Lagrange Multiplier Approach to Variational Problems and Applications

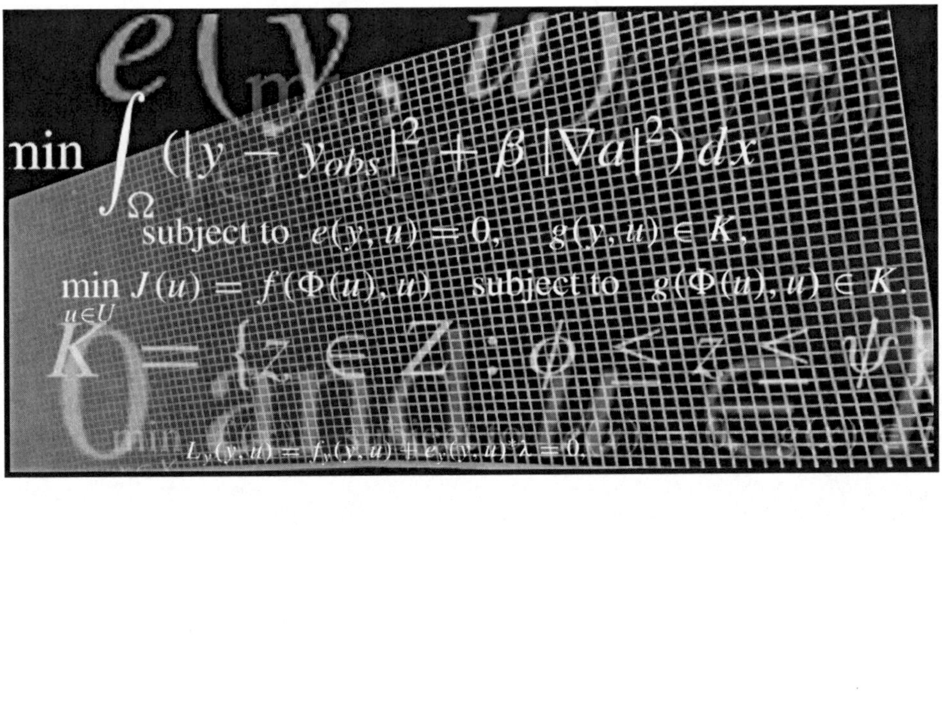

Advances in Design and Control

SIAM's Advances in Design and Control series consists of texts and monographs dealing with all areas of design and control and their applications. Topics of interest include shape optimization, multidisciplinary design, trajectory optimization, feedback, and optimal control. The series focuses on the mathematical and computational aspects of engineering design and control that are usable in a wide variety of scientific and engineering disciplines.

Series Volumes

Betts, John T., *Practical Methods for Optimal Control and Estimation Using Nonlinear Programming, Second Edition*
Shima, Tal and Rasmussen, Steven, eds., *UAV Cooperative Decision and Control: Challenges and Practical Approaches*
Speyer, Jason L. and Chung, Walter H., *Stochastic Processes, Estimation, and Control*
Krstic, Miroslav and Smyshlyaev, Andrey, *Boundary Control of PDEs: A Course on Backstepping Designs*
Ito, Kazufumi and Kunisch, Karl, *Lagrange Multiplier Approach to Variational Problems and Applications*
Xue, Dingyü, Chen, YangQuan, and Atherton, Derek P., *Linear Feedback Control: Analysis and Design with MATLAB*
Hanson, Floyd B., *Applied Stochastic Processes and Control for Jump-Diffusions: Modeling, Analysis, and Computation*
Michiels, Wim and Niculescu, Silviu-Iulian, *Stability and Stabilization of Time-Delay Systems: An Eigenvalue-Based Approach*
Ioannou, Petros and Fidan, Barış, *Adaptive Control Tutorial*
Bhaya, Amit and Kaszkurewicz, Eugenius, *Control Perspectives on Numerical Algorithms and Matrix Problems*
Robinett III, Rush D., Wilson, David G., Eisler, G. Richard, and Hurtado, John E., *Applied Dynamic Programming for Optimization of Dynamical Systems*
Huang, J., *Nonlinear Output Regulation: Theory and Applications*
Haslinger, J. and Mäkinen, R. A. E., *Introduction to Shape Optimization: Theory, Approximation, and Computation*
Antoulas, Athanasios C., *Approximation of Large-Scale Dynamical Systems*
Gunzburger, Max D., *Perspectives in Flow Control and Optimization*
Delfour, M. C. and Zolésio, J.-P., *Shapes and Geometries: Analysis, Differential Calculus, and Optimization*
Betts, John T., *Practical Methods for Optimal Control Using Nonlinear Programming*
El Ghaoui, Laurent and Niculescu, Silviu-Iulian, eds., *Advances in Linear Matrix Inequality Methods in Control*
Helton, J. William and James, Matthew R., *Extending H^∞ Control to Nonlinear Systems: Control of Nonlinear Systems to Achieve Performance Objectives*

Lagrange Multiplier Approach to Variational Problems and Applications

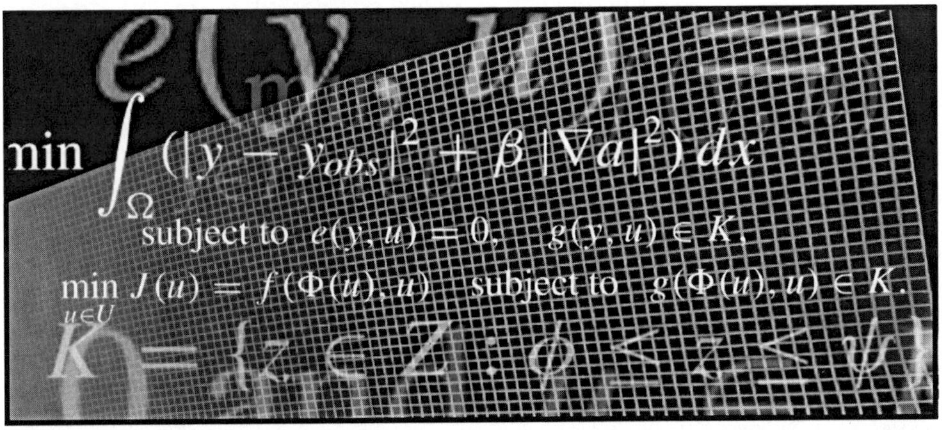

Kazufumi Ito
North Carolina State University
Raleigh, North Carolina

Karl Kunisch
University of Graz
Graz, Austria

siam®
Society for Industrial and Applied Mathematics
Philadelphia

10 9 8 7 6 5 4 3 2

Library of Congress Cataloging-in-Publication Data

Ito, Kazufumi.
Lagrange multiplier approach to variational problems and applications / Kazufumi Ito, Karl Kunisch.
p. cm. – (Advances in design and control ; 15)
Includes bibliographical references and index.
ISBN 978-0-898716-49-8 (pbk. : alk. paper)
1. Linear complementarity problem. 2. Variational inequalities (Mathematics). 3. Multipliers (Mathematical analysis). 4. Lagrangian functions. 5. Mathematical optimization. I. Kunisch, K. (Karl), 1952– II. Title.

QA402.5.I89 2008
519.3–dc22

2008061103

To our families:

Junko, Yuka, and Satoru

Brigitte, Katharina, Elisabeth, and Anna

Contents

Preface

The objective of this monograph is the treatment of a general class of nonlinear variational problems of the form

$$\begin{aligned} &\min_{y\in Y,\ u\in U} \quad f(y,u) \\ &\text{subject to}\ \ e(y,u)=0, \quad g(y,u)\in K, \end{aligned} \tag{0.0.1}$$

where $f: Y\times U \to \mathbb{R}$ denotes the cost functional, and $e: Y\times U \to W$ and $g: Y\times U \to Z$ are functionals used to describe equality and inequality constraints. Here Y, U, W, and Z are Banach spaces and K is a closed convex set in Z. A special choice for K is simple box constraints

$$K = \{z \in Z : \phi \le z \le \psi\}, \tag{0.0.2}$$

where Z is a lattice with ordering $\le$, and ϕ, ψ are elements in Z. Theoretical issues which will be treated include the existence of minimizers, optimality conditions, Lagrange multiplier theory, sufficient optimality considerations, and sensitivity analysis of the solutions to (0.0.1) with respect to perturbations in the problem data. These topics will be covered mainly in the first part of this monograph. The second part focuses on selected computational methods for solving the constrained minimization problem (0.0.1). The final chapter is devoted to the characterization of shape gradients for optimization problems constrained by partial differential equations.

Problems which fit into the framework of (0.0.1) are quite general and arise in application areas that were intensively investigated during the last half century. They include optimal control problems, structural optimization, inverse and parameter estimation problems, contact and friction problems, problems in image reconstruction and mathematical finance, and others. The variable y is often referred to as the state variable and u as the control or design parameter. The relationship between the variables y and u is described by e. It typically represents a differential or a functional equation. If e can be used to express the variable y as a function of u, i.e., $y = \Phi(u)$, then (0.0.1) reduces to

$$\min_{u\in U} J(u) = f(\Phi(u),u) \quad \text{subject to} \quad g(\Phi(u),u)\in K. \tag{0.0.3}$$

This is referred to as the reduced formulation of (0.0.1). In the more general case when y and u are both independent variables linked by the equation constraint $e(y,u) = 0$ it will be convenient at times to introduce $x = (y,u)$ in $X = Y \times U$ and to express (0.0.1) as

$$\min_{x\in X} \ f(x) \quad \text{subject to} \quad e(x) = 0, \ \ g(x)\in K. \tag{0.0.4}$$

From a computational perspective it can be advantageous to treat y and u as independent variables even though a representation of y in terms of u is available.

As an example consider the inverse problem of determining the diffusion parameter a in

$$-\nabla \cdot (a\, \nabla y) = f \text{ in } \Omega, \quad y|_{\partial\Omega} = 0 \tag{0.0.5}$$

from the measurement y_{obs}, where Ω is a bounded open set in $\mathbb{R}^2$ and $f \in H^{-1}(\Omega)$. A nonlinear least squares approach to this problem can be formulated as (0.0.1) by choosing $y \in Y = H_0^1(\Omega)$, $a \in U = H^1(\Omega) \cap L^\infty(\Omega)$, $W = H^{-1}(\Omega)$, $Z = L^2(\Omega)$ and considering

$$\begin{aligned} &\min \int_\Omega (|y - y_{obs}|^2 + \beta\, |\nabla a|^2)\, dx \\ &\text{subject to (0.0.5) and } \underline{a} \le a \le \bar{a}, \end{aligned} \tag{0.0.6}$$

where $0 < \underline{a} < \bar{a} < \infty$ are lower and upper bounds for the "control" variable a. In this example, given $a \in U$, the state $y \in Y$ can be uniquely determined by solving the elliptic boundary value problem (0.0.5) for y, and we obtain a problem of type (0.0.3).

The general approach that we follow in this monograph for the analytical as well as the numerical treatment of (0.0.1) is based on Lagrange multiplier theory. Let us subsequently suppose that Y, U, and W are Hilbert spaces and discuss the case $Z = U$ and $g(y, u) = u$, i.e., the constraint $u \in K$. We assume that f and e are C^1 and denote by f_y, f_u the Fréchet derivatives of f with respect to y and u, respectively. Let W^* be the dual space of W and let the duality product be denoted by $\langle \cdot, \cdot \rangle_{W^*,W}$. The analogous notation is used for Z and Z^*. We form the Lagrange functional

$$L(y, u, \lambda) = f(y, u) + \langle e(y, u), \lambda \rangle_{W,W^*}, \tag{0.0.7}$$

where $\lambda \in W^*$ is the Lagrange multiplier associated with the equality constraint $e(y, u) = 0$ which, for the present discussion, is supposed to exist. It will be shown that a minimizing pair (y, u) satisfies

$$\begin{aligned} &L_y(y, u) = f_y(y, u) + e_y(y, u)^*\lambda = 0, \\ &\langle f_u(y, u) + e_u(y, u)^*\lambda\, , v - u \rangle_{Z^*,Z} \ge 0 \text{ for all } v \in K, \\ &e(y, u) = 0 \text{ and } u \in K. \end{aligned} \tag{0.0.8}$$

In the case $K = Z$ the second equation of (0.0.8) results in the equality $f_y(y, u) + e_u(y, u) = 0$. In this case a first possibility for solving the system (0.0.8) for the unknowns (y, u, λ) is the use of a direct equation solver of Newton type, for example. Here the Lagrange multiplier λ is treated as an independent variable just like y and u. Alternatively, for (y, u) satisfying $e(y, u) = 0$ and λ satisfying $f_y(y, u) + e_u(y, u)^*\lambda = 0$ the gradient of $J(u)$ of (0.0.4) can be evaluated as

$$J_u = f_u(y, u) + e_u(y, u)^*\lambda. \tag{0.0.9}$$

Thus the combined step of determining $y \in Y$ for given $u \in K$ such that $e(y, u) = 0$ and finding $\lambda \in W^*$ satisfying $f_y(y, u) + e_u(y, u)^*\lambda = 0$ for $(y, u) \in Y \times U$ provides a possibility for evaluating the gradient of J at u. If in addition u has to satisfy a constraint of the form $u \in K$, the projected gradient is obtained by projecting $f_u(y,u) + e_u(y,u)^*\lambda$ onto K, and (projected) gradient-based iterative methods can be employed to solve (0.0.1).

In optimal control of differential equations, the multiplier λ is called the adjoint state and the second equation in (0.0.8) is called the optimality condition. Further the system (0.0.8) coincides with the celebrated Pontryagin maximum principle. The steps in the procedure explained above for obtaining the gradient are referred to as the forward equation step for the state equation (the third equation in (0.0.8)) and the backward equation step for the adjoint equation (the first equation). This terminology is motivated by time-dependent problems, where the third equation is an initial value problem and the first one is a problem with terminal time boundary condition. Lastly, if K is of type (0.0.2), then the second equation of (0.0.8) can be written as the complementarity condition

$$\begin{aligned} &f_u(y,u) + e_u(y,u)^*\lambda + \eta = 0, \\ &\eta = \max(0, \eta + (u - \psi)) + \min(0, \eta + (u - \phi)). \end{aligned} \tag{0.0.10}$$

Due to the nondifferentiability of the max and min operations classical Newton methods are not directly applicable to system (0.0.10). Active set methods provide a very efficient alternative. They turn out to be equivalent to semismooth Newton methods in function spaces. If appropriate structural conditions are met, including that f be quadratic and e affine, then such techniques are globally convergent. Moreover, under more general conditions they exhibit local superlinear convergence. Due to the practical relevance of problems with box constraints, a significant part of this monograph is devoted to the analysis of these methods.

A frequently employed alternative to the multiplier approach is given by the penalty method. To explain the procedure consider the sequence of minimization problems

$$\min_{y \in Y, u \in K} \ f(y,u) + \frac{c_k}{2}\,|e(y,u)|_W^2 \tag{0.0.11}$$

for an increasing sequence of penalty parameter c_k. That is, the equality constraint is eliminated through the quadratic penalty function. This requires to solve the unconstrained minimization problem (0.0.11) over $Y \times K$ for a sequence of $\{c_k\}$ tending to infinity. Under mild assumptions it can be shown that the sequence of minimizers (y_k, u_k) determined by (0.0.11) converges to a minimizer of (0.0.1) as $c_k \to \infty$. Moreover, (y_k, u_k) satisfies

$$\begin{aligned} &f_y(y_k,u_k) + e_y(y_k,u_k)^*(c_k\, e(y_k,u_k)) = 0, \\ &\langle f_u(y_k,u_k) + e_u(y_k,u_k)^*(c_k\, e(y_k,u_k)), v - u_k\rangle_{Z^*,Z} \ge 0. \end{aligned} \tag{0.0.12}$$

Comparing (0.0.12) with (0.0.8) it is natural to ask whether $c_k\, e(y_k,u_k)$ tends to a Lagrange multiplier associated to $e(y,u) = 0$ as $c_k \to \infty$. Indeed this can be shown under suitable conditions. Despite disadvantages due to slow convergence and possible ill-conditioning for solving (0.0.12) with large values of c_k, the penalty method is widely accepted in practice. This is due, in part, to the simplicity of the approach and the availability of powerful algorithms for solving (0.0.12) if $K = Z$, when the inequality in (0.0.12) becomes an equality.

A third methodology is given by duality techniques. They are based on the introduction of the dual functional

$$d(\lambda) = \inf_{y \in Y,\ u \in K} \ L(y,u,\lambda) \tag{0.0.13}$$

and the duality property

$$\sup_{\lambda \in Y} d(\lambda) = \inf_{y \in Y,\, u \in K} f(y,u) \quad \text{subject to } e(y,u) = 0. \tag{0.0.14}$$

The duality method for solving (0.0.1) requires minimizing $L(y, u, \lambda_k)$ over $(y, u) \in Y \times K$ and updating λ by means of

$$\lambda_{k+1} = \lambda_k + \alpha_k \mathcal{J} e(y_k, u_k), \tag{0.0.15}$$

where

$$(y_k, u_k) = \underset{y \in Y,\, u \in K}{\operatorname{argmin}}\ L(y, u, \lambda_k),$$

$\alpha_k > 0$ is an appropriately chosen step size and $\mathcal{J}$ denotes the Riesz mapping from W onto W^*. It can be argued that $e(y_k, u_k)$ is the gradient of $d(\lambda)$ at λ_k and (0.0.15) is in turn a steepest ascent method for the maximization of $d(\lambda)$ under appropriate conditions. Such methods are called primal-dual methods. Despite the fact that the method can be justified only under fairly restrictive convexity assumptions on $L(y, u, \lambda)$ with respect to (y, u), it provides an elegant use of Lagrange multipliers and is a basis for so-called augmented Lagrangian methods.

Augmented Lagrangian methods with $K = Z$ are based on the following problem which is equivalent to (0.0.1):

$$\begin{aligned} &\min_{y \in Y, u \in K} f(y,u) + \frac{c}{2}|e(y,u)|_W^2 \\ &\text{subject to} \quad e(y,u) = 0. \end{aligned} \tag{0.0.16}$$

Under rather mild conditions the quadratic term enhances the local convexity of $L(y, u, \lambda)$ in the variables (y, u) for sufficiently large $c > 0$. It helps the convergence of direct solvers based on the necessary optimality conditions (0.0.8). To carry this a step further we introduce the augmented Lagrangian functional

$$L_c(y, u, \lambda) = f(x, u) + \langle e(y, u), \lambda \rangle + \frac{c}{2}|e(y, u)|_W^2. \tag{0.0.17}$$

The first order augmented Lagrangian method is the primal-dual method applied to (0.0.17), i.e.,

$$\begin{aligned} (y_k, u_k) &= \underset{y \in Y,\, u \in K}{\operatorname{argmin}}\ L_c(y, u, \lambda_k), \\ \lambda_{k+1} &= \lambda_k + c\, \mathcal{J}\, e(y_k, u_k). \end{aligned} \tag{0.0.18}$$

Its advantage over the penalty method is attributed to the fact that local convergence of (y_k, u_k) to a minimizer (y, u) of (0.0.1) holds for all sufficiently large and fixed $c > 0$, without requiring that $c \to \infty$. As we noted, $L_c(y, u, \lambda)$ has local convexity properties under well-studied assumptions, and the primal-dual viewpoint is applicable to the multiplier update. The iterates (y_k, u_k, λ_k) converge linearly to the triple (y, u, λ) satisfying the first order necessary optimality conditions, and convergence can improve as $c > 0$ increases. Due to these attractive characteristics and properties, the method of multipliers and its subsequent Newton-like variants have been recognized as a powerful method for minimization problems with equality constraints. They constitute an important part of this book.

In (0.0.16) and (0.0.18) the constraint $u \in K$ remained as explicit constraint and was not augmented. To describe a possibility for augmenting inequalities we return to the general form $g(y,u) \in K$ and consider inequality constraints with finite rank, i.e., $Z = \mathbb{R}^p$ and $K = \{z \in \mathbb{R}^p : z_i \leq 0,\ 1 \leq i \leq p\}$. Then under appropriate conditions the formulation

$$\begin{aligned} &\min_{y\in Y,\, u\in U,\, q\in\mathbb{R}^p} \quad L_c(y,u,\lambda) + (\mu, g(y,u) - q) + \frac{c}{2}\,|g(y,u) - q|^2_{\mathbb{R}^p} \\ &\text{subject to } q \leq 0 \end{aligned} \tag{0.0.19}$$

is equivalent to (0.0.1). Here $\mu \in \mathbb{R}^p$ is the Lagrange variable associated with the inequality constraint $g(y,u) \leq 0$. Minimizing the functional in (0.0.19) over $q \leq 0$ results in the augmented Lagrangian functional

$$L_c(y,u,\lambda,\mu) = L_c(y,u,\lambda) + \frac{1}{2}\,|\max(0, \mu + c\,g(y,u))|^2_{\mathbb{R}^p} - \frac{c}{2}|\mu|^2_{\mathbb{R}^p}, \tag{0.0.20}$$

where equality and finite rank inequality constraints are augmented. The corresponding augmented Lagrangian method is

$$\begin{aligned} (y_k, u_k) &= \operatorname*{argmin}_{y\in Y,\, u\in K}\ L_c(y,u,\lambda_k,\mu_k), \\ \lambda_{k+1} &= \lambda_k + c\,\mathcal{J}\,e(y_k,u_k), \\ \mu_{k+1} &= \max(0, \mu_k + c\,g(y_k,u_k)). \end{aligned} \tag{0.0.21}$$

In the discussion of box-constrained problems we already pointed out the relevance of numerical methods for nonsmooth problems; see (0.0.10). In many applied variational problems, for example in mechanics, fluid flow, or image analysis, nonsmooth cost functionals arise. Consider as a special case the simplified friction problem

$$\min\ f(y) = \int_\Omega \frac{1}{2}\,(|\nabla y|^2 + |y|^2) - \tilde{f}\,y\,dx + g\int_\Gamma |y|\,ds \quad \text{over } y \in H^1(\Omega), \tag{0.0.22}$$

where Ω is a bounded domain with boundary Γ. This is an unconstrained minimization problem with $X = H^1(\Omega)$ and no control variables. Since the functional is not continuously differentiable, the necessary optimality condition $f_y(y) = 0$ is not applicable. However, with the aid of a generalized Lagrange multiplier theory a necessary optimality condition can be written as

$$\begin{aligned} &-\Delta y + y = \tilde{f}, \quad \frac{\partial y}{\partial \nu} = g\,\lambda \quad \text{on } \Gamma, \\ &|\lambda(x)| \leq 1 \text{ and } \lambda(x)y(x) = |\,y(x)| \quad \text{a.e. in } \Gamma. \end{aligned} \tag{0.0.23}$$

From a Lagrange multiplier perspective, problems with L^1-type cost functionals as in (0.0.22) and box-constrained problems are dual to each other. So it comes as no surprise that again semismooth Newton methods provide an efficient technique for solving (0.0.22) or (0.0.23). We shall analyze them and provide a theoretical basis for their numerical efficiency.

This monograph also contains the analysis of optimization problems constrained by partial differential equations which are singular in the sense that the state variable cannot be differentiated with respect to the control. This, however, does not preclude that the cost functional is differentiable with respect to the control and that an optimality principle can be derived. In terms of shape optimization problems this means that the cost functional is shape differentiable while the state is not differentiable with respect to the shape.

In summary, Lagrange multiplier theory provides a tool for the analysis of general constrained optimization problems with cost functionals which are not necessarily C^1 and with state equations which are in some sense singular. It also leads to a theoretical basis for developing efficient and powerful iterative methods for solving such problems. The purpose of this monograph is to provide a rather thorough analysis of Lagrange multiplier theory and to show its impact on the development of numerical algorithms for problems which are posed in a function space setting.

Let us give a short description of the book for those readers who do not intend to read it by consecutive chapters. Chapter 1 provides a variety of tools to establish existence of Lagrange multipliers and is called upon in all the following chapters. Here, as in other chapters, we do not attempt to give the most general results, nor do we strive for covering the complete literature. Chapter 2 is devoted to the sensitivity analysis of abstract constrained nonlinear programming problems and it essentially stands for itself. This chapter is of great importance, addressing continuity, Lipschitz continuity, and differentiability of the solutions to optimization and optimal control problems with respect to parameters that appear in the problem formulation. Such results are not only of theoretical but also of practical importance. The sensitivity equations have been a starting point for the development of algorithmic concepts for decades. Nevertheless, readers who are not interested in this topic at first may skip this chapter without missing technical results which might be needed for later chapters.

Chapters 3, 5, and 6 form a unit which is devoted to smooth optimization problems. Chapter 3 covers first order augmented Lagrangian methods for optimization problems with equality and inequality constraints. Here as in the remainder of the book, the inequality constraints that we have in mind typically represent partial differential equations. In fact, during the period in which this monograph was written, the terminology "PDE-constrained optimization" emerged. Inverse problems formulated as regularized least squares problems and optimal control problems for (partial) differential equations are primary examples for the theories that are discussed here. Chapters 5 and 6 are devoted to second order iterative solution methods for equality-constrained problems. Again the equality constraints represent partial differential equations. This naturally gives rise to the following situation: The variables with respect to which the optimization is carried out can be classified into two groups. One group contains the state variables of the differential equations and the other group consists of variables which represent the control or input variables for optimal control problems, or coefficients in parameter estimation problems. If the state variables are considered as functions of the independent controls, inputs, or coefficients, and the cost functional in the optimization problem is only considered as a functional of the latter, then this is referred to as the *reduced* formulation. Applying a second order method to the reduced functional we arrive at the Newton method for optimization problems with partial differential equations as constraints. If both state and control variables are kept as independent variables and the optimality system involving primal and adjoint variables, which are the Lagrange multipliers corresponding to the PDE-constraints, is derived, we arrive at the sequential quadratic programming (SQP) technique: It essentially consists of applying a Newton algorithm to the first order necessary optimality conditions. The Newton method for the reduced formulation and the SQP technique are the focus of Chapter 5. Chapter 6 is devoted to second order augmented Lagrangian techniques which are closely related, as we shall see, to SQP methods. Here the equation constraint is augmented in a penalty term, which has the effect of locally convexifying the optimization problem. Since augmented Lagrangians also involve Lagrange multipliers, there is, however, no necessity to let the penalty parameter tend to infinity and, in fact, we do not suggest doing so.

A second larger unit is formed by Chapters 4, 7, 8, and 9. Nonsmoothness, primal-dual active set strategy, and semismooth Newton methods are the keywords which characterize the contents of these chapters. Chapter 4 is essentially a recapture of concepts from convex analysis in a format that is used in the remaining chapters. A key result is the formulation of differential inclusions which arise in optimality systems by means of nondifferentiable equations which are derived from Yosida–Moreau approximations and which will serve as the basis for the primal-dual active set strategy. Chapter 7 is devoted to the primal-dual active set strategy and its global convergence properties for unilaterally and bilaterally constrained problems. The local analysis of the primal-dual active set strategy is achieved in the framework of semismooth Newton methods in Chapter 8. It contains the notion of Newton derivative and establishes local superlinear convergence of the Newton method for problems which do not satisfy the classical sufficient conditions for local quadratic convergence. Two important classes of applications of semismooth Newton methods are considered in Chapter 9: image restoration and deconvolution problems regularized by the bounded variation (BV) functional and friction and contact problems in elasticity.

Chapter 10 is devoted to a Lagrangian treatment of parabolic variational inequalities in unbounded domains as they arise in the Black–Scholes equation, for example. It contains the use of monotone techniques for analyzing parabolic systems without relying on compactness assumptions in a Gelfand-triple framework. In Chapter 11 we provide a calculus for obtaining the shape derivative of the cost functional in shape optimization problems which bypasses the need for using the shape derivative of the state variables of the partial differential equations. It makes use of the expansion technique that is proposed in Chapters 1 and 5 for weakly singular optimal control problems, and of the assumption that an appropriately defined adjoint equation admits a solution. This provides a versatile technique for evaluating the shape derivative of the cost functional using Lagrange multiplier techniques.

There are many additional topics which would fit under the title of this monograph which, however, we chose not to include. In particular, issues of discretization, convergence, and rate of convergence are not discussed. Here the issue of proper discretization of adjoint equations consistent with the discretization of the primal equation and the consistent time integration of the adjoint equations must be mentioned. We do not enter into the discussion of whether to discretize an infinite-dimensional nonlinear programming problem first and then to decide on an iterative algorithm to solve the finite-dimensional problems, or the other way around, consisting of devising an optimization algorithm for the infinite-dimensional problem which is subsequently discretized. It is desirable to choose a discretization and an iterative optimization strategy in such a manner that these two approaches commute. Discontinuous Galerkin methods are well suited for this purpose; see, e.g., [BeMeVe]. Another important area which is not in the focus of this monograph is the efficient solution of those large scale linear systems which arise in optimization algorithms. We refer the reader to, e.g., [BGHW, BGHKW], and the literature cited there. The solution of large scale time-dependent optimal control problems involving the coupled system of primal and adjoint equations, which need to be solved in opposite directions with respect to time, still offers a significant challenge, despite the advances that were made with multiple shooting, receding horizon, and time-domain decomposition techniques. From the point of view of optimization theory there are several topics as well into which one could expand. These include globalization strategies, like trust region methods, exact penalty methods, quasi-Newton methods, and a more abstract Lagrange multiplier theory than that presented in Chapter 1.

As a final comment we stress that for a full treatment of a variational problem in function spaces, both its infinite-dimensional analysis as well as its proper discretization and the relation between the two are indispensable. Proceeding from an infinite-dimensional problem directly to its disretization without such a treatment, important issues can be missed. For instance discretization without a well-posed analysis may result in the use of inappropriate inner products, which may lead to unnecessary ill-conditioning, which entails unnecessary preconditioning. Inconsiderate discretization may also result in the loss of structural properties, as for instance symmetry properties.

Chapter 1

Existence of Lagrange Multipliers

1.1 Problem statement and generalities

We consider the constrained optimization problem

$$\begin{aligned} &\min f(x) \\ &\text{subject to } x \in C \text{ and } g(x) \in K, \end{aligned} \tag{1.1.1}$$

where f is a real-valued functional defined on a real Banach space X; further C is a closed convex subset of X, g is a mapping from X into the real Banach space Z, and K is a closed convex cone in Z. Throughout the monograph it is understood that the vertex of a cone coincides with the origin. Further, in this chapter it is assumed that (1.1.1) admits a solution x^*. Here we refer to x^* as solution if it is feasible, i.e., x^* satisfies the constraints in (1.1.1), and f has a local minimum at x^*. It is assumed that

$$\begin{cases} f \text{ is Fréchet differentiable at } x^*, \text{ and} \\ g \text{ is continuously Fréchet differentiable at } x^* \end{cases} \tag{1.1.2}$$

with Fréchet derivatives denoted by $f'(x^*)$ and $g'(x^*)$, respectively. The set of feasible points for (1.1.1) will be denoted by $M = \{x \in C : g(x) \in K\}$. Moreover, the following notation will be used. The topological duals of X and Z are denoted by X^* and Z^*. For the subset A of X the polar or dual cone is defined by

$$A^+ = \{x^* \in X^* : \langle x^*, a \rangle \leq 0 \text{ for all } a \in A\},$$

where $\langle \cdot, \cdot \rangle$ $(= \langle \cdot, \cdot \rangle_{X^*,X})$ denotes the duality pairing between X^* and X. Further for $x \in X$ the conical hull of $C \setminus \{x\}$ is given by

$$C(x) = \{\lambda(c - x) : c \in C, \lambda \geq 0\},$$

and $K(z)$ with $z \in Z$ is defined analogously. The Lagrange functional $\mathcal{L} : X \times Z^* \to \mathbb{R}$ associated to (1.1.1) is defined by

$$\mathcal{L}(x, \lambda) = f(x) + \langle \lambda, g(x) \rangle_{Z^*,Z}.$$

Definition 1.1. *An element* $\lambda^* \in Z^*$ *is called a Lagrange multiplier for* (1.1.1) *at the solution* x^* *if* $\lambda^* \in K^+$, $\langle \lambda^*, g(x^*)\rangle = 0$, *and*

$$f'(x^*) + \lambda^* \circ g'(x^*) \in -C(x^*)^+.$$

Lagrange multipliers are of fundamental importance for expressing necessary as well as sufficient optimality conditions for (1.1.1) and for the development of algorithms to solve (1.1.1). If $C = X$, then the inclusion in Definition 1.1 results in the equation

$$f'(x^*) + \lambda^* \circ g'(x^*) = 0.$$

For the existence analysis of Lagrange multipliers one makes use of the following tangent cones which approximate the feasible set M at x^*:

$$T(M, x^*) = \left\{ x \in X : x = \lim_{n\to\infty} \frac{1}{t_n}(x_n - x^*),\, t_n \to 0^+,\, x_n \in M \right\},$$
$$L(M, x^*) = \{x \in X : x \in C(x^*) \text{ and } g'(x^*)x \in K(g(x^*))\}.$$

The cone $T(M, x^*)$ is called the sequential tangent cone (or Bouligand cone) and $L(M, x^*)$ the linearizing cone of M at x^*.

Proposition 1.2. *If* x^* *is a solution to* (1.1.1), *then* $f'(x^*)x \ge 0$ *for all* $x \in T(M, x^*)$.

Proof. Let $x \in T(M, x^*)$. Then there exist sequences $\{x_n\}_{n=1}^\infty$ in M and $\{t_n\}$ with $\lim_{n\to\infty} t_n = 0$ such that $x = \lim_{n\to\infty} \frac{1}{t_n}(x_n - x^*)$. By the definition of Fréchet differentiability there exists a sequence $\{r_n\}$ in Z such that

$$f'(x^*)x = \lim_{n\to\infty} \frac{1}{t_n} f'(x^*)(x_n - x^*) = \lim_{n\to\infty} \frac{1}{t_n}(f(x_n) - f(x^*) + r_n)$$

and $\lim_{n\to\infty} \frac{1}{|x_n - x^*|} r_n = 0$. By optimality of x^* we have

$$f'(x^*)x \ge \lim_{n\to\infty} |x| \frac{1}{|x_n - x^*|} r_n = 0$$

and the result follows. ☐

We now briefly outline the main steps involved in the abstract approach to prove the existence of Lagrange multipliers. For this purpose we assume that $C = X$, $K = \{0\}$ so that the set of feasible points is described by $M = \{x : g(x) = 0\}$. We assume that $g'(x^*) : X \to Z$ is surjective. Then by Lyusternik's theorem [Ja, We] we have

$$L(M, x^*) \subset T(M, x^*). \tag{1.1.3}$$

Consider the convex cone

$$B = \{(f'(x^*)x + r, g'(x^*)x) : r \ge 0, x \in X\} \subset \mathbb{R} \times Z.$$

Observe that $(0, 0) \in B$ and that due to Proposition 1.2 and (1.1.3) the origin $(0, 0)$ is a boundary point of B. Since $g'(x^*)$ is surjective, B has nonempty interior and hence by the Eidelheit separation theorem that we shall recall below there exists a closed hyperplane in $\mathbb{R} \times Z$ which supports B at $(0, 0)$, i.e., there exists $0 \neq (\alpha, \lambda^*) \in \mathbb{R} \times Z^*$ such that

$$\alpha(f'(x^*)x + r) + \langle \lambda^*, g'(x^*)x \rangle \geq 0 \text{ for all } (r, x) \in \mathbb{R}^+ \times X.$$

Setting $x = 0$ we have $\alpha \geq 0$. If $\alpha = 0$, then $\lambda^* g'(x^*) = 0$, which would imply $\lambda^* = 0$, which is impossible. Consequently $\alpha > 0$ and without loss of generality we can assume $\alpha = 1$. Hence λ^* is a Lagrange multiplier. For the general case of equality and inequality constraints the application of Lyusternik's theorem is replaced by stability results for solutions of systems of inequalities. The existence of a separating hyperplane will follow from a generalized open mapping theorem, which is covered in the following section.

The separation theorem announced above is presented next.

Theorem 1.3. *Let K_1 and K_2 be nontrivial, convex, and disjoint subsets of a normed linear space $\mathcal{X}$. If K_1 is closed and K_2 is compact, then there exists a closed hyperplane strictly separating K_1 and K_2, i.e., there exist $x^* \in \mathcal{X}$, $\beta \in R$, and $\epsilon > 0$ such that*

$$\langle x^*, x \rangle \leq \beta - \epsilon \text{ for all } x \in K_1 \quad \text{and} \quad \langle x^*, x \rangle \geq \beta + \epsilon \text{ for all } x \in K_2.$$

If K_1 is open and K_2 is arbitrary, then these inequalities still hold with $\epsilon = 0$.

Let us give a brief description of the sections of this chapter. Section 1.2 contains the open mapping theorem already alluded to above. Regularity properties which guarantee the existence of a Lagrange multiplier in general nonlinear optimal programming problems in infinite dimensions are analyzed in section 1.3 and applications to parameter estimation and optimal control problems are described in section 1.4. Section 1.5 is devoted to the derivation of a first order optimality system for a class of weakly singular optimal control problems and to several applications which are covered by this class. Finally in section 1.6 several further alternatives for deriving optimality systems are presented in an informal manner.

1.2 A generalized open mapping theorem

Throughout this section T denotes a bounded linear operator from X to Z. As in the previous section C stands for a closed convex set in X, and K stands for a closed convex cone in Z. For Theorem 1.4 below it is sufficient that K is a closed convex set. For $\rho > 0$ we set $X_\rho = \{x : |x| \leq \rho\}$ and Z_ρ is defined analogously. Recall that by the open mapping theorem, surjectivity of T implies that T maps open sets of X onto open sets of Z. As a consequence TX_1 contains Z_ρ for some $\rho > 0$. The following theorem generalizes this result.

Theorem 1.4. *Let $\bar{x} \in C$ and $\bar{y} \in K$, where K is a closed convex set. Then the following statements are equivalent:*

(a) $Z = TC(\bar{x}) - K(\bar{y})$,

(b) *there exists* $\rho > 0$ *such that* $Z_\rho \subset T((C - \{\bar{x}\}) \cap X_1) - (K - \{\bar{y}\}) \cap Z_1$,

where $C(\bar{x}) = \{\lambda(x - \bar{x}) : \lambda \geq 0, x \in C\}$ *and* $K(\bar{y}) = \{k - \lambda\bar{y} : \lambda \geq 0, k \in K\}$.

Proof. Clearly (b) implies (a). To prove the converse we first show that

$$C(\bar{x}) = \bigcup_{n \in \mathbb{N}} n((C - \bar{x}) \cap X_1). \tag{1.2.1}$$

The inclusion $\supset$ is obvious and hence it suffices to prove that $C(\bar{x})$ is contained in the set on the right-hand side of (1.2.1). Let $x \in C(\bar{x})$. Then $x = \lambda y$ with $\lambda \geq 0$ and $y \in C - \bar{x}$. Without loss of generality we can assume that $\lambda > 0$ and $|y| > 1$. Convexity of C implies that $\frac{1}{|y|}y \in (C - \bar{x}) \cap X_1$. Let $n \in \mathbb{N}$ be such that $n \geq \lambda|y|$. Then $x = (\lambda|y|)\frac{1}{|y|}y \in n((C - \bar{x}) \cap X_1)$, and (1.2.1) is verified.

For $\alpha > 0$ let $A_\alpha = \alpha T((C - \bar{x}) \cap X_1 - (K - \bar{y}) \cap Z_1)$. We will show that $0 \in \operatorname{int} \overline{A_1}$. In fact, from (1.2.1) and the analogous equality with C replaced by K it follows from (a) that

$$Z = \bigcup_{n \in \mathbb{N}} nT((C - \bar{x}) \cap X_1) \cup \bigcup_{n \in \mathbb{N}} (-n)((K - \bar{y}) \cap Z_1).$$

This implies hat

$$Z = \bigcup_{n \in \mathbb{N}} A_n = \bigcup_{n \in \mathbb{N}} \overline{A_n}. \tag{1.2.2}$$

Thus the complete space Z is the countable union of the closed sets $\overline{A_n}$. By the Baire category theorem there exists some $m \in \mathbb{N}$ such that $\operatorname{int} \overline{A_m} \neq \emptyset$. Let $a \in \operatorname{int} \overline{A_m}$. By (1.2.2) there exists $k \in \mathbb{N}$ with $-a \in \overline{A_k}$. Using $\overline{A_\alpha} = \alpha \overline{A_1}$ this implies that $-\frac{m}{k}a \in \overline{A_m}$. It follows that the half-open interval $\{\lambda(-\frac{m}{k}a) + (1 - \lambda)a : 0 \leq \lambda < 1\}$ belongs to $\operatorname{int} \overline{A_m}$ and consequently $0 \in \operatorname{int} \overline{A_m} = m \operatorname{int} \overline{A_1}$. Thus we have

$$0 \in \operatorname{int} \overline{A_1}. \tag{1.2.3}$$

Hence for some $\rho > 0$

$$Z_\rho \subset \frac{1}{2}\overline{A_1} = \overline{A_{\frac{1}{2}}} \subset A_{\frac{1}{2}} + Z_{\frac{\rho}{2}}, \tag{1.2.4}$$

and consequently for all $i = 0, 1, \ldots$

$$Z_{(\frac{1}{2})^i \rho} = \left(\frac{1}{2}\right)^i Z_\rho \subset \left(\frac{1}{2}\right)^i (A_{\frac{1}{2}} + Z_{\frac{\rho}{2}}) = A_{(\frac{1}{2})^{i+1}} + Z_{(\frac{1}{2})^{i+1}\rho}. \tag{1.2.5}$$

Now choose $y \in Z_\rho$. By (1.2.4) there exist $x_1 \in (C - \bar{x}) \cap X_1$, $y_1 \in (K - \bar{y}) \cap Z_1$, and $r_1 \in Z_{\frac{\rho}{2}}$ such that $y = T(\frac{1}{2}x_1) - \frac{1}{2}y_1 + r_1$.

Applying (1.2.5) with $i = 1$ to r_1 implies the existence of $x_2 \in (C - \bar{x}) \cap X_1$, $y_2 \in (K - \bar{y}) \cap Z_1$, and $r_2 \in Z_{(\frac{1}{2})^2 \rho}$ such that

$$y = T\left(\frac{1}{2}x_1 + \left(\frac{1}{2}\right)^2 x_2\right) - \left(\frac{1}{2}y_1 + \left(\frac{1}{2}\right) y_2\right) + r_2.$$

Repeated application of this argument implies that $y = Tu_n - v_n + r_n$, where

$$u_n = \sum_{i=1}^{n} \left(\frac{1}{2}\right)^i x_i \text{ with } x_i \in (C - \bar{x}) \cap X_1,$$

$$v_n = \sum_{i=1}^{n} \left(\frac{1}{2}\right)^i y_i \text{ with } y_i \in (K - \bar{y}) \cap Y_1,$$

and $r_n \in Y_{(\frac{1}{2})^n \rho}$. Observe that

$$u_n = \frac{1}{2}x_1 + \cdots + \left(\frac{1}{2}\right)^n x_n + \left(1 - \sum_{i=1}^{n}\left(\frac{1}{2}\right)^i\right) 0 \in (C - \bar{x}) \cap X_1$$

and

$$|u_n - u_{n+m}| \le \sum_{i=n+1}^{n+m} \left(\frac{1}{2}\right)^i \le \left(\frac{1}{2}\right)^n.$$

Consequently $\{u_n\}_{n=1}^{\infty}$ is a Cauchy sequence and there exists some $x \in (C - \bar{x}) \cap X_1$ such that $\lim_{n\to\infty} u_n = x$. Similarly there exists $v \in (K - \bar{y}) \cap Z_1$ such that $\lim_{n\to\infty} v_n = v$. Moreover $\lim_{n\to\infty} r_n = 0$ and continuity of T implies that $y = Tx - v$. □

1.3 Regularity and existence of Lagrange multipliers

In this section existence of a Lagrange multiplier for the optimization problem (1.1.1) will be established using a regular point condition which was proposed in the publications of Robinson for sensitivity analysis of generalized equations and later by Kurcyusz, Zowe, and Maurer for the existence of Lagrange multipliers.

Definition 1.5. *The element $\bar{x} \in M$ is called regular (or alternatively $\bar{x}$ satisfies the regular point condition) if*

$$0 \in \text{int}\,\{g'(\bar{x})(C - \bar{x}) - K + g(\bar{x})\}. \tag{1.3.1}$$

If (1.3.1) holds, then clearly

$$0 \in \text{int}\,\{g'(\bar{x})C(\bar{x}) - K(g(\bar{x}))\}. \tag{1.3.2}$$

Note that $g'(\bar{x})C(\bar{x}) - K(g(\bar{x}))$ is a cone. Consequently (1.3.2) implies that

$$g'(\bar{x})C(\bar{x}) - K(g(\bar{x})) = Z. \tag{1.3.3}$$

Finally (1.3.3) implies (1.3.1) by Theorem 1.4. Thus, the three conditions (1.3.1)–(1.3.3) are equivalent. Condition (1.3.1) is used in the work of Robinson (see, e.g., [Ro2]) on the stability of solutions to systems of inequalities, and (1.3.3) is the regularity condition employed in [MaZo, ZoKu] to guarantee the existence of Lagrange multipliers.

Remark 1.3.1. If $C = X$, then the so-called Slater condition $g'(\bar{x})h \in \text{int}\, K(g(\bar{x}))$ for some $h \in X$ implies (1.3.1). In case $C = X = \mathbb{R}^n$, $Z = \mathbb{R}^m$, $K = \{y \in \mathbb{R}^m : y_i = 0$ for $i = 1, \dots, k$, and $y_i \le 0$ for $i = k+1, \dots, m\}$, $g = (g_1, \dots, g_k, g_{k+1}, \dots, g_m)$, the constraint $g(x) \in K$ amounts to k equality and $m - k$ inequality constraints, and the regularity condition (1.3.3) becomes

$$\begin{cases} \text{the gradients } \{g_i'(\bar{x})\}_{i=1}^k \text{ are linearly independent and} \\ \text{there exists } x \in X \text{ such that} \\ g_i'(\bar{x})x = 0 \text{ for } i = 1, \dots, k, \\ g_i'(\bar{x})x < 0 \text{ for } i = k+1, \dots, m \text{ with } g_i(\bar{x}) = 0. \end{cases} \tag{1.3.4}$$

This regularity condition is referred to as the Mangasarian–Fromowitz constraint qualification.

Other variants of conditions which imply existence of Lagrange multipliers are named after F. John, Karush–Kuhn–Tucker (KKT), and Arrow, Hurwicz, Uzawa.

Theorem 1.6. *If the solution x^* of (1.1.1) is regular, then there exists an associated Lagrange multiplier $\lambda^* \in Z^*$.*

Proof. By Proposition 1.2 we have

$$f'(x^*)x \ge 0 \text{ for all } x \in T(M, x^*).$$

From Theorem 1 and Corollary 2 in [We], which can be viewed as generalizations of Lyusternik's theorem, it follows that $L(M, x^*) \subset T(M, x^*)$ and hence

$$f'(x^*)x \ge 0 \text{ for all } x \in L(M, x^*). \tag{1.3.5}$$

We define the set

$$B = \{(f'(x^*)x + r, g'(x^*)x - y) : r \ge 0, x \in C(x^*), y \in K(g(x^*))\} \subset \mathbb{R} \times Z.$$

This is a convex cone containing the origin $(0, 0) \in \mathbb{R} \times Z$. Due to (1.3.5) the origin is a boundary point of B. By the regular point condition and Theorem 1.4 with $T = g'(x^*)$ and $\bar{y} = g(x^*)$ there exists $\rho > 0$ such that

$$\{(\alpha, y) : \alpha \ge \max\{f'(x^*)x : x \in (C - x^*) \cap X_1\} \text{ and } y \in Z_\rho\} \subset B,$$

and hence int $B \ne \emptyset$. Consequently there exists a hyperplane in $\mathbb{R} \times Z$ which supports B at $(0, 0)$, i.e., there exists $(\alpha, \lambda^*) \ne (0, 0) \in \mathbb{R} \times Z^*$ such that

$$\alpha(f'(x^*)x + r) + \lambda^*(g'(x^*)x - y) \ge 0 \text{ for all } x \in C(x^*),$$
$$y \in K(g(x^*)), r \ge 0. \tag{1.3.6}$$

Setting $(r, x) = (0, 0)$ this implies that $\langle \lambda^*, y \rangle \le 0$ for all $y \in K(g(x^*))$ and consequently

$$\lambda^* \in K^* \quad \text{and} \quad \lambda^* g(x^*) = 0. \tag{1.3.7}$$

The choice $(x, y) = (0, 0)$ in (1.3.6) implies $\alpha \geq 0$. If $\alpha = 0$, then due to regularity of x^* we have $\lambda^* = 0$, which is impossible. Consequently $\alpha > 0$, and without loss of generality we can assume that $\alpha = 1$. Finally with $(r, y) = (0, 0)$, inequality (1.3.6) implies

$$f'(x^*) + \lambda^* g'(x^*) \in -C(x^*)^+.$$

This concludes the proof. $\square$

We close this subsection with two archetypical problems.

Equality-constrained problem. We consider

$$\begin{aligned} &\min f(x) \\ &\text{subject to } g(x) = 0. \end{aligned} \tag{1.3.8}$$

Let x^* denote a local solution at which (1.1.2) is satisfied and such that $g'(x^*) : X \to Z$ is surjective. Then there exists $\lambda^* \in Z^*$ such that the KKT condition

$$f'(x^*) + \lambda^* g'(x^*) = 0 \text{ in } X^*$$

holds. Surjectivity of $g'(x^*)$ is of course only a sufficient, not a necessary, condition for existence of a Lagrangian. This topic is taken up in Example 1.12 and below.

Inequality-constrained problem. This is the inequality-constrained problem with the so-called unilateral box, or simple constraints. We suppose that $Z = L^2(\Omega)$ with Ω a domain in $\mathbb{R}^n$ and consider

$$\begin{aligned} &\min f(x) \\ &\text{subject to } g(x(s)) \leq 0 \text{ for a.e. } s \in \Omega. \end{aligned} \tag{1.3.9}$$

Setting $C = X$ and $K = \{v \in L^2(\Omega) : v \leq 0\}$, this problem becomes a special case of (1.1.1). Let x^* denote a local solution at which (1.1.2) is satisfied and such that $g'(x^*) : X \to Z$ is surjective. Then there exists $\lambda^* \in L^2(\Omega)$ such that the KKT conditions

$$\begin{aligned} &f'(x^*) + \lambda^* g'(x^*) = 0 \text{ in } X^*, \\ &\lambda^*(s) \geq 0, \;\; g(x^*(s)) \leq 0, \;\; \lambda^*(s) g(x^*(s)) = 0 \text{ for a.e. } s \in \Omega \end{aligned} \tag{1.3.10}$$

hold. Again surjectivity of $g'(x^*)$ is a sufficient, but not a necessary, condition for (1.3.10) to hold.

For later use let us consider the case where the functionals $f'(x^*) \in X^*$ and $\lambda^* \in Z^*$ are represented by duality pairings, i.e.,

$$f'(x^*)(x) = \langle \mathcal{J}_X f'(x^*), x \rangle_{X^*,X}, \quad \lambda^*(z) = \langle \mathcal{J}_Z \lambda^*, z \rangle_{Z^*,Z}$$

for $x \in X$ and $z \in Z$. Here $\mathcal{J}_X$ is the canonical isomorphism from $\mathcal{L}(X, \mathbb{R})$ into the space of representations of continuous linear functionals on X with respect to the $\langle \cdot, \cdot \rangle_{X^*,X}$ duality pairing, and $\mathcal{J}_Z$ is defined analogously. In finite dimensions these isomorphisms are transpositions. With this notation $f'(x^*) + \lambda^* g'(x^*) = 0$ can be expressed as

$$\mathcal{J}_X f'(x^*) + g'(x^*)^* \mathcal{J}_Z \lambda^* = 0.$$

Henceforth the notation of the isomorphisms will be omitted.

1.4 Applications

This section is devoted to a discussion of examples in which the general existence result of the previous section is applicable, as well as to other examples in which it fails. Throughout Ω denotes a bounded domain in $\mathbb{R}^n$ with Lipschitz continuous boundary $\partial\Omega$. We use standard Hilbert space notation as introduced in [Ad], for example.

Differently from the previous sections we shall use J to denote the cost functional and f for inhomogeneities in equality constraints representing partial differential equations. Moreover the constraints $g(x) \in K$ from the previous sections will be split into equality constraints, denoted by $e(x) = 0$, and inequality constraints, $g(x) \in K$, with K a nontrivial cone.

Example 1.7. Consider the unilateral obstacle problem

$$\begin{cases} \min J(y) = \dfrac{1}{2}\displaystyle\int_\Omega |\nabla y|^2 dx - \int_\Omega f y dx \\ \\ \text{over } y \in H_0^1(\Omega) \text{ and } y(x) \le \psi(x) \text{ for a.e. } x \in \Omega, \end{cases} \tag{1.4.1}$$

where $f \in L^2(\Omega)$ and $\psi \in H_0^1(\Omega)$. This is a special case of (1.1.1) with $X = Z = H_0^1(\Omega)$, $K = \{\varphi \in H_0^1(\Omega) : \varphi(x) \le 0 \text{ for a.e. } x \in \Omega\}$, $C = X$, and $g(y) = y - \psi$. It is well known and simple to argue that (1.4.1) admits a unique solution $y^* \in H_0^1(\Omega)$. Moreover, since g' is the identity, every element of X satisfies the regular point condition. Hence by Theorem 1.6 there exists $\lambda^* \in H_0^1(\Omega)^*$ such that

$$\begin{aligned} &\langle \lambda^*, \varphi \rangle_{H_0^{-1}, H_0^1} \le 0 \text{ for all } \varphi \in K, \\ &\langle \lambda^*, y^* - \psi \rangle_{H_0^{-1}, H_0^1} = 0, \\ &-\Delta y^* + \lambda^* = f \text{ in } H^{-1}, \end{aligned}$$

where $H^{-1} = H_0^1(\Omega)^*$. Under well-known regularity assumptions on ψ and $\partial\Omega$, cf. [Fr, IK1], for example, $\lambda^* \in L^2(\Omega)$. This extra smoothness of the Lagrange multiplier does not follow from the general multiplier theory of the previous section.

The following result is useful to establish the regular point condition in diverse optimal control and parameter estimation problems. We require some notation. Let L be a closed convex cone in the real Hilbert space U which induces an ordering denoted according to $u \ge 0$ if $u \in L$. Let $q^* \in U^*$ be a nontrivial bounded linear functional on U, with $\pi : U \to \ker q^*$ the orthogonal projection. For $\varphi \in U$, $\mu \in \mathbb{R}$, and $\gamma \in \mathbb{R}^+$ define

$$g : U \to Z := U \times \mathbb{R} \times \mathbb{R}$$

by

$$g(u) = \left(u - \varphi, |u|^2 - \gamma^2, q^*(u) - \mu\right),$$

where $|\cdot|$ denotes the norm in U, and put $K = L \times \mathbb{R}^- \times \{0\} \subset Z$.

Proposition 1.8. *Assume that* $[\ker q^*]^\perp \cap L \ne \{0\}$ *and let* $h_0 \in [\ker q^*]^\perp \cap L$ *with* $q^*(h_0) = 1$. *If* $q^*(L) \subset \mathbb{R}^+$, $q^*(\varphi) < \mu$, *and* $|\pi(\varphi)|^2 + \mu^2 |h_0|^2 < \gamma^2$; *then the set*

$$M = \{u \in U : g(u) \in K\}$$

is nonempty and every element $u \in M$ is regular, i.e.,

$$0 \in \text{int}\ \{g(u) + g'(u)U - K\} \text{ for each } u \in M. \tag{1.4.2}$$

Proof. Note that $U = \ker q^* \oplus \text{span}\ \{h_0\}$ and that the orthogonal projection onto $[\ker q^*]^\perp$ is given by $q^*(u)h_0$ for $u \in U$. To show that M is nonempty, we define

$$\hat{u} = \varphi + (\mu - q^*(\varphi))\, h_0.$$

Observe that

$$\hat{u} - \varphi = (\mu - q^*(\varphi))\, h_0 \in L,$$
$$|\hat{u}|^2 = |\pi(\varphi)|^2 + \mu^2 |h_0|^2 < \gamma,$$

and $q^*(\hat{u}) = \mu$. Thus $\hat{u} \in M$. Next let $u \in M$ be arbitrary. We have to verify that

$$0 \in \text{int}\{ (u - \varphi + h - L, |u|^2 - \gamma^2 + 2(u, h) - \mathbb{R}^-, q^*(h)) : h \in U\} \subset Z, \tag{1.4.3}$$

where $(\cdot, \cdot)$ denotes the inner product in U. Put

$$\delta = \min \left(\frac{\gamma^2 - |\pi(\varphi)|^2 - \mu^2 |h_0|^2}{1 + 2\gamma + 2\gamma |h_0|}, \frac{\mu - q^*(\varphi)}{\| q^* \| + 1} \right),$$

and define $B = \{(\tilde{u}, \tilde{r}, \tilde{s}) \in Z : |(\tilde{u}, \tilde{r}, \tilde{s})|_Z \leq \delta\}$. Without loss of generality we endow the product space with the supremum norm in this proof. We shall show that

$$B \subset \{(u - \varphi + h_1 + \sigma h_0 - L, |u|^2 - \gamma^2 + 2(u, h_1 + \sigma h_0) - \mathbb{R}^-, \tag{1.4.4}$$
$$\sigma q^*(h_0)) : h_1 \in \ker\ q^*, \sigma \in \mathbb{R}\},$$

which implies (1.4.3). Let $(\tilde{u}, \tilde{r}, \tilde{s}) \in B$ be chosen arbitrarily. We put $\sigma = \tilde{s}$ and verify that there exists $(h_1, l, r^-) \in \ker q^* \times L \times \mathbb{R}^-$ such that

$$\left(u - \varphi + h_1 + \tilde{s} h_0 - l, |u|^2 - \gamma^2 + 2(u, h_1 + \tilde{s} h_0) - r^-\right) = (\tilde{u}, \tilde{r}). \tag{1.4.5}$$

This will imply the claim. Observe that by the choice of δ we find $\mu - q^*(\varphi) - q^*(\tilde{x}) + \tilde{s} \geq 0$. Hence l is defined by

$$l = (\mu - q^*(\varphi) - q^*(\tilde{u}) + \tilde{s})\, h_0 \in L.$$

Choosing $h_1 = \pi(\phi - u + \tilde{u})$ we obtain equality in the first coordinate of (1.4.5). For the second coordinate in (1.4.5) we observe that

$$\begin{aligned}
&|u|^2 - \gamma^2 + 2(u, h_1 + \tilde{s} h_0) \\
&= |\pi u|^2 + \mu^2 |h_0|^2 - \gamma^2 + 2(u, \pi(\varphi - u + \tilde{u}) + \tilde{s} h_0) \\
&\leq |\pi u|^2 + \mu^2 |h_0|^2 - \gamma^2 + |\pi u|^2 + |\pi \varphi|^2 - 2|\pi u|^2 + 2\gamma\, (|\tilde{u}|^2 + |\tilde{s}|\, |h_0|) \\
&\leq \mu^2 |h_0|^2 - \gamma^2 + |\pi \varphi|^2 + 2\delta\gamma \Big(1 + |h_0| \Big) \leq -\delta
\end{aligned}$$

by the definition of δ. Hence there exists $r^- \in \mathbb{R}^-$ such that equality holds in the second coordinate of (1.4.5). ☐

Remark 1.4.1. If the constraint $q^*(u) = \mu$ is not present in the definition of M, then the conclusion of Proposition 1.8 holds provided that $|\varphi| < \gamma$.

Example 1.9. We consider a least squares formulation for the estimation of the potential c in

$$\begin{cases} -\Delta y + cy = f & \text{in } \Omega, \\ \qquad\quad y = 0 & \text{on } \partial\Omega \end{cases} \tag{1.4.6}$$

from an observation $z \in L^2(\Omega)$, where $f \in L^2(\Omega)$ and $\Omega \subset \mathbb{R}^n$, with $n \leq 4$. For this purpose we set

$$M = \{c \in L^2(\Omega) : c(x) \geq 0, |c| \leq \gamma\}$$

for some $\gamma > 0$ and consider

$$\begin{cases} \min J(c) = \frac{1}{2}|y(c) - z|^2 + \frac{\alpha}{2}|c|^2 \\ \\ \text{subject to } c \in M \text{ and } y(c) \text{ solution to (4.6).} \end{cases} \tag{1.4.7}$$

The Lax–Milgram theorem implies the existence of a variational solution $y(c) \in H_0^1(\Omega)$ for every $c \in M$. Here we use the fact that $H^1(\Omega) \subset L^4(\Omega)$ and $cy \in L^{4/3}(\Omega)$ for $c \in L^2(\Omega)$, $y \in H^1(\Omega)$, for $n \leq 4$. Proposition 1.8 with $L = \{c \in L^2(\Omega) : c(x) \geq 0\}$ and

$$g(c) = \left(c, |c|^2 - \gamma^2\right)$$

is applicable and implies that every element of M is a regular point. Using subsequential weak limit arguments one can argue the existence of a solution c^* to (1.4.7). To apply Theorem 1.6 Fréchet differentiability of $c \to J(c)$ at c^* must be verified. This will follow from Fréchet differentiability of $c \to y(c)$ at every $c \in M$, which in turn can be argued by means of the implicit function theorem, for example. The Fréchet derivative of $c \to y(c)$ at $c \in M$ in direction $h \in L^2(\Omega)$ denoted by $y'(c)h$ satisfies

$$\begin{cases} -\Delta y'(c)h + cy'(c)h = -hy(c) & \text{in} \quad \Omega, \\ \qquad\qquad\quad y'(c)h = 0 & \text{on} \quad \partial\Omega. \end{cases} \tag{1.4.8}$$

With these preliminaries we obtain a Lagrange multiplier $(\mu_1, \mu_2) \in -L \times \mathbb{R}^+$ such that the following first order necessary optimality system holds:

$$\begin{cases} \left(y(c^*) - z, y'(c^*)h\right)_{L^2} + \alpha\,(c^*, h) - (\mu_1, h) + 2\mu_2(c^*, h) = 0 \\ \qquad\qquad\qquad\qquad\qquad\qquad\qquad \text{for all } h \in L^2(\Omega), \\ -(\mu_1, c^*) + \mu_2\left(|c^*|^2 - \gamma\right) = 0, \\ (\mu_1, \mu_2) \in L \times \mathbb{R}^+. \end{cases} \tag{1.4.9}$$

The first equation in (1.4.9) can be simplified by introducing $p(c^*)$ as the variational solution to the adjoint equation given by

$$\begin{cases} -\Delta p + c^* p = -\,(y(c^*) - z) & \text{in} \quad \Omega, \\ \qquad\qquad p = 0 & \text{on} \quad \partial\Omega. \end{cases} \tag{1.4.10}$$

Using (1.4.8) and (1.4.10) in (1.4.9) we obtain the optimality system

$$\begin{cases} p(c^*)\, y(c^*) + \alpha c^* - \mu_1 + 2\mu_2 c^* = 0, \\ (\mu_1, \mu_2) \in -L \times \mathbb{R}^+, \left(c^*,\ \left|c^*\right|^2 - \gamma^2\right) \in L \times \mathbb{R}^-, \\ (\mu_1, c^*) = 0,\ \mu_2 \left((\left|c^*\right|^2 - \gamma^2\right) = 0. \end{cases} \tag{1.4.11}$$

Example 1.10. We revisit Example 1.9 but this time the state variable is considered as an independent variable. This results in the following problem with equality and inequality constraints:

$$\begin{cases} \min J(y, c) = \frac{1}{2}|y - z|^2 + \frac{\alpha}{2}|c|^2 \\ \\ \text{subject to } e(y, c) = 0 \text{ and } g(c) \in L \times \mathbb{R}^-, \end{cases} \tag{1.4.12}$$

where g is as in Example 1.9 and $e : H_0^1(\Omega) \times L^2(\Omega) \to H^{-1}(\Omega)$ is defined by

$$e(y, c) = -\Delta y + cy - f.$$

Note that $cy \in L^{\frac{4}{3}}(\Omega) \subset H^{-1}(\Omega)$ for $c \in L^2(\Omega)$, $y \in H^1(\Omega)$ if $n \le 4$. Let $(c^*, y^*) \in L^2(\Omega) \times H_0^1(\Omega)$ denote a solution of (1.4.12). Fréchet differentiability of J, e, and g is obvious for this formulation with y as independent variable. The regular point condition now has to be established for the constraint

$$\begin{pmatrix} e \\ g \end{pmatrix} \in K = \{0\} \times L \times \mathbb{R}^- \subset H^{-1}(\Omega) \times L^2(\Omega) \times \mathbb{R}.$$

The second and third components are treated as in Example 1.9 by means of Proposition 1.8 and the first coordinate is incorporated by an existence argument for (1.4.6). The details are left to the reader. Consequently there exists a Lagrange multiplier $(p, \mu_1 \mu_2) \in H_0^1(\Omega) \times L \times \mathbb{R}^+$ such that the derivative of

$$\begin{aligned} &\mathcal{L}(y, c, p, \mu_1, \mu_2) \\ &\quad = J(y, c) + \langle e(y, c), p\rangle_{H^{-1}, H_0^1} + (\mu_1, -c) + \mu_2\left(\left|c^*\right|^2 - \gamma^2\right) \end{aligned}$$

with respect to (y, c) is zero at $(y^*, c^*, p, \mu_1, \mu_2)$. This implies that p is the variational solution in $H_0^1(\Omega)$ to

$$\begin{cases} -\Delta p + c^* p = -(y^* - z) & \text{in } \Omega, \\ \qquad\qquad p = 0 & \text{on } \partial\Omega \end{cases}$$

and

$$py^* + \alpha c^* - \mu_1 + 2\mu_2 c^* = 0.$$

In addition, of course,

$$\begin{aligned} &(\mu_1, \mu_2) \in -L \times \mathbb{R}^+, \left(c^*, \left|c^*\right|^2 - \gamma^2\right) \in L \times \mathbb{R}^-, \\ &(\mu_1, c^*) = 0, \quad \mu_2\left(\left|c^*\right|^2 - \gamma\right) = 0. \end{aligned}$$

Thus the first order optimality condition derived in Example 1.9 above and the present one coincide, and the adjoint variable of Example 1.9 becomes the Lagrange multiplier of the equality constraint $e(y, c) = 0$.

Example 1.11. We consider a least squares formulation for the estimation of the diffusion coefficient a in

$$\begin{cases} -(ay_x)_x + cy = f \text{ in } \Omega = (0, 1), \\ y(0) = y(1) = 0, \end{cases} \tag{1.4.13}$$

where $f \in L^2(0, 1)$ and $c \in L^2(\Omega)$, $c \geq 0$, are fixed. We assume that a is known outside of an interval $I = (\beta, \gamma)$ with $0 < \beta < \gamma < 1$ and that it coincides there with the values of a known function $\nu \in H^1(0, 1)$, which satisfies $\min\{\nu(x) : x \in (0, 1)\} > 0$. The set of admissible parameters is then defined by

$$M = \Big\{ a \in H^1(I) : a \geq \nu, \quad |a - \hat{a}|_{H^1(I)} \leq \gamma, \quad a(\beta) = \nu(\beta), \\ a(\gamma) = \nu(\gamma), \text{ and } \int_\beta^\gamma a\,dx = m \Big\}.$$

Here $\hat{a} \in H^1(I)$ is a fixed reference parameter with $\hat{a}(\beta) = \nu(\beta), \hat{a}(\gamma) = \nu(\gamma)$, and $m = \int_\beta^\gamma \hat{a}\,dx$. Recall that $H^1(I) \subset C(\bar{I})$ and hence $a(\beta)$ and $a(\gamma)$ are well defined. The coefficient a appearing in (1.4.13) coincides with ν on the complement of I and with some element of M on I. Consider the least squares formulation

$$\begin{cases} \min \frac{1}{2} |y(a) - z|^2_{L^2(0,1)} + \frac{\alpha}{2} |a|^2_{H^1(I)} \\ \\ \text{subject to } a \in M \text{ and } y(a) \text{ solution to (1.4.13)}, \end{cases} \tag{1.4.14}$$

where $z \in L^2(0, 1)$ is given. It is simple to argue the existence of a solution a^* to (1.4.14). Note that minimizing over M implies that the mean of the coefficient a is known. This, together with some additional technical assumptions, implies that the solution a^* is (locally) stable with respect to perturbations of z [CoKu]. To argue the existence of Lagrange multipliers associated to the constraints defining M one considers the set of shifted parameters $\tilde{M} = \{a - \hat{a} : a \in M\}$, i.e.,

$$\tilde{M} = \Big\{ a \in H^1(I) : a \geq \nu - \hat{a}, |a|_{H^1(I)} \leq \gamma, \int_\beta^\gamma a\,dx = 0, \quad a(\beta) = a(\gamma) = 0 \Big\}.$$

We apply Proposition 1.8 with $U = H_0^1(I)$, and $(|u|^2_{L^2(I)} + |u_x|^2_{L^2(I)})^{1/2}$ as norm, $L = \{a \in U : a \geq 0\}$, $q^*(a) = \int_\beta^\gamma a\,dx$, $\varphi = \nu - \hat{a} \in U$, and $\mu = 0$. The mapping $g : U \to U \times \mathbb{R} \times \mathbb{R}$ is given by

$$g(a) = \left(a - \varphi, |a|^2_{H^1(I)} - \gamma^2, q^*(a) \right).$$

Let us assume that $\int_\beta^\gamma \nu\,dx < m$ and $|\nu - \hat{a}|_{H^1(I)} < \gamma$. Then we have $q^*(L) \subset \mathbb{R}^+$, $q^*(\varphi) < m - \int_\beta^\gamma \hat{a} dx = \mu$, and

$$|\pi\varphi|_{H^1(I)} = |\pi(\nu - \hat{a})|_{H^1(I)} \leq |\nu - \hat{a}|_{H^1(I)} < \gamma.$$

It remains to ascertain that $[\ker q^*]^\perp \cap L$ is nonempty. Let ψ be the unique solution of

$$\begin{cases} -\psi_{xx} + \psi = 1 \text{ on } (\beta, \gamma), \\ \psi(\beta) = \psi(\gamma) = 0. \end{cases}$$

By the maximum principle $\psi \in L$. A short calculation shows that $\psi \in [\ker q^*]^\perp$ and hence $h_0 := q^*(\psi)^{-1}\psi$ satisfies $h_0 \in [\ker q^*]^\perp \cap L$ and $q^*(h_0) = 1$.

Next we present examples where the general existence result of Section 1.3 is not applicable. Different techniques to be developed in the following sections will guarantee, however, the existence of a Lagrange multiplier. In these examples we shall again use e for equality and g for inequality constraints.

Example 1.12. Consider the problem

$$\begin{cases} \min x_1^2 + x_3^2 \\ \text{subject to} \quad e(x_1, x_2, x_3) = 0, \end{cases} \tag{1.4.15}$$

where

$$e(x_1, x_2, x_3) = \begin{pmatrix} x_1 - x_2^2 - x_3 \\ x_2^2 - x_3^2 \end{pmatrix}.$$

Here $X = \mathbb{R}^3$, $Z = \mathbb{R}^2$, e of Section 1.3 is e here, and $K = \{0\}$. Note that $x^* = (0, 0, 0)$ is a solution of (1.4.15) and that $\nabla e(x^*)$ is not surjective. Hence x^* is not a regular point for the constraint $e(x) = 0$. Nevertheless $(0, \lambda_2)$ with $\lambda_2 \subset \mathbb{R}$ arbitrary is a Lagrange multiplier for the constraint in (1.4.15).

Note that in case $C = X$ and $K = \{0\}$, the last requirement in Definition 1.1 becomes

$$f'(x^*) + e'(x^*)^*\lambda^* = 0 \text{ in } X^*,$$

where $e'(x^*)^* : Z^* \to X^*$ is the adjoint of $e'(x^*)$. Hence if $e'(x^*)$ is not surjective and if it has closed range, then $\ker e'(x^*)^*$ is not empty and the Lagrange multiplier is not unique.

Example 1.13. Here we consider optimal control of the equation with elliptic nonlinearity (Bratu problem)

$$\begin{cases} -\Delta y + \exp(y) = u \text{ in } \Omega, \\ \frac{\partial y}{\partial n} = 0 \text{ on } \Gamma \text{ and } y = 0 \text{ on } \partial\Omega \setminus \Gamma, \end{cases} \tag{1.4.16}$$

where Ω is a bounded domain in $\mathbb{R}^n$, $n \le 3$, with smooth boundary $\partial\Omega$ and Γ is a connected, strict subset of $\partial\Omega$. Let $H^1_\Gamma(\Omega) = \{y \in H^1(\Omega) : y = 0 \text{ on } \partial\Omega \setminus \Gamma\}$. The following lemma establishes the concept of solution to (1.4.16). Its proof is given at the end of this section.

Lemma 1.14. *For every $u \in L^2(\Omega)$ the variational problem*

$$(\nabla y, \nabla v) + \langle \exp y, v \rangle_{H^1_\Gamma(\Omega)^*, H^1_\Gamma(\Omega)} = (u, v) \textit{ for } v \in H^1_\Gamma(\Omega)$$

has a unique solution $y = y(u) \in H^1_\Gamma(\Omega)$ and there exists a constant C such that

$$|y|_{H^1_\Gamma \cap L^\infty} \le C(|u|_{L^2(\Omega)} + C) \textit{ for all } u \in L^2(\Omega) \tag{1.4.17}$$

and

$$|y(u_1) - y(u_2)|_{H^1_\Gamma \cap L^\infty} \le C|u_1 - u_2|_{L^2(\Omega)} \textit{ for all } u_i \in L^2(\Omega), i = 1, 2. \tag{1.4.18}$$

The optimal control problem is given by

$$\begin{cases} \min J(y,u) = \frac{1}{2}\,|y-z|^2_{L^2(\Omega)} + \frac{\alpha}{2}\,|u|^2_{L^2(\Omega)} \\ \text{subject to } (y,u) \in \big(H^1_\Gamma(\Omega) \cap L^\infty(\Omega)\big) \times L^2(\Omega) \\ \text{and } (y,u) \text{ solution to (1.4.17),} \end{cases} \tag{1.4.19}$$

where $z \in L^2(\Omega)$. To consider this problem as a special case of (1.1.1) we set $X = Y \times L^2(\Omega)$, where $Y = H^1_\Gamma(\Omega) \cap L^\infty(\Omega)$, $Z = H^1_\Gamma(\Omega)^*$, $K = \{0\}$, and define $e : X \to Z^*$ as the mapping which assigns to $(y, u) \in Y \times U$ the functional

$$v \to (\nabla y, \nabla v) + (\exp(y), v) - (u, v).$$

The control component of any minimizing sequence $\{(y_n, u_n)\}$ for J is bounded. Together with Lemma 1.14 it is therefore simple to argue that (1.4.19) admits a solution $(y^*, u^*) \in Y \times U$. Clearly J and e are continuously Fréchet differentiable. If (y^*, u^*) was a regular point, then this would require surjectivity of $e'(y^*, u^*) : X \to H^1_\Gamma(\Omega)^*$. For $(\delta y, \delta u) \in X$, $e'(y^*, u^*)(\delta y, \delta u)$ is the functional defined by

$$v \to (\nabla \delta y, \nabla v) + (\exp(y^*)\delta y - \delta u, v), \qquad v \in H^1_\Gamma(\Omega).$$

Since $-\Delta + \exp(y^*)$ is an isomorphism from $H^1_\Gamma(\Omega)$ to $H^1_\Gamma(\Omega)^*$ and $H^1_\Gamma(\Omega)$ is not contained in $L^\infty(\Omega)$, if $n > 1$, it follows that $e'(y^*, u^*) : X \to H^1_\Gamma(\Omega)^*$ is not surjective if $n > 1$.

Example 1.15. We consider the least squares problem for the estimation of the vector-valued convection coefficient u in

$$\begin{cases} -\Delta y + u \cdot \nabla y = f \text{ in } \Omega, \\ y = 0 \text{ on } \partial\Omega, \ \operatorname{div} u = 0, \end{cases} \tag{1.4.20}$$

from data $z \in L^2(\Omega)$. Here Ω is a bounded domain in $\mathbb{R}^n$, $n \le 3$, with smooth boundary and $f \in L^2(\Omega)$. To cast the problem in abstract form we choose $U = \{u \in L^2_n(\Omega) : \operatorname{div} u = 0\}$, $X = \big(H^1_0(\Omega) \cap L^\infty(\Omega)\big) \times U$, $Z = H^{-1}(\Omega)$, $K = \{0\}$ and define $e : X \to Z$ as the mapping which assigns to $(y, u) \in X$ the functional

$$v \to (\nabla y, \nabla v) - (uy, \nabla v) - (f, v) \quad \text{for } v \in H^1_0(\Omega).$$

Note that $e(y, u)$ is not well defined for $(y, u) \in H^1_0(\Omega) \times U$, since $u \cdot \nabla y \in L^1(\Omega)$ only. The regularized least squares problem is given by

$$\begin{cases} \min J(y,u) = \frac{1}{2}\,|y-z|^2_{L^2(\Omega)} + \frac{\alpha}{2}\,|u|^2_{L^2_n(\Omega)} \\ \text{subject to } e(y,u) = 0, \ \operatorname{div} u = 0, \quad (y,u) \in X. \end{cases} \tag{1.4.21}$$

Observe that $(u \cdot \nabla y, y) = (\nabla \cdot (uy), y) = 0$. Using this fact and techniques similar to those of the proof of Lemma 1.14 (compare [Tr, Section 2.3] and [IK14, IK15]) it can be shown that for every $u \in U$ there exists a unique solution $y = y(u) \in X$ of (1.4.20) and for every bounded subset B of U there exists a constant $k(B)$ such that

$$|y(u)|_X \le k(B)|u|_{L^2_n} \text{ for all } u \in B. \tag{1.4.22}$$

Extracting a subsequence of a minimizing sequence to (1.4.21) it is simple to argue the existence of a solution $(y^*, u^*) \in X$ to (1.4.21). Clearly e is Fréchet differentiable at (y^*, u^*) and for $\delta y \in H^1_0(\Omega) \cap L^\infty(\Omega)$, $e_y(y^*, u^*)\delta y \in H^{-1}$ is the functional defined by

$$\langle e_y(y^*, u^*)\delta y, v\rangle_{H^{-1},H^1_0} = (\nabla\delta y, \nabla v) + (u^* \cdot \nabla\delta y, v) \text{ with } v \in H^1_0(\Omega).$$

Note that $e_y(y^*, u^*)$ is well defined on $H^1_0(\Omega) \cap L^\infty(\Omega)$. But as a consequence of the bilinear term $u^* \cdot \nabla\delta y$ which is only in $L^1(\Omega)$, $e_y(y^*, u^*)$ is not defined on $H^1_0(\Omega)$. The operator $e'(y^*, u^*)$ from X to $H^{-1}(\Omega)$ is not surjective if $n > 1$, and hence (y^*, u^*) does not satisfy the regular point condition.

We now turn to the proof of Lemma 1.14 for which we require the following result from [Tr].

Lemma 1.16. *Let $\varphi : (k_1, h_1) \to \mathbb{R}$ be a nonnegative, nonincreasing function and suppose that there are positive constants r, K, and β, with $\beta > 1$, such that*

$$\varphi(h) \le K(h-k)^{-r}\varphi(k)^\beta \textit{ for } k_1 < k < h < h_1.$$

If $\hat{k} := K^{\frac{1}{r}} 2^{\frac{\beta}{\beta-1}} \varphi(k_1)^{\frac{\beta-1}{r}}$ satisfies $k_1 + \hat{k} < h_1$, then $\varphi(k_1 + \hat{k}) = 0$.

***Proof of Lemma* 1.14.** Let us first argue the existence of a solution $y = y(u) \in$ of

$$(\nabla y, \nabla v) + (e^y, v)_{H^1_\Gamma(\Omega)^*, H^1_\Gamma(\Omega)} = (u, v) \text{ for all } v \in H^1_\Gamma(\Omega). \tag{1.4.23}$$

Since $(\exp(s_1) - \exp(s_2))(s_1 - s_2) \ge 0$ for $s_1, s_2 \in \mathbb{R}$, it follows that $-\Delta\phi + \exp(\phi) - \gamma I$ defines a maximal monotone operator from a subset of $H^1_\Gamma(\Omega)$ to $H^1_\Gamma(\Omega)^*$ for some $\gamma > 0$ [Ba], and hence (1.4.23) has a unique variational solution $y = y(u) \in H^1_\Gamma(\Omega)$ for every $u \in L^2(\Omega)$; see [IK15] for details. Since

$$|\nabla(y(u_1) - y(u_2))|^2 \le (u_1 - u_2, y(u_1) - y(u_2))$$

and $\partial\Omega \setminus \Gamma$ is nonempty, there exists an embedding constant C such that

$$|y(u_1) - y(u_2)|_{H^1_\Gamma(\Omega)} \le C\,|u_1 - u_2|_{L^2(\Omega)} \tag{1.4.24}$$

for u_1, u_2 in $L^2(\Omega)$, and

$$|y(u)|_{H^1_\Gamma} \le C(|u|_{L^2(\Omega)} + C) \text{ for all } u \in L^2(\Omega).$$

Throughout the remainder of the proof C will denote a generic constant, independent of $u \in L^2(\Omega)$. To verify (1.4.17) it remains to obtain an $L^\infty(\Omega)$ bound for $y = y(u)$. The proof is based on a generalization of well-known $L^\infty(\Omega)$ estimates due to Stampacchia and Miranda [Tr] for linear variational problems to the nonlinear problem (1.4.23). Let us aim first for a pointwise (a.e.) upper bound for y. For $k \in (0, \infty)$ we set $y_k = (y-k)^+$ and $\Omega_k = \{x \in \Omega : y_k > 0\}$. Note that $y_k \in H^1_\Gamma(\Omega)$ and $y_k \geq 0$. Using (1.4.23) we find

$$(\nabla y_k, \nabla y_k) = (\nabla y, \nabla y_k) = (u, y_k) - (e^y, y_k) \leq (u, y_k),$$

and hence

$$|\nabla y_k|^2_{L^2} \leq |u|_{L^2} |y_k|_{L^2}. \tag{1.4.25}$$

By Hölder's inequality and a well-known embedding result

$$|y_k|_{L^2} = \left(\int_{\Omega_k} y_k^2 \right)^{\frac{1}{2}} \leq |y_k|_{L^6} |\Omega_k|^{\frac{1}{3}} \leq C |\nabla y_k|_{L^2} |\Omega_k|^{\frac{1}{3}}.$$

Here we used the assumption that $n \leq 3$. Employing this estimate in (1.4.25) implies that

$$|\nabla y_k|_{L^2} \leq C |\Omega_k|^{\frac{1}{3}} |u|_{L^2}. \tag{1.4.26}$$

We denote by h and k arbitrary real numbers satisfying $0 < k < h < \infty$, and we find

$$|y_k|^4_{L^4} = \int_{\Omega_k} (y-k)^4 > \int_{\Omega_h} (y-k)^4 \geq |\Omega_h| (h-k)^4,$$

which, combined with (1.4.26), gives

$$|\Omega_h| \leq \hat{C} (h-k)^{-4} |\Omega_k|^{\frac{4}{3}} |u|^4_{L^2}, \tag{1.4.27}$$

where the constant $\hat{C}$ is independent of h, k, and u. It will be shown that Lemma 1.16 is applicable to (1.4.27) with $\varphi(k) = |\Omega_k|$, $\beta = \frac{4}{3}$ and $K = \hat{C} |u|^4_{L^2}$. The conditions on k_1 and h_1 can easily be satisfied. In fact, in our case $k_1 = 0$, $h_1 = \infty$, and $\hat{k} = \hat{C}^{\frac{1}{4}} |u|_{L^2} 2^{\frac{\beta}{\beta-1}} |\Omega_0|^{\frac{\beta-1}{4}}$. The condition $k_1 + \hat{k} < h_1$ is satisfied since

$$\hat{k} = \hat{C}^{\frac{1}{4}} |u|_{L^2} 2^{\frac{\beta}{\beta-1}} |\Omega_0|^{\frac{\beta-1}{4}} < \hat{C}^{\frac{1}{4}} |u|_{L^2} 2^{\frac{\beta}{\beta-1}} |\Omega|^{\frac{\beta-1}{4}} < \infty.$$

We conclude that $|\Omega_{\hat{k}}| = 0$ and hence $y \leq \hat{k}$ a.e. in Ω. A uniform lower bound on y can be obtained in an analogous manner by considering $y_k = (-(k+y))^+$. We leave the details to the reader. This concludes the proof of (1.4.17). To verify (1.4.18) the H^1 estimate for $y(u_1) - y(u_2)$ is already clear from (1.4.24) and it remains to verify the $L^\infty(\Omega)$ estimate. Let us set $y_i = y(u_i)$, $z = y_1 - y_2$, $z_k = (z-k)^+$, and $\Omega_k = \{x \in \Omega : z_k > 0\}$ for $k \in (0, \infty)$. We obtain

$$|\nabla z_k|^2_{L^2} = (\nabla z, \nabla z_k) = (u_1 - u_2, z_k) - (e^{y_1} - e^{y_2}, z_k) \leq (u_1 - u_2, z_k).$$

Proceeding as above with y and y_k replaced by z and z_k the desired pointwise upper bound for $y_1 - y_2$ is obtained. For the lower bound we define $z_k = (-(k+z))^+$ for $k \in (0, \infty)$ and $\Omega_k = \{x \in \Omega : z_k > 0\} = \{x : k + y_1(x) < y_2(x)\}$. It follows that

$$|\nabla z_k|^2_{L^2} = -(\nabla(y_1 - y_2), \nabla z_k) = (e^{y_1} - e^{y_2}, z_k) - (u_1 - u_2, z_k) \leq -(u_1 - u_2, z_k).$$

From this inequality we obtain the desired uniform pointwise lower bound on $y_1 - y_2$. □

1.5 Weakly singular problems

We consider the optimization problem with equality and inequality constraints of the type

$$\begin{cases} \min J(y,u) \\ \text{subject to } e(y,u)=0, \quad u\in C\subset U, \end{cases} \tag{1.5.1}$$

with $J: Y\times U\to\mathbb{R}$, $e: Y_1\times U\to W$, where Y, U, W are Hilbert spaces and Y_1 is a Banach space that is densely embedded in Y. Further C is a closed convex subset of U. Below W^* will denote the dual of W. Let (y^*, u^*) denote a local solution to (1.5.1) and let $V(y^*)\times V(u^*)\subset Y_1\times U$ denote a neighborhood of (y^*, u^*) such that $J(y^*, u^*)\le J(y,u)$ for all $(y,u)\in V(y^*)\times(V(u^*)\cap C)$ satisfying $e(y,u)=0$. It is assumed throughout that J is Fréchet differentiable in a neighborhood in the $Y\times U$ topology of (y^*, u^*) and that the Fréchet derivative is locally Lipschitz continuous. Further e is assumed to be Fréchet differentiable at (y^*, u^*) with Fréchet derivative

$$e'(y^*, u^*)(\delta y, \delta u) = e_y(y^*, u^*)\delta y + e_u(y^*, u^*)\delta u.$$

In particular $e_y(y^*, u^*)\in\mathcal{L}(Y_1, W)$. Since Y_1 is dense in Y, one may consider $e_y(y^*, u^*)$ as densely defined linear operator with domain in Y. To distinguish this operator from $e_y(y^*, u^*)$ defined on Y_1 we shall denote it in this section by

$$G: D(G)\subset Y\to W$$

and we assume that

(H1) $G^*: D(G^*)\subset W^*\to Y^*$ is densely defined.

Then necessarily G is closable [Ka, p. 168]. Its closure will be denoted by the same symbol. In addition the following assumptions will be required:

(H2) $J_y(y^*, u^*)\in Rg\, G^*$.

Condition (H2) is a regularity assumption. It implies the existence of a solution $\lambda^*\in D(G^*)$ to the adjoint equation

$$G^*\lambda + J_y(y^*, u^*) = 0,$$

which is a Lagrange multiplier associated to the constraint $e(y,u)=0$.

(H3) There exists a dense subset D of C with the following property: For every $u\in D$ there exists $t_u>0$ such that for all $t\in[0,t_u]$ there exists $y(t)\in Y_1$ satisfying $e(y(t), u^*+t(u-u^*))=0$ and

$$\lim_{t\to 0^+}\frac{1}{t}|y(t)-y^*|_Y^2 = 0. \tag{1.5.2}$$

(H4) For every $u\in D$ and $y(\cdot)$ as in (H3), e is directionally differentiable at every element of $\{(y^*+s(y(t)-y^*), u^*+st(u-u^*)) : s\in[0,1], t\in[0,t_u]\}$ in all directions $(\tilde{y},\tilde{u})\in Y_1\times U$ and

$$\lim_{t\to 0^+}\frac{1}{t}\left\langle\int_0^1 [e'(y^*+s(y(t)-y^*), u^*+stv)-e'(y^*, u^*)](y(t)-y^*, tv)ds, \lambda^*\right\rangle_{W,W^*} = 0,$$

where $v = u-u^*$.

Note that (H4) is satisfied if (H3) holds with Y replaced by Y_1 and if $e : Y_1 \to W$ is Fréchet differentiable with locally Lipschitzian derivative.

Our assumptions do not require surjectivity of $e'(y^*, u^*) : Y_1 \times U \mapsto W$ which is required by (1.3.1) nor that $e'(y^*, u^*)$ is well defined on all of $Y \times U$. Below we shall give several examples which illustrate the applicability of the assumptions. For now, we argue that if they are satisfied, then a Lagrange multiplier with respect to the equality constraint exists.

Theorem 1.17. *Let* (y^*, u^*) *be a local solution of* (1.5.1) *and assume that* (H1)–(H4) *hold. Then*

$$\begin{cases} e(y^*, u^*) = 0 & \textit{in } W & \textit{(primal equation)}, \\ G^*\lambda^* + J_y(y^*, u^*) = 0 & \textit{in } Y^* & \textit{(adjoint equation)}, \\ \left\langle e_u(y^*, u^*)^*\lambda^* + J_u(y^*, u^*), u - u^* \right\rangle_{U^*,U} \geq 0 & \textit{for all } u \in C & \textit{(optimality)}. \end{cases} \tag{1.5.3}$$

The system of equations (1.5.3) is referred to as an optimality system.

Proof. Let $u \in D$, set $v = u - u^*$, choose t_u according to (H3), and assume that $t \in (0, t_u]$. Due to (H3) and (H4)

$$\begin{aligned} 0 = e(y(t), u^* + tv) - e(y^*, u^*) &= G(y(t) - y^*) + t\, e_u(y^*, u^*)\, v \\ &+ \int_0^1 [e'(y^* + s(y(t) - y^*), u^* + stv) - e'(y^*, u^*)](y(t) - y^*, tv)ds. \end{aligned} \tag{1.5.4}$$

(H2) implies the existence of a solution λ^* to the adjoint equation. Observe that by (1.5.4) and the fact that u^* is a local solution to (1.5.1)

$$\begin{aligned} 0 \leq J(y(t), u^* + tv) - J(y^*, u^*) &= J'(y^*, u^*)(y(t) - y^*, tv) \\ &+ \int_0^1 \left[J'(y^* + s(y(t) - y^*), u^* + stv) - J'(y^*, u^*)\right](y(t) - y^*, tv)ds \\ &+ \langle G^*\lambda^*, y(t) - y^*\rangle_{Y^*,Y} + t\langle e_u(y^*, u^*)v, \lambda^*\rangle_{W,W^*} \\ &+ \int_0^1 \left\langle \left[e'(y^* + s(y(t) - y^*), u^* + stv) - e'(y^*, u^*)\right](y(t) - y^*, tv), \lambda^* \right\rangle_{W,W^*} ds. \end{aligned}$$

By the second equation in (1.5.3), local Lipschitz continuous differentiability of J, (H3), (H4), and the fact that $J'(y^*, u^*)(y(t) - y^*, tv) = J_y(y^*, u^*)(y(t) - y^*) + tJ_u(y^*, u^*)v$ we obtain

$$0 \leq \lim_{t \to 0^+} \frac{1}{t} J(y(t), u^* + tv) - J(y^*, u^*)) = \langle J_u(y^*, u^*)(u^*) + e_u(y^*, u^*)^*\lambda^*, u - u^*\rangle_{U^*,U}.$$

Since u is arbitrary in D and D is dense in C it follows that

$$\langle J_u(y^*, u^*)(u^*) + e_u(y^*, u^*)^*\lambda^*, u - u^*\rangle_{U^*,U} \geq 0 \text{ for all } u \in C.$$

This ends the proof. $\square$

We next give several examples which demonstrate the applicability of hypotheses (H1)–(H4) and the necessity to allow for two spaces Y_1 and Y with $Y_1 \subsetneq Y$. The typical situation that we have in mind is $Y_1 = Y \cap L^\infty(\Omega)$ with Y a Hilbertian function space over Ω.

Example 1.18. Consider first the finite-dimensional equality-constrained optimization problem

$$\begin{cases} \min y_1^2 + u^2, \\ y_1 - y_2^2 = u, \\ y_2^3 = u^2 \end{cases} \tag{1.5.5}$$

for which $(y^*, u^*) = (0, 0, 0)$ is the solution. Here $Y = \mathbb{R}^2$, $U = \mathbb{R}^1$, $W = \mathbb{R}^1$, and

$$e(y, u) = \begin{pmatrix} y_1 - y_2^2 - u \\ y_2^3 - u^2 \end{pmatrix}.$$

Note that with $(y_1, y_2, u) = (x_1, x_2, x_3)$ this problem coincides with Example 1.12. We recall that $e'(y^*, u^*)$ is not surjective and the theory of Section 1.3 assuring the existence of a Lagrange multiplier is therefore not applicable. However, $G^* = \left(\begin{smallmatrix} 1 & 0 \\ 0 & 0 \end{smallmatrix}\right)$ and the adjoint equation

$$G^* \begin{pmatrix} \lambda_1 \\ \lambda_2 \end{pmatrix} = \begin{pmatrix} 0 \\ 0 \end{pmatrix}$$

has infinitely many solutions. Thus (H1), (H2) are satisfied. As for (H3) note that $\left(\begin{smallmatrix} y_1(u) \\ y_2(u) \end{smallmatrix}\right) = \left(\begin{smallmatrix} u+u^{\frac{4}{3}} \\ u^{\frac{2}{3}} \end{smallmatrix}\right)$ defines a solution branch to $e(y, u) = 0$ for which (H3) is satisfied. It is simple to verify (H4). Hence (1.5.5) is an example for an optimization problem where all hypotheses (H1)–(H4) are satisfied and Theorem 1.17 is applicable.

Example 1.19. We consider the optimal control problems with distributed control

$$\min J(y, u) = \frac{1}{2}|y - z|^2_{L^2(\Omega)} + \frac{\alpha}{2}|u|^2_{L^2(\Omega)} \tag{1.5.6}$$

subject to

$$\begin{cases} -\Delta y + \exp(y) = u \text{ in } \Omega, \\ \frac{\partial y}{\partial n} = 0 \text{ on } \Gamma, \\ y = 0 \text{ on } \partial\Omega \setminus \Gamma, \end{cases} \tag{1.5.7}$$

where $\alpha > 0$, $z \in L^2(\Omega)$, Ω is a bounded domain in $\mathbb{R}^n$, $n \le 3$, with smooth boundary $\partial\Omega$, and Γ is a connected, strict subset of $\partial\Omega$. The concept of solution to (1.5.7) is the variational one of Lemma 1.14. To consider this problem in the general setting of this section the control variable u is chosen in $U = L^2(\Omega)$. We set $Y = H^1_\Gamma(\Omega) = \{y \in H^1(\Omega) : y = 0 \text{ on } \partial\Omega\backslash\Gamma\}$, $Y_1 = H^1_\Gamma(\Omega) \cap L^\infty(\Omega)$, and $W = (H^1_\Gamma(\Omega))^*$. Moreover $e : Y_1 \times U \to W$ is defined by assigning to $(y, u) \in Y_1 \times U$ the functional on W^* given by

$$v \to (\nabla y, \nabla v) + (\exp y, v) - (u, v) \text{ for } v \in H^1_\Gamma(\Omega).$$

Let $(y^*, u^*) \in Y_1$ denote a solution to (1.5.6), (1.5.7). For $(y, u) \in Y_1 \times L^2(\Omega)$ the Fréchet derivative $e_y(y, u)\delta y$ of e with respect to y in direction $\delta y \in H^1_\Gamma(\Omega)$ is given by the functional $v \to (\nabla \delta y, \nabla v) + ((\exp y)\delta y, v)$. Clearly $e_y(y, u) : Y \to W$ is symmetric and (H1), (H2) are satisfied. (H3) is a direct consequence of Lemma 1.14. To verify (H4) note that $e'(y, u)(\delta y, \delta u)$ is the functional defined by

$$v \to (\nabla \delta y, \nabla v) + (\exp(y)\delta y, v) - (\delta u, v)_\Omega \text{ for } v \in H^1_\Gamma(\Omega).$$

If $(y, u) \in Y_1 \times U$, then $e'(y, u)$ is well defined on $Y \times U$. (H4) requires us to consider

$$\lim_{t\to 0^+} \frac{1}{t} \int_0^1 \int_\Omega (\exp(y^* + s(y(t) - y^*)) - \exp y^*)(y(t) - y^*)\lambda^* dx ds, \tag{1.5.8}$$

where $y(t)$ is the solution to (1.5.7) with $u = u^* + tv$, $v \in U$. Note that $\lambda^* \in L^\infty$ and that $\{|y(t)|_{L^\infty} : t \in [0, 1]\}$ is bounded by Lemma 1.14. Moreover $|y(t) - y^*|_{Y_1} \le c\, t |v|_{L^2}$ for a constant c independent of $t \in [0, 1]$, and thus the pointwise local Lipschitz property of the exponential function implies that the limit in (1.5.8) is zero. (H4) now easily follows.

The considerations of this example remain correct for cost functionals J that are much more general than the one in (1.5.6). In fact, it suffices that J is weakly lower semicontinuous from $Y \times U$ to $\mathbb{R}$ and radially unbounded with respect to u, i.e., $\lim_{n\to\infty} |u_n|_{L^2} = \infty$ implies that $\limsup_{n\to\infty} \inf_{y\in Y} J(y, u_n) = \infty$. This guarantees existence of a solution (y^*, u^*). The general regularity assumptions and (H1)–(H4) are satisfied if $J : Y \times U \to \mathbb{R}$ is continuous and Fréchet differentiable in a neighborhood of (y^*, u^*) with locally Lipschitz continuous derivative.

Example 1.20. Consider the nonlinear optimal boundary control problem

$$\min J(y, u) = \frac{1}{2}|y - z|^2_{L^2(\Omega)} + \frac{\alpha}{2}|u|^2_{H^s(\Gamma)} \tag{1.5.9}$$

subject to

$$\begin{cases} -\Delta y + \exp y = f \text{ in } \Omega, \\ \frac{\partial y}{\partial n} = u \text{ on } \Gamma, \\ y = 0 \text{ on } \partial\Omega \setminus \Gamma, \end{cases} \tag{1.5.10}$$

where $\alpha > 0$, $z \in L^2(\Omega)$, $f \in L^\infty$ are fixed, Ω is a bounded domain in $\mathbb{R}^n$, with smooth boundary $\partial\Omega$, and Γ is a nonempty, connected, strict subset of $\partial\Omega$. Further s is a real number strictly larger than $\frac{n-3}{2}$ if $n \ge 3$, and $s = 0$ if $n < 3$. Differently from Example 1.19 the dimension n of Ω is now allowed to be arbitrary. This example can be treated within the general framework of this book by setting Y, Y_1, W as in Example 1.19. The control space U is chosen to be $H^s(\Gamma)$. To verify (H1)–(H4) one proceeds as in Example 1.19 by utilizing the following lemma. Its proof is given in [IK15] and is similar to that of Lemma 1.14.

Lemma 1.21. *The variational problem*

$$(\nabla y, \nabla v) + (\exp y, v) = (f, v) + (u, v)_\Gamma, \quad v \in H^1_\Gamma(\Omega),$$

has a unique solution $y = y(u) \in H^1_\Gamma(\Omega) \cap L^\infty(\Omega)$ *for every* $u \in H^s(\Gamma)$*, and there exists a constant c such that*

$$|y|_{H^1_\Gamma \cap L^\infty} \le c(|u|_{H^s(\Gamma)} + c) \text{ for all } u \in H^s(\Gamma). \tag{1.5.11}$$

Moreover, c can be chosen such that

$$|y(u_1) - y(u_2)|_{H^1_\Gamma(\Omega)\cap L^\infty} \le c|u_1 - u_2|_{H^s} \text{ for all } u_i \in H^s(\Gamma), i = 1, 2. \tag{1.5.12}$$

Example 1.22. We reconsider (1.4.21) from Example 1.15 as a special case of (1.5.1). For this purpose we set $Y = H^1_0(\Omega)$, $Y_1 = H^1_0(\Omega) \cap L^\infty(\Omega)$, $W = H^{-1}(\Omega)$, and e as in Example 1.15. Let $(y^*, u^*) \in Y_1 \times U$ denote a solution of (1.4.21). Clearly e is Fréchet differentiable at (y^*, u^*) and the partial derivative $G = e_y(y^*, u^*) \in \mathcal{L}(Y_1, W)$ is the functional characterized by

$$\langle e_y(y^*, u^*)\delta y, v\rangle_{W,W^*} = (\nabla\delta y, \nabla v) + (u^* \cdot \nabla\delta y, v), \; v \in W^* = H^1_0(\Omega).$$

As a consequence of the quadratic term $u^* \cdot \nabla\delta y$ which is only in $L^1(\Omega)$, G is not defined on all of $Y = H^1_0(\Omega)$. As an operator from Y_1 to W, the operator G is not surjective. Considered as an operator with domain in Y, its adjoint is given by

$$G^* w = -\Delta w - \nabla \cdot (u^* w).$$

The domain of G^* contains Y_1 and hence G^* is densely defined. Moreover its range contains $L^2(\Omega)$ and thus (H1) as well as (H2) are satisfied. Let $U(u^*) \subset U$ be a bounded neighborhood of u^*. Since for every $u \in U(u^*)$

$$(\nabla(y(u) - y^*), \nabla v) - (u(y(u) - y^*), \nabla v) = ((u - u^*)y^*, \nabla v) \text{ for all } v \in H^1_0(\Omega),$$

it follows that there exists a constant $k > 0$ such that

$$|y(u) - y^*|_{H^1} \le k|u - u^*|_{L^2_n} \text{ for all } u \in U(u^*), \tag{1.5.13}$$

and (H3) follows. The validity of (H4) is a consequence of (1.5.13) and the fact that λ^* is the unique variational solution in $H^1_0(\Omega)$ to

$$-\Delta\lambda^* - \nabla \cdot (u^*\lambda^*) = -(y^* - z)$$

and hence an element of $L^\infty(\Omega)$.

Remark 1.5.1. Comparing Examples 1.19 and 1.20 with Example 1.22 we observe that the linearization $e'(y, u)$, with $(y, u) \in Y_1 \times U$, is well defined on $Y \times U$ for Examples 1.19 and 1.20 but it is only defined with domain strictly contained in $Y \times U$ for Example 1.22. For none of these examples is e defined on all of $Y \times U$.

Example 1.23. Here we consider the nonlinear optimal control problem with nonlinearity of blowup type:

$$\begin{cases} \min \frac{1}{2}|\nabla(y - z)|^2_{L^2_2(\Omega)} + \frac{\alpha}{2}|u|^2_{L^2(\Gamma)} \text{ subject to} \\ -\Delta y - \exp y = f \text{ in } \Omega, \\ \frac{\partial y}{\partial n} = u \text{ on } \Gamma, \\ y = 0 \text{ on } \partial\Omega \setminus \Gamma, \end{cases} \tag{1.5.14}$$

where $\alpha > 0$, $z \in H^1_\Gamma(\Omega)$, $f \in L^2(\Omega)$, Ω is a smooth bounded domain in $\mathbb{R}^2$, and $\Gamma \subset \partial\Omega$ is a connected strict subset of $\partial\Omega$. Since Ω is assumed to be a two-dimensional domain we have the following property of the exponential function: for every $p \in [1, \infty)$,

$$\{|\exp y|_{L^p} : y \in B\} \text{ is bounded} \tag{1.5.15}$$

provided that B is a bounded subset of $H^1_0(\Omega)$ [GT, p. 155]. The variational form of the boundary value problem in (1.5.14) is given by

$$(\nabla y, \nabla v) - (\exp\, y, v) = (f, v) + (u, v)_\Gamma \text{ for all } v \in H^1_\Gamma(\Omega), \tag{1.5.16}$$

where $H^1_\Gamma(\Omega)$ is defined in Example 1.19. To argue existence of a solution to (1.5.14) let $\{(y_n, u_n)\}$ be a minimizing sequence with weak limit $(y^*, u^*) \in H^1_0(\Omega) \times L^2(\Omega)$. Due to (1.5.15) and the radial unboundedness of the cost functional with respect to the $H^1_\Gamma(\Omega) \times L^2(\Gamma)$-norm the set $\{|\exp y_n|_{L^p} : n \in \mathbb{N}\}$ is bounded for every $p \in [1, \infty)$ and $\{|\exp y_n|_{W^{1,p}} : n \in \mathbb{N}\}$ is bounded for every $p \in [1, 2)$. Since $W^{1,p}(\Omega)$ is compactly embedded in $L^2(\Omega)$ for every $p \in (1, 2)$ it follows that for a subsequence of $\{y_n\}$, denoted by the same symbol, $\lim \exp(y_n) = \exp y^*$ in $L^2(\Omega)$. It is now simple to argue that (y^*, u^*) is a solution to (1.5.14). Let us discuss then the validity of (H1)–(H4) with $Y = Y_1 = H^1_\Gamma(\Omega)$, $W = (H^1_\Gamma(\Omega))^*$, with the obvious choice for J, and with $e : Y \times U \to W$ the mapping assigning to $(y, u) \in Y \times U$ the functional $v \to (\nabla y, \nabla v) - (\exp y, v) - (f, v) - (u, v)_\Gamma$ for $v \in H^1_\Gamma(\Omega)$. From (1.5.15) it follows that e is well defined and its Fréchet derivative at (y, u) in direction $(\delta y, \delta u)$ is characterized by

$$(e'(y, u)(\delta y, \delta u), v) = (\nabla \delta y, \nabla v) - (\exp(y)\delta y, v) - (\delta u, v)_\Gamma \text{ for } v \in H^1_\Gamma(\Omega).$$

The operator $G = e_y(y^*, u^*)$ can be expressed as

$$G(\delta y) = -\Delta \delta y - \exp(y^*)\delta y.$$

Note that $G \in \mathcal{L}(Y, W^*)$, and G is self-adjoint with compact resolvent. In particular, (H1) is satisfied. The spectrum of G consists of eigenvalues only. It will be assumed that

$$0 \text{ is not an eigenvalue of } G. \tag{1.5.17}$$

Due to the regularity assumption for z (note that it would suffice to have $\Delta z \in (H^1_\Gamma)^*$), (1.5.17) implies that (H2) holds. To argue the validity of (H3) and (H4) one can rely on the implicit function theorem. Let B be a bounded open neighborhood of y^* in $H^1_\Gamma(\Omega)$. Using (1.5.15) one argues the existence of a constant $\kappa > 0$ such that

$$|\exp y - \exp \bar{y}|_{L^4} \le \kappa |y - \bar{y}|_{H^1} \text{ for all } y, \bar{y} \in B.$$

It follows that e is continuous on $B \times U$ and its partial derivative $e_y(y, u)$ is Lipschitz continuous with respect to $(y, u) \in B \times U$. The implicit function theorem implies the existence of a neighborhood $U(u^*)$ of u^* such that for every $u \in U(u^*)$ there exists a solution $y(u^*)$ of (1.5.16) depending continuously on u. Since $e(y, u)$ is Lipschitz continuous with respect to u it follows, moreover, that there exists $L > 0$ such that

$$|y(u) - y^*|_{H^1} \le L|u - u^*|_{L^2(\Gamma)} \text{ for all } u \in U(u^*).$$

(H3) and (H4) are a direct consequence of this estimate.

The methodology utilized to consider this example can also be applied to Examples 1.19 and 1.20 provided that Ω is restricted to being two-dimensional. This is essential for (1.5.15) to hold. For Example 1.23 it is essential for the cost functional to be radially unbounded with respect to the $H^1_\Gamma(\Omega)$-norm for the y-component to guarantee that minimizing sequences are bounded. For Examples 1.19 and 1.20 the a priori bound on the y-component of minimizing sequences can be obtained through the state equation.

1.6 Approximation, penalty, and adapted penalty techniques

As in the previous section we consider problems of type

$$\begin{cases} \min J(y,u) \\ \text{subject to } e(y,u)=0, \quad u \in C \subset U. \end{cases} \tag{1.6.1}$$

We describe techniques that in specific situations can be more powerful than the general theory of Section 1.4 to obtain optimality systems of type

$$\begin{cases} e(y^*,u^*)=0, \\ \\ e_y^*(y^*,u^*)\lambda^* + J_y(y^*,u^*)=0, \\ \\ \langle e_u^*(y^*,u^*)\lambda^* + J_u(y^*,u^*), u-u^*\rangle_{U^*,U} \geq 0 \quad \text{for all} \quad u \in C, \end{cases} \tag{1.6.2}$$

where (y^*,u^*) denotes a solution to (1.6.1). We shall proceed formally here, referring the reader to [Ba, Chapter 3.2.2] and [Lio1] for details.

1.6.1 Approximation techniques

If J and/or e are not smooth, then one can aim for obtaining an optimality system of the form (1.6.2) by introducing families of smooth operators e_ε and smooth functionals J_ε, $\varepsilon > 0$, and replacing (1.6.1) by

$$\begin{cases} \min J_\varepsilon(y,u) \\ \text{subject to } e_\varepsilon(y,u)=0, \text{ and } u \in C. \end{cases} \tag{1.6.3}$$

Let $(y_\varepsilon, u_\varepsilon)$, $\varepsilon > 0$, denote solutions to (1.6.3). Under relatively mild conditions it can be argued that $\{(y_\varepsilon,u_\varepsilon)\}_{\varepsilon>0}$ contains accumulation points and that they are solutions of (1.6.2). But they need not coincide with (y^*,u^*). To overcome this difficulty a penalty term is added to the cost in (1.6.3), resulting in the adapted penalty formulation

$$\begin{cases} \min J_\varepsilon(y,u) + \dfrac{1}{2}\,|u-u^*|^2 \\ \\ \text{subject to } e_\varepsilon(y,u)=0 \text{ and } u \in C. \end{cases} \tag{1.6.4}$$

It can be shown that under appropriate assumptions $(y_\varepsilon,u_\varepsilon)$ converges to (y^*,u^*) and that there exists λ^* such that (1.6.2) holds. This approach can be used for optimal control of semilinear elliptic equations, variational inequalities [Ba, BaK1, BaK2], and state-dependent parameter estimation problems [BKR], for example.

1.6.2 Penalty techniques

This technique is well suited for systems with multiple steady states or finite explosion property. The procedure is explained by means of an example. We consider the nonlinear parabolic control system

$$\begin{cases} y_t - \Delta y - y^3 = u \quad \text{in } Q = (0,T)\times\Omega, \\ y = 0 \quad \text{on } \Sigma = (0,T)\times\partial\Omega, \\ y(0,\cdot) = y_0 \quad \text{in } \Omega, \end{cases} \tag{1.6.5}$$

where $y = y(t,x)$, $T > 0$, and y_0 is a given initial condition. With (1.6.5) we associate the optimal control problem

$$\begin{cases} \min J(y,u) = \frac{1}{6}\,|y-z|^6_{L^6(Q)} + \frac{\alpha}{2}\,|u|^2_{L^2(Q)} \\ \text{subject to } u \in C, \quad y \in H^{1,2}(Q) \text{ and (1.6.5)}, \end{cases} \tag{1.6.6}$$

where $\alpha > 0$, $z \in L^6(Q)$, C is a closed convex set in $L^2(Q)$, and $H^{1,2}(Q) = \{\varphi \in L^2(Q) : \varphi_t, \varphi_{x_i}, \varphi_{x_i x_j} \in L^2(Q)\}$. Realizing the state equation by a penalty term leads to

$$\begin{cases} \min J_\varepsilon(y,u) = J(y,u) + \frac{1}{\varepsilon}\,\left|y_t - \Delta y - y^3 - u\right|^2_{L^2(Q)} \\ \text{subject to } u \in C, \quad y \in H^{1,2}(Q), \quad y|_\Sigma = 0,\ y(0,\cdot) = y_0. \end{cases} \tag{1.6.7}$$

It can be shown that for every $\varepsilon > 0$ there exists a solution $(y_\varepsilon, u_\varepsilon)$ to (1.6.7). Considering the behavior of the pair $(y_\varepsilon, u_\varepsilon)$ as $\varepsilon \to 0^+$ the question of convergence to a particular solution (y^*, u^*) of (1.6.6) again arises. As before this can be realized by means of adaptation terms:

$$\begin{cases} \min J_\varepsilon(y,u) + \frac{1}{2}\,\left|y - y^*\right|^2_{L^2(Q)} + \frac{1}{2}\,\left|u - u^*\right|^2_{L^2(Q)} \\ \text{subject to } u \in C, \quad y \in H^{1,2}(Q), \quad y|_\Sigma = 0,\ y(0,\cdot) = y_0. \end{cases} \tag{1.6.8}$$

Let us denote solutions to (1.6.8) by $(y^\varepsilon, u^\varepsilon)$. It is fairly straightforward to derive optimality conditions for (1.6.8): Setting

$$\lambda_\varepsilon = -\frac{1}{\varepsilon}\left(y^\varepsilon_t - \Delta y^\varepsilon - (y^\varepsilon)^3 - u^\varepsilon\right) \in L^2(Q) \tag{1.6.9}$$

we find

$$\begin{cases} \lambda^\varepsilon_t - \Delta\lambda^\varepsilon - 3(y^\varepsilon)^2\lambda_\varepsilon = -(y^\varepsilon - z)^5 - (y^\varepsilon - y^*) \quad \text{in } Q, \\ \lambda^\varepsilon = 0 \quad \text{on } \Sigma, \\ \lambda^\varepsilon(T,\cdot) = 0 \quad \text{in } Q \end{cases} \tag{1.6.10}$$

and

$$\int_Q (\alpha u^\varepsilon - \lambda^\varepsilon)(u - u^\varepsilon)dQ + \int_Q (u^\varepsilon - u^*)(u - u^\varepsilon)dQ \geq 0 \quad \text{for all } u \in C, \tag{1.6.11}$$

where λ^ε must be interpreted as a weak solution in $L^2(Q)$ of (1.6.10) (i.e., the inner product with smooth test function is taken, and partial integration bringing all differentiations on the test function is carried out). It can be verified that $u^\varepsilon \to u^*$ in $L^2(Q)$, $y^\varepsilon \rightharpoonup y^*$ weakly in $H^{1,2}(Q)$ and that there exists $\lambda^* \in W^{2,1;6/5}$, the weak limit of λ^ε, such that the following optimality system for (1.6.6) is satisfied:

$$\begin{cases} y_t - \Delta y - y^3 = u \quad \text{in } Q, \\ -\lambda_t - \Delta\lambda - 3y^2\lambda = -(y-z)^5 \quad \text{in } Q, \\ y = \lambda = 0 \quad \text{on } \Sigma, \\ y(0,\cdot) = y^0, \quad \lambda(T,\cdot) = 0 \quad \text{in } \Omega, \end{cases} \tag{1.6.12}$$

where $W^{2,1;6/5} = \left\{\varphi \in L^{6/5}(Q) : \varphi_t, \varphi_{x_i}, \varphi_{x_i x_j} \in L^{6/5}(Q)\right\}$; see [Lio1, Chapter I.3].

Chapter 2

Sensitivity Analysis

2.1 Generalities

In this chapter we discuss the sensitivity of solutions to parameterized mathematical programming problems of the form

$$\begin{aligned} &\min_{x \in C} \quad f(x, p) \quad \text{subject to} \\ &e(x, p) = 0, \quad g(x, p) \le 0, \ \text{and} \ \ell(x, p) \in K, \end{aligned} \tag{2.1.1}$$

where C is a closed convex set in X, and further $e : X \times P \to W$ represents an equality constraint, $g : X \times P \to \mathbb{R}^m$ a finite-dimensional inequality constraint, and $\ell : X \times P \to Z$ an infinite-dimensional affine constraint. The cost f as well as the equality and inequality constraints are allowed to depend on a parameter $p \in P$. Unless otherwise specified P is a normed linear space, $X,\ W,\ Z$ are real Hilbert spaces, and K is a closed convex cone with vertex at 0 in Z. Such problems arise in inverse, control problems, and variational inequalities, and some examples will be discussed in Section 2.6.

The cone K induces a natural ordering $\le$ by means of $z_1 \le z_2$ if $z_1 - z_2 \in K$. We denote by K^+ the dual cone given by

$$K^+ = \{z \in Z^* : \langle z, \tilde{z} \rangle \le 0 \text{ for all } \tilde{z} \in K\},$$

where $\langle \cdot, \cdot \rangle$ denotes the duality pairing between Z and Z^*. Suppose that x_0 is the solution to (2.1.1) for the nominal parameter $p = p_0$. The objective of this section is to investigate

- continuity and Lipschitz continuity of the solution mapping

  $$p \in P \to (x(p), \lambda(p), \mu(p), \eta(p)) \in X \times W^* \times \mathbb{R}^{m,+} \times K^+,$$

  where for p in a neighborhood of p_0, $x(p) \in X$ denotes the local solution of (2.1.1) in a neighborhood of x_0, and $\lambda,\ \mu,\ \eta$ are the Lagrange multipliers associated with constraints described by $e,\ g,\ \ell$ in (2.1.1);

- differentiability of the minimum value function $F(p) = f(x(p), p)$ at p_0; and

- directional differentiability of the solution mapping.

Due to the local nature of the analysis, f, e, and g need only be defined in a neighborhood of (x_0, p_0). We assume throughout this chapter that they are twice continuously Fréchet differentiable with respect to x and that their first and second derivatives are continuous in a neighborhood of (x_0, p_0). It is assumed that ℓ is affine in x for every $p \in P$ and that the first derivative $\ell'(p_0)$ with respect to x is continuous in a neighborhood of p_0.

In order to ensure the existence of a Lagrange multiplier at x_0 we assume that

(H1) x_0 is a regular point, i.e.,

$$0 \in \operatorname{int} \left\{ \begin{pmatrix} e'(x_0, p_0) \\ g'(x_0, p_0) \\ \ell'(p_0) \end{pmatrix} (C - x_0) + \begin{pmatrix} 0 \\ \mathbb{R}^m_+ \\ -K \end{pmatrix} + \begin{pmatrix} 0 \\ g(x_0, p_0) \\ \ell(x_0, p_0) \end{pmatrix} \right\}, \tag{2.1.2}$$

where the interior is taken in $W \times R^m \times Z$ and primes denote the Fréchet derivatives with respect to x. We define the Lagrange functional $\mathcal{L} : X \times P \times W^* \times \mathbb{R}^m \times Z^* \to R$ by

$$\mathcal{L}(x, p, \lambda, \mu, \eta) = f(x, p) + \langle \lambda, e(x, p) \rangle + \langle \mu, g(x, p) \rangle_{R^m} + \langle \eta, \ell(x, p) \rangle.$$

With (2.1.2) holding, it follows from Theorem 1.6 that there exists a Lagrange multiplier $(\lambda_0, \mu_0, \eta_0) \in W^* \times \mathbb{R}^{m,+} \times K^+$ such that

$$\begin{aligned} &\langle \mathcal{L}'(x_0, p_0, \lambda_0, \mu_0, \eta_0), c - x_0 \rangle \geq 0 \quad \text{for all } c \in C, \\ &e(x_0, p_0) = 0, \\ &\langle \mu_0, g(x_0, p_0) \rangle = 0, \quad g(x_0, p_0) \leq 0, \quad \mu_0 \geq 0, \\ &\langle \eta_0, \ell(x_0, p_0) \rangle = 0, \quad \ell(x_0, p_0) \in K, \quad \eta_0 \in K^+. \end{aligned} \tag{2.1.3}$$

To express (2.1.3) in a more compact form we recall that the subdifferential $\partial \psi_C$ of the indicator function

$$\psi_C(x) = \begin{cases} 0 & \text{if } x \in C, \\ \infty & \text{if } x \notin C \end{cases}$$

of a closed convex set C in a Banach space X is given by

$$\partial \psi_C(x) = \begin{cases} \{y \in X^* : \langle y, c - x \rangle_{X^*, X} \leq 0 \text{ for all } c \in C\} & \text{if } x \in C, \\ \{\,\} & \text{if } x \notin C. \end{cases}$$

The set $\partial \psi_C(x)$ is also referred to as the normal cone to C at x. For convenience we also specify

$$\partial \psi_{K^+}(\eta) = \begin{cases} \{z \in Z : \langle z^* - \eta, z \rangle_{X^*, X} \leq 0 \text{ for all } z^* \in K^+\} & \text{if } \eta \in K^+, \\ \{\,\} & \text{if } \eta \notin K^+. \end{cases}$$

Note that (2.1.3) is equivalent to

$$0 \in \begin{cases} \mathcal{L}'(x_0, p_0, \lambda_0, \mu_0, \eta_0) + \partial\psi_C(x_0), \\ e(x_0, p_0), \\ -g(x_0, p_0) + \partial\psi_{\mathbb{R}^{m,+}}(\mu_0), \\ -\ell(x_0, p_0) + \partial\psi_{K^+}(\eta_0). \end{cases} \tag{2.1.4}$$

In fact this follows from the following characterization of the normal cone to K^+.

Lemma 2.1. *Suppose that K is a closed convex cone in Z. Then*

$$z \in \partial\psi_{K^+}(\eta) \textit{ if and only if } z \in K, \ \ \eta \in K^+, \textit{ and } \ \langle \eta, z\rangle = 0.$$

Proof. Assume that $z \in \partial\psi_{K^+}(\eta)$. Then $\partial\psi_{K^+}(\eta)$ is nonempty, $\eta \in K^+$, and

$$\langle z^* - \eta, z\rangle \le 0 \quad \text{for all } z^* \in K^+. \tag{2.1.5}$$

Hence with $z^* = 2\eta$ this implies $\langle \eta, z\rangle \le 0$ and for $z^* = \frac{\eta}{2}$ we have $\langle \eta, z\rangle \ge 0$, and therefore $\langle \eta, z\rangle = 0$. From (2.1.5) it follows that $\langle z^*, z\rangle \le 0$ for all $z^* \in K^+$, and hence the geometric form of the Hahn–Banach theorem implies that $z \in K$.

Conversely, if $z \in K$, $\eta \in K^+$, and $\langle \eta, z\rangle = 0$, then $\langle z^* - \eta, z\rangle \le 0$ for all $z^* \in Z^*$ and thus $z \in \partial\psi_{K^+}(\eta)$. ☐

Let $A : X \to X^*$ be the operator representation of the Hessian $\mathcal{L}''(x_0, p_0, \lambda_0, \mu_0, \eta_0)$ of $\mathcal{L}$ such that

$$\langle Ax_1, x_2\rangle = \mathcal{L}''(x_0, p_0, \lambda_0, \mu_0, \eta_0)(x_1, x_2)$$

and define operators $E : X \to W$, $G : X \to \mathbb{R}^m$, and $L : X \to Z$ by

$$E = e'(x_0, p_0), \quad G = g'(x_0, p_0), \quad L = \ell'(p_0).$$

Without loss of generality one can assume that the coordinates of the inequality constraints $g(x, p_0) \le 0$ and the associated Lagrange multiplier μ_0 are arranged so that $\mu_0 = (\mu_0^+, \mu_0^0, \mu_0^-)$ and $g = (g^+, g^0, g^-)$, with $g^+ : X \to \mathbb{R}^{m_1}$, $g^0 : X \to \mathbb{R}^{m_2}$, $g^- : X \to \mathbb{R}^{m_3}$, $m = m_1 + m_2 + m_3$, and

$$\begin{aligned} g^+(x_0, p_0) &= 0, & \mu_0^+ &> 0, \\ g^0(x_0, p_0) &= 0, & \mu_0^0 &= 0, \\ g^-(x_0, p_0) &< 0, & \mu_0^- &= 0. \end{aligned} \tag{2.1.6}$$

We further define

$$G_+ = g^+(x_0, p_0)', \quad G_0 = g^0(x_0, p_0)',$$

and

$$E_+ = \begin{pmatrix} E \\ G_+ \end{pmatrix} : X \to W \times \mathbb{R}^{m_1}.$$

Note that the coordinates denoted by superscripts $+$ behave locally like equality constraints. Define $\mathcal{E}(z) : X \times \mathbb{R} \to (W \times \mathbb{R}^{m_1}) \times \mathbb{R}^{m_2} \times Z$ for $z \in Z$ by

$$\mathcal{E}(z) = \begin{pmatrix} E_+ & 0 \\ G_0 & 0 \\ L & z. \end{pmatrix}.$$

The adjoint $\mathcal{E}^*(z)$ is given by

$$\mathcal{E}^*(z) = \begin{pmatrix} E_+^* & G_0^* & L^* \\ 0 & 0 & \langle \cdot, z \rangle_Z \end{pmatrix}.$$

The following conditions will be used.

(H2) There exists a $\kappa > 0$ such that $\langle Ax, x \rangle_{X^*,X} \geq \kappa\, |x|_X^2$ for all $x \in \ker(E_+)$.

(H3) $\mathcal{E}(z)$ is surjective at $z = \ell(x_0, p_0)$.

(H4) There exists a neighborhood $\tilde{U} \times \tilde{N}$ of (x_0, p_0) and a positive constant ν such that

$$|f(x,p) - f(x,q)| + |e(x,p) - e(x,q)|_W + |g(x,p) - g(x,q)|_{\mathbb{R}^m} \\ + |\ell(x,p) - \ell(x,q)|_Z \leq \nu\, |p - q|_P$$

for all (x, p) and (x, q) in $\tilde{U} \times \tilde{N}$.

Condition (H2) is a second order sufficient optimality condition for (2.1.1) at x_0 and (H3) implies that there exists a neighborhood $\mathcal{O}$ of $\ell(x_0, p_0)$ in Z such that $\mathcal{E}(z)$ is surjective for all $z \in \mathcal{O}$.

The chapter is organized as follows. In Sections 2.2 and 2.3 we discuss the basic results for establishing Lipschitz continuity of the solution mapping. The implicit function theory for the generalized equation of the form (2.1.4) is discussed in Section 2.2. In Section 2.3 stability results for solutions to mathematical programming problems including (2.1.1) are addressed. Sufficient conditions for stability of the solutions are closely related to sufficient optimality conditions for local minimizers. Section 2.3 therefore also contains first and second order sufficient conditions for local optimality. Sections 2.4 and 2.5 are devoted to establishing Lipschitz continuity and directional differentiability of the solution mapping. For the latter we employ the assumption of polyhedricity of a closed convex cone K which, together with appropriate additional conditions, implies that the directional derivative $(\dot{x}, \dot{\lambda}, \dot{\mu}, \dot{\eta})$ in direction q satisfies

$$0 \in \begin{cases} \mathcal{L}'_p(x_0, p_0, \lambda_0, \mu_0, \eta_0)q + A\dot{x} + E^*\dot{\lambda} + G_+^*\dot{\mu}^+ + G_0^*\dot{\mu}^0 + L^*\dot{\eta}, \\ -e_p^+(x_0, p_0)q - E_+\dot{x}, \\ -g_p^0(x_0, p_0)q - G_0\dot{x} + \partial\psi_{\mathbb{R}^{m_2,+}}(\dot{\mu}^0), \\ -\ell_p(x_0, p_0)q - L\dot{x} + \partial\psi_{\hat{K}^+}(\dot{\eta}), \end{cases} \tag{2.1.7}$$

where $e^+ = \binom{e}{g^+}$ and $\hat{K}^+$ is the dual cone of $\hat{K} = \overline{\cup_{\lambda>0}\lambda(K - \ell(x_0, p_0))} \cap [\eta_0]^\perp$. Section 2.6 contains some applications to optimal control of ordinary differential equations. This chapter is strongly influenced by the work of S. M. Robinson and also by that of W. Alt.

2.2 Implicit function theorem

In this section we discuss an implicit function theorem for parameterized generalized equations of the form

$$0 \in F(x, p) + \partial\psi_C(x), \tag{2.2.1}$$

where $F : X \times P \to X^*$, with X^* the strong dual space of X, and C is a closed convex set in X. We assume that X and P are normed linear spaces unless specified otherwise. Suppose $x_0 \in X$ is a solution to (2.2.1) for a reference parameter $p = p_0 \in P$. Let $F(x, p_0)$ be Fréchet differentiable at x_0 and define the linearized form by

$$\mathcal{T}x = F(x_0, p_0) + F'(x_0, p_0)(x - x_0) + \partial\psi_C(x). \tag{2.2.2}$$

Definition 2.2 (Strong Regularity). *The generalized equation* (2.2.1) *is called strongly regular at x_0 with associated Lipschitz constant ρ if there exist neighborhoods of V of* 0 *and U of x_0 in X such that $(\mathcal{T}^{-1}V) \cap U$, the intersection of U with the restriction of $\mathcal{T}^{-1}$ to V, is single valued and Lipschitz continuous from V to U with modulus ρ.*

Note that if $C = X$, the strong regularity assumption coincides with $F'(x_0, p_0)^{-1}$ being a bounded linear operator, which is the common condition for the implicit function theorem. The following result is taken from Robinson's work [Ro3].

Theorem 2.3 (Generalized Implicit Function Theorem). *Let P be a topological space, and assume that F is Fréchet differentiable with respect to x and both F and F' are continuous at (x_0, p_0). If* (2.2.1) *is strongly regular at x_0 with associated Lipschitz constant ρ, then for every $\epsilon > 0$ there exists neighborhoods N_ϵ of p_0 and U_ϵ of x_0, and a single-valued function $x : N_\epsilon \to U_\epsilon$ such that for every $p \in N_\epsilon$, $x(p)$ is the unique solution in U_ϵ of* (2.2.1)*. Furthermore, for each $p, q \in N_\epsilon$*

$$|x(p) - x(q)|_X \le (\rho + \epsilon)\,|F(p, x(q)) - F(q, x(q))|_{X^*}. \tag{2.2.3}$$

Proof. For $\epsilon > 0$ we choose a $\delta > 0$ such that $\rho\delta < \epsilon/(\rho + \epsilon)$. By strong regularity, there exists neighborhoods V of 0 and U of x_0 in X such that $(\mathcal{T}^{-1}V) \cap U$ is single valued and Lipschitz continuous from V to U with modulus ρ. Let

$$h(x, p) = F(x_0, p_0) + F'(x_0, p_0)(x - x_0) - F(x, p)$$

and choose a neighborhood N_ϵ of p_0 and a closed ball U_ϵ of radius r about x_0 contained in U so that for each $p \in N_\epsilon$ and $x \in U_\epsilon$ we have for $h(x, p) \in V$

$$|F'(x, p) - F'(x_0, p_0)| \le \delta$$

and

$$\rho\,|F(x_0, p) - F(x_0, p_0)| \le (1 - \rho\delta)r.$$

For any $p \in N_\epsilon$ define an operator Φ_p from $U_\epsilon \to U$ by

$$\Phi_p(x) = \mathcal{T}^{-1}(h(x, p)) \cap U. \tag{2.2.4}$$

Note that $x \in U_\epsilon \cap \Phi_p(x)$ if and only if $x \in U_\epsilon$ and (2.2.1) holds. We show that Φ_p is a strict contraction and Φ_p maps U_ϵ into itself. For $x_1, x_2 \in U_\epsilon$ we find, using $h'(x, p) = F'(x_0, p_0) - F'(x, p)$, that

$$|\Phi_p(x_1) - \Phi_p(x_2)| \le \rho\, |h(x_1, p) - h(x_2, p)|$$

$$= \rho | \int_0^1 h'(x_1 + t\,(x_1 - x_2), p)(x_1 - x_2)\, dt| \le \rho\delta\, |x_1 - x_2|,$$

where $\rho\,\delta < 1$. Since $x_0 = \mathcal{T}^{-1}(0) \cap U$ we have

$$|\Phi_p(x_0) - x_0| \le \rho\, |h(x_0, p)| = \rho\, |F(x_0, p) - F(x_0, p_0)| \le (1 - \rho\delta)r,$$

and thus for $x \in U_\epsilon$

$$|\Phi_p(x) - x_0| \le |\Phi_p(x) - \Phi_p(x_0)| + |\Phi_p(x_0) - x_0| \le \rho\delta\, |x - x_0| + (1 - \rho\delta)r \le r.$$

By the Banach fixed point theorem Φ_p has a unique fixed point $x(p)$ in U_ϵ and for each $x \in U_\epsilon$ we have

$$|x(p) - x| \le (1 - \rho\delta)^{-1} |\Phi_p(x) - x|. \tag{2.2.5}$$

It follows from our earlier observation that $x(p)$ is the unique solution to (2.2.1) in U_ϵ. To verify (2.2.3) with $p, q \in N_\epsilon$ we use (2.2.5) with $x = x(q)$ and obtain

$$|x(p) - x(q)| \le (1 - \rho\delta)^{-1} |\Phi_p(x(q)) - x(q)|.$$

Since $x(q) = \Phi_q(x(q))$ we find

$$|\Phi_p(x(q)) - x(q)| \le \rho\, |h(x(q), p) - h(x(q), q)| = \rho\, |F(x(q), p) - F(x(q), q)|,$$

and hence

$$|x(p) - x(q)| \le \rho(1 - \rho\delta)^{-1} |F(x(q), p) - F(x(q), q)|.$$

Observing that $\rho(1 - \rho\delta)^{-1} \le \rho + \epsilon$ the desired estimate (2.2.3) follows. □

As a consequence of the previous theorem, Lipschitz continuity of $F(x, p)$ in p implies local Lipschitz continuity of $x(p)$ at p_0.

Corollary 2.4. *Assume in addition to the conditions of Theorem 2.3 that P is a normed linear space and that for some $\nu > 0$*

$$|F(x, p) - F(x, q)| \le \nu\, |p - q|$$

for $p, q \in N_\epsilon$ and $x \in U_\epsilon$. Then $x(p)$ is Lipschitz on N_ϵ with modulus $\nu\,(\rho + \epsilon)$.

It should be remarked that the condition of strong regularity is the weakest possible condition which can be imposed on the values of a function F and its derivative at a point x_0, so that for each perturbation satisfying the hypothesis of Theorem 2.3, a function $x(\cdot)$ will exist having the properties stated in Theorem 2.3. To see this, one has only to consider a function $F : X \to X^*$ which is Fréchet differentiable at x_0 and satisfies

$$0 \in F(x_0) + \partial\psi_C(x_0).$$

Let P be a neighborhood of the origin in X^*, and let

$$F(x, p) = F(x_0, p_0) + F'(x_0, p_0)(x - x_0) - p$$

with $p_0 = 0$. Choose $\epsilon > 0$. If there exist neighborhoods N_ϵ and V_ϵ and a function $x(\cdot)$ satisfying the properties stated in Theorem 2.3, then with

$$\mathcal{T}x = F(x_0, p_0) + F'(x_0, p_0)(x - x_0) + \partial\psi_C(x)$$

we see that the restriction to N_ϵ of $\mathcal{T}^{-1}V$ is a singled-valued, Lipschitz continuous function. Therefore the generalized equation $0 \in F(x, p_0) + \partial\psi_C(x)$ is strongly regular at x_0.

One of the important consequences of Theorem 2.3 is the following theorem on parametric sensitivity analysis.

Theorem 2.5. *Under the conditions of Theorem* 2.3 *and Corollary* 2.4 *there exists for each* $\epsilon > 0$ *a function* $\alpha_\epsilon(p) : N_\epsilon \to R^+$ *satisfying* $\lim_{p\to p_0} \alpha_\epsilon(p) = 0$, *such that*

$$|x(p) - \Phi_p(x_0)| \le \alpha_\epsilon(p)\,|p - p_0|,$$

where $\Phi_p(x_0)$ *is the unique solution in* U_ϵ *of the linear generalized equation*

$$0 \in F(x_0, p) + F'(x_0, p_0)(x - x_0) + \partial\psi_C(x). \tag{2.2.6}$$

Proof. From the proof of Theorem 2.3 it follows that $x(p) = \Phi_p(x(p))$ and thus by strong regularity

$$\begin{aligned}
|x(p) - \Phi_p(x_0)| &\le \rho\,|h(x(p), p) - h(x_0, p)| \\
&\le \rho \left| \int_0^1 h'(x_0 + \theta\,(x(p) - x_0))\,d\theta \right| |x(p) - x_0| \\
&\le \rho\nu(\rho + \epsilon)|p - p_0| \left| \int_0^1 h'(x_0 + \theta\,(x(p) - x_0))\,d\theta \right|.
\end{aligned}$$

Since $h'(x, p) = F'(x_0, p_0) - F'(x, p)$ and F' is continuous,

$$\left| \int_0^1 e'(x_0 + \theta\,(x(p) - x_0))\,d\theta \right| \to 0$$

as $p \to p_0$, which completes the proof. $\square$

In the case $C = X$ the generalized equation (2.2.1) reduces to the nonlinear equation $F(x, p) = 0$ and

$$\Phi_p(x_0) = F'(x_0, p_0)^{-1}(F(x_0, p_0) - F(x_0, p)).$$

Thus, Theorem 2.5 shows that if $F(x_0, \cdot)$ is Fréchet differentiable at p_0, then so is $x(p)$ and

$$x'(p_0) = -F'(x_0, p_0)^{-1} \frac{\partial F}{\partial p}(x_0, p_0). \tag{2.2.7}$$

In many applications, one might find (2.2.6) significantly easier to solve than (2.2.1). In fact, the necessary optimality condition (2.1.3) is of the form (2.2.1), and (2.2.6) corresponds to a quadratic programming problem, as we will discuss in Section 2.4. Thus Theorem 2.5 can provide a relatively cheap way to find a good approximation to $x(p)$ for p near p_0.

2.3 Stability results

Throughout this section, let X, Y be real Banach spaces and let C be closed convex subsets of X. Concerning K it suffices in this section to assume that it is a closed convex set in Y, unless it is specifically assumed that it is a closed convex cone in Y. For p in the metric space P with metric $\delta(\cdot, \cdot)$ consider the parameterized minimization problem

$$\min f(x, p) \quad \text{subject to } x \in C, \quad g(x, p) \in K. \tag{2.3.1}$$

This class of problems contains (2.1.1) as a special case. To facilitate the analysis of (2.3.1) let $\Sigma(p)$ denote the set of feasible points, i.e.,

$$\Sigma(p) = \{x \in C : g(x, p) \in K\}.$$

For a given $p_0 \in P$ and $x_0 \in \Sigma(p_0)$ a local minimizer of (2.3.1), define the set $\Sigma_r(p)$ by

$$\Sigma_r(p) = \Sigma(p) \cap \bar{B}(x_0, r),$$

where $r > 0$ and $\bar{B}(x_0, r)$ is the closure of the open ball with radius r at x_0 in X. We further define the value function $\mu_r(p)$ by

$$\mu_r(p) = \inf \{f(x, p) | x \in \Sigma_r(p)\}. \tag{2.3.2}$$

Throughout this section it is assumed that $f : X \times P \to \mathbb{R}$ and $g : X \times P \to Y$ are continuously Fréchet differentiable with respect to x and that their first derivatives are continuous in a neighborhood of (x_0, p_0). Recall from Definition 1.5 that $x_0 \in \Sigma(p_0)$ is a regular point if

$$0 \in \text{int } \{g(x_0, p_0) + g_x(x_0, p_0)(C - x_0) - K\}. \tag{2.3.3}$$

We use a generalized inverse function theorem for set-valued functions due to Robinson [Ro3] for the stability analysis of (2.3.1).

For a nonempty set S in X and $x \in X$ we define

$$d(x, S) = \inf\{|y - x| : \quad y \in S\},$$

and $d(x, S) = \infty$ if $S = \emptyset$. For nonempty sets S and $\bar{S}$ in X the Hausdorff distance is defined by

$$d_H(S, \bar{S}) = \max \left\{ \sup\{d(x, \bar{S}) : x \in S\}, \ \sup\{d(y, S) : y \in \bar{S}\} \right\}.$$

We further set $B_X = \{x \in X : |x| < 1\}$. For a set-valued function $F : X \to Y$ we define the preimage of F by $F^{-1}(y) = \{x \in X : y \in F(x)\}$. A set-valued function is called closed and convex if its graph is closed and convex.

Theorem 2.6 (Generalized Inverse Function Theorem). *Let X and Y be normed linear spaces and F a closed convex set-valued function from X to Y. Let $y_0 \in F(x_0)$ and suppose that for some $\eta > 0$,*

$$y_0 + \eta\, B_Y \subset F(x_0 + B_X).$$

Then for $x \in x_0 + B_X$ and any $y \in y_0 + \operatorname{int}(\eta\, B_Y)$,

$$d(x, F^{-1}(y) \cap (x_0 + B_X)) \le (\eta - |y - y_0|)^{-1}(1 + |x - x_0|)d(y, F(x)).$$

Proof. Let $x \in x_0 + B_X$ and $y - y_0 \in \operatorname{int}(\eta\, B_Y)$. For $y \in F(x)$ the claim is clearly satisfied and therefore we assume $y \notin F(x)$. Let $\delta > 0$ be arbitrary and choose $y_\delta \in F(x)$ such that

$$0 < |y_\delta - y| < d(y, F(x)) + \delta. \tag{2.3.4}$$

Let $\alpha = \eta - |y - y_0| > 0$. Choose any $\epsilon \in (0, \alpha)$ and set

$$y_\epsilon = y + (\alpha - \epsilon)|y - y_\delta|^{-1}(y - y_\delta).$$

Then $|y_\epsilon - y_0| \le |y - y_0| + (\alpha - \epsilon) < \eta$, and

$$y_\epsilon \in y_0 + \operatorname{int}(\eta B_Y).$$

Thus by assumption there exists

$$x_\epsilon \in x_0 + B_X \text{ with } y_\epsilon \in F(x_\epsilon).$$

For $\frac{1}{1+(\alpha-\epsilon)|y-y_\delta|^{-1}} \in (0, 1)$, we find, using convexity of F and the fact that $y_\delta \in F(x)$,

$$y = (1 - \lambda)y_\delta + \lambda\, y_\epsilon \in (1 - \lambda)F(x) + \lambda\, F(x_\epsilon) \subset F((1 - \lambda)x + \lambda\, x_\epsilon). \tag{2.3.5}$$

Since x and x_ϵ are contained in $x_0 + B_X$ we have $(1 - \lambda)x + \lambda\, x_\epsilon \in x_0 + B_X$. Together with (2.3.5) this implies

$$d(x, F^{-1}(y) \cap (x_0 + B_X)) \le |x - (1 - \lambda)x + \lambda\, x_\epsilon| = \lambda\, |x - x_\epsilon|.$$

However, $|x - x_\epsilon| \le |x - x_0| + |x_0 - x_\epsilon| \le |x - x_0| + 1$ and

$$\lambda = \frac{|y - y_\delta|}{|y - y_\delta| + \alpha - \epsilon} < \frac{|y - y_\delta|}{\alpha - \epsilon}.$$

Hence by (2.3.4)

$$d(x, F^{-1}(y) \cap (x_0 + B_X)) \le (\eta - |y - y_0| - \epsilon)^{-1}(1 + |x - x_0|)\,(d(y, F(x)) + \delta),$$

and letting $\epsilon \to 0^+$, $\delta \to 0^+$ the claim follows. $\square$

The proof of the stability result below relies on the following set-valued fixed point lemma. For a metric space $(\mathcal{X}, d)$ let $\mathcal{F}(\mathcal{X})$ denote the collection of all nonempty closed subsets of $\mathcal{X}$.

Lemma 2.7. *Let $(\mathcal{X}, d)$ be a complete metric space and let $\Phi\colon \mathcal{X} \to \mathcal{F}(\mathcal{X})$ be a set-valued mapping. Suppose that there exists $x_0 \in \mathcal{X}$ and constants $r > 0$, $\alpha \in (0,1)$, $\epsilon \in (0,r)$ such that for all x_1, x_2 in the closed ball $\bar{B}(x_0, r)$ the sets $\Phi(x_1)$, $\Phi(x_2)$ are nonempty,*

$$d_H(\Phi(x_1), \Phi(x_2)) \le \alpha\, d(x_1, x_2),$$

and

$$d(x_0, \Phi(x_0)) \le (1-d)(r-\epsilon).$$

Then there exists $\bar{x} \in \bar{B}(x_0, r)$ such that $\bar{x} \in \Phi(\bar{x})$ and $d(x_0, \bar{x}) \le (1-\alpha)^{-1} d(x_0, \Phi(x_0)) + \epsilon$. If moreover

$$d_H(\Phi(\mathcal{S}_1), \Phi(\mathcal{S}_2)) \le \alpha\, d_H(\mathcal{S}_1, \mathcal{S}_2) \tag{2.3.6}$$

for all nonempty closed subsets $\mathcal{S}_1$ and $\mathcal{S}_2$ of $\bar{B}(x_0, r)$, then the sequence $S_k = \Phi(S_{k-1})$, with $S_0 = \{x_0\}$, converges in the Hausdorff metric to a set $\bar{S}$ satisfying $\bar{S} = \Phi(\bar{S})$.

Proof. Let $\delta = \alpha(x_0, \Phi(x_0))$ and set $\gamma = \epsilon\alpha^{-1}(1-\alpha)$. By induction one proves the existence of a sequence $x_{k+1} \in \Phi(x_k)$, $k = 0, 1, \dots$, such that $x_k \in \bar{B}(x_0, r)$ and

$$\begin{aligned} d(x_{k+1}, x_k) &\le \alpha^k\delta + (1-2^{-(k+1)})\gamma\, \alpha^{k+1}, \\ d(x_{k+ü1}, x_0) &\le (1-\alpha^{k+1})(r-\epsilon) + \gamma \sum_{i=1}^{k+1} (1-2^{-i})\alpha^i. \end{aligned} \tag{2.3.7}$$

In fact, choose $x_1 \in \Phi(x_0)$ with $d(x_0, x_1) \le \delta + \frac{\gamma}{2}\alpha \le (1-\alpha)(r-\epsilon) + \frac{\gamma}{2}\alpha \le r$. Hence $x_1 \in \bar{B}(x_0, r)$ and (2.3.7) holds for $k = 0$. Now suppose that (2.3.7) holds with k replaced by $k-1$. Then $d(x_0, x_k) < r$ and hence $\Phi(x_k)$ is nonempty and there exists $x_{k+1} \in \Phi(x_k)$ satisfying $d(x_{k+1}, x_k) \le d(\Phi(x_k), x_k) + 2^{-(k+1)}\gamma\, \alpha^{k+1}$. Consequently $d(x_{k+1}, x_k) \le d_H(\Phi(x_k), \Phi(x_{k-1})) + 2^{-(k+1)}\gamma\,\alpha^{k+1} \le \alpha\ d(x_k, x_{k-1}) + 2^{-(k+1)}\gamma\,\alpha^{k+1} \le \alpha^k\delta + (1 - 2^{-(k+1)})\gamma\,\alpha^{k+1}$, and $d(x_{k+1}, x_0) \le d(x_{k+1}, x_k) + d(x_k, x_0) \le \gamma \sum_{i=1}^{k+1}(1-2^{-i})\alpha^i + (1-\alpha^{k+1})(r-\epsilon)$, and (2.3.7) holds for all k. For $k \ge 0$ and for every $m \ge 1$ we have

$$\begin{aligned} d(x_{k+m}, x_k) &\le \sum_{i=0}^{m-1} d(x_{k+i+1}, x_{k+i}) \\ &\le \sum_{i=0}^{m-1} \left(\alpha^{k+i}\delta + (1-2^{-(k+i+1)})\gamma\, \alpha^{k+i+1}\right) \\ &\le \alpha^k(1-\alpha)^{-1}(\delta + \gamma\,\alpha), \end{aligned}$$

and consequently $\{x_k\}$ is a Cauchy sequence in $\bar{B}(x_o, r)$. Since $\mathcal{X}$ is complete there exists $\bar{x} \in \bar{B}(x_0, r)$ with $\lim_{k\to\infty} x_k = \bar{x}$ and $d(\bar{x}, x_0) \le \frac{\delta}{1-\alpha} + \epsilon$. To verify the fixed point property, note that for every k we have

$$\begin{aligned} d(\bar{x}, \Phi(\bar{x})) &\le d(\bar{x}, x_{k+1}) + d(x_{k+1}, \Phi(\bar{x})) \\ &\le d(\bar{x}, x_{k+1}) + d_H(\Phi(x_k), \Phi(\bar{x})) \\ &\le d(\bar{x}, x_{k+1}) + \alpha\, d(x_k, \bar{x}). \end{aligned} \tag{2.3.8}$$

Passing to the limit with respect to k implies that $\bar{x} \in \Phi(\bar{x})$.

We now assume that Φ also satisfies (2.3.6). Since the set of closed subsets of $\bar{B}(x_o, r)$ is complete with respect to the Hausdorff metric, the existence of a set $\bar{S}$ with $\lim S_k = \bar{S}$ in the Hausdorff metric follows. To argue invariance of $\bar{S}$ under Φ note that

$$\begin{aligned} d_H(\bar{S}, \Phi(\bar{S})) &= \lim_{k\to\infty} d_H(S_k, \Phi(\bar{S})) \\ &= \lim_{k\to\infty} d_H(\Phi(S_{k-1}), \Phi(\bar{S})) \le \lim_{k\to\infty} d_H(S_{k-1}, \bar{S}) = 0 \end{aligned}$$

and hence $\bar{S} = \Phi(\bar{S})$. ☐

The following result is an important consequence of Theorem 2.6.

Theorem 2.8 (Stability). *Let $x_0 \in \Sigma(p_0)$ be a regular point. Then for every $\epsilon > 0$ there exist neighborhoods U_ϵ of x_0, N_ϵ of p_0 such that for $p \in N_\epsilon$ the set $\Sigma(p)$ is nonempty and for each $x \in U_\epsilon \cap C$*

$$d(x, \Sigma(p)) \le (M + \epsilon)\, d(0, \{g(x, p) - K : x \in C\}), \tag{2.3.9}$$

where a constant $M > 0$ is independent of $\epsilon > 0$.

Proof. Let F be the closed convex set-valued mapping from X into Y given by

$$F(x) = \begin{cases} g'(x_0, p_0)(x - x_0) + g(x_0, p_0) - K & \text{for } x \in C, \\ \{\} & \text{for } x \notin C. \end{cases}$$

The regular point condition implies that there exists an $\eta > 0$ such that

$$\eta\, B_Y \subset F(x_0 + B_X).$$

Here we use Theorem 1.4 and the discussion below Definition 1.5. For $\delta > 0$ and $r > 0$ let $\rho = (\eta - 2\delta\, r)^{-1}(1 + r)$ where we assume that

$$2\delta\, r < \eta \quad \text{and} \quad 2\rho\delta < 1.$$

Let

$$h(x, p) = g(x_0, p_0) + g'(x_0, p_0)(x - x_0) - g(x, p)$$

and choose a neighborhood of N of p_0 and a closed ball $\bar{B}(x_0, r)$ of radius r about x_0 such that for $x \in U = C \cap \bar{B}(x_0, r)$ and $p \in N$

$$|g'(x, p) - g'(x_0, p_0)| \leq \delta$$

and

$$|g(x_0, p) - g(x_0, p_0)| \leq \delta\, r.$$

As in the proof of Theorem 2.3 it can be shown that

$$|h(x_1, p) - h(x_2, p)| \leq \delta |x_1 - x_2|$$

for all $x_1, x_2 \in U$ and $p \in N$. In particular, if $x \in U$ and $p \in N$, then

$$|h(x, p)| \leq |h(x_0, p_0)| + |h(x, p) - h(x_0, p_0)| \leq \delta\, r + \delta\, r < \eta.$$

For any $p \in N$ define the set-valued closed convex mapping Φ_p from $U \to C$ by

$$\Phi_p(x) = F^{-1}(h(x, p)). \tag{2.3.10}$$

Note that $x \in \Phi_p(x)$ implies that $g(x, p) \in K$. We argue that Lemma 2.7 is applicable with $\Phi = \Phi_p$ and $\mathcal{X} = C$ endowed with the metric induced by X. For every $x \in U$ the set $\Phi_p(x)$ is a closed convex subset of X. (Hence, if X is a reflexive Banach space, then the metric projection is well defined and Lemma 2.7 will be applicable with $\epsilon = 0$.) For every $x_1,\ x_2 \in U$ we have by Theorem 2.6

$$d_H(\Phi_p(x_1), \Phi_p(x_2))| \leq \rho\, |h(x_1, p) - h(x_2, p)| < \rho\delta\, |x_1 - x_2|,$$

where we used that $h(x_i, p) \in F(x_i)$ for $i = 1, 2$. Moreover

$$|\Phi_p(x_0) - x_0| \leq \rho\, |g(x_0, p) - g(x_0, p_0)| \leq \rho\delta r,$$

and for every pair of sets $\mathcal{S}_1$ and $\mathcal{S}_2$ in U we have

$$d_H(\Phi_p(\mathcal{S}_1), \Phi_p(\mathcal{S}_2)) \leq \rho\, \delta\, d_H(\mathcal{S}_1, \mathcal{S}_2).$$

Hence all conditions of Lemma 2.7 are satisfied with $\alpha = 2\rho\delta$ and $\epsilon = \frac{r(1-2\alpha)}{1-\alpha}$. Consequently there exists a Cauchy sequence $\{x_k\}$ in C such that $x_{k+1} \in \Phi_p(x_k)$ and $\bar{x} = \lim_{k\to\infty} x_k \in C$ satisfies $\bar{x} \in \Phi_p(\bar{x})$. Moreover, if $S_k = \Phi_p(S_{k-1})$ with $S_0 = \{x_0\}$, then $d_H(S_m, S_k) \to 0$ as $m \geq k \to \infty$ and thus $\bar{S} = \lim_{k\to\infty} S_k$ satisfies $\bar{S} = \Phi_p(\bar{S})$ and $\bar{S} \subset \Sigma_p$. Let

$$(x, \tilde{y}) \in M = \{(x, \tilde{y}) \in U \times Y : g(x_0, p_0) + g'(x_0, p_0)(x - x_0) + \tilde{y} \in K\}.$$

For all $\bar{x} \in \bar{S}$ we have by Theorem 2.6

$$d(x, F^{-1}(h(\bar{x}, p)) \leq \rho\, d(h(\bar{x}, p), F(x)) \leq \rho\, |h(\bar{x}, p) + \tilde{y}|$$

$$\leq \rho\, (|g(x, p) - g(x_0, p_0) - g'(x_0, p_0)(x - x_0) - \tilde{y}| + |h(\bar{x}, p) - h(x, p)|\,).$$

Since $|h(x, p) - h(\bar{x}, p)| \leq \delta\, |x - \bar{x}|$,

$$d(x, \bar{S}) \leq \frac{\rho}{1 - \rho\delta}\, |g(x, p) - g(x_0, p_0) - g'(x_0, p_0)(x - x_0) - \tilde{y}|$$

for all $(x, \tilde{y}) \in M$. Now, for $x \in C$ and $y \in g(x, p) - K$ we let $\tilde{y} = g(x, p) - g(x_0, p_0) - g'(x_0, p_0)(x - x_0) - y$. Then $(x, \tilde{y}) \in M$ and hence

$$d(x, \Sigma(p)) \leq d(x, \bar{S}) \leq \frac{\rho}{1 - \rho\delta}\, |y|.$$

This holds for all $y \in g(x, p) - K$. Since $\frac{\rho}{1-\rho\delta} = \frac{1+r}{\eta - \delta(1+3r)}$, we can select for every $\epsilon > 0$ neighborhoods $N = N_\epsilon$ and $U = U_\epsilon$ such that $\frac{\rho}{1-\rho\delta} \leq \eta^{-1} + \epsilon$. This implies the claim with $M = \eta^{-1}$. □

Theorem 2.9 (Lipschitz Continuity of Value Function). *Let $x_0 \in \Sigma(p_0)$ be a local minimizer. Assume moreover that x_0 is regular and that there exists a neighborhood $U \times N$ of (x_0, p_0) such that*

$$|f(x, p) - f(\tilde{x}, p_0)| \leq L_f\, (|x - \tilde{x}| + \delta(p, p_0)) \tag{2.3.11}$$

for $x,\ \tilde{x} \in U$ and $p \in N$ and

$$|g(x, p) - g(x, p_0)| \leq L_g\, \delta(p, p_0) \tag{2.3.12}$$

for $(x, p) \in U \times N$. Then there exist constants $r > 0$ and L_r and a neighborhood $\tilde{N}$ of p_0 such that

$$|\mu_r(p) - \mu_r(p_0)| \leq L_r\, \delta(p, p_0)$$

for all $p \in \tilde{N}$.

Proof. Since x_0 is a local minimizer of (2.3.1) at p_0 there exists $s > 0$ such that

$$f(x_0, p_0) \leq f(x, p_0) \quad \text{for all } \ x \in \Sigma_s(p_0).$$

Let $C_s = C \cap \bar{B}(x_0, s)$. By Theorem 1.4 and the discussion following Definition 1.5 the regular point condition (2.3.3) also holds with C replaced by C_s. Applying Theorem 2.8 with $C = C_s$ and $\epsilon = 1$ there exist neighborhoods U_1 of x_0 and N_1 of p_0 such that $\Sigma_s(p)$ is nonempty and

$$d(x, \Sigma_s(p)) \leq (M + 1)\, d(0, \{g(x, p) - K : x \in C_s\}) \tag{2.3.13}$$

for each $x \in U_1$ and $p \in N_1$.

We now choose $r > 0$ and a neighborhood $\tilde{N}$ of p_0 such that

$$r \leq s,\ \ \bar{B}(x_0, r) \subset U_1 \cap U,\ \ \tilde{N} \subset N_1 \cap N, \ \ \text{and } 2(M + 1)L_g\, \delta(p, p_0) < r \text{ for all } p \in \tilde{N}. \tag{2.3.14}$$

By (2.3.13) and (2.3.12) we obtain for all $p \in \tilde{N}$

$$d(x_0, \Sigma_s(p)) \leq (M + 1)\, |g(x_0, p) - g(x_0, p_0)| \leq (M + 1)L_g\, \delta(p, p_0).$$

Thus there exists a $x(p) \in \Sigma_s(p)$ such that

$$|x(p) - x_0| \le 2(M+1)L_g\,\delta(p, p_0) < r$$

and therefore $x(p) \in \Sigma_r(p)$ and $f(x(p), p) \ge \mu_r(p)$. From (2.3.11)

$$|f(x(p), p) - f(x_0, p_0)| \le L_r\,\delta(p, p_0),$$

where $L_r = L_f(2(M+1)L_g + 1)$. Combining these inequalities, we have

$$\mu_r(p) \le f(x(p), p) \le f(x_0, p_0) + L_r\,\delta(p, p_0) = \mu_r(p_0) + L_r\,\delta(p, p_0) \tag{2.3.15}$$

for all $p \in \tilde{N}$.

Conversely, let $p \in \tilde{N}$ and $x \in \Sigma_r(p)$. From (2.3.13) and (2.3.12)

$$d(x, \Sigma_s(p_0)) \le (M+1)\,|g(x, p) - g(x_0, p_0)|.$$

Hence there exists $\bar{x} \in \Sigma_s(p_0)$ such that

$$|x - \bar{x}| \le 2(M+1)L_g\,\delta(p, p_0) < r, \tag{2.3.16}$$

and thus $f(x_0, p_0) \le f(\bar{x}, p_0)$. From (2.3.11) we deduce that $|f(x, p) - f(\bar{x}, p_0)| \le L_r\,\delta(p, p_0)$, and hence

$$\mu_r(p_0) = f(x_0, p_0) \le f(\bar{x}, p_0) \le f(x, p) + L_r\,\delta(p, p_0) \tag{2.3.17}$$

for each $x \in \Sigma_r(p)$. The desired estimate follows from (2.3.15) and (2.3.17). □

In the above proof we followed the work of Alt [Alt3]. Next, we establish Hölder continuity of the local minimizers $x(p)$ of (2.3.1). Theorem 2.9 provides Lipschitz continuity of the value function under the regular point condition. The following example shows that the local solutions to the perturbed problems may behave rather irregularly.

Example 2.10. Let $X = Y = R^2$, $P = R$, $C = R^2$, $K = \{(y_1, y_2) : y_1 \ge 0,\ y_2 \ge 0\}$, $f(x_1, x_2, p) = x_1 + p\,x_2$, and $g(x_1, x_2, p) = (x_1, x_2)$. For $p_0 = 0$ a local solution to (2.3.1) is given by $x_0 = (0, 1)$ and $\mu_r(p_0) = 0$. Now let $r > 0$ be arbitrary. Then for $p \ne p_0$ the perturbed problem

$$\min f(x, p), \quad x \in \Sigma_r(p) \tag{2.3.18}$$

has a unique solution

$$x(p) = \begin{cases} (0, \max(1-r, 0)) & \text{if } p > 0, \\ (0, 1+r) & \text{if } p < 0 \end{cases}$$

and

$$\mu_r(p) = \begin{cases} p\max(1-r, 0) & \text{if } p > 0, \\ p(1+r) & \text{if } p < 0. \end{cases}$$

This shows that $\mu_r(p)$ is Lipschitz continuous but the distance $|x(p) - x_0|$ is not continuous with respect to p at $p_0 = 0$. The reason for this behavior of $x(p)$ is that the unperturbed problem does not depend on x_2 and all $(0, x_2)$ with $x_2 \geq 0$ are optimal for $p = p_0$. On the other hand if we let $f(x_1, x_2, p) = x_1 + px_2 + \frac{1}{2}(x_2 - 1)^2$, then for $r \geq 1$ and $|p| \leq 1$ the solution to (2.3.18) is given by

$$x(p) = (0, 1 - p).$$

Thus the distance $|x(p) - x_0|$ is continuous in p.

This example suggests that some kind of local invertibility should be assumed on f in order to ensure continuity of the solutions. We define a condition closely related to sufficient optimality conditions:

There exist $\kappa > 0$, $\beta \geq 1$, $\gamma > 0$, and neighborhoods $\hat{U}$ of x_0 and $\hat{N}$ of p_0 such that

$$f(x, p) \geq f(x_0, p_0) + \kappa\, |x - x_0|^\beta - \gamma\, \delta(p, p_0) \tag{2.3.19}$$

for all $p \in \hat{N}$ and $x \in \Sigma(p) \cap \hat{U}$.

The following result shows that this condition is implied by sufficient optimality conditions at x_0.

Theorem 2.11. *Assume that x_0 is regular, that* (2.3.11)–(2.3.12) *hold, and that there exist constants $\kappa > 0$, $\beta \geq 1$ and neighborhood $\tilde{U}$ of x_0 such that*

$$f(x, p_0) \geq f(x_0, p_0) + \kappa\, |x - x_0|^\beta \quad \textit{for all } x \in \Sigma(p_0) \cap \tilde{U}. \tag{2.3.20}$$

Then condition (2.3.19) *holds.*

Proof. We select $r, s > 0$ and a neighborhood N_1 of p_0 as in (2.3.14) in the proof of Theorem 2.9. We can assume that $\bar{B}(x_0, s) \in \tilde{U}$. Set $\hat{U} = \bar{B}(x_0, r)$ and choose $x \in \Sigma_r(p)$, $p \in N$. From (2.3.16) there exists an $\bar{x} \in \Sigma_s(p_0)$ such that

$$|x - \bar{x}| \leq 2(M + 1)L_g\, \delta(p, p_0),$$

and due to (2.3.17)

$$f(x, p) \geq f(\bar{x}, p_0) - L_r\, \delta(p, p_0).$$

By (2.3.20)

$$f(\bar{x}, p_0) \geq f(x_0, p_0) + \tilde{\kappa}\, |\bar{x} - x_0|^\beta.$$

Note that

$$\begin{aligned} |x - x_0|^\beta &\leq (|\bar{x} - x_0| + |\bar{x} - x|)^\beta \\ &\leq |\bar{x} - x_0|^\beta + \beta(|\bar{x} - x_0| + |\bar{x} - x|)^{\beta-1}\, |x - \bar{x}| \leq |\bar{x} - x_0|^\beta + \beta(3s)^{\beta-1}\, |x - \bar{x}|. \end{aligned}$$

Thus,

$$\begin{aligned} f(x, p) &\geq f(x_0, p_0) + \kappa\, |x - x_0|^\beta - L_r\, \delta(p, p_0) \\ &\geq f(x_0, p_0) + \kappa\, |x - x_0|^\beta - \gamma\, \delta(p, p_0), \end{aligned}$$

where $\gamma = L_r + \kappa\beta(3s)^{\beta-1}2(M + 1)L_g$ and (2.3.19) follows with $\hat{N} = \tilde{N}$. ☐

Condition (2.3.20) can be verified under the following second order sufficient optimality condition due to Maurer and Zowe [MaZo] .

Theorem 2.12. *Assume that K is a closed convex cone in Y and that $x_0 \in \Sigma(p_0)$ is a regular point, and let $\mathcal{L}(x,\lambda) = f(x, p_0) + \langle \lambda, g(x, p_0)\rangle_{Y^*,Y}$ be the Lagrangian for* (2.3.1) *at x_0 with Lagrange multiplier λ_0. If there exist constants $\omega > 0$, $\bar{\beta} > 0$ such that*

$$\mathcal{L}''(x_0, \lambda_0, p_0)(h,h) \geq \omega\, |h|^2 \text{ for all } h \in \mathcal{S}, \tag{2.3.21}$$

where

$$\mathcal{S} = L(\Sigma(p_0), x_0) \cap \{h \in X : \langle \lambda_0, g'(x_0, p_0)h\rangle_{Y^*,Y} \geq -\bar{\beta}\,|h|\},$$

then there exist $\kappa > 0$ and a neighborhood $\tilde{U}$ of x_0 such that (2.3.20) *holds with $\beta = 2$.*

For convenience we recall the definition of the linearizing cone: $L(\Sigma(p_0), x_0) = \{x \in C(x_0) : g'(x_0)x \in K(g(x_0))\}$, where $K(g(x_0)) = \{\lambda(y - g(x_0)) : y \in K, \lambda \geq 0\}$.

Proof. All quantities are evaluated at p_0 and this dependence will therefore be suppressed. First we show that every $x \in \Sigma = \Sigma(p_0)$ can be represented as

$$x - x_0 = h(x) + z(x) \text{ with } h(x) \in L(\Sigma, x_0) \text{ and } |z(x)| = o(|x - x_0|), \tag{2.3.22}$$

where $\frac{o(s)}{s} \to 0$ as $s \to 0^+$. Let $x \in \Sigma$ and expand g at x_0:

$$g(x) - g(x_0) = g'(x_0)(x - x_0) + r(x, x_0) \quad \text{with } |r(x, x_0)| = o(|x - x_0|).$$

By the generalized open mapping theorem (see Theorem 1.4), there exist $\alpha > 0$ and $k(x) \in K(g(x_0))$ such that for some $z(x) \in \alpha\, |r(x, x_0)|\, (B_X \cap C(x_0))$

$$r(x, x_0) = g'(x_0)z(x) - k(x).$$

If we put $h(x) = x - x_0 + z(x)$, then $h(x) \in C(x_0)$, $|z(x)| = o(|x - x_0|)$, and

$$g'(x_0)h(x) = g'(x_0)(x - x_0) + r(x, x_0) + k(x)$$

$$= g(x) - g(x_0) + k(x) \in K - g(x_0) + k(x) \subset K(g(x_0)),$$

i.e., $h(x) \in L(\Sigma, x_0)$. Next, for $x \in \Sigma$

$$f(x) \geq f(x) + \langle \lambda_0, g(x)\rangle = \mathcal{L}(x, \lambda_0) = \mathcal{L}(x_0, \lambda_0) + \frac{1}{2}\mathcal{L}''(x - x_0, x - x_0) + r(x - x_0), \tag{2.3.23}$$

where $|r(x - x_0)| = o(|x - x_0|^2)$, and we used $\langle \lambda_0, g(x)\rangle \leq 0$ for $x \in \Sigma$, and $\mathcal{L}'(x_0, \lambda_0) = 0$. Now put $B = \mathcal{L}''(x_0, \lambda_0)$ and $S = L(\Sigma) \cap \{h \in X : \langle \lambda_0, g'(x_0)h\rangle \geq -\bar{\beta}\,|h|\}$ in Lemma 2.13. It implies the existence of $\delta_0 > 0$ and $\gamma > 0$ such that

$\mathcal{L}''(x_0, \lambda_0)(x - x_0, x - x_0) \geq \delta_0\, |x - x_0|^2$ for all $x - x_0 = h(x) + z(x) \in \Sigma - x_0$ with $|z(x)| \leq \gamma\, |h(x)|$ and $\langle \lambda_0, g'(x_0)h\rangle \geq -\bar{\beta}\,|h|$.

Choose $\delta \in (0, 1)$ satisfying $\delta/(1-\delta) < \gamma$ and $\rho > 0$ such that $|z(x)| \le \delta|x - x_0|$ for $|x - x_0| \le \rho$. Then for $|x - x_0| \le \rho$

$$|h(x)| \ge |x - x_0| - |z(x)| \ge (1-\delta)|x - x_0|$$

and thus

$$|z(x)| \le \frac{\delta}{1-\delta}\,|h(x)| < \gamma\,|h(x)|. \tag{2.3.24}$$

Hence

$$\mathcal{L}''(x_0, \lambda_0)(x - x_0, x - x_0) \ge \delta_0\,|x - x_0|^2 \text{ for all } x - x_0 = h(x) + z(x) \in \Sigma - x_0$$
$$\text{with } |x - x_0| \le \rho \text{ and } \langle \lambda_0, g'(x_0)h(x)\rangle \ge -\bar{\beta}\,|h(x)|.$$

By (2.3.23) there exists $\bar{\rho} \in (0, \rho]$ such that

$$\begin{aligned} &f(x) \ge f(x_0) + \tfrac{\delta_0}{4}|x - x_0|^2 \text{ for all } x - x_0 = h(x) + z(x) \in \Sigma - x_0 \\ &\text{with } |x - x_0| \le \bar{\rho} \text{ and } \langle \lambda_0, g'(x_0)h(x)\rangle \ge -\bar{\beta}\,|h(x)|. \end{aligned} \tag{2.3.25}$$

For the case $x - x_0 = h(x) + z(x) \in \Sigma - x_0$ with $|x - x_0| \le \bar{\rho}$ and $\langle \lambda_0, g'(x_0)h(x)\rangle < -\bar{\beta}\,|h(x)|$ we find

$$\begin{aligned} f(x) - f(x_0) &= f'(x_0)(x - x_0) + \bar{r}(x - x_0) \\ &= -\langle \lambda_0, g'(x_0)h(x)\rangle - \langle \lambda_0, g'(x_0)z(x)\rangle + \bar{r}(x - x_0) \\ &\ge \bar{\beta}|h(x)| + r_1(x - x_0), \end{aligned}$$

where $r_1(x - x_0) = -\langle \lambda_0, g'(x_0)z(x)\rangle + \bar{r}(x - x_0)$ and $r_1(x - x_0) = o(|x - x_0|)$. Together with (2.3.24) this implies

$$f(x) \ge f(x_0) + \bar{\beta}(1-\delta)|x - x_0| + r_1(x - x_0) \text{ for } x - x_0 = h(x) + z(x) \in \Sigma - x_0$$
$$\text{with } |x - x_0| \le \bar{\rho} \text{ and } \langle \lambda_0, g'(x_0)h(x)\rangle < -\bar{\beta}\,|h(x)|.$$

Combined with (2.3.25) we arrive at the desired conclusion. ☐

Lemma 2.13. *Let S be a subset of a normed linear space X and let $B : X \times X \to R$ be a bounded symmetric quadratic form satisfying for some $\omega > 0$*

$$B(h, h) \ge \omega\,|h|^2 \quad \text{for all } h \in S.$$

Then there exist $\delta_0 > 0$ and $\gamma > 0$ such that

$$B(h + z, h + z) \ge \delta_0\,|h + z|^2 \quad \text{for all } h \in S,\ z \in X \text{ with } |z| \le \gamma\,|h|.$$

Proof. Let $b = \|B\|$ and choose $\gamma > 0$ such that $\delta = \omega - 2b\gamma - b\gamma^2 > 0$. Then for all $z \in X$ and $h \in S$ satisfying $|z| \le \gamma\,|h|$

$$B(h + z, h + z) \ge \omega\,|h|^2 - 2b|h||z| - b|z|^2 \ge \delta\,|h|^2.$$

Since $|h+z| \le |h| + |z| \le (1+\gamma)\,|h|$ we get

$$B(h+z, h+z) \ge \delta_0\,|h+z|^2$$

with $\delta_0 = \delta/(1+\gamma)^2$. □

Remark 2.3.1. An example which illustrates the benefit of including the set $\{h \in X : \langle \lambda_0, g'(x_0, p_0)h\rangle_{Y^*,Y} \ge -\bar{\beta}\,|h|\}$ into the definition of the set $\mathcal{S}$ on which the Hessian must be positive definite is given by the one-dimensional example with

$$f(x) = -x^3 - x,\ \ g(x) = x,\ \ K = R^-, \ \text{ and } C = R.$$

In this case $\Sigma = R^-, x_0 = 0, f(x_0) = 0, f'(x_0) = -1$, and $f''(x_0) = 0$. We further find that $L(\Sigma, x_0) = K$ and that the Lagrange multiplier is given by $\lambda_0 = 1$. Consequently for $\bar{\beta} \in (0,1)$ we have that $\mathcal{S} = L(\Sigma, x_0) \cap \{h \in R : \lambda_0 g'(x_0)h \ge -\bar{\beta}\,|h|\} = \{0\}$. Hence (2.3.21) is trivially satisfied. Let us note that f is bounded below as $f(x) \ge f(x_0) + \max(|x|, 2|x|^2)$ for $x \in K$. A similar argument does not hold if f is replaced by $f(x) = -x^3$. In this case $x_0 = 0,\ \lambda_0 = 0,\ f''(x_0) = 0,\ \mathcal{S} = K$, and (2.3.21) is not satisfied. Note that in this case we have lack of strict complementarity; i.e., the constraint is active and simultaneously $\lambda_0 = 0$.

Theorem 2.14. *Suppose that $x_0 \in \Sigma(p_0)$ is a regular point and that for some $\beta > 0$*

$$f'(x_0)h \ge \beta|h| \ \textit{ for all } \ h \in L(\Sigma(p_0), x_0).$$

Then there exists $\alpha > 0$ and a neighborhood $\widetilde{U}$ of x_0 such that

$$f(x) \ge f(x_0) + \alpha|x - x_0| \ \textit{ for all } \ x \in \Sigma(p_0) \cap \widetilde{U}.$$

Proof. For any $x \in \Sigma(p_0)$ we use the representation

$$x - x_0 = h(x) + z(x)$$

from (2.3.22) in the proof of Theorem 2.12. By expansion we have

$$f(x) - f(x_0) = f'(x_0)h(x) + f'(x_0)z(x) + r(x, x_0)$$

with $|r(x, x_0)| = o(|x - x_o|)$. The first order condition implies

$$f(x) - f(x_0) \ge \beta|h(x)| + r_1(x, x_0),$$

where $r_1(x, x_0) = f'(x_o)z(x) + r(x, x_o)$ and $|r_1(x, x_0)| = o(|x - x_0|)$. Choose a neighborhood $\widetilde{U}$ of x_0 such that $|z(x)| \le \frac{1}{2}|x - x_0|$ and $|r_1(x, x_0)| \le \frac{\beta}{4}|x - x_0|$ for $x \in \Sigma(p_0) \cap \widetilde{U}$. Then $|h(x)| \le \frac{1}{2}|x - x_0|$ and $f(x) - f(x_0) \ge \frac{\beta}{4}|x - x_0|$ for $x \in \widetilde{U}$. □

Note that for $C = X$ we have $f'(x_0) + g(x_0, p_0)^*\lambda_0 = 0$ and the second order condition of Theorem 2.12 involves the set of directions $h \in L(\Sigma(p_0), x_0)$ for which the first order condition is violated.

Following [Alt3] we now show that (2.3.19) implies stability of local minimizers.

Theorem 2.15. *Suppose that the local solution $x_0 \in \Sigma(p_0)$ is a regular point and that* (2.3.11), (2.3.12), *and* (2.3.19) *are satisfied. Then there exist real numbers $r > 0$ and $L > 0$ and a neighborhood N of p_0 such that the value function μ_r is Lipschitz continuous at p_0 and for each $p \in N$ the following statements hold:*

(a) *For every sequence $\{x_n\}$ in $\Sigma_r(p)$ with the property that $\lim_{n\to\infty} f(x_n, p) = \mu_r(p)$ it follows that $x_n \in B(x_0, r)$ for all sufficiently large n.*

(b) *If there exists a point $x(p) \in \Sigma_r(p)$ with $\mu_r(p) = f(x(p), p)$, then $x(p) \in B(x_0, r)$, i.e., $x(p)$ is a local minimizer of* (2.3.1) *and*

$$|x(p) - x_0| \leq L\,\delta(p, p_0)^{\frac{1}{\beta}}.$$

Proof. We choose $r > 0$ and a neighborhood $\tilde{N}$ of p_0 as in (2.3.14) in the proof of Theorem 2.9. Without loss of generality we can assume that $\tilde{N} \times \bar{B}(x_0, r) \subset \hat{N} \times \hat{U}$ with $\hat{N}, \hat{U}$ determined from (2.3.19) and that

$$\kappa\, r^{\beta} > (L_r + \gamma)\delta(p, p_0) \tag{2.3.26}$$

for all $p \in \tilde{N}$, where L_r is determined in Theorem 2.9. Thus μ_r is Lipschitz continuous at p_0 by Theorem 2.9. Let $p \in \tilde{N}$ be arbitrary and let $\{x_n\}$ in $\Sigma(p)$ be a sequence satisfying $\lim_{n\to\infty} f(x_n, p) = \mu_r(p)$. By (2.3.19) and Lipschitz continuity of μ_r

$$\begin{aligned}\kappa\, |x_n - x_0|^{\beta} &\leq |f(x_n, p) - f(x_0, p_0)| + \gamma\delta(p, p_0)\\ &\leq |f(x_n, p) - \mu_r(p)| + |\mu_r(p) - \mu_r(p_0)| + \gamma\,\delta(p, p_0)\\ &\leq |f(x_n, p) - \mu_r(p)| + (L_r + \gamma)\,\delta(p, p_0).\end{aligned}$$

This estimate together with (2.3.26) imply $|x_n - x_0| < r$ for all sufficiently large n. This proves (a).

Similarly, for $x(p) \in \Sigma_r(p)$ satisfying $\mu_r(p) = f(x(p), p)$ we have

$$\kappa\, |x(p) - x_0|^{\beta} \leq (L_r + \gamma)\,\delta(p, p_0).$$

Thus, $|x(p) - x_0| < r$,

$$|x(p) - x_0| \leq \left(\frac{L_r + \gamma}{\kappa}\,\delta(p, p_0)\right)^{\frac{1}{\beta}}$$

and (b) follows. □

2.4 Lipschitz continuity

In this section we investigate Lipschitz continuous dependence of the solution and the Lagrange multipliers with respect to the parameter p in (2.1.1). Throughout this section and the next we assume that $C = X$ and that the requirements on the function spaces specified below (2.1.1) are met.

Theorem 2.16. *Assume* (H1)–(H4) *hold at the local solution* x_0 *of* (2.1.1). *Then there exist neighborhoods* $N = N(p_0)$ *of* p_0 *and* $U = U(x_0, \lambda_0, \mu_0, \eta_0)$ *of* $(x_0, \lambda_0, \mu_0, \eta_0)$ *and a constant* M *such that for all* $p \in N$ *there exists a unique* $(x_p, \lambda_p, \mu_p, \eta_p) \in U$ *satisfying*

$$0 \in \begin{cases} \mathcal{L}'(x, p, \lambda, \mu, \eta), \\ -e(x, p), \\ -g(x, p) + \partial\psi_{R^{m,+}}(\mu), \\ -\ell(x, p) + \partial\psi_{K^+}(\eta), \end{cases} \tag{2.4.1}$$

and

$$|(x(p), \lambda(p), \mu(p), \eta(p)) - (x(q), \lambda(q), \mu(q), \eta(q))| \le M\,|p - q| \tag{2.4.2}$$

for all $p, q \in N$. *Moreover, there exists a neighborhood* $\tilde{N} \subset N$ *of* p_0 *such that* $x(p)$ *is a local solution of* (2.1.1) *if* $p \in \tilde{N}$.

For the proof we shall employ the following lemma.

Lemma 2.17. *Consider the quadratic programming problem in a Hilbert space* H

$$\min J(x) = \frac{1}{2}\langle Ax, x\rangle + \langle \tilde{a}, x\rangle$$
$$\text{subject to } Ex = \tilde{b} \text{ in } \tilde{Y} \text{ and } Gx - \tilde{c} \in \tilde{K} \subset \tilde{Z}, \tag{2.4.3}$$

where $\tilde{Y}$, $\tilde{Z}$ *are Banach spaces,* $E \in \mathcal{L}(H, \tilde{Y})$, $G \in \mathcal{L}(H, \tilde{Z})$, *and* $\tilde{K}$ *is a closed convex cone in* $\tilde{Z}$. *If*

(a) $\langle A\cdot, \cdot\rangle$ *is a symmetric bounded quadratic form on* $H \times H$ *and there exists an* $\omega > 0$ *such that* $\langle Ax, x\rangle \ge \omega\,|x|_H^2$ *for all* $x \in \ker(E)$, *and*

(b) *for all* $(\tilde{b}, \tilde{c}) \in \tilde{Y} \times \tilde{Z}$

$$\tilde{S}(\tilde{b}, \tilde{c}) = \{x \in H : Ex = \tilde{b},\quad Gx - \tilde{c} \in K\} \text{ is nonempty,}$$

then there exists a unique solution to (2.4.3) *satisfying*

$$\langle Ax + \tilde{a}, v - x\rangle \ge 0 \quad \text{for all } v \in \tilde{S}(\tilde{b}, \tilde{c}).$$

Moreover, if x *is a regular point, then there exists* $(\tilde{\lambda}, \tilde{\eta}) \in \tilde{Y}^* \times \tilde{Z}^*$ *such that*

$$0 \in \begin{pmatrix} \tilde{a} \\ \tilde{b} \\ \tilde{c} \end{pmatrix} + \begin{pmatrix} A & E^* & G^* \\ -E & 0 & 0 \\ -G & 0 & 0 \end{pmatrix} \begin{pmatrix} x \\ \tilde{\lambda} \\ \tilde{\eta} \end{pmatrix} + \begin{pmatrix} 0 \\ 0 \\ \partial\psi_{\tilde{K}^+}(\tilde{\eta}) \end{pmatrix}.$$

Proof. Since $\tilde{K}$ is closed and convex, $\tilde{S}(\tilde{b}, \tilde{c})$ is closed and convex as well. Since $\tilde{S}(\tilde{b}, \tilde{c})$ is nonempty, there exists a unique $w \in \operatorname{range} E^*$ such that $Ew = \tilde{b}$. Moreover, every $x \in \tilde{S}(\tilde{b}, \tilde{c})$ can be expressed as $x = w + y$, where $y \in \ker(E)$. From (a) it follows that J is coercive and bounded below on $\tilde{S}(\tilde{b}, \tilde{c})$. Hence there exists a bounded minimizing sequence $\{x_n\}$ in $\tilde{S}(\tilde{b}, \tilde{c})$ such that $\lim_{n\to\infty} J(x_n) \to \inf_{x \in \tilde{S}(\tilde{b}, \tilde{c})} J(x)$. Boundedness of

x_n and weak sequential closedness of $\tilde{S}(\tilde{b}, \tilde{c})$ imply the existence of a subsequence of x_n that converges weakly to some x in $\tilde{S}(\tilde{b}, \tilde{c})$. Since J is weakly lower semicontinuous, $J(x) \le \liminf J(x_n)_{n\to\infty}$. Hence x minimizes J over $\tilde{S}(\tilde{b}, \tilde{c})$.

For $v \in \tilde{S}(\tilde{b}, \tilde{c})$ and $t \in (0, 1)$ we have $x + t(v - x) \in \tilde{S}(\tilde{b}, \tilde{c})$. Thus,

$$0 \le J(x + t\,(v - x)) - J(x) = t\,\langle Ax + \tilde{a}, v - x\rangle + \frac{t^2}{2}\langle A(v - x), v - x\rangle$$

and letting $t \to 0^+$ we have

$$\langle Ax + \tilde{a}, v - x\rangle \ge 0 \quad \text{for all } v \in \tilde{S}(\tilde{b}, \tilde{c}).$$

The last assertion follows from Theorem 1.6. $\square$

***Proof of Theorem* 2.16.** The proof of the first assertion of the theorem is based on the implicit function theorem of Robinson for generalized equations; see Theorem 2.3. It requires us to verify the strong regularity condition for the linearized form of (2.4.1) which is given by

$$0 \in \begin{cases} \mathcal{L}'(x_0, p_0, \lambda_0, \mu_0, \eta_0) + A(x - x_0) + E^*(\lambda - \lambda_0) + G^*(\mu - \mu_0) + L^*(\eta - \eta_0), \\ -e(x_0, p_0) - E(x - x_0), \\ -g(x_0, p_0) - G(x - x_0) + \partial\psi_{R^{m,+}}(\mu), \\ -\ell(x_0, p_0) - L(x - x_0) + \partial\psi_{K^+}(\eta), \end{cases}$$

where the operators A, E, G, and L are defined in the introduction to this chapter.

If we define the multivalued operator $\mathcal{T}$ from $X \times W \times R^m \times Z$ to itself by

$$\mathcal{T}\begin{pmatrix} x \\ \lambda \\ \mu \\ \eta \end{pmatrix} = \begin{pmatrix} A & E^* & G^* & L^* \\ -E & 0 & 0 & 0 \\ -G & 0 & 0 & 0 \\ -L & 0 & 0 & 0 \end{pmatrix}\begin{pmatrix} x \\ \lambda \\ \mu \\ \eta \end{pmatrix} + \begin{pmatrix} f'(x_0, p_0) - Ax_0 \\ Ex_0 \\ -g(x_0, p_0) + Gx_0 \\ -\ell(x_0, p_0) + Lx_0 \end{pmatrix} + \begin{pmatrix} 0 \\ 0 \\ \partial\psi_{R^{m,+}}(\mu) \\ \partial\psi_{K^+}(\eta) \end{pmatrix},$$

then the linearization can be expressed as

$$0 \in \mathcal{T}(x, \lambda, \mu, \eta).$$

Since the constraint associated with $g^-(x, p_0) \le 0$ is inactive in a neighborhood of x_0, strong regularity of $\mathcal{T}$ at $(x_0, \lambda_0, \mu_0, \eta_0)$ easily follows from strong regularity of the multivalued mapping T from $X \times (W \times R^{m_1}) \times R^{m_2} \times Z$ into itself at $(x_0, (\tilde{\lambda}_0, \mu_0^+), \mu_0^0, \eta_0)$ defined by

$$T\begin{pmatrix} x \\ \tilde{\lambda} \\ \mu^0 \\ \eta \end{pmatrix} = \begin{pmatrix} A & E_+^* & G_0^* & L^* \\ -E_+ & 0 & 0 & 0 \\ -G_0 & 0 & 0 & 0 \\ -L & 0 & 0 & 0 \end{pmatrix}\begin{pmatrix} x \\ \tilde{\lambda} \\ \mu^0 \\ \eta \end{pmatrix} + \begin{pmatrix} f'(x_0, p_0) - Ax_0 \\ E_+x_0 \\ G_0x_0 \\ -\ell(x_0, p_0) + Lx_0 \end{pmatrix} + \begin{pmatrix} 0 \\ 0 \\ \partial\psi_{R^{m_2,+}}(\mu^0) \\ \partial\psi_{K^+}(\eta) \end{pmatrix}.$$

Strong regularity of T requires us to show that there exist neighborhoods of $\hat{V}$ of 0 and $\hat{U}$ of $(x_0, (\tilde{\lambda}_0, \mu_0^+), \mu_0^0, \eta_0)$ in $X \times (W \times R^{m_1}) \times R^{m_2} \times Z$ such that $T^{-1}(\hat{V}) \cap \hat{U}$ is single valued and Lipschitz continuous from $\hat{V}$ to $\hat{U}$.

We now divide the proof in several steps.

Existence. We show the existence of a solution to

$$(\alpha, \beta, \gamma, \delta) \in T(x, \tilde{\lambda}, \mu^0, \eta)$$

for $(\alpha, \beta, \gamma, \delta) \in \hat{V}$. Observe that this is equivalent to

$$0 \in \begin{pmatrix} a \\ b \\ c \\ d \end{pmatrix} + \begin{pmatrix} A & E_+^* & G_0^* & L^* \\ -E_+ & 0 & 0 & 0 \\ -G_0 & 0 & 0 & 0 \\ -L & 0 & 0 & 0 \end{pmatrix} \begin{pmatrix} x \\ \tilde{\lambda} \\ \mu^0 \\ \eta \end{pmatrix} + \begin{pmatrix} 0 \\ 0 \\ \partial \psi_{R^{m_2},+}(\mu^0) \\ \partial \psi_{K^+}(\eta) \end{pmatrix}, \tag{2.4.4}$$

where

$$(a, b, c, d) = f'(x_0, p_0) - Ax_0 - \alpha,\ E_+x_0 - \beta,\ G_0x_0 - \gamma,\ -\ell(x_0, p_0) + Lx_0 - \delta). \tag{2.4.5}$$

To solve (2.4.4) we introduce the quadratic optimization problem with linear constraints:

$$\begin{gathered} \min \frac{1}{2}\langle Ax, x\rangle + \langle a, x\rangle \\ \text{subject to } E_+x = b,\ G_0x \le c, \text{ and } Lx - d \in K. \end{gathered} \tag{2.4.6}$$

We verify the conditions in Lemma 2.17 for (2.4.6) with $\tilde{Y} = W \times R^{m_1}$, $\tilde{Z} = R^{m_2} \times Z$, $E = E_+$, $G = (G_0, L)$, and $\tilde{K} = R^{m_2,-} \times K$. Let S be the feasible set of (2.4.6):

$$S(\alpha, \beta, \gamma, \delta) = \{x \in X : E_+x = b,\quad G_0x \le c,\quad Lx - d \in K\},$$

where the relationship between $(\alpha, \beta, \gamma, \delta)$ and (a, b, c, d) is given by (2.4.5). Clearly, $x_0 \in S(0)$ and moreover x_0 is regular point. From Theorem 2.8 it follows that there exists a neighborhood V of 0 such that for all $(\alpha, \beta, \delta, \gamma) \in V$ the set $S(\alpha, \beta, \gamma, \delta)$ is nonempty. Lemma 2.17 implies the existence of a unique solution x to (2.4.6) for every $(\alpha, \beta, \delta, \gamma) \in V$. By (H1)–(H2) and Theorems 2.12, 2.11, and 2.14 the solution x is Hölder continuous with exponent $\frac{1}{2}$ with respect to $(\alpha, \beta, \gamma, \delta) \in V$. Next we show that $x = x(\alpha, \beta, \gamma, \delta)$ is a regular point for (2.4.6), i.e.,

$$0 \in \operatorname{int} \left\{ \begin{pmatrix} 0 \\ G_0x - c \\ Lx - d \end{pmatrix} + \begin{pmatrix} E_+ \\ G_0 \\ L \end{pmatrix} (X - x) + \begin{pmatrix} 0 \\ R^{m_2,+} \\ -K \end{pmatrix} \right\}.$$

This is equivalent to

$$0 \in \operatorname{int} \left\{ \begin{pmatrix} \beta \\ \gamma \\ \ell(x_0, p_0) + \delta \end{pmatrix} + \begin{pmatrix} E_+ \\ G_0 \\ L \end{pmatrix} (X - x_0) + \begin{pmatrix} 0 \\ R^{m_2,+} \\ -K \end{pmatrix} \right\}.$$

Since x_0 is a regular point, there exists a neighborhood $\tilde{V} \subset V$ of 0 such that this inclusion holds for all $(\alpha, \beta, \gamma, \delta) \in \tilde{V}$. Hence the existence of a solutions to (2.4.4) follows from Lemma 2.17.

Uniqueness. To argue uniqueness, assume that $(x_i, \tilde{\lambda}_i, \mu_i^0, \eta_i)$, $i = 1, 2$, are solutions to (2.4.4). It follows that

$$\begin{aligned} &\langle A(x_1 - x_2) + E_+^*(\tilde{\lambda}_1 - \tilde{\lambda}_2) + G_0^*(\mu_1^0 - \mu_2^0) + L^*(\eta_1 - \eta_2), x_1 - x_2 \rangle \le 0, \\ &E_+(x_1 - x_2) = 0, \\ &\langle G_0(x_1 - x_2),\ \mu_1^0 - \mu_2^0 \rangle \ge 0, \\ &\langle L(x_1 - x_2), \eta_1 - \eta_2 \rangle \ge 0. \end{aligned}$$

Using (H2) we find

$$\kappa\,|x_1 - x_2|^2 \le \langle A(x_1 - x_2), x_1 - x_2 \rangle \le -\langle G_0^*(\mu_1^0 - \mu_2^0), x_1 - x_2 \rangle - (L^*(\eta_1 - \eta_2), x_1 - x_2) \le 0,$$

which implies that $x_1 = x_2 = x$. To show uniqueness of the remaining components, we observe that

$$\begin{aligned} &E_+^*(\tilde{\lambda}_1 - \tilde{\lambda}_2) + G_0^*(\mu_1^0 - \mu_2^0) + L^*(\eta_1 - \eta_2) = 0, \\ &\langle \eta_1 - \eta_2, L(x - x_0) + \ell(x_0, p_0) + \delta \rangle = 0, \end{aligned}$$

or equivalently

$$\mathcal{E}^*(z)(\tilde{\lambda}_1 - \tilde{\lambda}_2, \mu_1^0 - \mu_2^0 + \eta_1 - \eta_2) = 0,$$

where $z = L(x - x_0) + \ell(x_0, p_0) + \delta$ and $\mathcal{E}$ is defined below (2.1.6). Due to continuous dependence of x on $(\alpha, \beta, \gamma, \delta)$ we can assure that $z \in \mathcal{O}$ for all $(\alpha, \beta, \gamma, \delta) \in \tilde{V}$. Hence (H3) implies that $(\tilde{\lambda}_1 - \tilde{\lambda}_2, \mu_1^0 - \mu_2^0, \eta_1 - \eta_2) = 0$.

Continuity. By (H3) the operator $\mathcal{E}^*(\ell(x_0, p_0))$ has closed range and injective. Hence there exists $\epsilon > 0$ such that

$$|\mathcal{E}^*(\ell(x_0, p_0))\Lambda| \ge \epsilon\,|\Lambda|$$

for all $\Lambda = (\tilde{\lambda}, \mu^0, \eta)$. Since $\|\mathcal{E}^*(\ell(x_0, p_0)) - \mathcal{E}^*(z)\| \le |z - \ell(x_0, p_0)|$, there exists a neighborhood of 0 in $X \times (W \times R^{m_1}) \times R^{m_2} \times Z$, again denoted by $\tilde{V}$, such that

$$|\mathcal{E}^*(z_\delta)\Lambda| \ge \frac{\epsilon}{2}\,|\Lambda| \tag{2.4.7}$$

for all $(\alpha, \beta, \gamma, \delta) \in \tilde{V}$, where $z_\delta = \ell(x_0, p_0) + Lx - Lx_0 + \delta$. Any solution $(x, \tilde{\lambda}, \mu^0, \eta)$ satisfies

$$\mathcal{E}^*(z_\delta)(\tilde{\lambda}, \mu^0, \eta) = (-Ax - a, 0).$$

It thus follows from (2.4.7) that there exists a constant k_1 such that for all $(\alpha, \beta, \gamma, \delta) \in \tilde{V}$ the solution $(x, \tilde{\lambda}, \tilde{\mu}, \eta)$ satisfies

$$|(x, \tilde{\lambda}, \mu^0, \eta)|_{X \times (W \times R^{m_1}) \times R^{m_2} \times Z} \le k_1.$$

Henceforth let $(\alpha_i, \beta_i, \gamma_i, \delta_i) \in \tilde{V}$ and let $(x_i, \tilde{\lambda}_i, \mu_i^0, \eta_i)$ denote the corresponding solution to (2.4.4) for $i = 1, 2$. Then

$$\mathcal{E}^*(z_{\delta_1}) \begin{pmatrix} \tilde{\lambda}_1 - \tilde{\lambda}_2 \\ \mu_1^0 - \mu_2^0 \\ \eta_1 - \eta_2 \end{pmatrix} = \begin{pmatrix} \alpha_1 - \alpha_2 + A(x_2 - x_1) \\ \langle \eta_2, L(x_2 - x_1) + \delta_2 - \delta_1 \rangle \end{pmatrix}.$$

By (2.4.7) this implies that there exists a constant k_2 such that

$$|(\tilde{\lambda}_1 - \tilde{\lambda}_2, \mu_1^0 - \mu_2^0, \eta_1 - \eta_2)|_{(W \times R^{m_1}) \times R^{m_2} \times Z} \le k_2 \left(|\alpha_1 - \alpha_2| + |\delta_1 - \delta_2| + |x_1 - x_2|\right). \quad (2.4.8)$$

Now, from the first equation in (2.4.4) we have

$$\langle A(x_1 - x_2) + E_+^*(\tilde{\lambda}_1 - \tilde{\lambda}_2) + G_0^*(\mu_1^0 - \mu_2^0) + L^*(\eta_1 - \eta_2) + a_1 - a_2, x_1 - x_2 \rangle \le 0. \quad (2.4.9)$$

We also find

$$\begin{aligned} \langle \mu_1^0 - \mu_2^0, G_0(x_1 - x_2) \rangle &= \langle \mu_1^0, c_1 - c_2 \rangle + \langle \mu_1^0, c_2 - G_0 x_2 \rangle \\ &\quad - \langle \mu_2^0, G_0 x_1 - c_1 \rangle - \langle \mu_2^0, c_1 - c_2 \rangle \ge \langle \mu_1^0 - \mu_2^0, c_1 - c_2 \rangle \end{aligned} \quad (2.4.10)$$

and similarly

$$\langle \eta_1 - \eta_2, L(x_1 - x_2) \rangle \ge \langle \eta_1 - \eta_2, d_1 - d_2 \rangle. \quad (2.4.11)$$

Representing $x_1 - x_2 = v + w$ with $v \in \ker(R_+)$ and $w \in \operatorname{range}(E_+^*)$, one obtains

$$E_+(x_1 - x_2) = E_+ w = b_1 - b_2.$$

Thus

$$|w| \le k_3 \, |b_1 - b_2| \quad (2.4.12)$$

for some k_3, and from (H2) and (2.4.9)–(2.4.11)

$$\begin{aligned} \kappa \, |v|^2 \le \langle Av, v \rangle &= \langle A(x_1 - x_2), x_1 - x_2 \rangle - 2 \langle Av, w \rangle - \langle Aw, w \rangle \\ &\le -\langle \tilde{\lambda}_1 - \tilde{\lambda}_2, b_1 - b_2 \rangle - \langle \mu_1^0 - \mu_2^0, c_1 - c_2 \rangle - \langle \eta_1 - \eta_2, d_1 - d_2 \rangle \\ &\quad - \langle a_1 - a_2, v + w \rangle - 2 \langle Av, w \rangle - \langle Aw, w \rangle. \end{aligned}$$

Let $\Lambda = (\tilde{\lambda}_1 - \tilde{\lambda}_2, \mu_1^0 - \mu_2^0, \eta_1 - \eta_2)$. Then

$$\kappa \, |v|^2 \le |\Lambda| |(\beta_1 - \beta_2, \gamma_1 - \gamma_2, \delta_1 - \delta_2)| + |\alpha_1 - \alpha_2|(|v| + |w|) + \|A\|(2|v||w| + |w|^2).$$

It thus follows from (2.4.8), (2.4.12) that there exists a constant k_4 such that

$$|x_1 - x_2| \le k_4 \, |(\alpha_1 - \alpha_2, \beta_1 - \beta_2, \gamma_1 - \gamma_2, \delta_1 - \delta_2)|$$

for all $(\alpha_i, \beta_i, \gamma_i, \delta_i) \in \tilde{V}$. We apply (2.4.8) once again to obtain Lipschitz continuity of $(x, \tilde{\lambda}, \mu^0, \eta)$ with respect to $(\alpha, \beta, \gamma, \delta)$ in a neighborhood of the origin. Consequently $\mathcal{T}$

is strongly regular at $(x_0, \lambda_0, \mu_0, \eta_0)$, Robinson's theorem is applicable, and (2.4.1), (2.4.2) follow.

Local solution. We show that there exists a neighborhood $\tilde{N}$ of p_0 such that for $p \in \tilde{N}$ the second order sufficient optimality (2.3.21) is satisfied at $x(p)$ so that $x(p)$ is a local solution of (2.1.1) by Theorem 2.12. Due to (H2) and the smoothness properties of f, e, g we can assume that

$$\mathcal{L}''(x(p), p, \lambda(p), \eta(p))(h, h) \geq \frac{\kappa}{2}\,|h|^2 \quad \text{for all } h \in \ker\, E_+ \tag{2.4.13}$$

if $p \in N(p_0)$. Let us define $E_p = (e'(x(p), p), g'_+(x(p), p))$ for $p \in N(p_0)$. Due to surjectivity of E_{p_0} and smoothness properties of e, g there exists a neighborhood $\tilde{N} \subset N(p_0)$ of p_0 such that E_p is surjective for all $p \in \tilde{V}$. Lemma 2.13 implies the existence of $\delta_0 >$ and $\gamma > 0$ such that

$$\mathcal{L}''(x(p), p, \lambda(p), \eta(p))(h + z, h + z) \geq \delta_0\,|h + z|^2$$

for all $h \in \ker\, E_+$ and $z \in X$ satisfying $|z| \leq \gamma\,|h|$. The orthogonal projection onto $\ker\, E_p$ is given by $P_{\ker\, E_p} = I - E_p^*(E_p E_p^*)^{-1} E_p$. We can select $\tilde{N}$ so that

$$|E_p^*(E_p E_p^*)^{-1} E_p - E_{p_0}^*(E_{p_0} E_{p_0}^*)^{-1} E_{p_0}| \leq \frac{\gamma}{1 + \gamma}$$

for all $p \in \tilde{V}$. For $x \in \ker\, E_p$, we have $x = h + z$ with $h \in \ker\, E_+$ and $z \in (\ker\, E_+)^\perp$ and $|x|^2 = |h|^2 + |z|^2$. Thus,

$$|z| \leq |E_p^*(E_p E_p^*)^{-1} E_p x - E_{p_0}^*(E_{p_0} E_{p_0}^*)^{-1} E_{p_0} x| \leq \frac{\gamma}{1 + \gamma}(|h| + |z|)$$

and hence $|z| \leq \gamma\,|h|$. From (2.4.13) this implies

$$\mathcal{L}''(x(p), p, \lambda(p), \eta(p)) \geq \delta_0\,|x|^2 \quad \text{for all } x \in \ker\, E_p$$

and, since $L(\Sigma(p), p) \subset \ker\, E_p$, the second order sufficient optimality (2.3.21) is satisfied at $x(p)$. $\square$

The first part of the proof of Theorem 2.16 contains a Lipschitz continuity result for linear complementarity problems. In the following corollary we reconsider this result in a modified form which will be used for the convergence analysis of sequential quadratic programming problems.

Throughout the following discussion p_0 in (2.1.1) (with $C = X$) is fixed and therefore its notation is suppressed. For $(x, \lambda, \mu) \in X \times W^* \times \mathbb{R}^m$ let

$$\mathcal{L}(x, \lambda, \mu) = f(x) + \langle \lambda, e(x) \rangle + \langle \mu, g(x) \rangle;$$

i.e., the equality and finite rank inequality constraints are realized in the Lagrangian term. For $(x, \lambda, \mu) \in X \times W^* \times \mathbb{R}^m$, $(\bar{x}, \bar{\lambda}, \bar{\mu}) \in X \times W^* \times \mathbb{R}^m$, and $(a, b, c, d) \in X \times W \times \mathbb{R}^n \times Z$

consider

$$\begin{cases} \min \frac{1}{2} \mathcal{L}''(\bar{x}, \bar{\lambda}, \bar{\mu})(x - \bar{x}, x - \bar{x}) + \langle f'(\bar{x}) + a, x - \bar{x} \rangle, \\ e(\bar{x}) + e'(\bar{x})(x - \bar{x}) = b, \\ g(\bar{x}) + g'(\bar{x})(x - \bar{x}) \le c, \\ \ell(\bar{x}) + L(x - \bar{x}) - d \in K. \end{cases} \tag{2.4.14}$$

Let $(\bar{x}, \bar{\lambda}, \bar{\mu}) \in X \times W^* \times \mathbb{R}^m$ and define the operator $\mathcal{G}$ from $X \times W^* \times \mathbb{R}^m \times Z^*$ to $X \times W \times \mathbb{R}^m \times Z$ by

$$\mathcal{G}(\bar{x}, \bar{\lambda}, \bar{\mu})(x, \lambda, \mu, \eta) = \begin{pmatrix} f'(\bar{x}) \\ -e(\bar{x}) \\ -g(\bar{x}) \\ -\ell(\bar{x}) \end{pmatrix} + \begin{pmatrix} \mathcal{L}''(\bar{x}, \bar{\lambda}, \bar{\mu})(x - \bar{x}) + e'(\bar{x})^*\lambda + g'(\bar{x})^*\mu + L^*\eta \\ -e'(\bar{x})(x - \bar{x}) \\ -g'(\bar{x})(x - \bar{x}) \\ -L(x - \bar{x}) \end{pmatrix}.$$

Note that

$$0 \in \begin{pmatrix} a \\ b \\ c \\ d \end{pmatrix} + \mathcal{G}(\bar{x}, \bar{\lambda}, \bar{\mu})(x, \lambda, \mu, \eta) + \begin{pmatrix} 0 \\ 0 \\ \partial\psi_{\mathbb{R}^{m,+}}(\mu) \\ \partial\psi_{K^+}(\eta) \end{pmatrix} \tag{2.4.15}$$

is the first order optimality system for (2.4.14). The following modification of (H2) will be used

$(\widetilde{\mathrm{H2}})$ there exists $\kappa > 0$ such that $\langle \mathcal{L}''(x_0, \lambda_0, \mu_0) x, x \rangle_X \ge \kappa |x|_X^2$ for all $x \in \ker(E_+)$.

Corollary 2.18. *Assume that* (H1), $(\widetilde{\mathrm{H2}})$, (H3) *hold at a local solution of* (2.1.1) *and that the second derivatives of f, e, and g are Lipschitz continuous in a neighborhood of x_0. Then there exist neighborhoods $U(\xi_0)$ of $\xi_0 = (x_0, \lambda_0, \mu_0, \eta_0)$ in $X \times W^* \times \mathbb{R}^m \times Z^*$, $\hat{U}(x_0, \lambda_0, \mu_0)$ of (x_0, λ_0, μ_0), and V of the origin in $X \times W \times \mathbb{R}^m \times Z$ and a constant $\tilde{K}$, such that for all $(\bar{x}, \bar{\lambda}, \bar{\mu}) \in \hat{U}(x_0, \lambda_0, \bar{\mu}_0))$ and $q = (a, b, c, d) \in V$, there exists a unique solution $\xi = \xi(\bar{x}, \bar{\lambda}, \bar{\mu}, q) = (x, \lambda, \mu, \eta) \in U(\xi_0)$ of* (2.4.15)*, and x is a local solution of* (2.4.14)*. Moreover, for every pair $q_1, q_2 \in V$ and $(\bar{x}_1, \bar{\lambda}_1, \bar{\mu}_1), (\bar{x}_2, \bar{\lambda}_2, \bar{\mu}_2) \in \hat{U}(x_0, \lambda_0, \bar{\mu}_0)$ we have*

$$|\xi(\bar{x}_1, \bar{\lambda}_1, \bar{\mu}_1, q_1) - \xi(\bar{x}_2, \bar{\lambda}_2, \bar{\mu}_2, q_2)| \le \tilde{K}(|(\bar{x}_1, \bar{\lambda}_1, \bar{\mu}_1) - (\bar{x}_2, \bar{\lambda}_2, \bar{\mu}_2)| + |q_1 - q_2|).$$

Proof. From the first part of the proof of Theorem 2.16 it follows that

$$0 \in \begin{pmatrix} a \\ b \\ c \\ d \end{pmatrix} + \mathcal{G}(x_0, \lambda_0, \mu_0)(x, \lambda, \mu, \eta) + \begin{pmatrix} 0 \\ 0 \\ \partial\psi_{\mathbb{R}^{m,+}}(\mu) \\ \partial\psi_{K^+}(\eta) \end{pmatrix}$$

admits a unique solution $\xi = (x, \lambda, \mu, \eta)$ which depends on (a, b, c, d) Lipschitz continuously provided that (a, b, c, d) is sufficiently small. The proof of the corollary can be completed by observing that the estimates in the proof of Theorem 2.16 also hold uniformly if (x_0, λ_0, μ_0) is replaced by $(\bar{x}, \bar{\lambda}, \bar{\mu})$ in a sufficiently small neighborhood $\hat{U}(x_0, \lambda_0, \mu_0)$ of (x_0, λ_0, μ_0), and (a, b, c, d) is in a sufficiently small neighborhood V of the origin in $X \times W \times \mathbb{R}^m \times Z$.

As an alternative to reconsidering the proof of Theorem 2.16 one can apply Theorem 2.16 to (2.1.1) with $P = (X \times W^* \times \mathbb{R}^m) \times (X \times W \times \mathbb{R}^m \times Z)$, $p = (x, \lambda, \mu, a, b, c, d)$, $p_0 = (x_0, \lambda_0, \mu_0, 0)$, and f, e, g, ℓ replaced by

$$\tilde{f}(x, \bar{x}, \bar{\lambda}, \bar{\mu}, a) = \tfrac{1}{2}\mathcal{L}''(\bar{x}, \bar{\lambda}, \bar{\mu})(x - \bar{x}, x - \bar{x}) + \langle f'(\bar{x}) + a, x - \bar{x}\rangle,$$

$$\tilde{e}(x; \bar{x}, b) = e(\bar{x}) + e'(\bar{x})(x - \bar{x}) - b,$$

$$\tilde{g}(x; \bar{x}, c) = g(\bar{x}) + g'(\bar{x})(x - \bar{x}) - c,$$

$$\tilde{\ell}(x; \bar{x}, d) = \ell(\bar{x}) + L(x - \bar{x}) - d.$$

Clearly (H1)–(H3) hold for $\tilde{e}, \tilde{g}, \tilde{\ell}$ at (x_0, p_0) with $p_0 = (x_0, \lambda_0, \mu_0, 0)$. Lipschitz continuity of $\tilde{f}, \tilde{e}, \tilde{g}, \tilde{\ell}$ with respect to $(\bar{x}, \bar{\lambda}, \bar{\mu}, a, b, c, d)$ is a consequence of the general regularity requirements on f, e, g, ℓ made at the beginning of the chapter and the assumption that the second derivatives of f and e are Lipschitz continuous in a neighborhood of x_0. This implies (H4) for $\tilde{e}, \tilde{f}, \tilde{g}, \tilde{\ell}$. $\square$

2.5 Differentiability

In this section we discuss the differentiability properties of the solution $\xi(p) = (x(p), \lambda(p), \mu(p), \eta(p))$ of the optimality system (2.1.3) with respect to p. Throughout it is assumed that $p \in N$ so that (2.1.3) has the unique solutions $\xi(p)$ in U, where N and U are specified in Theorem 2.16. We assume that $C = X$.

Definition 2.19. *A function ϕ from P to a normed linear space $\tilde{X}$ is said to have a directional derivative at $p_0 \in P$ if*

$$\lim_{t \to 0^+} \frac{\phi(p_0 + tq) - \phi(p_0)}{t}$$

exists for all $q \in P$.

Consider the linear generalized equation

$$0 \in \begin{cases} \mathcal{L}'(x_0, p, \lambda_0, \mu_0, \eta_0) + A(x - x_0) + E^*(\lambda - \lambda_0) + G^*(\mu - \mu_0) + L^*(\eta - \eta_0), \\ -e(x_0, p) - E(x - x_0), \\ -g(x_0, p) - G(x - x_0) + \partial\psi_{R^{m,+}}(\mu), \\ -\ell(x_0, p) - L(x - x_0) + \partial\psi_{K^+}(\eta). \end{cases} \tag{2.5.1}$$

In the proof of Theorem 2.16 strong regularity of (2.1.4) at $\xi(0) = (x_0, \lambda_0, \mu_0, \eta_0)$ was established. Therefore it follows from Theorem 2.5 that there exist neighborhoods of p_0

and $\xi(p_0)$ which, without loss of generality, we again denote by N and U, such that (2.5.1) admits a unique solution $\hat{\xi}(p) = (\hat{x}(p), \hat{\lambda}(p), \hat{\mu}(p), \hat{\eta}(p))$ in U for every $p \in N$. Moreover we have

$$|\xi(p) - \hat{\xi}(p)| \le \alpha(p)\,|p - p_0|, \tag{2.5.2}$$

where $\alpha(p) : N \to R^+$ satisfies $\alpha(p) \to 0$ as $p \to p_0$. Thus if $\hat{\xi}$ has a directional derivative at p_0, then so does $\xi(p)$ and the directional derivatives coincide. We henceforth concentrate on the analysis of directional differentiability of $\hat{\xi}(p)$. From (2.5.2) and Lipschitz continuity if $p \to \xi(p)$ at p_0, it follows that for every $q \in P$

$$\limsup_{t>0} \left| \frac{\hat{\xi}(p_0 + t\,q) - \hat{\xi}(p_0)}{t} \right| < \infty.$$

Hence there exist weak cluster points of $\frac{\hat{\xi}(p_0+t\,q)-\hat{\xi}(p_0)}{t}$ as $t \to 0^+$. Note that these weak cluster points coincide with those of $\frac{\xi(p_0+t\,q)-\xi(p_0)}{t}$ as $t \to 0^+$. We will show that under appropriate conditions, these weak cluster points are strong limit points and we derive equations that they satisfy. The following definition and additional hypotheses will be used.

Definition 2.20. *A closed convex set C of a Hilbert space H is called polyhedric at $z \in H$ if*

$$\overline{\cup_{\lambda>0}\lambda(C - Pz) \cap [z - Pz]^\perp} = \overline{\cup_{\lambda>0}\lambda(C - Pz)} \cap [z - Pz]^\perp,$$

where P denotes the metric projection onto C and $[z - Pz]^\perp$ stands for the orthogonal complement of the subspace spanned by $x - Pz \in H$. Moreover C is called polyhedric if C is polyhedric at every $z \in H$.

(H5) *$e(x_0, \cdot)$, $\ell(x_0, \cdot)$, $f'(x_0, \cdot)$, $e'(x_0, \cdot)$, $g'(x_0, \cdot)$, and $\ell'(x_0, \cdot)$ are directionally differentiable at p_0.*

(H6) *K is polyhedric at $\ell(x_0, p_0) + \eta_0$.*

(H7) *There exists $\nu > 0$ such that $\langle Ax, x\rangle \ge \nu\,|x|^2$ for all $x \in \ker\, E$.*

(H8) *$\binom{E}{L} : X \to W \times Z$ is surjective.*

Since every element $z \in Z$ can be decomposed uniquely as $z = z_1 + z_2$ with $z_1 = P_K z$ and $z_2 = P_{K^+} z$ and $\langle z_1, z_2\rangle = 0$ (see [Zar]*),* (H.6) *is equivalent to*

$$\overline{\cup_{\lambda>0}\lambda(K - \ell(x_0, p_0)) \cap [\eta_0]^\perp} = \overline{\cup_{\lambda>0}\lambda(K - \ell(x_0, p_0))} \cap [\eta_0]^\perp.$$

Recall the decomposition of the inequality constraint with finite rank and the notation that was introduced in (2.1.6). Due to the complementarity condition and continuous dependency of $\xi(p)$ on p one can assume that

$$g^+(x(p), p) = 0, \quad g^-(x(p), p) < 0, \quad \mu^+(p) > 0, \quad \mu^-(p) = 0 \tag{2.5.3}$$

for $p \in N$. This also holds with $(x(p), \mu(p))$ replaced by $(\hat{x}(p), \hat{\mu}(p))$.

Theorem 2.21. *Let* (H1)–(H5) *hold and let* $(\dot{x}, \dot{\lambda}, \dot{\mu}, \dot{\eta})$ *denote a weak cluster point of* $\frac{\xi(p_0+tq)-\xi(p_0)}{t}$ *as* $t \to 0^+$. *Then* $(\dot{x}, \dot{\lambda}, \dot{\mu}, \dot{\eta})$ *satisfies*

$$0 \in \begin{cases} \mathcal{L}'_p(x_0, p_o, \lambda_0, \mu_0, \eta_0)q + A\dot{x} + E^*\dot{\lambda} + G^*\dot{\mu} + L^*\dot{\eta}, \\ -e^+_p(x_0, p_0)q - E_+\dot{x}, \\ -g^0_p(x_0, p_0) - G_0\dot{x} + \partial\psi_{R^{m_2,+}}(\dot{\mu}^0), \\ \dot{\mu}^-, \\ \langle \dot{\eta}, \ell(x_0, p_0)\rangle + \langle \eta_0, L\dot{x} + \ell_p(x_0, p_0)q\rangle. \end{cases} \tag{2.5.4}$$

Proof. As described above, we can restrict ourselves to a weak cluster point of $\frac{\hat{\xi}(p_0+tq))-\hat{\xi}(p_0)}{t}$. We put $p_n = p_0 + t_n q$. By (2.5.1)

$$0 = \mathcal{L}'(x_0, p_n, \lambda_0, \mu_0, \eta_0) - \mathcal{L}'(x_0, p_0, \lambda_0, \mu_0, \eta_0)$$
$$+A(\hat{x}(p_n) - x_0) + E^*(\hat{\lambda}(p_n) - \lambda_0) + G^*(\hat{\mu}(p_n) - \mu_0) + L^*(\hat{\eta}(p_n) - \eta_0) = 0,$$
$$0 = -e(x_0, p_n) + e(x_0, p_0) - E(\hat{x}(p_n) - x_0) = 0.$$

Dividing these two equations by $t_n > 0$ and letting $t_n \to 0^+$ we obtain the first two equations in (2.5.4). For the third equation note that $\dot{\mu}^0 \geq 0$ since $\hat{\mu}^0(p_n) \geq 0$ and $\mu^0(p_0) = 0$. Since $g^0(x_0, p_0) = 0$ we have from (2.5.1)

$$\langle g^0(x_0, p_n) - g^0(x_0, p_0) + G^0(\hat{x}(p_n) - x_0), z - (\hat{\mu}^0(p_n) - \mu^0(p_0))\rangle \geq 0$$

for all $z \in R^{m_2,+}$. Dividing this inequality by $t_n > 0$ and letting $t_n \to 0^+$ we obtain the third inclusion. The fourth equation is obvious. For the last equation we recall that

$$\langle \hat{\eta}(p_n), \ell(x_0, p_n) + L(\hat{x}(p_n) - x_0)\rangle = \langle \eta_0, \ell(x_0, p_0)\rangle = 0.$$

Thus,

$$\langle \hat{\eta}(p_n) - \eta_0, \ell(x_0, p_n)\rangle + \langle \hat{\eta}(p_n), L(\hat{x}(p_n) - x_0)\rangle + \langle \eta_0, \ell(x_0, p_n) - \ell(x_0, p_0)\rangle = 0,$$

which implies the last equality. □

Define the value function $V(p)$

$$V(p) = \inf_{x \in c} \{f(x, p) : e(x, p) = 0,\ g(x, p) \leq 0,\ \ell(x, p) \in K\}.$$

Then we have the following corollary.

Corollary 2.22 (Sensitivity of Cost). *Let* (H.1)–(H.4) *hold and assume that* e, g, ℓ *are continuously differentiable in the sense of Fréchet at* (x_0, p_0). *Then the Gâteaux derivative* *of the value function μ at p_0 exists and is given by*

$$V'(p_0) = \mathcal{L}_p(x_0, p_0, \lambda_0, \mu_0, \eta_0)$$
$$= f_p(x_0, p_0) + \langle \lambda_0, e_p(x_0, p_0)\rangle + \langle \mu_0, g_p(x_0, p_0)\rangle + \langle \eta_0, \ell_p(x_0, p_0)\rangle.$$

Proof. Note that

$$V(p) = f(x(p), p) = \mathcal{L}(x(p), p, \lambda(p), \mu(p), \eta(p)).$$

For $q \in P$ let $\xi(t) = t^{-1}(x(p_0 + t\,q) - x(p_0))$, $t > 0$. Then there exists a subsequence t_n such that $\lim_{n\to\infty} t_n = 0$ and $\xi(t_n) \to (\dot{x}, \dot{\lambda}, \dot{\mu}, \dot{\eta})$. Now, let $p_n = p_0 + t_n\,q$ and

$$V(p_n) - V(p_0) = \mathcal{L}(x(p_n), p_n, \lambda(p_n), \mu(p_n), \eta(p_n))$$
$$-\mathcal{L}(x(p_n), p_0, \lambda(p_n), \mu(p_n), \eta(p_n)) + \mathcal{L}(x(p_n), p_0, \lambda(p_n), \mu(p_n), \eta(p_n))$$
$$-\mathcal{L}(x(p_n), p_0, \lambda(p_0), \mu(p_0), \eta(p_0)) + \mathcal{L}(x(p_n), p_0, \lambda(p_0), \mu(p_0), \eta(p_0))$$
$$-\mathcal{L}(x(p_0), p_0, \lambda(p_0), \mu(p_0), \eta(p_0)).$$

Thus,

$$\tfrac{V(p_n)-V(p_0)}{t_n} = \langle \mathcal{L}'(x_0, p_0, \lambda_0, \mu_0, \eta_0), \dot{x}\rangle + \langle \mathcal{L}(x_0, p_0, \lambda_0, \mu_0, \eta_0)_p, q\rangle$$
$$+\langle \dot{\lambda}, e(x_0, p_0)\rangle + \langle \dot{\mu}, g(x_0, p_0)\rangle + \langle \dot{\eta}, \ell(x_0, p_0)\rangle + O(t_n).$$

Clearly $\langle \dot{\lambda}, e(x_0, p_0)\rangle = 0$ and considering separately the cases according to (2.5.3) we find $\langle \dot{\mu}, g(x_0, p_0)\rangle = 0$ as well. Below we verify that

$$\langle \eta_0, L\dot{x} + \ell_p(x_0, p_0)q\rangle = 0. \tag{2.5.5}$$

From Theorem 2.21 this implies that $\langle \dot{\eta}, \ell(x_0, p_0)\rangle = 0$. To verify (2.5.5) note that from (2.5.1) we have

$$\langle \hat{\eta}(t_n), \ell(x_0, p_0) - \ell(x_0, p_n) - L(\hat{x}(t_n) - x_0)\rangle \le 0$$

and consequently $\langle \eta_0, \ell_p(x_0, p_0)q + L\,\dot{x}\rangle \ge 0$. Moreover

$$\langle \eta_0, \ell(x_0, p_n) - \ell(x_0, p_0) + L(\hat{x}(t_n) - x_0)\rangle \le 0$$

and therefore $\langle \eta_0, \ell_p(x_0, p_0)q + L\,\dot{x}\rangle \le 0$ and (2.5.5) holds. It follows that

$$\lim_{n\to\infty} \frac{V(p_n) - V(p_0)}{t_n} = \langle \mathcal{L}_p(x_0, p_0, \lambda_0, \mu_0, \eta_0), q\rangle.$$

Since the limit does not depend on the sequence $\{t_n\}$ the desired result follows. $\square$

In order to prove strong differentiability of $x(p)$ we use a result due to Haraux [Har] on directional differentiability of the metric projection onto a closed convex set.

Theorem 2.23. *Let C be a closed convex set in a Hilbert space H with metric projection P from H onto C. Let ψ be an H-valued function and assume that C is polyhedric at $\psi(0)$. Set*

$$\hat{K}(\psi(0)) = \overline{\cup_{\lambda>0}\lambda(C - P\psi(0))} \cap [\psi(0) - P\psi(0)]^{\perp},$$

and denote by $P_{\hat{K}(\psi(0))}$ the projection onto $\hat{K}(\psi(0))$. If there exists a sequence t_n such that $\lim_{n\to\infty} t_n \to 0^+$ and $\lim_{n\to\infty} \frac{\psi(t_n)-\psi(0)}{t_n} =: \dot{\psi}$ exists in H, then

$$\lim_{t_n\to 0^+} \frac{P\psi(t_n) - P\psi(0)}{t_n} = P_{\hat{K}(\psi(0))}\dot{\psi}.$$

Proof. Let $\gamma(t) = \frac{P\psi(t)-P\psi(0)}{t}$. Since the metric projection onto a closed convex set in a Hilbert space is a contraction, $\gamma(t_n)$ has a weak limit which is denoted by γ. Since

$$\langle \psi(t_n) - P\psi(t_n), P\psi(0) - P\psi(t_n)\rangle \le 0$$

and $P\psi(t_n) = t_n\,\gamma(t_n) + P\psi(0)$, we obtain

$$\langle \psi(t_n) - t_n\,\gamma(t_n) - P\psi(0), -t_n\gamma(t_n)\rangle \le 0.$$

This implies

$$\left\langle t_n\,\frac{\psi(t_n) - \psi(0)}{t_n} + \psi(0) - t_n\,\gamma(t_n) - P\psi(0), -t_n\gamma(t_n)\right\rangle \le 0$$

and hence

$$t_n^2\left\langle \gamma(t_n), \gamma(t_n) - \frac{\psi(t_n) - \psi(0)}{t_n}\right\rangle \le t_n\langle \gamma(t_n), \psi(0) - P\psi(0)\rangle$$
$$= \langle P\psi(t_n) - P\psi(0), \psi(0) - P\psi(0)\rangle \le 0 \tag{2.5.6}$$

for all n. Since the norm is weakly lower semicontinuous, (2.5.6) implies that

$$\langle \gamma, \gamma - \dot{\psi}\rangle \le 0. \tag{2.5.7}$$

Employing (2.5.6) again we have

$$0 \ge \langle \gamma(t_n), \psi(0) - P\psi(0)\rangle \ge t_n\left\langle \gamma(t_n), \gamma(t_n) - \frac{\psi(t_n) - \psi(0)}{t_n}\right\rangle$$

for $t_n > 0$. Since $\{\gamma(t_n)\}$ is bounded, this implies

$$\langle \gamma, \psi(0) - P\psi(0)\rangle = 0,$$

and thus $\gamma \in \hat{K}(\psi(0))$.

Let $w \in \cup_{\lambda>0} \lambda(C - P\psi(0)) \cap [\psi(0) - P\psi(0)]^{\perp}$ be arbitrary. Then there exist $\lambda > 0$ and $u \in C$ such that $w = \lambda(u - P\psi(0))$ and $\langle \psi(0) - P\psi(0), u - P\psi(0) \rangle = 0$. Since

$$\langle \psi(t_n) - t_n \gamma(t_n) - P\psi(0), u - P\psi(0) - t_n\gamma(t_n) \rangle \le 0,$$

we find for $\delta_n = \gamma(t_n) - \gamma$ that

$$\left\langle t_n \frac{\psi(t_n) - \psi(0)}{t_n} + \psi(0) - P\psi(0) - t_n\gamma - t_n\delta_n, u - P\psi(0) - t_n\gamma - t_n\delta_n \right\rangle \le 0.$$

Thus,

$$\left\langle t_n \frac{\psi(t_n) - \psi(0)}{t_n} - t_n\gamma - t_n\delta_n, u - P\psi(0) - t_n\gamma - t_n\delta_n \right\rangle \le \langle \psi(0) - P\psi(0), t_n\delta_n \rangle$$

and therefore

$$\left\langle \frac{\psi(t_n) - \psi(0)}{t_n} - \gamma, u - P\psi(0) \right\rangle \le \langle \delta_n, u - P\psi(0) \rangle + \langle \psi(0) - P\psi(0), \delta_n \rangle + t_n M$$

for some constant M. Letting $t_n \to 0^+$ we have

$$\langle \dot{\psi} - \gamma, u - P\psi(0) \rangle \le 0.$$

Since C is polyhedric at $\psi(0)$ it follows that

$$\langle \dot{\psi} - \gamma, w \rangle \le 0 \quad \text{for all } w \in \hat{K}(\psi(0)). \tag{2.5.8}$$

Combining (2.5.7) and (2.5.8) one obtains

$$\langle \gamma - \dot{\psi}, \gamma - w \rangle \le 0 \quad \text{for all } w \in \hat{K}(\psi(0)),$$

and thus $\gamma = P_{\hat{K}(\psi(0))}\dot{\psi}$

To show strong convergence of $\gamma(t_n)$ observe that from (2.5.7) and (2.5.8)

$$\langle \dot{\psi} - \gamma, \gamma \rangle = 0.$$

Hence

$$\langle \dot{\psi}, \gamma \rangle = |\gamma|^2 \le \liminf |\gamma(t_n)|^2 \le \limsup \left\langle \frac{\psi(t_n) - \psi(0)}{t_n}, \gamma(t_n) \right\rangle = \langle \dot{\psi}, \gamma \rangle,$$

which implies that $\lim |\gamma(t_n)|^2 = |\gamma|^2$ and completes the proof. □

Theorem 2.24 (Sensitivity Equation). *Let* (H1)–(H8) *hold. Then the solution mapping $p \to \xi(p) = (x(p), \lambda(p), \mu(p), \eta(p))$ is directionally differentiable at p_0, and the directional derivative $(\dot{x}, \dot{\lambda}, \dot{\mu}, \dot{\eta})$ at p_0 in direction $q \in P$ satisfies*

$$0 \in \begin{cases} \mathcal{L}_p(x_0, p_0, \lambda_0, \mu_0, \eta_0)q + A\dot{x} + E^*\dot{\lambda} + G_+^*\dot{\mu}^+ G_0^*\dot{\mu}^0 + L^*\dot{\eta}, \\ -e_p^+(x_0, p_0)q - E_+\dot{x}, \\ -g_p^0(x_0, p_0)q - G_0\dot{x} + \partial\psi_{R^{m,+}}(\dot{\mu}^0), \\ -\ell_p(x_0, p_0)q - L\dot{x} + \partial\psi_{\hat{K}^+}(\dot{\eta}). \end{cases} \tag{2.5.9}$$

Proof. Let $\{t_n\}$ be a sequence of real numbers with $\lim_{n\to\infty} t_n = 0^+$ and $w-\lim\ t_n^{-1}(\xi(p_0+t_n q) - \xi(p_0)) = (\dot{x}, \dot{\lambda}, \dot{\mu}, \dot{\eta})$. Then $(\dot{x}, \dot{\lambda}, \dot{\mu}, \dot{\eta})$ is also a weak cluster point of $w-\lim\ t_n^{-1}(\hat{\xi}(p_0+t_n q) - \xi(p_0))$. The proof is now given in several steps.

We first show that $\frac{\hat{x}(t_n)-x(0)}{t_n}$ converges strongly in X as $n\to\infty$ and (2.5.9) holds for all weak cluster points. Let $p_n = p_0 + t_n\, q$ and define

$$\phi(t_n) = \mathcal{L}'(x_0, p_n, \lambda_0, \mu_0, \eta_0) + f'(x_0, p_0) - Ax_0.$$

From (2.5.1) we have

$$0 \in \begin{cases} \mathcal{L}'(x_0, p_n, \lambda_0, \mu_0, \eta_0) + f'(x_0, p_0) + A(\hat{x}(t_n) - x_0) + E^*(\hat{\lambda}(t_n)) \\ \qquad +G^*(\hat{\mu}(t_n)) + L^*(\hat{\eta}(t_n)), \\ -e(x_0, p_n) - E(\hat{x}(t_n) - x_0), \\ -g(x_0, p_n) - G(\hat{x}(t_n) - x_0) + \partial\psi_{R^{m,+}}(\hat{\mu}(t_n)), \\ -\ell(x_0, p_n) - L(\hat{x}(t_n) - x_0) + \partial\psi_{K^+}(\hat{\eta}(t_n)). \end{cases} \tag{2.5.10}$$

By (H8) there exists a unique $w(t_n) \in \text{range}\,(E^*, L^*)$ such that

$$\begin{pmatrix} E \\ \\ L \end{pmatrix} w(t_n) = \begin{pmatrix} -e(x_0, p_n) + Ex_0 \\ \\ -\ell(x_0, p_n) + Lx_0 \end{pmatrix}.$$

Due to (H5) and (H8) there exists $\dot{w} \in X$ such that $\lim\ t_n^{-1}(w(t_n) - w(0)) = \dot{w}$. If we define $y(t_n) = \hat{x}(t_n) - w(t_n)$, then using Lemma 2.7 one verifies that $y(t_n) \in \mathcal{C}$, where

$$\mathcal{C} = \{x \in X : Ex = 0 \text{ and } Lx \in K\}.$$

Observe that

$$\langle E^*\hat{\lambda}(t_n) + L^*\hat{\eta}(t_n), c - y_n(t_n)\rangle \le 0 \quad \text{for all } c \in \mathcal{C}$$

and thus

$$\langle Ay_n(t) + \phi(t_n) + Aw(t_n) + G^*\hat{\mu}(t_n), c - y(t_n)\rangle \ge 0 \quad \text{for all } c \in \mathcal{C}. \tag{2.5.11}$$

If we equip $\ker\ E$ with the inner product

$$((x, y)) = \langle Ax, y\rangle \quad \text{for } x, y \in \ker\ E,$$

then $\ker\ E$ is a Hilbert space by (H7). Let $A_P = P_{\ker\ E}\, A\, P_{\ker\ E}$ and define

$$\psi(t_n) = -A_P^{-1} P_{\ker\ E}(\phi(t_n) + Aw(t_n) + G^*\mu(t_n)) \in \ker\ E.$$

Then $\lim_{t_n\to 0^+} \frac{\psi(t_n)-\psi(0)}{t_n} =: \dot{\psi}$ exists, and (2.5.11) is equivalent to

$$((y(t_n) - \psi(t_n), c - y(t_n))) \ge 0 \quad \text{for all } c \in \mathcal{C}.$$

Thus $y(t_n) = P_C\psi(t_n)$, where P_C denotes the metric projection in $\ker E$ onto P_C with respect to the inner product $((\cdot,\cdot))$. For any $h \in \ker E$ we find

$$((h, \psi(0) - P_C\,\psi(0))) = \langle h, -f'(x_0, p_0) + A\,x_o - A\,w(0) - G^*\mu_0 - A\,y(0)\rangle$$

$$= \langle h, -f'(x_0, p_0) - G^*\mu_0 - E^*\lambda_0\rangle = \langle h, L^*\eta_0\rangle,$$

and therefore

$$[\psi(0) - P_C\psi(0)]^\perp = \{h : \langle \eta_0, L\,h\rangle = 0\}. \tag{2.5.12}$$

Here the orthogonal complement is taken with respect to the $((\cdot,\cdot))$—the inner product on $\ker E$. Moreover we have $L(y(0) = \ell(x_0, p_0)$. Hence from Lemma 2.25 below we have $\overline{\cup_{\lambda>0}\lambda(C - P_C\,\psi(0)) \cap [\psi(0) - P_C\,\psi(0)]^\perp} = \overline{\cup_{\lambda>0}\lambda(C - P_C\,\psi(0))} \cap [\psi(0) - P_C\,\psi(0)]^\perp$, i.e., C is polyhedric with respect to $\psi(0)$. Theorem 2.23 therefore implies that $\lim_{t_n\to o^+}(t_n^{-1})(y(t_n) - y(0)) =: \dot y$ exists and satisfies

$$((\dot\psi - \dot y, v - \dot y)) \geq 0 \text{ for all } v \in \hat{C}(\psi(0)). \tag{2.5.13}$$

Moreover $\lim_{t_n\to 0}(t_n^{-1})(x(t_n) - x(0))$ exists and will be denoted by $\dot x$.

To verify that $(\dot x, \dot\lambda, \dot\mu, \dot\eta)$ satisfies (2.5.9) it suffices to prove the last inclusion. From (2.5.13) we have

$$\langle -\mathcal{L}'_p(x_0, p_0, \lambda_0, \mu_0, \eta_0)q - A\,\dot x - G^*\dot\mu, v - \dot y\rangle \leq 0$$

and from the first equation of (2.5.4)

$$\langle L^*\dot\eta, v - \dot y\rangle = \langle \dot\eta, L\,v - \ell_p(x_0, p_0)q\rangle \leq 0 \tag{2.5.14}$$

for all $v \in \hat{C}(\psi(0)) = \overline{\cup_{\lambda>0}\lambda(C - P_C\,\psi(0))} \cap [\psi(0) - P_C\,\psi(0)]^\perp$. Here we used the facts that $\dot\phi = \mathcal{L}'_p(x_0, \lambda_0, \mu_0, \eta_o)q$ and $L\,\dot w = -\ell_p(x_0, p_0)q$. From (2.5.12), (H8), and $L\,y(0) = \ell(x_0, p_0)$ we have

$$L\,\hat{C}(\psi(0)) = \cup_{\lambda>0}\lambda(K - \ell(x_0, p_0) \cap [\eta_0]^\perp =: \hat K$$

and therefore by (2.5.14)

$$\langle \dot\eta, v - L\,\dot x - \ell\,p(x_0, p_0)q\rangle \leq 0 \text{ for all } v \in \hat K. \tag{2.5.15}$$

Recall from (2.5.5) that

$$\langle \eta_0, \ell_p(x_o, p_0)q + L\,\dot x\rangle = 0.$$

This further implies $\ell_p(x_0, p_0)q + L\,\dot x \in \hat K$, and from (2.5.15) we deduce $\dot\eta \in \hat K^+$. Consequently $\dot\eta \in \partial\,\psi_{\hat K}(\ell_p(x_0, p_0)q + L\,\dot x)$, which is equivalent to $\ell\,p(x_o, p_0)q + L\,\dot x \in \partial\,\psi_{\hat K^+}(\dot\eta)$ and implies the last inclusion in (2.5.9).

Next, we show that the weak cluster points of $\lim_{t\to 0^+} \frac{\hat{\xi}(t)-\xi(0)}{t}$ are unique. Let $(\dot{x}_i, \dot{\lambda}_i, \dot{\mu}_i, \dot{\eta}_i)$, $i = 1, 2$, be two weak cluster points. Then by (2.5.9)

$$\begin{aligned} 0 &= \langle A(\dot{x}_1 - \dot{x}_2), \dot{x}_1 - \dot{x}_2\rangle + \langle \dot{\mu}_1 - \dot{\mu}_2, G(\dot{x}_1 - \dot{x}_2)\rangle + \langle \dot{\eta}_1 - \dot{\eta}_2, L(\dot{x}_1 - \dot{x}_2)\rangle \\ &= \langle A(\dot{x}_1 - \dot{x}_2), \dot{x}_1 - \dot{x}_2\rangle + \langle \dot{\mu}_1^0, -g_p^0(x_0, p_0)q - G_0\dot{x}_2\rangle + \langle \dot{\mu}_2^0, -g_p^0(x_0, p_0)q - G_0\dot{x}_1\rangle \\ &\quad + \langle \dot{\eta}_1, -\ell_p(x_0, p_0)q - L\dot{x}_2\rangle + \langle \dot{\eta}_2, -\ell_p(x_0, p_0)q - L\dot{x}_1\rangle \\ &\geq \langle A(\dot{x}_1 - \dot{x}_2), \dot{x}_1 - \dot{x}_2\rangle, \end{aligned}$$

and by (H2) this implies $\dot{x}_1 = \dot{x}_2$. From (2.5.9) we have

$$\mathcal{E}^*(\ell(x_0, p_0)) \begin{pmatrix} \dot{\tilde{\lambda}}_1 - \dot{\tilde{\lambda}}_2 \\ \dot{\mu}_1^0 - \dot{\mu}_2^0 \\ \dot{\eta}_1 - \dot{\eta}_2 \end{pmatrix} = 0$$

and hence (H3) implies $(\dot{\lambda}_1, \dot{\mu}_1, \dot{\eta}_1) = (\dot{\lambda}_2, \dot{\mu}_2, \dot{\eta}_2)$. Consequently $\frac{\hat{\xi}(t)-\xi(0)}{t}$ has a unique weak limit as $t \to 0^+$. Finally we show that the unique weak limit is also strong. For the x-component this was already verified. Note that from (2.5.1)

$$\mathcal{E}^*(\ell(x_0, p_0)) \begin{pmatrix} \tilde{\hat{\lambda}}(t) - \tilde{\lambda}_0 \\ \hat{\mu}^0 - \mu_0^0 \\ \hat{\eta}(t) - \eta_0 \end{pmatrix} = \begin{pmatrix} -\mathcal{L}'(x_0, p_0 + t\,q, \lambda_0, \mu_0, \eta_0) - A(\hat{x}(t) - x_0) \\ \\ \langle \hat{\eta}(t), -\ell(x_0, p + t\,q) + \ell(x_0, p_0) - L(\hat{x(t)} - x_0)\rangle \end{pmatrix}.$$

Dividing this equation by t and noticing that the right-hand side converges strongly as $t \to 0^+$, it follows from (H3) that $\lim_{t\to 0^+} t^{-1}(\tilde{\lambda}(t) - \lambda_0, \tilde{\mu}(t) - \mu_0, \tilde{\eta}(t) - \eta_0)$ converges strongly as well. $\square$

Lemma 2.25. *Assume that* (H6) *and* (H8) *hold and let* $y(0) \in \mathcal{C}$ *be such that* $L\,y(0) = \ell(x_0, p_0)$. *Then we have*

$$\begin{aligned} &\overline{\cup_{\lambda>0}\lambda(\mathcal{C} - y(0)) \cap \{h \in \ker E : \langle \eta_0, L\,h\rangle = 0\}} \\ &= \overline{\cup_{\lambda>0}\lambda(\mathcal{C} - y(0))} \;\cap\; \{h \in \ker E : \langle \eta_0, L\,h\rangle = 0\}. \end{aligned} \tag{2.5.16}$$

Proof. It suffices to verify that the set on the right-hand side of (2.5.16) is contained in the set on the left. For this purpose let $y \in \overline{\cup_{\lambda>0}\lambda(\mathcal{C} - y(0))} \cap \{h \in \ker E : \langle \eta_0, L\,h\rangle = 0\}$ and decompose $y = y_1 + y_2$ with $y_1 \in \ker\binom{E}{L}$ and $y_2 \in (\ker\binom{E}{L})^\perp$. We need only consider y_2. Since $L\,y_2 \in \overline{\cup_{\lambda\to 0}\lambda(K - \ell(x_0, p_0))} \cap [\eta_0]^\perp$ and therefore $L\,y_2 \in \overline{\cup_{\lambda>0}\lambda(K - \ell(x_0, p_0)) \cap [\eta_0]^\perp}$ by (H6), hence there exist sequences $\{\lambda_n\}$ and $\{k_n\}$ with $\lambda_n > 0, k_n \in K, \langle k_n, \eta_0\rangle = 0$, and $\lim \lambda_n(k_n - \ell(x_0, p_0)) = L\,y_2$. Let us define $c_n = \tilde{c}_n + y(0)$, where $\tilde{c}_n$ is the unique element in $\binom{E}{L}^*$ satisfying $\binom{E}{L}\tilde{c}_n = \binom{0}{k_n - L\,y(0)}$. Then we have $E(c_n - y(0)) = 0$, $\langle \eta_0, L\,c_n\rangle = \langle \eta_0, k_n\rangle = 0$ and hence the sequence $\lambda_n(c_n - y(0))$ is contained in the set on the left-hand side of (2.5.16). Moreover $\lim_{n\to\infty}\binom{E}{L}(\lambda_n(c_n - y(0)) = \binom{E}{L}y_2$, and since both $\lambda_n(c_n - y(0))$ and y_2 are contained in the range of $\binom{E}{L}^*$ we find that $\lim_{n\to\infty}\lambda_n(c_n - y(0)) = y_2$. $\square$

2.6 Application to optimal control of an ordinary differential equation

We consider optimal control of the linear ordinary differential equation with a terminal constraint

$$\min \int_0^T \hat{f}(y(t), u(t), \alpha)dt$$

subject to

$$\begin{aligned} &\dot{y}(t) = A(\alpha)y(t) + B(\alpha)u(t) + h(t) \text{ on } (0, T), \\ &y(0) = 0, \\ &|y(T) - y^d| \leq \delta, \\ &u \leq z \quad \text{a.e. on } (0, T), \end{aligned} \tag{2.6.1}$$

where $T > 0$, $y(t) \in \mathbb{R}^n$, $u(t) \in \mathbb{R}^n$, $\alpha \in \mathbb{R}^r$, $y^d \in \mathbb{R}^n$, $\delta > 0$, $z \in L^2$, $A(\alpha) \in \mathbb{R}^{n\times n}$, $B(\alpha) \in \mathbb{R}^{n\times m}$, $\hat{f}: \mathbb{R}^n \times \mathbb{R}^m \times \mathbb{R}^r \to \mathbb{R}$, and $h \in L^2(0, T)$. All function spaces of this section are considered on the interval $(0, T)$ with Euclidean image space of appropriate dimension. The perturbation parameter is given by $p = (\alpha, y^d, h, z) \in P = \mathbb{R}^r \times \mathbb{R}^n \times L^2 \times L^2$ with reference parameter $p_0 = (\alpha_0, y_0^d, h_0, z_0)$. To identify (P_p) as a special case of (2.1.1) we set

$$X = H_L^1 \times L^2,\ W = L^2,\ Z = L^2,\ K = \{\phi \in L^2 : \phi \leq 0 \text{ a.e.}\},$$

where $H_L^1 = \{\phi \in H^1 : \phi(0) = 0\}$, $x = (y, u)$, and

$$\begin{aligned} f(y, u, p) &= \int_0^T \hat{f}(y, u, \alpha)dt, \\ e(y, u, p) &= \dot{y} - A(\alpha)y - B(\alpha)u - h, \\ g(y, u, p) &= \frac{1}{2}\left(|y(T) - y^d|^2 - \delta^2\right), \\ \ell(y, u, p) &= u - z. \end{aligned}$$

The linearizations E, G, and L are

$$\begin{aligned} E(y_0, u_0, p_0)(h, v) &= \dot{h} - A(\alpha_0)h - B(\alpha_0)v, \\ G(y_0, u_0, p_0)(h, v) &= \langle y_0(T) - y_0^d, h(T)\rangle, \\ L(y_0, u_0, p_0)(h, v) &= v. \end{aligned}$$

The following assumptions will imply conditions (H1)–(H8) of this chapter.

(A1) There exists a solution (y_0, u_0) of (2.6.1).

(A2) $A(\cdot)$ and $B(\cdot)$ are Lipschitz continuous and directionally differentiable at α_0.

(A3) $(y, u) \to f(y, u, \alpha)$ is twice continuously differentiable in a neighborhood of $(y_0, u_0, \alpha_0) \in H^1_L \times L^2 \times \mathbb{R}^r$.

(A4) There exists $\epsilon > 0$ such that for the Hessian with respect to (y, u)

$$(h, v)\, \hat{f}''(y_0, u_0, \alpha_0)\, (h, v)^T \geq \epsilon |v|^2$$

for all $(h, v) \in \mathbb{R}^n \times \mathbb{R}^m$.

(A5) There exist a neighborhood $\hat{V}$ of $(y_0, u_0, \alpha_0) \in H^1_L \times L^2 \times \mathbb{R}^r$ and a constant κ such that $|f(y, u, \alpha_1) - f(y, u, \alpha_2)| \leq \kappa |\alpha_1 - \alpha_2|$ for all $(y, u, \alpha_i) \in \hat{V}$, $i = 1, 2$, and $f'(y_0, u_0, \cdot)$ is directionally differentiable at α_0, where the prime denotes the derivative with respect to (y, u).

(A6) $\langle y_0(T) - y^d_0, \int_0^T e^{A(\alpha_0)(T-s)}\, B(\alpha_0)\, (z_0 - u_0) ds \rangle \neq 0$.

With (A3) holding, the regularity requirements made at the beginning of the chapter are satisfied. The regular point condition (H1) will be satisfied if for every $(\phi, \rho, \psi) \in L^2 \times \mathbb{R} \times L^2$ there exist $(h, v) \in H^1_L \times L^2$ and $(r^+, r, k) \in \mathbb{R}^+ \times \mathbb{R} \times K$ such that

$$\begin{cases} \dot{h} - A(\alpha_0)h - B(\alpha_0)v = \phi, \\ \langle y_0(T) - y^d_0, h(T)\rangle + r^+ + r\, g(y_0, u_0, p_0) = \rho, \\ v - k + r(u_0 - z_0) = \psi. \end{cases} \tag{2.6.2}$$

From the first and third equations we have

$$v = \psi + k + r(z_0 - u_0),$$

$$h(t) = \rho_1(t) + \int_0^t e^{A(\alpha_0)(t-s)} B(\alpha_0)(k + r(z_0 - u_0))dt,$$

where $\rho_1(t) = \int_0^t e^{A(\alpha_0)(t-s)}(B(\alpha_0)\psi + \phi)ds$. The second equation in (2.6.2) is equivalent to

$$\left\langle y_0(T) - y^\alpha_0, \int_0^T e^{A(\alpha_0)(T-s)} B(\alpha_0)(k + r(z_0 - u_0))ds \right\rangle + r^+ + r\, g(y_0, u_0, p_0)$$
$$= \rho - \rho_1(T).$$

If $g(y_0, u_0, p_0) = 0$, then, using (A6), the desired solution for (2.6.2) is obtained by setting $r^+ = k = 0$ and

$$r = \left\langle y_0(t) - y^d_o, \int_0^T e^{A(\alpha_0)(T-s)} B(\alpha_0)(z_0 - u_0)ds \right\rangle^{-1} (\rho - \rho_1(T)).$$

If $g(y_0, u_0, p_0) < 0$ and $\rho - \rho_1 \geq 0$, then the choice $r^+ = \rho - \rho_1, r = k = 0$ gives the desired solution. Finally if $g(y_0, u_0, p_0) < 0$ and $\rho - \rho_1 < 0$, then $r = (\rho - \rho_1)g(y_0, u_0, p_0) > 0$,

$r^+ = 0$, and $k = r(u_0 - u) \in K$ give a solution for (2.6.2). Thus the regular point condition holds and implies the existence of a Lagrange multiplier $(\lambda_0, \mu_0, \eta_0) \in L^2 \times \mathbb{R} \times L^2$. For the Lagrangian functional

$$\mathcal{L}(y, u, p_0, \lambda_0, \mu_0, \eta_0) = \int_0^T \hat{f}(y, u, \alpha_0)dt + \langle \lambda_0, \dot{y} - A(\alpha_0)y - B(\alpha_0)u - h_0 \rangle$$
$$+ \frac{\mu_0}{2}(|y(T) - y_0^d| - \delta^2|) - \langle \eta_0, u - z_0 \rangle,$$

the Hessian with respect to (y, u) at (y_0, u_0) in directions $(h, v) \in H_L^1 \times L^2$ satisfies

$$\mathcal{L}''(y_0, u_0, p_0, \lambda_0, \mu_0, \eta_0)((h, v), (h, v))$$
$$= \int_0^T (h(t), v(t))\hat{f}''(y_0, u_0, \alpha_0)(h(t), v(t))^T dt + \mu_0 |h(T)|^2 \geq \epsilon |v|_{L^2}^2$$

by (A4). Let $k_1 > 0$ be chosen such that $|h|_{H_L^1} \leq k_1 |v|_{L^2}$ for all $(h, v) \in H_L^1 \times L^2$ in ker E. Then there exists a constant $k_2 > 0$ such that

$$\mathcal{L}''(y_0, u_0, p_0, \lambda_0, \mu_0, \eta_0)((h, v), (h, v)) \geq k_2(|h|_{H_L^1}^2 + |v|_{L^2}^1)$$

for all $(h, v) \in \ker E$ and (H2), (H7) hold.

We turn to the verification of (H3). The case $g(y_0, u_0, \alpha_0) < 0$ is simple and we therefore consider the case $g(y_0, u_0, \alpha_0) = 0$. For arbitrary $(\phi, \rho, \psi) \in L^2 \times \mathbb{R} \times L^2$ existence of a solution $(h, v, r) \in H_1^1 \times L^2 \times \mathbb{R}$ to

$$\begin{cases} \dot{h} - A(\alpha_0)h - B(\alpha_0)v = \phi, \\ \langle y_0(T) - y_0^d, h(T) \rangle = \rho, \\ v + r(z_0 - u_0) = \psi \end{cases} \tag{2.6.3}$$

must be shown. From the first and third equations

$$v = \psi - r(z_0 - u_0)$$

and

$$h(t) = \rho_1(t) - r \int_0^t e^{A(\alpha_0)(t-s)} B(\alpha_0)(z_0 - u_0)ds.$$

From the second equation of (2.6.3) and (H7) we find

$$r = \left\langle y_0(T) - y_0^d, \int_0^T e^{A(\alpha_0)(T-s)} B(\alpha_0)(z_0 - u_0)ds \right\rangle^{-1}$$
$$(\langle y_0(T) - y_0^d, \rho_1(T) \rangle - \rho),$$

and (H3) holds. Conditions (H4) and (H5) follow from (A2) and (A5). The cone of a.e. nonpositive functions in L^2 is polyhedric [Har] and hence (H6) holds. Finally (H8) is simple to check. We note that (A5) and (A6) are not needed if (2.6.1) is considered without terminal constraint.

Chapter 3

First Order Augmented Lagrangians for Equality and Finite Rank Inequality Constraints

3.1 Generalities

This chapter is devoted to first order augmented Lagrangian methods for constrained problems of the type

$$\begin{cases} \min f(x) \text{ over } x \in X \\ \text{subject to } e(x) = 0, \;\; g(x) \leq 0, \;\; \ell(x) \in K, \end{cases} \tag{3.1.1}$$

where $f : X \to \mathbb{R}$, $e : X \to W$, $g : X \to \mathbb{R}^m$, $\ell : X \to Z$ with X, W, and Z real Hilbert spaces, K a closed convex cone with vertex at 0 in Z, and $\mathbb{R}^m$ endowed with the natural ordering $x \leq 0$ if $x_i \leq 0$ for $i = 1, \ldots, m$. A sequence of unconstrained problems with the property that their solutions converge to the solution of (3.1.1) will be formulated. As we shall see, first order augmented Lagrangian techniques are related to Lagrangian or duality methods, as well as to the penalty technique. They arise from considering the Lagrangian functional for (3.1.1) and adding a proper penalty term. Differently from pure penalty methods, however, the penalty parameter is not taken to infinity. Rather the penalty term with fixed weight is used to enhance convexity of the Lagrangian functional in the neighborhood of a local minimum and to speed up convergence of pure Lagrangian methods. Affine inequality constraints with infinite-dimensional image space can be added to the problem formulation (3.1.1). Such constraints, however, are not augmented in this chapter. Central to the analysis is the augmentability property, which is also of interest in its own right. It has received a considerable amount of attention for finite-dimensional optimization problems; see, e.g., [Hes1]. Augmentability of (3.1.1) means that a properly defined penalty term involving the constraints in (3.1.1) can be added to the Lagrangian functional for (3.1.1) such that the resulting augmented Lagrangian functional has the property that its Hessian is positive definite on all of X under minimal assumptions on (3.1.1). We shall present a convergence theory for the first order augmented Lagrangian method applied to (3.1.1) without strict complementarity assumption. This means that we do not assume that $\mu_i^* = 0$ implies $g_i(x^*) < 0$, where x^* denotes a local solution of (3.1.1) and μ_i^* is the ith coordinate of the Lagrange multiplier associated to the inequality constraint $g(x) \leq 0$. As an application we shall discuss the fact that in the context of parameter estimation problems the first order augmented Lagrangian method can be considered as a hybrid method combining the output-matching and equation error methods. In the context of parameter estimation or, equally well, of optimal control problems, the first order augmented Lagrangian method lends itself to vectorization and parallelization in a natural way; see [KuTa].

It will be convenient throughout this section to *identify* the Hilbert spaces X, W, and Z with their duals. As a consequence the Lagrange multiplier associated to the equality constraint $e(x) = 0$ is sought in W and that for $\ell(x) \in K$ in Z.

Let us now fix those assumptions which are assumed throughout this chapter: There exists $x^* \in X$ and $(\lambda^*, \mu^*, \eta^*) \in W \times \mathbb{R}^p \times Z$ such that

$$\begin{cases} f, e, \text{ and } g \text{ are twice continuously Fréchet differentiable} \\ \text{in a neighborhood of } x^*, \end{cases} \tag{3.1.2}$$

x^* is a stationary point of (3.1.1) with Lagrange multiplier (λ^*, μ^*), i.e.,

$$\begin{cases} (f'(x^*), h)_X + \left(\lambda^*, e'(x^*)h\right)_W + \left(\mu^*, g'(x^*)h\right)_{\mathbb{R}^m} + \left(\eta^*, \ell'(x^*)h\right)_Z = 0 \\ \text{for all } \ h \in X, \end{cases} \tag{3.1.3}$$

and

$$\begin{cases} e(x^*) = 0, \ \left(\mu^*, g(x^*)\right)_{\mathbb{R}^m} = 0, \ \mu^* \geq 0, \ g(x^*) \leq 0, \\ \left(\eta^*, \ell(x^*)\right)_Z = 0, \ \eta^* \in K^+, \ \ell(x^*) \in K. \end{cases} \tag{3.1.4}$$

Moreover we assume that

$$e'(x^*) : X \to W \text{ is surjective.} \tag{3.1.5}$$

Above $(\cdot, \cdot)_W$ and $(\cdot, \cdot)_Z$ denote the inner products in W and Z, respectively, and $(\cdot, \cdot)_{\mathbb{R}^m}$ is the usual inner product in $\mathbb{R}^m$. If x^* is a local solution of (3.1.1), if further (3.1.2) holds and x^* is a regular point in the sense of Definition 1.5, then there exists a Lagrange multiplier $(\lambda^*, \mu^*, \eta^*)$ such that (3.1.3), (3.1.4) hold.

Introducing the Lagrange functional $\mathcal{L} : X \times W \times \mathbb{R}^m \times Z \to \mathbb{R}$ by

$$\mathcal{L}(x, \lambda, \mu, \eta) = f(x) + \left(\lambda, e(x)\right)_W + \left(\mu, g(x)\right)_{\mathbb{R}^m} + \left(\eta, \ell(x)\right)_Z,$$

(3.1.3) can be expressed as

$$\mathcal{L}'(x^*, \lambda^*, \mu^*, \eta^*)h = 0 \text{ for all } h \in X.$$

Here and below the prime denotes differentiation with respect to x. With (3.1.2) holding, $\mathcal{L}''(x^*, \lambda^*, \mu^*, \eta^*) : X \times X \to \mathbb{R}$ exists as symmetric bilinear form given by

$$\mathcal{L}''(x^*, \lambda^*, \mu^*, \eta^*)(h, k) = f''(x^*)(h, k) + \left(\lambda^*, e''(x^*)(h, k)\right)_W + \left(\mu^*, g''(x^*)(h, k)\right)_{\mathbb{R}^m}$$

for all $(h, k) \in X \times X$. The index set of the coordinates of the finite rank inequality constraints is decomposed as

$$\begin{aligned} I_1 &= \{i : g_i(x^*) = 0, \ \mu_i^* = 0\}, \\ I_2 &= \{i : g_i(x^*) = 0, \ \mu_i^* < 0\}, \\ I_3 &= \{i : g_i(x^*) < 0\}. \end{aligned}$$

Clearly $I_1 \cup I_2 \cup I_3 = \{1, \ldots, m\}$. If I_1 is empty, then we say that strict complementarity holds. Let us denote the cardinality of I_i by m_i. Without loss of generality we assume that $m_i \geq 1$ for $i = 1, 2, 3$ and that $I_1 = 1, \ldots, m_1$, $I_2 = m_1 + 1, \ldots, m_1 + m_2$, and $I_3 = m_1 + m_2 + 1, \ldots, m$.

The following second order sufficient optimality condition will be of central importance:

There exists a constant $\gamma > 0$ such that

$$\begin{cases} \mathcal{L}''(x^*, \lambda^*, \mu^*, \eta^*)(h, h) \geq \gamma |h|_X^2 \text{ for all } h \in \mathcal{C}, \\ \text{where } \mathcal{C} = \{h \in X : e'(x^*)h = 0,\ g_i'(x^*)h \leq 0 \\ \text{for } i \in I_1,\ g_i'(x^*)h = 0 \text{ for } i \in I_2\}. \end{cases} \tag{3.1.6}$$

In case X is finite-dimensional, (3.1.6) is well known to imply that x^* is a strict local solution to (3.1.1); see, e.g., [Be, p. 71]. For the infinite-dimensional case this will be proved in Theorem 3.4 below.

Section 3.2 is devoted to augmentability of problem (3.1.1). This means that a functional is associated to (3.1.1) with the property that if x^* is a local minimizer of (3.1.1), then it is also a minimizer of that functional where the essential constraints are eliminated and at most simple affine constraints remain. Here this is achieved on the basis of Lagrangian functionals and augmentability is obtained without the use of a strict complementarity assumption. Section 3.3 contains the foundations for the first order augmented Lagrangian algorithm based on a duality framework. The convergence analysis for the first order algorithm is given in Section 3.4. Section 3.5 contains an application to parameter estimation problems.

3.2 Augmentability and sufficient optimality

We consider a modification of (3.1.1) where the equality and finite rank inequality constraints of (3.1.1) are augmented:

$$\begin{aligned} &\min f_c(x, u) = f(x) + \frac{c}{2}|e(x)|_W^2 + \frac{c}{2}|g(x) + u|_{\mathbb{R}^m}^2 \text{ over } (x, u) \in X \times \mathbb{R}^m \\ &\text{subject to } e(x) = 0,\ g(x) + u = 0,\ u \geq 0,\ \ell(x) \in K, \end{aligned} \tag{3.2.1}$$

where $c > 0$ will be appropriately chosen. Observe that x^* is a local solution to (3.1.1) if and only if $(x^*, u^*) = (x^*, -g(x^*))$ is a (local) solution to (3.2.1). Moreover, x^* is stationary for (3.1.1) with Lagrange multiplier $(\lambda^*, \mu^*, \eta^*)$ if and only if $(x^*, -g(x^*))$ is stationary for (3.2.1) with Lagrange multiplier $(\lambda^*, \mu^*, \mu^*, \eta^*)$, i.e., (3.1.3)–(3.1.4) hold with $g(x^*) = -u^*$. Associated to (3.2.1) we introduce the functional

$$\mathcal{L}_c(x, u, \lambda, \mu, \eta) = \mathcal{L}(x, \lambda, \mu, \eta) + \frac{c}{2}|e(x)|_W^2 + \frac{c}{2}|g(x) + u|_{\mathbb{R}^m}^2. \tag{3.2.2}$$

Its Hessian $\mathcal{L}_c''$ at (x^*, u^*) in direction $((h,k),(h,k)) \in (X \times \mathbb{R}^m)$ is given by

$$\begin{aligned}
&\mathcal{L}_c''(x^*, u^*, \lambda^*, \mu^*, \eta^*)((h,k),(h,k)) \\
&\quad = \mathcal{L}''(x^*, \lambda^*, \mu^*, \eta^*)(h,h) + c|e'(x^*)h|^2 + c|g'(x^*)h + k|^2 \\
&\quad = \mathcal{L}''(x^*, \lambda^*, \mu^*, \eta^*)(h,h) + c|e'(x^*)h|^2 + c\sum_{i\in I_2} |g'(x^*)h|^2 \\
&\qquad + c\sum_{i\in I_1\cup I_3} |g_i(x^*)h + k_i|^2 + c\sum_{i\in I_2}(|k_i|^2 + 2k_i g_i'(x^*)h).
\end{aligned} \tag{3.2.3}$$

By the Riesz representation theorem there exists for every $i \in \{1,\dots,m\}$ a unique $l_i \in X$ such that $g_i'(x^*)h = (l_i, h)_X$ for every $h \in X$, where $(\cdot,\cdot)_X$ denotes the inner product in X. The operator $E : X \to W \times \mathbb{R}^{m_2}$, defined by

$$Eh = \big(e'(x^*)h, (l_{m_1+1}, h)_X, \dots, (l_{m_1+m_2}, h)_X\big),$$

combines the linearized equality constraint and the active inequality constraints, which act like equalities in the sense that the associated Lagrange multipliers are positive. The essential technical result which will imply augmentability of the constraints in (3.1.1) is given next.

Proposition 3.1. *If* (3.1.2)–(3.1.6) *hold, then there exist constants* $\bar{c} > 0$ *and* $\tau \in (0, \gamma]$ *such that*

$$\begin{aligned}
H((h,\hat{k}),(h,\hat{k})) &:= \mathcal{L}''(x^*, \lambda^*, \mu^*, \eta^*)(h,h) + c|Eh|^2 + c\sum_{i=1}^{m_1} |(l_i, h)_X + \hat{k}_i|^2 \\
&\geq \tau(|h|^2 + |\hat{k}^2_{\mathbb{R}^{m_1}}|)
\end{aligned}$$

for all $c \geq \bar{c}$, $h \in X$, *and* $\hat{k} \in \mathbb{R}^{m_1}_+$.

Proof. *Step* 1. An equivalent characterization of $\mathcal{C}$ is derived by eliminating linear dependencies in the definition of $\mathcal{C}$. The cone in (3.1.6) can be equivalently expressed as

$$\mathcal{C} = \{h \in X : Eh = 0 \text{ and } (l_i, h)_X \leq 0 \text{ for all } i \in I_1\}. \tag{3.2.4}$$

We next show that

$$\begin{cases}
\text{there exist } \tilde{m}_2 \leq m_2 \text{ and vectors } n_i,\, i = m_1+1, \dots, m_1+\tilde{m}_2, \\
\text{such that } \tilde{E} : X \to Y \times \mathbb{R}^{\tilde{m}_2} \text{ defined by} \\
\tilde{E}h = \big(e'(x^*)h, (n_{m_1+1}, h)_X, \dots, (n_{m_1+\tilde{m}_2}, h)_X\big) \\
\text{is surjective and } \ker E = \ker \tilde{E}.
\end{cases} \tag{3.2.5}$$

To verify (3.2.5) let $n_i = P_{\ker e'} l_i$, where $P_{\ker e'}$ denotes the projection onto $\ker e'(x^*)$. Choose $\tilde{I}_2 \subset I_2$ such that $\{n_i\}_{i\in\tilde{I}_2}$ are linearly independent and $\operatorname{span}\{n_i\}_{i\in\tilde{I}_2} = \operatorname{span}\{n_i\}_{i\in I_2}$. Possibly after reindexing we have $\{n_i\}_{i\in\tilde{I}_2} = \{n_i\}_{m_1+1}^{m_1+\tilde{m}_2}$. We show that $\ker E = \ker \tilde{E}$. Let $h \in \ker E$. Then $e'(x^*)h = 0$. For every $i \in \tilde{I}_2$ let $l_i = n_i + l_i^\perp$, where $l_i^\perp \in (\ker e'(x^*))^\perp$. Then

$(n_i, h)_X = (l_i, h)_X = 0,\ i \in \tilde{I}_2$, and therefore $h \in \ker \tilde{E}$. The converse, $\ker \tilde{E} \subset \ker E$, is proved similarly. To verify surjectivity of $\tilde{E}$, observe that the adjoint $\tilde{E}^* : W \times \mathbb{R}^{\tilde{m}_2} \to X$ of $\tilde{E}$ is given by $\tilde{E}^*(w, r) = e'(x^*)^* w + \sum_{i \in \tilde{I}_2} r_i n_i$. If $\tilde{E}^*(w, r) = 0$, then, using $\sum_{i \in \tilde{I}_2} r_i n_i \in \ker e'(x^*)$ and $e'(x^*)^* w \in (\ker e'(x^*))^\perp$, we have $\sum_{i \in \tilde{I}_2} r_i n_i = 0$ and $e'(x^*)^* w = 0$. Therefore $(w, r) = 0$, $\ker \tilde{E}^* = 0$, and range $\tilde{E} = W \times \mathbb{R}^{m_2}$, by the closed range theorem. Thus (3.2.5) holds.

Finally we consider the set of vectors $\{n_i\}_{i \in I_1}$ with $n_i = P_{\tilde{E}} l_i$. After possible rearrangement of indices there exists a subset $\tilde{I}_1 = \{1, \ldots, \tilde{m}_1\} \subset I_1$ such that $\{n_i\}_{i \in \tilde{I}_1}$ are linearly independent in $\ker E$. The cone C can therefore equivalently be expressed as

$$C = \left\{h \in X : h \in \ker E \text{ and } (n_i, h)_X \le 0 \text{ for all } i \in \tilde{I}_1\right\}.$$

To exclude trivial cases we assume throughout that $\tilde{I}_1$ and $\tilde{I}_2$ are nonempty.

Step 2. Here we characterize the polar cone

$$C_1^* = \left\{h \in \ker E : (h, \tilde{h})_X \le 0 \text{ for all } \tilde{h} \in C_1\right\}$$

of the closed convex cone

$$C_1 = \left\{h \in \ker E : (n_i, h)_X \le 0 \text{ for all } i \in \tilde{I}_1\right\}.$$

Denoting by P and P^* the canonical projections in $\ker E$ onto C_1 and C_1^*, respectively, every element $h \in \ker E$ can be uniquely expressed as $h = h_1 + h_2$ with $(h_1, h_2)_X = 0$ and $h_1 = Ph \in C_1,\ h_2 = P^* h \in C_1^*$ (see [Zar, p. 256]). Moreover

$$C_1 = \bigcap_{i \in \tilde{I}_1} C_i \text{ with } C_i = \{h \in \ker E : (n_i, h)_X \le 0\}$$

and

$$C_1^* = \left(\bigcap C_i\right)^* = \overline{co \bigcup C_i^*} = co \bigcup C_i^*$$

(see [Zar]), with *co* denoting convex closure. It follows that

$$C_1^* = \left\{\sum_{i=1}^{\tilde{m}_1} \alpha_i n_i : (\alpha_1, \ldots, \alpha_{\tilde{m}_1}) \in \mathbb{R}_+^{\tilde{m}_1}\right\}. \tag{3.2.6}$$

Step 3. Let K_1 be chosen such that

$$K_1 \sum_{i=1}^{\tilde{m}_1} |\alpha_i|^2 \le \left|\sum_{i=1}^{\tilde{m}_1} \alpha_i n_i\right|^2$$

for every $(\alpha_1, \ldots, \alpha_{\tilde{m}_1}) \in \mathbb{R}^{\tilde{m}_1}$. For arbitrary $h = \sum_{i=1}^{\tilde{m}_1} \alpha_i n_i \in C_1^*$ we have

$$|h|^2 = \sum_{i=1}^{\tilde{m}_1} \alpha_i (n_i, h)_X \le \sum_{I_1^+} \alpha_i (n_i, h)_X,$$

where $I_1^+ = \{i \in \tilde{I}_1 : (n_i, h)_X > 0 \text{ and } \alpha_i > 0\}$. Consequently

$$|h|^2 \le \left(\sum_{I_1^+} \alpha_i^2\right)\left(\sum_{I_1^+} (n_i, h)_X^2\right)^{1/2} \le K_1^{-1/2}|h|\left(\sum_{I_1^+} (n_i, h)_X^2\right)^{1/2}$$

and

$$K_1|h|^2 \le \sum_{I_1^+} (n_i, h)_X^2 \text{ for every } h \in C_1^*. \tag{3.2.7}$$

Step 4. Every $h \in X$ can be uniquely decomposed into mutually orthogonal elements as $h = h_1 + h_2 + h_3$, where $h_1 \in C_1 \subset \ker E$, $h_2 \in C_1^* \subset \ker E$, and $h_3 \in \ker E^\perp$.

By Step 2 there exists a vector $(\alpha_1, \ldots, \alpha_{\tilde{m}_1}) \in \mathbb{R}^{\tilde{m}_1}$ such that $h_2 = \sum_{i=1}^{\tilde{m}_1} \alpha_i n_i$. Note that $(n_i, h_1)_X \le 0$ for all $i \in \tilde{I}_1$. Therefore, using the fact that $(h_1, h_2)_X = \sum_{i=1}^{\tilde{m}_1} \alpha_i (h_1, n_i)_X = 0$, it follows that

$$\begin{cases} \text{if } \alpha_i > 0 \text{ for some } i = 1, \ldots, \tilde{m}_1, \text{ then} \\ (n_i, h_1)_X = (P_{\tilde{E}} l_i, h_1)_X = (l_i, h_1)_X = 0. \end{cases} \tag{3.2.8}$$

We shall use Step 3 with $I_1^+ = \{i \in \tilde{I}_1 : (n_i, h_2)_X > 0 \text{ and } \alpha_i > 0\}$. Let $\hat{k} \in \mathbb{R}_+^{m_1}$ and $c > 0$. Then by (3.1.6) and (3.2.8) we have

$$\begin{aligned} H((h, \hat{k}), (h, \hat{k})) &= \mathcal{L}''(x^*, \lambda^*, \mu^*, \eta^*)(h, h) + c|Eh_3|^2 \\ &\quad + c\sum_{I_1^+} |(l_i, h_3)_X + (l_i, h_2)_X + \hat{k}_i|^2 + c\sum_{I_1 \setminus I_1^+} |(l_i, h)_X + \hat{k}_i|^2 \\ &\ge \gamma|h_1|^2 + 2\mathcal{L}''(x^*, \lambda^*, \mu^*, \eta^*)(h_1, h_2 + h_3) \\ &\quad + \mathcal{L}''(x^*, \lambda^*, \mu^*, \eta^*)(h_2 + h_3, h_2 + h_3) \\ &\quad + c|Eh_3|^2 + \frac{c_1}{2}\sum_{I_1^+}\left((n_i, h_2)_X + \hat{k}_i\right)^2 - c_1\sum_{I_1^+}(l_i, h_3)_X^2 \\ &\quad + \frac{c_2}{2}\sum_{I_1 \setminus I_1^+} |\hat{k}_i|^2 - c_2\sum_{I_1 \setminus I_1^+}(l_i, h)_X^2 \end{aligned}$$

for all $0 < c_1, c_2 \le c$ to be chosen below. Here we used $(a+b)^2 \ge \frac{1}{2}a^2 - b^2$ and the fact that $(l_i, h_2) = (n_i, h_2)$ for $i \in I_1$. To estimate Eh_3 from below we make use of the fact that $\tilde{E}$ is an isomorphism from $(\ker E)^\perp$ to $W \times \mathbb{R}^{\tilde{m}_2}$. Hence there exists $K_2 > 0$ such that $K_2|h_3| \le |\tilde{E}h_3| \le |Eh_3|$ for all $h_3 \in (\ker E)^\perp$. Setting $L = \|\mathcal{L}''(x^*, \lambda^*, \mu^*)\|$, $c_3 = \min(c_1, c_2)$, and $c_4 = \max(c_1, c_2)$ we obtain

$$\begin{aligned} H((h, \hat{k}), (h, \hat{k})) &= \gamma|h_1|^2 - 2L|h_1|(|h_2| + |h_3|) - L(|h_2| + |h_3|)^2 \\ &\quad + cK_2^2|h_3|^2 + \frac{c_1}{2}\sum_{I_1^+}(n_i, h_2)_X^2 + \frac{c_3}{2}\sum_{I_1}\hat{k}_i^2 \\ &\quad - 3c_4\sum_{I_1}(l_i, h_3)_X^2 - 3c_2\sum_{I_1 - I_1^+}\left((l_i, h_1)_X^2 + (l_i, h_2)_X^2\right), \end{aligned}$$

where we used the fact that $(n_i, h_2)_X \ge 0$ for $i \in I_1^+$.

Setting $|l|^2 = \sum_{I_1} |l_i|^2$ and using (3.2.7) we find

$$
\begin{aligned}
H((h,\hat{k}),(h,\hat{k})) &\geq \gamma|h_1|^2 + \frac{c_1}{2}K_1|h_2|^2 + cK_2^2|h_3|^2 + \frac{c_3}{2}\sum_{I_1}\hat{k}_i^2 \\
&\quad - 2L|h_1|(|h_2|+|h_3|) - 2L(|h_2|^2+|h_3|^2) \\
&\quad - 3c_2|l|^2|h_1|^2 - 3c_2|l|^2|h_2|^2 - 3c_4|l|^2|h_3|^2 \\
&\geq \gamma|h_1|^2 + \frac{c_1}{2}K_1|h_2|^2 + cK_2^2|h_3|^2 + \frac{c_3}{2}\sum_{I_1}\hat{k}_i^2 \\
&\quad - 2L\left(\frac{\gamma}{4}\right)^{1/2}\left(\frac{\gamma}{4}\right)^{-1/2}|h_1||h_2| - 2L\left(\frac{\gamma}{4}\right)^{1/2}\left(\frac{\gamma}{4}\right)^{-1/2}|h_1||h_3| \\
&\quad - 2L|h_2|^2 - 2L|h_3|^2 - 3c_2|l|^2|h_1|^2 - 3c_2|l|^2|h_2|^2 - 3c_4|l|^2|h_3|^2 \\
&\geq |h_1|^2\left(\gamma - \frac{\gamma}{2} - 3c_2|l|^2\right) \\
&\quad + |h_2|^2\left(\frac{c_1}{2}K_1 - \frac{4}{\gamma}L^2 - 2L - 3c_2|l|^2\right) \\
&\quad + |h_3|^2\left(cK_2^2 - \frac{4}{\gamma}L^2 - 2L - 3c_4|l|^2\right) + \frac{c_3}{2}\sum_{I_1}\hat{k}_i^2.
\end{aligned}
$$

We now choose the constants in the order c_2, c_1, and $\bar{c}$ such that the coefficients of $|h_1|^2$, $|h_2|^2$, and $|h_3|^2$, with c replaced by $\bar{c}$, are positive. This implies the claim for every $c \geq \bar{c}$. □

It is worthwhile to point out the following corollary to Proposition 3.1 which is well known in finite-dimensional spaces.

Corollary 3.2. *Let $E \in \mathcal{L}(X,W)$ be surjective and let $A \in \mathcal{L}(X)$ be self-adjoint and coercive on* $\ker E$; *i.e., there exists $\gamma > 0$ such that $(Ax,x) \geq \gamma|x|^2$ for all $x \in \ker E$. Then there exist constants $\tau > 0$ and $\bar{c} > 0$ such that $((A + cE^*E)x, x)_X \geq \tau|x|^2$ for all $x \in X$ and $c \geq \bar{c}$.*

This follows from Proposition 3.1 with $A = \mathcal{L}''(x^*)$.

Proposition 3.3. *Let* (3.1.2)–(3.1.6) *hold. Then there exists $\sigma > 0$ such that*

$$
\mathcal{L}''_{\bar{c}}(x^*,u^*,\lambda^*,\mu^*,\eta^*)((h,k),(h,k)) + \sum_{i\in I_2}\mu_i^* k_i \geq \sigma(|h|^2 + |k^2|)
$$

for all $h \in X$ with $|h| \leq \tilde{\mu}/(2\bar{c}\sup_{i\in I_2}|l_i|)$, $\tilde{\mu} = \min_{I_2}\mu_i^*$ and all $k \in \mathbb{R}^m$ with $k_i \geq 0$ for $i \in I_1 \cup I_2$, and $\bar{c}$ is as introduced in Proposition 3.1.

Proof. Let $A = \mathcal{L}''_{\bar{c}}(x^*,u^*,\lambda^*,\mu^*,\eta^*)((h,k),(h,k)) + \sum_{i\in I_2}\mu_i^* k_i$. Then from (3.2.3) and the definition of H

$$
A = H((h,\hat{k}),(h,\hat{k})) + \bar{c}\sum_{I_3}\left|(l_i,h)_X + k_i\right|^2 + \bar{c}\sum_{I_2}\left\{|k_i|^2 + 2(l_i,h)_X k_i\right\} + \sum_{I_2}\mu_i^* k_i,
$$

where $\hat{k}$ denotes the first m_1 coordinates of the vector k. By Proposition 3.1 we have for every $\varepsilon > 1$

$$
\begin{aligned}
A &\geq \tau(|h|^2 + |\hat{k}|^2_{\mathbb{R}^{m_1}}) + \bar{c} \sum_{I_3} \Big((l_i, h)^2_X (1 - \varepsilon^2) + |k_i|^2 (1 - \varepsilon^{-2}) \Big) \\
&\quad + \bar{c} \sum_{I_2} |k_i|^2 - 2\bar{c} \sum_{I_2} \big| (l_i, h)_X \big| k_i + \sum_{I_2} \mu_i^* k_i \\
&\geq \tau(|h|^2 + |\hat{k}|^2_{\mathbb{R}^{m_1}}) + \bar{c}(1 - \varepsilon^{-2}) \sum_{I_2 \cup I_3} |k_i|^2 \\
&\quad + \bar{c}(1 - \varepsilon^2)|h|^2 \sum_{I_3} |l_i|^2 - 2\bar{c} \sum_{I_2} \big| (l_i, h)_X \big| k_i + \sum_{I_2} \mu_i^* k_i .
\end{aligned}
$$

Now choose $\varepsilon > 1$ such that $\frac{\tau}{2} = (\varepsilon^2 - 1)\tau \sum_{I_3} |l_i|^2$. Then using the constraint for h we find that

$$
\begin{aligned}
A &\geq \frac{1}{2}\tau(|h|^2 + |\hat{k}|^2_{\mathbb{R}^{m_1}}) + \bar{c}(1 - \varepsilon^{-2}) \sum_{I_2 \cup I_3} k_i^2 + \Big(\tilde{\mu} - 2\bar{c} \sup_{I_2} |l_i||h| \Big) \sum_{I_2} k_i \\
&\geq \frac{1}{2}\tau(|h|^2 + |\hat{k}|^2_{\mathbb{R}^{m_1}}) + \bar{c}(1 - \varepsilon^{-2}) \sum_{I_2 \cup I_3} k_i^2 .
\end{aligned}
$$

This proves the claim for $c = \bar{c}$. For arbitrary $c \geq \bar{c}$ the assertion follows from the form of $\mathcal{L}_c''(x^*, u^*, \lambda^*, \mu^*)$. □

Theorem 3.4. *Assume that* (3.1.2)–(3.1.6) *hold. Then there exist constants* $\bar{\sigma} > 0$, $\bar{c} > 0$ *and a neighborhood* $U(x^*, -g(x^*)) = U(x^*, u^*)$ *of* (x^*, u^*) *such that*

$$
\begin{aligned}
&\mathcal{L}_c(x, u, \lambda^*, \mu^*, \eta^*) + \big(\mu^*, u\big)_{\mathbb{R}^m} \\
&= f(x) + \big(\lambda^*, e(x)\big)_W + \big(\mu^*, g(x) + u\big)_{\mathbb{R}^m} + \big(\eta^*, \ell(x)\big)_Z + \frac{c}{2}|e(x)|^2_W + \frac{c}{2}|g(x) + u|^2_{\mathbb{R}^m} \\
&\geq f(x^*) + \bar{\sigma}(|x - x^*|^2 + |u - u^*|^2)
\end{aligned}
\tag{3.2.9}
$$

for all $c \geq \bar{c}$ *and* $(x, u) \in U(x^*, u^*)$ *with* $u_i \geq 0$ *for* $i \in I_1 \cup I_2$.

Proof. By (3.1.4) we have $\big(\mu^*, u^*\big) = -\big(\mu^*, g(x^*)\big) = 0$. Moreover

$$
\begin{aligned}
&\mathcal{L}_c(x^*, u^*, \lambda^*, \mu^*, \eta^*) + \big(\mu^*, u^*\big)_{\mathbb{R}^m} = f(x^*), \\
&(\mathcal{L}_c)_x(x^*, u^*, \lambda^*, \mu^*, \eta^*) = (\mathcal{L}_c)_u(x^*, u^*, \lambda^*, \mu^*, \eta^*) = 0,
\end{aligned}
$$

where $(\mathcal{L}_c)_x$ and $(\mathcal{L}_c)_u$ denote the partial derivatives of $\mathcal{L}_c$ with respect to x and u. Consequently

$$
\begin{aligned}
&\mathcal{L}_c(x^*, u^*, \lambda^*, \mu^*, \eta^*) + \big(\mu^*, u^*\big)_{\mathbb{R}^m} = f(x^*) + \big(\mu^*, u - u^*\big)_{\mathbb{R}^m} \\
&+ \mathcal{L}_c''(x^*, u^*, \lambda^*, \mu^*, \eta^*)((x - x^*, u - u^*), (x - x^*, u - u^*)) + o(|x - x^*|^2 + |u - u^*|^2).
\end{aligned}
$$

The definitions of I_1 and I_3 imply that $\mu_i^* = 0$ for $i \in I_1 \cup I_3$. Thus Proposition 3.3 implies the existence of a neighborhood $U(x^*, u^*)$ of (x^*, u^*) and a constant $\bar{\sigma} > 0$ such that (3.2.9) holds for all $(x, u) \in U(x^*, u^*)$ with $u_i \geq 0$ for $i \in I_1 \cup I_2$. $\square$

The conclusion of Theorem 3.4 is referred to as augmentability of problem (3.1.1). This means that a functional, which in our case is $\mathcal{L}_c(x, u, \lambda^*, \mu^*, \eta^*) + \big(\mu^*, u^*\big)_{\mathbb{R}^m}$, can be found with the property that it has (x^*, u^*) as a strict local minimizer under the simple constraint $u \geq 0$, while the explicit nonlinear constraints are eliminated. Let us note that if (3.1.6) holds with C replaced by the larger set

$$\hat{C} = \{h \in X : e'(x^*)h = 0,\ g_i'(x^*)h \leq 0 \text{ for } i \in I_1 \cup I_2\},$$

then the conclusion of Theorem 3.4 holds with $\mathcal{L}_c(x, u, \lambda^*, \mu^*, \eta^*) + \big(\mu^*, u^*\big)_{\mathbb{R}^m}$ replaced by $\mathcal{L}_c(x, u, \lambda^*, \mu^*, \eta^*)$. In case X is finite-dimensional, augmentability is analyzed in detail in [Hes1], for example. The proof of augmentability in [Hes1] depends in essential manner on compactness of the unit ball. In [Be, p. 161 ff], the augmentability analysis relies on the strict complementarity assumption, i.e., $I_1 = \emptyset$ is assumed.

We obtain, as immediate consequence of Theorem 3.4, that (3.1.6) provides a second order sufficient optimality condition.

Corollary 3.5. *Assume that* (3.1.2)–(3.1.6) *hold. Then there exists a neighborhood $U(x^*)$ of x^* such that*

$$f(x) \geq f(x^*) + \bar{\sigma}|x - x^*|^2$$

for all $x \in U(x^*)$ satisfying $e(x) = 0$, $g(x) \leq 0$, and $\ell(x) \in K$.

Theorem 3.4 can be used as the basis to define augmented cost functionals in terms of x only with the property that x^* is a uniform strict local unconstrained minimum. The two choices we present differ in the way in which the inequality constraints are treated. The first choice uses the classical cutoff penalty functional

$$\tilde{g}(x) = \max(g(x), 0), \tag{3.2.10}$$

and the second is the Bertsekas penalty functional

$$\hat{g}(x, \mu, c) = \max\left(g(x), -\frac{\mu}{c}\right), \tag{3.2.11}$$

where $\mu \in \mathbb{R}^m$ and $c > 0$. In each of the two cases the max operation acts coordinatewise. To motivate the choice (3.2.11) we use the Lagrangian $\mathcal{L}_c(x, u, \lambda, \mu, \eta)$ introduced in (3.2.2) for (3.2.1) and consider

$$\begin{cases} \min \mathcal{L}_c(x, u, \lambda, \mu, \eta) = f_c(x, u) + \big(\lambda, e(x)\big)_W \\ \quad +\big(\mu, g(x) + u\big)_{\mathbb{R}^m} + \big(\eta, \ell(x)\big)_Z \text{ over } x \in X, u \geq 0. \end{cases} \tag{3.2.12}$$

Carrying out the constraint minimization with respect to u with x and μ fixed results in

$$u = u(x, \mu) = \max\left(0, -\left(g(x) + \frac{\mu}{c}\right)\right),$$

and consequently

$$g(x) + u(x, \mu) = \max\left(g(x), -\frac{\mu}{c}\right) = \hat{g}(x, \mu, c). \tag{3.2.13}$$

Corollary 3.6. *Let* (3.1.2)–(3.1.6) *hold. Then there exists a neighborhood $U(x^*)$ of x^* such that*

$$\begin{aligned} f(x)+&\left(\lambda^*, e(x)\right)_W + \left(\mu^*, \tilde{g}(x)\right)_{\mathbb{R}^m} + \left(\eta^*, \ell(x)\right)_Z + \frac{c}{2}|e(x)|_W^2 + \frac{c}{2}|\tilde{g}(x)|^2 \\ &\geq f(x^*) + \bar{\sigma}|x - x^*|^2 \end{aligned}$$

for all $x \in U(x^*)$, and $c \geq \bar{c}$, where $\tilde{g}(x) = \max(g(x), 0)$.

Proof. Setting

$$u = \max(-g(x), 0) \tag{3.2.14}$$

we have $g(x) + u = \tilde{g}(x)$ and $u \geq 0$. Recall the definition of $U = U(x^*, -g(x^*))$ from Theorem 3.4. Determine a neighborhood $U(x^*)$ such that $x \in U(x^*)$ implies $(x, -g(x)) \in U$ and $g_i(x) \leq 0$ if $g_i(x^*) < 0$. It is simple to argue that $x \in U(x^*)$ implies $(x, u) \in U$, where u is defined in (3.2.14). The claim now follows from Theorem 3.4. ☐

Corollary 3.7. *Let* (3.1.2)–(3.1.6) *hold and let $r > 0$. Then there exist constants $\delta > 0$ and $\tilde{c} = \tilde{c}(r) \geq \bar{c}$ such that*

$$\begin{aligned} f(x)+&\left(\lambda^*, e(x)\right)_X + \left(\mu^*, \hat{g}(x, \mu, c)\right)_{\mathbb{R}^m} + \left(\eta^*, \ell(x)\right)_Z + \frac{c}{2}|e(x)|_W^2 + \frac{c}{2}|\hat{g}(x, \mu, c)|^2 \\ &\geq f(x^*) + \bar{\sigma}|x - x^*|^2 \end{aligned}$$

for all $c \geq \tilde{c}$, $x \in B_\delta = \{x \in X : |x - x^*| \leq \delta\}$, and $\mu \in B_r^+ = \{\mu \in \mathbb{R}_+^m : |\mu| \leq r\}$, where $\hat{g}(x, \mu, c) = \max(g(x), -\frac{\mu}{c})$.

Proof. Let $\varepsilon > 0$ be such that $g_i(x^*) \leq -\varepsilon$ for all $i \in I_3$ and such that $|x - x^*| < \varepsilon$ and $|u - u^*| < \varepsilon$ implies $(x, u) \in U(x^*, u^*)$, where $U(x^*, u^*)$ is given in Theorem 3.4. Determine $\delta \in (0, \varepsilon)$ such that $|x - x^*| \leq \delta$ implies $|g(x) - g(x^*)| < \frac{\varepsilon}{2}$ and choose $\tilde{c} \geq \frac{2r}{\varepsilon}$. For $c \geq \tilde{c}$ and $\mu \in B_r^+$ we define

$$u = \max\left(0, -\left(\frac{\mu}{c} + g(x)\right)\right). \tag{3.2.15}$$

Then $g(x) + u = \hat{g}(x)$ and $u \geq 0$. To verify the claim it suffices to show that $x \in B_\delta$ and $\mu \in B_r^+$ imply $|u - u^*| < \varepsilon$, where u is defined in (3.2.15). If $i \in I_1 \cup I_2$, then $|u_i - u_i^*| \leq |g_i(x^*) - g_i(x)| + \frac{\mu_i}{c}$. For $i \in I_3$ we have $\frac{\mu_i}{c} + g_i(x) \leq \frac{\mu_i}{c} + |g_i(x^*) - g_i(x)| + g_i(x^*) < \frac{r}{c} + \frac{\varepsilon}{2} - \varepsilon < 0$ and consequently $|u_i - u_i^*| = |g_i(x^*) - g_i(x) - \frac{\mu_i}{c}|$. Summing over $i = 1, \ldots, m$ we find $|u - u^*| \leq 2|g(x^*) - g(x)|^2 + \frac{2}{c^2}|\mu|^2 < \varepsilon^2$. ☐

Remark 3.2.1. Note that

$$x \to \left(\mu, \hat{g}(x, \mu, c)\right) + \frac{c}{2}|\hat{g}(x, \mu, c)|^2$$

is C^1 if g is C^1. This follows from the identity

$$\big(\mu, \hat{g}(x, \mu, c)\big)_{\mathbb{R}^m} + \frac{c}{2}|\hat{g}(x, \mu, c)|^2_{\mathbb{R}^m} = \frac{1}{2c}\left(|\max(0, \mu + cg(x))|^2_{\mathbb{R}^m} - |\mu|^2_{\mathbb{R}^m}\right). \quad (3.2.16)$$

Here, as throughout, the norm on $\mathbb{R}^m$ is the Euclidean one. On the other hand,

$$x \to \big(\mu, \tilde{g}(x, \mu, c)\big) + \frac{c}{2}|\tilde{g}(x, \mu, c)|^2$$

is not C^1.

3.3 The first order augmented Lagrangian algorithm

Here we describe the first order augmented Lagrangian algorithm which is a hybrid method combining the penalty technique and a Lagrangian method. Considering for a moment equality constraints only, the penalty method for (3.1.1) consists in minimizing

$$f(x) + \frac{c_k}{2}|e(x)|^2$$

for a sequence of penalty parameters c_k tending to ∞. The Lagrangian method relies on minimizing

$$f(x) + (\lambda_k, e(x))$$

and updating λ_k as a maximizer of the dual problem associated to (3.1.1), which will be defined below. The first order augmented Lagrangian method provides a systematic technique for the multiplier update. Its convergence analysis will be given in the next section.

Let x^* be a local solution of (3.1.1) with (3.1.2)–(3.1.6) holding. The algorithm will require startup values $(\lambda_0, \mu_0) \in W \times \mathbb{R}^m_+$ for the Lagrange multipliers. We set $r = |\mu^*| + (|\lambda_0 - \lambda^*|^2 + |\mu_0 - \mu^*|^2)^{1/2}$ and choose ε and $\tilde{c} = \tilde{c}(r, \varepsilon) \geq \bar{c}$ as in the proof of Corollary 3.6. Recall that ε was chosen such that $g_i(x^*) \leq -\varepsilon$ for all indices i in the set of inactive indices I_3 and such that $|x - x^*| < \varepsilon$, $|u - u^*| < \varepsilon$ imply $(x, u) \in U(x^*, -g(x^*))$ with $U(x^*, -g(x^*))$ given in Theorem 3.4. Finally, let $\{c_n\}_{n=1}^{\infty}$ be a monotonically nondecreasing sequence of penalty parameters with $c_1 > \tilde{c}$, and set $\sigma_n = c_n - \tilde{c}$. It is not required that $\lim_{n\to\infty} c_n = \infty$, as would be the case for penalty methods.

Algorithm ALM.

1. Choose $(\lambda_0, \mu_0) \in W \times \mathbb{R}^m_+$.

2. Let x_n be a solution to

$$\min \mathcal{L}_n(x) \text{ over } x \in X \text{ with } \ell(x) \in K. \quad (P_n)$$

3. Update the Lagrange multipliers

$$\lambda_n = \lambda_{n-1} + \sigma_n e(x_n), \quad \mu_n = \mu_{n-1} + \sigma_n \hat{g}(x_n, \mu_{n-1}, c_n),$$

where

$$\mathcal{L}_n(x) = f(x) + \big(\lambda_{n-1}, e(x)\big)_W + \big(\mu_{n-1}, \hat{g}(x, \mu_{n-1}, c_n)\big)_{\mathbb{R}^m}$$
$$+ \frac{c_n}{2}|e(x)|_W^2 + \frac{c_n}{2}|\hat{g}(x, \mu_{n-1}, c_n)|^2.$$

Observe that $\mu_n = \max(\mu_{n-1} + \sigma_n g(x_n), \mu_{n-1}(1 - \frac{\sigma_n}{c_n}))$ and consequently $\mu_n \in \mathbb{R}^n_+$ for each $n = 1, 2, \dots$. Existence of solutions to (P_n) will be discussed in Section 3.4. It will be shown that (P_n) has a solution in the interior of B_δ (see Corollary 3.6) for all n sufficiently large, or for all n if c_1 is sufficiently large, and that these solutions converge to x^*, while (λ_n, μ_n) converges to (λ^*, μ^*).

To motivate the Lagrange multiplier update in step 3 of the algorithm and to justify calling this a first order algorithm, we consider problem (3.2.1) without the infinite rank inequality constraint. Introducing Lagrange multipliers for the equality constraints $e(x) = 0$ and $g(x) + u = 0$ we obtained the augmented Lagrange functional $\overline{\mathcal{L}}_c$ in (3.2.12). Carrying out the minimization with respect to u and utilizing (3.2.13) suggests introducing the modified augmented Lagrangian functional

$$\hat{\mathcal{L}}_c(x, \lambda, \mu) = f(x) + \big(\lambda, e(x)\big)_W + \big(\mu, \hat{g}(x, \mu, c)\big)_{\mathbb{R}^m} + \frac{c}{2}|e(x)|_W^2 + \frac{c}{2}|\hat{g}(x, \mu, c)|_{\mathbb{R}^m}^2. \tag{3.3.1}$$

Since $\hat{\mathcal{L}}_c(x^*, \lambda^*, \mu^*) = f(x^*)$ we find

$$\hat{\mathcal{L}}_c(x^*, \lambda, \mu) \le \hat{\mathcal{L}}_c(x^*, \lambda^*, \mu^*) \le \hat{\mathcal{L}}_c(x, \lambda^*, \mu^*) \text{ for all } x \in B_\delta \text{ and } \mu \ge 0, \tag{3.3.2}$$

whenever $c \ge \tilde{c}$. The first inequality follows from the fact that $e(x^*) = \hat{g}(x^*, \mu) = 0$ for $\mu \ge 0$ and the second one from Corollary 3.6. From the saddle point property (3.3.2) for $\hat{\mathcal{L}}_c(x, \lambda, \mu)$ we deduce that

$$\sup_{\lambda, \mu \ge 0} \inf_x \hat{\mathcal{L}}_c(x, \lambda, \mu) \le \hat{\mathcal{L}}_c(x^*, \lambda^*, \mu^*) \le \inf_x \sup_{\lambda, \mu} \hat{\mathcal{L}}_c(x, \lambda, \mu). \tag{3.3.3}$$

For the following discussion we assume strict complementarity, i.e., $I_1 = \emptyset$. Then $x \to \hat{\mathcal{L}}_c(x, \lambda, \mu)$ is twice continuously differentiable for (x, λ, μ) in a neighborhood of (x^*, λ^*, μ^*) and

$$\hat{\mathcal{L}}_c''(x^*, \lambda^*, \mu^*)(h, h) = \hat{\mathcal{L}}''(x^*, \lambda^*, \mu^*)(h, h) + c|Eh|_W^2 \text{ for } h \in X.$$

Thus by (3.1.6) and Corollary 3.2, $\hat{\mathcal{L}}_c''(x^*, \lambda^*, \mu^*)$ is positive definite on X for all c sufficiently large. Moreover $(x, \lambda, \mu) \to \hat{\mathcal{L}}_c''(x, \lambda, \mu)$ is continuous in a neighborhood $U(x^*) \times U(\lambda^*, \mu^*)$ of (x^*, λ^*, μ^*) and $\hat{\mathcal{L}}_c''(x, \lambda, \mu) \ge \bar{\tau}|x|^2$ for $\bar{\tau} > 0$ independent of $x \in X$ and $(\lambda, \mu) \in W \times \mathbb{R}^m$. It follows that

$$\min_{x \in U(x^*)} \hat{\mathcal{L}}_c''(x, \lambda, \mu)$$

admits a unique solution $x(\lambda, \mu)$ for every $(\lambda, \mu) \in U(\lambda^*, \mu^*)$ and as a consequence of Theorem 2.24 $(\lambda, \mu) \to x(\lambda, \mu)$ is differentiable. Consequently the locally defined dual functional

$$d(\lambda, \mu) = \min_{x \in U(x^*)} \hat{\mathcal{L}}_c''(x, \lambda, \mu)$$

is differentiable for (λ, μ) in a neighborhood of (λ^*, μ^*). For the gradient ∇d at (λ, μ) in direction $(\delta_\lambda, \delta_\mu)$ we obtain using (3.2.16)

$$\begin{aligned}\nabla d(\lambda, \mu)(\delta_\lambda, \delta_\mu) = &\big(f'(x) + e'(x)^*\lambda + g'(x)^* \max(0, \mu + cg(x)), x_\lambda(\delta_\lambda) + x_\mu(\delta_\mu)\big)\\ &+ \big(\delta_\lambda, e(x)\big) + \big(\delta_\mu, \hat{g}(x, \mu, c)\big),\end{aligned}$$

where $x = x(\lambda, \mu)$.

Utilizing the first order optimality condition this implies that the gradient of $d(\lambda, \mu)$ is given by

$$\nabla d(\lambda, \mu) = (e(x(\lambda, \mu)), \hat{g}(x(\lambda, \mu), \mu, c)) \in W \times \mathbb{R}^m. \tag{3.3.4}$$

Thus the multiplier updates in step 3 of Algorithm ALM are steepest ascent directions for the dual problem

$$\sup_{\lambda, \mu \geq 0} d(\lambda, \mu).$$

In view of (3.3.3) the first order augmented Lagrangian algorithm is an iterative algorithm combining minimization in the primal direction in step 2 and maximization in the dual direction on every level of the iteration.

Remark 3.3.1. In this chapter we identify X and W with their dual spaces and consider $e'(x)^*$ as an operator from W to X. If, alternatively, $e'(x)^*$ is considered as an operator from W^* to X^*, then the Lagrange multiplier is taken as an element of W^* and the Lagrange functional is defined as $\mathcal{L}(x, \tilde{\lambda}) = f(x) + \langle \tilde{\lambda}, e(x) \rangle_{W^*,W}$. The relation between λ and $\tilde{\lambda}$ is given by $\mathcal{I}\lambda = \tilde{\lambda}$, with $\mathcal{I}$ the canonical isomorphism from W to W^*. If, for example, $W = H^{-1}(\Omega)$, then $W^* = H_0^1(\Omega)$ and $\mathcal{I} = (-\Delta)^{-1}$. The Lagrange multiplier update in step 3 of Algorithm ALM is then given by $\lambda_n = \lambda_{n-1} + \sigma_n \mathcal{I}\, e(x_n)$.

As already mentioned the augmented Lagrangian algorithm ALM is also a hybrid method combining Lagrange multiplier and penalty methods. Solving (P_n) without the penalty terms we obtain the former, for $\lambda_n = 0$ and c_n satisfying $\lim_{n\to\infty} c_n = \infty$ we obtain the latter. The choice of c_n for Algorithm ALM in practice is an important one. While no general purpose techniques appear to be available, guidelines for the choice include choosing them large enough such that the augmented Lagrangian has positive definite Hessian in the sense of Proposition 3.1. Large c_n will improve the convergence estimates, as we shall see in the following section, and at the same time they can be the cause for ill-conditioning of the auxiliary problems (P_n). In fact, for large c_n the Hessian of the cost functional $\mathcal{L}_n(x)$ can have eigenvalues on significantly different scales. For further discussion on the choice of the parameter c we refer the reader to [Be].

Let us make a few historical comments on the first order augmented Lagrangian algorithm. It was originated by Hestenes [Hes2] and Powell [Pow] and extended among others by Bertsekas [Be] and in [PoTi, Roc2]. The infinite-dimensional case has received less attention. In this respect we refer the reader to the work of Polyak and Tret'yakov [PoTr] and Fortin and Glowinski [FoGl]. Both consider the case of augmenting equality constraints only. In [FoGl] augmented Lagrangian methods are also developed systematically for solving nonlinear partial differential equations.

3.4 Convergence of Algorithm ALM

We shall use the following notation:

$$\hat{L}_c(x,\mu) = f(x) + \big(\lambda^*, e(x)\big) + \big(\mu^*, \hat{g}(x,\mu,c)\big) + \big(\eta^*, \ell(x)\big) + \frac{\bar{c}}{2}|e(x)|^2 + \frac{\bar{c}}{2}|\hat{g}(x,\mu,c)|^2.$$

Further we set $r = |\mu^*| + (|\lambda^0 - \lambda^*|^2 + |\mu^0 - \mu^*|^2)^{1/2}$. Corollary 3.7 then guarantees the existence of $\delta > 0$ and $\tilde{c} = \tilde{c}(r)$ such that

$$\hat{L}_c(x,\mu) - f(x^*) \geq \bar{\sigma}|x - x^*|^2 \tag{3.4.1}$$

for all $x \in B_\delta$, $\mu \in B_r^+$, and $c \geq \tilde{c}$. We also assume that B_δ is contained in the region of applicability of (3.4.1) and that $\{c_n\}_{n=1}^\infty$ is a monotonically nondecreasing sequence with $\tilde{c} < c_1$. Then we have the following convergence properties of Algorithm ALM from arbitrary initializations $(\lambda_0, \mu_0) \in W \times \mathbb{R}^m_+$ in the case that suboptimal solutions are chosen in step 2.

Theorem 3.8. *Assume that* (3.1.2)–(3.1.5), (3.4.1) *hold and that* $\tilde{x}_n \in B_\delta$ *satisfy*

$$\mathcal{L}_n(\tilde{x}_n) \leq \mathcal{L}_n(x^*) \text{ for each } n = 1, \dots. \tag{3.4.2}$$

If (λ_n, μ_n) *are defined as in step* 3 *of Algorithm* ALM *with* x_n *replaced by* $\tilde{x}_n$, *then for* $n \geq 1$ *we have with* $\sigma_n = c_n - \bar{c}$

$$\bar{\sigma}|\tilde{x}_n - x^*|^2 + \frac{1}{2\sigma_n}(|\lambda_n - \lambda^*|^2 + |\mu_n - \mu^*|^2) \leq \frac{1}{2\sigma_n}(|\lambda_{n-1} - \lambda^*|^2 + |\mu_{n-1} - \mu^*|^2). \tag{3.4.3}$$

This implies that $\mu_n \in B_r^+$ *for all* $n \geq 1$ *and*

$$|\tilde{x}_n - x^*|^2 \leq \frac{1}{2\bar{\sigma}\sigma_n}(|\lambda_0 - \lambda^*|^2 + |\mu_0 - \mu^*|^2) \tag{3.4.4}$$

and

$$\sum_{n=1}^{\infty} \sigma_n|\tilde{x}_n - x^*|^2 \leq \frac{1}{2\bar{\sigma}}(|\lambda_0 - \lambda^*|^2 + |\mu_0 - \mu^*|^2). \tag{3.4.5}$$

Proof. We proceed by induction and assume that the claim has been verified up to $n-1$. For $n = 1$ the result can be obtained by the general arguments given below. For the induction step we observe that (3.4.3), which is assumed to hold up to $n - 1$, implies

$$|\mu_{n-1}| \leq |\mu^*| + (|\lambda_0 - \lambda^*|^2 + |\mu_0 - \mu^*|^2)^{1/2} = r.$$

Consequently $\mu_{n-1} \in B_r^+$ and (3.4.1) with $\mu = \mu_{n-1}$ is applicable. Using the fact that

$$[\max(0, \mu_{n-1,i} + c_n g_i(x^*))]^2 - \mu^2_{n-1,i} \leq 0$$

and (3.2.16) we find

$$\mathcal{L}_n(x^*) = f(x^*) + \frac{1}{2c_n}\sum_{i=1}^{m}\{[\max(0, \mu_{n-1,i} + c_n g_i(x^*))]^2 - \mu_{n-1,i}^2\} \le f(x^*). \quad (3.4.6)$$

Next we rearrange terms in $\mathcal{L}_n(\tilde{x}_n)$ and obtain

$$\begin{aligned}
\mathcal{L}_n(\tilde{x}_n) &= f(\tilde{x}_n) + \big(\lambda^*, e(\tilde{x}_n)\big) + \big(\lambda_{n-1} - \lambda^*, e(\tilde{x}_n)\big) + \big(\mu^*, \hat{g}(\tilde{x}_n, \mu_{n-1}, c_n)\big) \\
&\quad + \big(\mu_{n-1} - \mu^*, \hat{g}(\tilde{x}_n, \mu_{n-1}, c_n)\big) + \frac{\bar{c}}{2}|e(\tilde{x}_n)|^2 + \frac{1}{2}(c_n - \bar{c})|e(\tilde{x}_n)|^2 \\
&\quad + \frac{\bar{c}}{2}|\hat{g}(\tilde{x}_n, \mu_{n-1}, c_n)|^2 + \frac{1}{2}(c_n - \bar{c})|\hat{g}(\tilde{x}_n, \mu_{n-1}, c_n)|^2 \\
&= \hat{L}(\tilde{x}_n, \mu_{n-1}) + \frac{1}{2\sigma_n}(|\lambda_n - \lambda^*|^2 - |\lambda_{n-1} - \lambda^*|^2) \\
&\quad + \frac{1}{2\sigma_n}(|\mu_n - \mu^*|^2 - |\mu_{n-1} - \mu^*|^2) - \big(\eta^*, \ell(\tilde{x}_n)\big)_Z.
\end{aligned}$$

Since $\eta^* \in K^+$ and $\ell(\tilde{x}_n) \in K$ we have $\big(\eta^*, \ell(\tilde{x}_n)\big) \le 0$. This fact together with the above equality, (3.4.2), and (3.4.6) implies

$$\begin{aligned}
&\hat{L}(\tilde{x}_n, \mu_{n-1}) + \frac{1}{2\sigma_n}(|\lambda_n - \lambda^*|^2 - |\lambda_{n-1} - \lambda^*|^2) \\
&\qquad + \frac{1}{2\sigma_n}(|\mu_n - \mu^*|^2 - |\mu_{n-1} - \mu^*|^2) \le \mathcal{L}_n(\tilde{x}_n) \le f(x^*).
\end{aligned}$$

Finally (3.4.1) implies that

$$\bar{\sigma}\,|\tilde{x}_n - x^*|^2 + \frac{1}{2\sigma_n}|\lambda_n - \lambda^*|^2 + \frac{1}{2\sigma_n}|\mu_n - \mu^*|^2 \le \frac{1}{2\sigma_n}|\lambda_{n-1} - \lambda^*|^2 + \frac{1}{2\sigma_n}|\mu_{n-1} - \mu^*|^2,$$

which establishes (3.4.3). Estimates (3.4.4) and (3.4.5) follow from (3.4.3). □

The following conditions will be utilized to guarantee existence of local solutions to the auxiliary problems (P_n):

$$\begin{cases} f : X \to \mathbb{R} \text{ and } g_i : X \to \mathbb{R},\ i = 1, \dots, m, \text{ are weakly lower} \\ \text{semicontinuous,} \\ e : X \to W \text{ maps weakly convergent sequences to} \\ \text{weakly convergent sequences.} \end{cases} \quad (3.4.7)$$

Further (P_n^C) denotes problem (P_n) with the additional constraint that x is contained in the closed ball B_δ. We refer to x_n as the solution to (P_n^C) if $\mathcal{L}_n(x_n) \le \mathcal{L}_n(x)$ for all $x \in B_\delta$.

Proposition 3.9. *If* (3.1.2)–(3.1.5), (3.4.1), *and* (3.4.7) *hold, then* (P_n^C) *admits a solution* x_n *for every* $n = 1, 2, \dots$. *Moreover, there exists* n_0 *such that every solution* x_n *of* (P_n^C)

satisfies $x_n \in \operatorname{int} B_\delta$ *if* $n \geq n_0$. *If* $\frac{1}{c_1 - \bar{c}}(|\lambda_0 - \lambda^*|^2 + |\mu_0 - \mu^*|^2)$ *is sufficiently small, then every solution* x_n *of* (P_n^C) *is in* $\operatorname{int} B_\delta$ *for every* $n = 1, 2, \ldots$.

Thus for n_0 or c sufficiently large the solutions to (P_n^C) are local solutions of the unconstrained problems (P_n).

Proof. Let $\{x_n^i\}_{i\in\mathbb{N}}$ be a minimizing sequence for (P_n^C). There exists a weakly convergent subsequence $\{x_n^{i_k}\}_{k\in\mathbb{N}}$ with weak limit $x_n \in B_\delta$. Condition (3.4.7) implies weak lower semicontinuity of $\mathcal{L}_n$ and hence $\mathcal{L}_n(x_n) \leq \liminf_{i_k} \mathcal{L}_n(x_n^{i_k})$ and x_n is a solution of (P_n^C).

In particular (3.4.2) holds with $\tilde{x}_n$ replaced by x_n for $n = 1, 2, \ldots$. Consequently $\lim_n x_n = x^*$ and $x_n \in \operatorname{int} B_\delta$ for all n sufficiently large. Alternatively, if $\frac{1}{c_1 - \bar{c}}(|\lambda_0 - \lambda^*|^2 + |\mu_0 - \mu^*|^2)$ is sufficiently small, then $x_n \in \operatorname{int} B_\delta$ for all $n = 1, 2, \ldots$ by (3.4.4). □

For the solutions x_n obtained in Proposition 3.9 the conclusions of Theorem 3.8 hold, in particular $\lim_n x_n = x^*$ and the associated Lagrange multipliers $\{\lambda_n\}$ and $\{\mu_n\}$ are bounded.

To investigate convergence of the Lagrange multipliers we introduce $M : X \to W \times \mathbb{R}^{m_1+m_2} \times Z$ defined by

$$M = (e'(x^*), g'_{ac}(x^*), \ell'(x^*)),$$

where g_{ac} denotes the first $m_1 + m_2$ coordinates of g. The following additional hypothesis will be needed:

There exists a constant $\kappa > 0$ such that

$$\begin{aligned} \|f'(x^*) - f'(x)\| &\leq \kappa |x^* - x|, \\ \|e'(x^*) - e'(x)\| &\leq \kappa |x^* - x|, \\ \|g'(x^*) - g'(x)\| &\leq \kappa |x^* - x| \end{aligned} \tag{3.4.8}$$

for all $x \in B_\delta$, and

$$M \text{ is surjective.} \tag{3.4.9}$$

We point out that (3.4.9) is more restrictive than the regular point condition in Definition 1.5 of Chapter 1. Below we set $\mu_{n,ac} = col(\mu_{n,1}, \ldots, \mu_{n,m_1+m_2})$ and $\mu_{ac}^* = col(\mu_1^*, \ldots, \mu_{m_1+m_2}^*)$. Without loss of generality we shall assume that $\bar{\sigma} \leq 1$.

Theorem 3.10. *Let* (3.1.2)–(3.1.5), (3.4.1), *and* (3.4.8)–(3.4.9) *hold and let* x_n *be a solution of* (P_n) *in* $\operatorname{int} B_\delta$, $\tilde{n} \geq 1$. *Then there exists a constant* K *independent of* n *such that*

$$|\lambda_n - \lambda^*| + |\mu_n - \mu^*| + |\eta_n - \eta^*| \leq \frac{K}{\sqrt{\bar{\sigma}\sigma_n}}(|\lambda_{n-1} - \lambda^*| + |\mu_{n-1} - \mu^*|) \text{ for all } n \geq 1.$$

Here η_n *denotes the Lagrange multiplier associated with the constraint* $\ell \in K$ *in* (P_n).

Sufficient conditions for $x_n \in B_\delta$ were given in Proposition 3.9 above.

Proof. Optimality of x_n and x^* for (P_n) and (3.1.1) (see (3.1.3)) imply

$$\big(f'(x_n), h\big) + \big(\lambda_{n-1} + c_n e(x_n), e'(x_n)h\big) + \big(\max(0, \mu_{n-1} + c_n g(x_n)), g'(x_n)h\big)$$
$$+\big(\eta_n, \ell'(x_n)h\big) = 0$$

and

$$\big(f'(x^*), h\big) + \big(\lambda^*, e'(x^*)h\big) + \big(\mu^*, g'(x^*)h\big) + \big(\eta^*, \ell'(x^*)h\big) = 0$$

for all $h \in X$. If we set $\tilde{\lambda}_n = \lambda_{n-1} + c_n e(x_n)$ and $\tilde{\mu}_n = \max(0, \mu_{n-1} + c_n g(x_n))$, we obtain the following equation in X:

$$e'(x^*)^*(\tilde{\lambda}_n - \lambda^*) + g'_{ac}(x^*)^*(\tilde{\mu}_{n,ac} - \mu^*_{ac}) + \ell'(x^*)^*(\eta_n - \eta^*) = f'(x^*) - f'(x_n)$$
$$+ (e'(x^*)^* - e'(x_n)^*)\tilde{\lambda}_n + (g'_{ac}(x^*)^* - g'_{ac}(x_n)^*)\mu_{n,ac}.$$

Here we used the fact that $\tilde{\mu}_{n,i} = 0$ for $i \in I_3$ by the choice of δ and $\tilde{c}$ (see the proof of Corollary 3.7), and we set $\mu_{n,ac} = \text{col}\ (\mu_{n,1}, \ldots, \mu_{n,m_1+m_2})$. Note that $M^* : W \times \mathbb{R}^{m_1+m_2} \to X$ is given by $M^* = \text{col}\ (e'(x^*)^*, g'_{ac}(x^*)^*)$. Hence we find

$$MM^*(\tilde{\lambda}_n - \lambda^*, \tilde{\mu}_{n,ac} - \mu^*_{ac}, \eta_n - \eta^*) = M\big(\tilde{f}'(x^*) - \tilde{f}'(x_n)$$
$$+ (e'(x^*)^* - e'(x_n)^*)\tilde{\lambda}_n + (g'_{ac}(x^*)^* - g'_{ac}(x_n)^*)\tilde{\mu}_{n,ac}\big). \quad (3.4.10)$$

Since $\tilde{\lambda}_n - \lambda^* = \bar{c}e(x_n)$ and $\tilde{\mu}_n - \mu_n = \bar{c}\hat{g}(x_n, \mu_{n-1}, c_n)$ we have

$$\begin{aligned} |\tilde{\lambda}_n - \lambda_n| &= \bar{c}|e(x_n) - e(x^*)|, \\ |\tilde{\mu}_{n,ac} - \mu_{n,ac}| &\le \bar{c}|g_{ac}(x_n) - g_{ac}(x^*)|. \end{aligned} \quad (3.4.11)$$

Thus by (3.4.8) and (3.4.3), (3.4.4) of Theorem 3.8 the sequences $\{\lambda_n\}$ and $\{\mu_n\}$ are uniformly bounded. Consequently by (3.4.9) and (3.4.10) there exists a constant $\bar{K}$ such that $|\tilde{\lambda}_n - \lambda^*| + |\tilde{\mu}_{n,ac} - \mu^*_{ac}| \le \bar{K}|x_n - x^*|$, and further by (3.4.11) a constant K can be chosen such that

$$|\lambda_n - \lambda^*| + |\mu^n_{ac} - \mu^*_{ac}| + |\eta_n - \eta^*| \le \bar{K}|x_n - x^*| \text{ for all } n = 1, 2, \ldots. \quad (3.4.12)$$

The choice of δ implies that $g_i(x_n) \le \frac{\mu_{n-1,i}}{c_n}$ for all $i \in I_3$ and therefore

$$\mu^n_i = \frac{\bar{c}}{c_n}\mu^{n-1}_i \text{ for } i \in I_3,\ n \ge 1. \quad (3.4.13)$$

From (3.4.3), with $\tilde{x}_n$ replaced by x_n, (3.4.12), and (3.4.13), and using $\bar{\sigma} \le 1$ the theorem follows. $\square$

Corollary 3.11. *Under the assumptions of Theorem* 3.10 *we have*

$$|\lambda_n - \lambda^*| + |\mu_n - \mu^*| + |\eta_n - \eta^*| \le \left(\frac{K}{\sqrt{\bar{\sigma}}}\right)^n \prod_{i=1}^{n} \frac{1}{\sqrt{\sigma_i}}(|\lambda_0 - \lambda^*| + |\mu_0 - \mu^*|)$$

for all $n \ge 1$.

Remark 3.4.1. If the surjectivity requirement for M is replaced by assuming that M^* is surjective, then the proof of Theorem 3.10 implies the existence of a constant K such that

$$|P_M(\lambda_n - \lambda^*, \mu_{n,ac} - \mu^*_{ac}, \eta_n - \eta^*)|_{W \times \mathbb{R}^{m_1+m_2} \times Z} \le K|x_n - x^*|,$$

where $P_M = M(M^*M)^{-1}M^*$ denotes the orthogonal projection of $W \times \mathbb{R}^{m_1+m_2} \times Z$ onto the range of M which is closed, since M^* is surjective. Since $x_n \to x^*$ in X this implies convergence of $P_M(\lambda_n, \mu_{n,ac}, \eta_n)$ to $P_M(\lambda^*, \mu^*_{ac}, \eta^*)$.

Remark 3.4.2. If a constraint of the type $x \in C$, with C a closed convex set in X, appears in (3.1.1), then Theorem 3.8 and Proposition 3.9 remain valid if Algorithm ALM is modified such that in (P_n) the functional $\mathcal{L}_n(x)$ is minimized over $x \in C$ and $\tilde{x}_n$ appearing in (3.4.2) satisfies $\tilde{x}_n \in C$. The stationarity condition (3.1.3) has to be replaced by

$$f'(x^*)(x - x^*) + \big(\lambda^*, e'(x^*)(x - x^*)\big)_W + \big(\mu^*, g'(x^*)(x - x^*)\big)_{\mathbb{R}^m} \ge 0 \qquad (3.1.3')$$

for all $x \in C$, in this case.

3.5 Application to a parameter estimation problem

We consider a least squares formulation for the estimation of the coefficient a in

$$\begin{cases} -\text{div } (a \text{ grad } y) = f \text{ in } \Omega, \\ \qquad\qquad\quad y = 0 \text{ in } \partial\Omega, \end{cases} \qquad (3.5.1)$$

where Ω is a bounded domain in $\mathbb{R}^n$, with Lipschitz continuous boundary $\partial\Omega$ if $n \ge 2$, and $f \in H^{-1}(\Omega)$, $f \ne 0$. Let Q be a Hilbert space that embeds compactly in $L^\infty(\Omega)$ and let $N : Q \to \mathbb{R}$ denote a seminorm on Q with $\ker(N) \subset \text{span}\{1\}$ and such that $N(a)$ defines a norm on $Q/\mathbb{R} = \{a \in Q : (1, a)_Q = 0\}$ that is equivalent to the norm induced by Q. For $z \in H^1_0(\Omega)$, $\alpha > 0$, and $\beta > 0$ we consider

$$\begin{cases} \min \frac{1}{2}|y - z|^2_{H^1_0(\Omega)} + \frac{\alpha}{2}N^2(a) \\ \text{subject to } e(a, y) = 0,\ a(x) \ge \beta, \end{cases} \qquad (3.5.2)$$

where $e : Q \times H^1_0(\Omega) \to H^{-1}(\Omega)$ is given by

$$e(a, y) = \text{div } (a \text{ grad } y) + f.$$

This is a special case of (3.1.1) with $X = Q \times H^1_0(\Omega)$, $W = H^{-1}(\Omega)$, $g = 0$, and f the cost functional in (3.5.2). The affine constraint $a(x) \ge \beta$ can be considered as an explicit constraint as discussed in Remark 3.4.2 by setting $C = \{a \in Q : a(x) \ge \beta\}$. Denote the solution to (3.5.1) as a function of a by $y(a)$. To guarantee the existence of a solution for (3.5.2) we require the following assumption:

$$\begin{cases} \text{either } N \text{ is a norm on } Q \\ \text{or } \inf\{|y(a) - z|_{H^1} : a \text{ is constant }\} < |z|_{H^1}. \end{cases} \qquad (3.5.3)$$

Lemma 3.12. *If* (3.5.3) *holds, then there exists a solution* (a^*, y^*) *of* (3.5.2).

Proof. Let $\{(a_n, y_n)\}_{n=1}^\infty$ denote a minimizing sequence for (3.5.2). If N is a norm, then a subsequence argument implies the existence of a solution to (3.5.2). Otherwise we decompose a_n as $a_n = a_n^1 + a_n^2$ with $a_n^2 \in Q/\mathbb{R}$ and $a_n^1 \in (Q/\mathbb{R})^\perp$. Since $\{(a_n, y_n)\}_{n=1}^\infty$ is a minimizing sequence, $\{N(a_n^2)\}_{n=1}^\infty$ and consequently $\{|a_n^2|_{L^\infty}\}_{n=1}^\infty$ are bounded. We have

$$a_n^1 \int_\Omega |\nabla y_n|^2\, dx = -\int_\Omega a_n^2 |\nabla y_n|^2\, dx + \int_\Omega f y_n\, dx.$$

Since $a \geq \beta$ implies $a^1 \geq \beta$, it follows that $\{a_n^1 |\nabla y_n|_{L^2}^2\}_{n=1}^\infty$, is bounded. If $a_n^1 \to \infty$, then $|\nabla y_n| = |\nabla y(a_n)| \to 0$ for $n \to \infty$. In this case

$$|z| \leq \lim_{n\to\infty} |y(a_n) - z|_{H^1(\Omega)}^2 + \beta N^2(a_n) \leq \inf\{|y(a) - z|_{H^1} : a = \text{constant}\} < |z|_{H^1},$$

which is impossible. Therefore $\{a_n^1\}_{n=1}^\infty$ and as a consequence $\{a_n\}_{n=1}^\infty$ are bounded, and a subsequence of $\{(a_n, y_n)\}$ converges weakly to a solution of (3.5.2). $\square$

Using Theorem 1.6 one argues the existence of Lagrange multipliers $(\lambda^*, \eta^*) \in H^{-1}(\Omega) \times Q^*$ such that the Lagrangian

$$\mathcal{L}(a, y, \lambda^*, \eta^*) = \frac{1}{2}|y - z|_{H_0^1(\Omega)}^2 + \frac{\alpha}{2} N^2(a) + \langle \lambda^*, e(a, y)\rangle_{H_0^1, H^{-1}} + \langle \eta^*, \beta - a\rangle_{Q^*, Q}$$

is stationary at (a^*, y^*), and $\langle \eta^*, \beta - a^*\rangle_{Q^*, Q} = 0$, $\langle \eta^*, h\rangle_{Q^*, Q} \geq 0$ for all $h \geq 0$. Hence (3.1.2)–(3.1.5) hold. Here (3.1.3)–(3.1.4) must be modified to include the affine inequality constraint with infinite-dimensional image space. We turn to the augmentability condition (3.4.1) and introduce the augmented Lagrangian

$$\mathcal{L}_c(a, y, \lambda^*, \eta^*) = \mathcal{L}(a, y, \lambda^*, \eta^*) + \frac{c}{2}|e(a, y)|_{H^{-1}}^2$$

for $c \geq 0$.

Henceforth we assume that (3.5.2) admits a solution (a_0, y_0) for $\alpha = 0$. Such a solution exists if z is attainable, i.e., if there exists $a_0 \in Q$, $a_0 \geq \beta$, such that $y(a_0) = y$ and in this case $y_0 = y(a_0)$. Alternatively, if the set of coefficients is further constrained by a norm bound $|a|_Q \leq \gamma$, for example, then existence of a solution to (3.5.2) with $\alpha = 0$ is also guaranteed.

Proposition 3.13. *Let* (a_0, y_0) *denote a solution to* (3.5.2) *with* $\alpha = 0$, *let* (a^*, y^*) *be a solution for* $\alpha > 0$, *and choose* $\gamma \geq |a^*|_Q$. *Then, if* $|y_0 - z|_{H^1}$ *is sufficiently small, there exist positive constants* c_0 *and* σ *such that*

$$\mathcal{L}_c(a, u; \lambda^*, \eta^*) \geq \frac{1}{2}|y^* - z|_{H^1}^2 + \frac{\alpha}{2} N^2(a^*) + \sigma(|a - a^*|_Q^2 + |y - y^*|_{H_0^1}^2)$$

for all $c \geq c_0$ *and* $(a, y) \in C \times H_0^1(\Omega)$ *with* $|a|_Q \leq \gamma$.

This proposition implies that (3.4.1) is satisfied and that Theorem 3.8 and Proposition 3.9 are applicable with $B_\delta = \{a \in Q : a(x) \geq \beta, |a|_Q \leq \gamma\} \times H_0^1(\Omega)$.

Proof. For $(a, y) \in C \times H_0^1(\Omega)$ we have

$$\mathcal{L}_c(a, y, \lambda^*, \eta^*) - \mathcal{L}_c(a^*, y^*, \lambda^*, \eta^*) = \nabla\mathcal{L}(a^*, y^*, \lambda^*, \eta^*)$$
$$- \big(h\nabla\lambda^*, \nabla v\big)_{L^2} + \frac{1}{2}|v|^2_{H_0^1} + \frac{\alpha}{2}N^2(h) + \frac{c}{2}|e(a, h)|^2_{H^{-1}},$$

where $h = a - a^*$, $v = y - y^*$, and $\nabla\mathcal{L}$ denotes the gradient of $\mathcal{L}$ with respect to (a, y). First order optimality implies that

$$\mathcal{L}_c(a, y, \lambda^*, \eta^*) - \mathcal{L}_c(a^*, y^*, \lambda^*, \eta^*) \geq -\big(h\nabla\lambda^*, \nabla v\big)_{L^2}$$
$$+ \frac{1}{2}|v|^2_{H_0^1} + \frac{\alpha}{2}N^2(h) + \frac{c}{2}|e(a, h)|^2_{H^{-1}} =: D.$$

Introduce $P = \nabla \cdot (-\Delta)^{-1}\nabla$ and note that P can be extended to an orthogonal projection on $L^2(\Omega)^n$. We find that

$$\begin{aligned}
|e(a, y)|^2_{H^{-1}} &= \big(P(a\nabla y - a^*\nabla y^*), a\nabla y - a^*\nabla y^*\big)_{L^2} \\
&= \big(P(a\nabla v + h\nabla y^*), h\nabla y^* + a\nabla v\big)_{L^2} \\
&\geq \frac{1}{2}\big(P(h\nabla y^*), h\nabla y^*\big)_{L^2} - \big(P(a\nabla v), a\nabla v\big)_{L^2} \\
&\geq \frac{1}{2}\big(P(h\nabla y^*), h\nabla y^*\big)_{L^2} - |a|^2_{L^\infty}|v|^2_{H_0^1} \\
&\geq \frac{1}{2}\big(P(h\nabla y^*), h\nabla y^*\big)_{L^2} - \gamma^2 k_1^2 |v|^2_{H_0^1},
\end{aligned}$$

where k_1 denotes the embedding constant of Q into $L^\infty(\Omega)$. Henceforth we consider the case that $\ker N = \operatorname{span}\{1\}$ and use the decomposition introduced in the proof of Lemma 3.12. We have

$$|e(a, y)|^2 \geq \frac{1}{2}|h_1|^2|y^*|^2_{H_0^1} - |h_1|\,|h_2|_{L^\infty}|y^*|^2_{H_0^1} - \gamma^2 k_1^2 |v|_{H_0^1}.$$

Since Q embeds continuously into $L^\infty(\Omega)$ and N is a norm equivalent to the norm on $Q/\mathbb{R}$, there exists a constant k_2 such that $|h_2|_{L^\infty} \leq k_2 N(h)$ for all $h = h_1 + h_2 \in Q$. Consequently

$$|e(a, y)|^2 \geq \frac{1}{2}|h_1|^2|y^*|^2_{H_0^1} - k_2|h_1|\,N(h_2)\,|y^*|^2_{H_0^1} - \gamma^2 k_1^2 |v|_{H_0^1} \tag{3.5.4}$$

for all $(a, y) \in C \times H_0^1(\Omega)$. Next note that $\mathcal{L}_y(a^*, u^*; \lambda^*, \eta^*) = 0$ implies that $-\nabla(a^*\nabla\lambda^*) = -\Delta(y^* - y)$ in $H^{-1}(\Omega)$, and hence

$$\big|\big(h\nabla\lambda^*, \nabla v\big)_{L^2}\big| \leq \frac{1}{\beta}(|h_1| + k_2 N(h_2))|y^* - z|_{H_0^1}|v|_{H_0^1}. \tag{3.5.5}$$

Setting $c = \delta\alpha$ with $\delta > 0$ and observing that $|f|_{H^{-1}} \leq k_3|y^*|_{H^1}$ for a constant k_3 depending only on β,

$$D = \frac{1}{2}(1 - \delta\alpha\gamma^2 k_1^2)|v|^2_{H_0^1} - \frac{1}{\beta}(|h_1| + k_2 N(h_2))|y^* - z||v|_{H_0^1}$$
$$+ \frac{\alpha}{2}\left(N^2(h_2) + \frac{\delta}{2}|y^*|^2_{H_0^1}|h_1|^2 - \frac{k_2}{k_3^2}\delta|h_1|N(h_2)\,|f|^2_{H^{-1}}\right).$$

There exist $\delta > 0$ and $k_4 > 0$ such that

$$D \geq k_4(|v|^2_{H^1_0} + |h_1|^2 + N^2(h_2)) - \frac{1}{\beta}(|h_1| + k_2 N(h_2))\, |y^* - z|_{H^1_0} |v|_{H^1_0}.$$

Since $|h_1| + N(h_2)$ defines an equivalent norm on Q, the claim follows with $c_0 = \delta\alpha$. □

It is worthwhile to comment on the auxiliary problems of Algorithm ALM. These involve the minimization of

$$\mathcal{L}_n(a, y) = \frac{1}{2}|y - z|^2_{H^1_0} + \frac{\alpha}{2}N^2(a) + \langle \lambda_{n-1}, A(a)y + f\rangle_{H^1_0, H^{-1}} + \frac{c_n}{2}|A(a)y + f|^2_{H^{-1}} \quad (3.5.6)$$

over $(a, y) \in C \times H^1_0(\Omega)$, where $A(a)y = \text{div}\,(a \text{ grad } y)$. The resulting optimality conditions are given by

$$\begin{aligned}&\frac{\alpha}{2}(N^2(a_n))'(a - a_n)\\ &\qquad + (-\nabla\lambda_{n-1}\, \nabla\, y_n + c_n\, \nabla\Delta^{-1}(A(a_n)y_n + f)\, \nabla y_n, a - a_n)_{L^n} \geq 0 \quad (3.5.7)\end{aligned}$$

for all $a \geq a_n$ and

$$-\Delta(y_n - z) + A(a_n)\lambda_{n-1} + A(a_n)(-\Delta)^{-1}(A(a_n)y_n + f) = 0. \quad (3.5.8)$$

The Lagrange multiplier is updated according to

$$\lambda_n = \lambda_{n-1} + \sigma_n(-\Delta)^{-1}(A(a_n)y_n + f). \quad (3.5.9)$$

To simplify (3.5.7), (3.5.8) for numerical realization one can proceed iteratively solving (3.5.7) with y_n replaced by y_{n-1} for a_n and then (3.5.8) for y_n. A good initialization for the variable λ is given by $\lambda_0 = 0$. In fact, for small residue problems $|y^* - z|_{H^1_0}$ is small and then $\frac{\partial}{\partial y}\mathcal{L}(a^*, y^*, \lambda^*, \eta^*) = 0$ implies that λ^* is small. Setting $\lambda_0 = 0$ and $y_0 = z$ the coefficient a_1 is determined from

$$\min_{a \geq \beta} \frac{\alpha}{2}N^2(a) + \frac{C_1}{2}|A(a)z + f|^2_{H^{-1}}. \quad (3.5.10)$$

Thus the first step of the augmented Lagrangian algorithm for estimating a in (3.5.1) involves a regularized equation error step. Recall that the equation error method for parameter estimation problems consists in solving the hyperbolic equation $-\text{div}\,(a \text{ grad } z) = f$ for a. This estimate is improved during the subsequent iterations. The first order augmented Lagrangian method can therefore be considered as a hybrid method combining the output least squares and the equation error approach to parameter estimation.

Let us close this section with some brief comments. If the H^1 least squares criterion is replaced by an L^2-criterion, then the first negative Laplacian in (3.5.8) must be replaced by the identity operator. For the H^1-criterion the same number of differentiations are applied to y in the least squares and the equation error term in (3.5.8). If one would approach the problem of estimating a in (3.5.1) by first discretizing and then applying an optimization strategy, the discrete analogue of $(-\Delta)^{-1}$ can be interpreted as preconditioning. We assumed that Q embeds compactly in $L^\infty(\Omega)$. This is the case, for example, if $Q = H^1(\Omega)$ when $n = 1$, or $Q = H^2(\Omega)$ when $n = 2$ or 3, or if $Q \subset L^\infty(\Omega)$ is finite-dimensional.

Chapter 4

Augmented Lagrangian Methods for Nonsmooth, Convex Optimization

4.1 Introduction

The class of optimization problems that motivates this chapter is given by

$$\min f(x) + \varphi(\Lambda x) \quad \text{over } x \in C, \tag{4.1.1}$$

where X, H are real Hilbert spaces, C is a closed convex subset of X, and $\Lambda \in \mathcal{L}(X, H)$. In this chapter we identify H with its dual space and we distinguish between X and its dual X^*. Further $f : X \to \mathbb{R}$ is a continuously differentiable, convex function, and $\varphi : H \to (-\infty, \infty]$ is a proper, lower semicontinuous, convex but not necessarily differentiable function. This problem class encompasses a wide variety of optimization problems including variational inequalities of the first and second kind [Glo]. Our formulation here is slightly more general than the one in [Glo, EkTe], since we allow the additional constraint $x \in C$.

For example, one can formulate regularized inverse scattering problems in the form (4.1.1):

$$\min \int_\Omega \frac{\mu}{2} |\nabla u|^2 + g\, |\nabla u|\, dx + \frac{1}{2} \int_\Omega \left| \int_\Omega k(x, y) u(y)\, dy - z(x) \right|^2 dx \tag{4.1.2}$$

over $u \in H^1(\Omega)$ and $u \geq 0$, where $k(x, y)$ denotes a scattering kernel. The problem consists in recovering the original image u defined on a domain Ω from scattered and noisy data $z \in L^2(\Omega)$. Here $\mu > 0$ and $g > 0$ are fixed and should be adjusted to the statistics of the noise. If $\mu = 0$, then this problem is equivalent to the image enhancement algorithm in [ROF] based on minimization of the BV-seminorm $\int_\Omega |\nabla u|\, ds$. In this example $\varphi(v) = g \int_\Omega |v|\, ds$, which is nondifferentiable. Several additional applications are treated at the end of this chapter.

We develop a Lagrange multiplier theory to deal with the nonsmoothness of φ. To briefly explain the approach let $x, \lambda \in H$ and $c > 0$, and define the family of generalized Yosida–Moreau approximations $\varphi_c(x, \lambda)$ by

$$\varphi_c(x, \lambda) = \inf_{u \in H} \left\{ \varphi(x - u) + (\lambda, u)_H + \frac{c}{2} |u|_H^2 \right\}. \tag{4.1.3}$$

This is equivalent to an augmented Lagrangian approach. In fact, note that (4.1.1) is equivalent to

$$\begin{aligned} &\min f(x) + \varphi(\Lambda x - u) \\ &\text{subject to } x \in C \quad \text{and} \quad u = 0 \text{ in } H. \end{aligned} \tag{4.1.4}$$

Treating the equality constraint $u = 0$ in (4.1.4) by the augmented Lagrangian method results in the minimization problem

$$\min_{x \in C, u \in H} f(x) + \varphi(\Lambda x - u) + (\lambda, u)_H + \frac{c}{2} |u|_H^2, \tag{4.1.5}$$

where $\lambda \in H$ is a multiplier and c is a positive scalar penalty parameter. Equivalently, problem (4.1.5) is written as

$$\min_{x \in C} L_c(x, \lambda) = f(x) + \varphi_c(\Lambda x, \lambda). \tag{4.1.6}$$

It will be shown that $\varphi_c(u, \lambda)$ is continuously Fréchet differentiable with respect to $u \in H$. Moreover, if $x_c \in C$ denotes the solution to (4.1.6), then it satisfies

$$\begin{aligned} &\langle f'(x_c) + \Lambda^* \lambda_c, x - x_c \rangle_{X^*, X} \geq 0 \quad \text{for all } x \in C, \\ &\lambda_c = \varphi_c'(\Lambda x_c, \lambda_c). \end{aligned}$$

It will further be shown that under appropriate conditions the pair $(x_c, \lambda_c) \in C \times H$ has a (strong-weak) cluster point $(\bar{x}, \bar{\lambda})$ as $c \to \infty$ such that $\bar{x} \in C$ is the minimizer of (4.1.1) and that $\bar{\lambda} \in H$ is a Lagrange multiplier in the sense that

$$\langle f'(\bar{x}) + \Lambda^* \bar{\lambda}, x - \bar{x} \rangle_{X^*, X} \geq 0 \quad \text{for all } x \in C \tag{4.1.7}$$

with the complementarity condition

$$\bar{\lambda} = \varphi_c'(\Lambda \bar{x}, \bar{\lambda}) \quad \text{for each } c > 0. \tag{4.1.8}$$

System (4.1.7)–(4.1.8) for the pair $(\bar{x}, \bar{\lambda})$ is a necessary and sufficient optimality condition for problem (4.1.1). We analyze iterative algorithms of Uzawa and augmented Lagrangian type for finding the optimal pair $(\bar{x}, \bar{\lambda})$ and present a convergence analysis. It will be shown that condition (4.1.8) is equivalent to the complementarity condition $\bar{\lambda} \in \partial \varphi(\Lambda \bar{x})$. Thus the frequently employed differential inclusion $\bar{\lambda} \in \partial \varphi(\Lambda \bar{x})$ is replaced by the nonlinear equation (4.1.8).

This chapter is organized as follows. In Sections 4.2–4.3 we present the basic convex analysis and duality theory in Banach spaces. Section 4.4 is devoted to the generalized Yosida–Moreau approximation which is the basis for the remainder of the chapter. Conditions for the existence of Lagrange multiplier for (4.1.1) and optimality systems are derived in Section 4.5. Section 4.6 is devoted to the augmented Lagrangian algorithm. Convergence of both the augmented Lagrangian and the Uzawa methods are proved. Section 4.7 contains a large number of concrete applications.

4.2 Convex analysis

In this section we present standard results from convex analysis and duality theory in Banach spaces following, in part, [BaPe, EkTu]. Throughout X denotes a real Banach space.

Definition 4.1. (1) *A functional* $F : X \to (-\infty, \infty]$ *is called convex if*

$$F((1-\lambda)\,x_1 + \lambda\,x_2) \le (1-\lambda)\,F(x_1) + \lambda\,F(x_2)$$

for all $x_1,\ x_2 \in X$ *and* $0 \le \lambda \le 1$. *It is called proper if it is not identically* ∞.

(2) *A functional* $F : X \to (-\infty, \infty]$ *is said to be lower semicontinuous (l.s.c.) at* $x \in X$ *if*

$$F(x) \le \liminf_{y \to x} F(y).$$

A functional F *is l.s.c. if it is l.s.c. at all* $x \in X$.

(3) *A functional* $F : X \to (-\infty, \infty]$ *is called weakly lower semicontinuous (w.l.s.c.) at* x *if*

$$F(x) \le \liminf_{n \to \infty} F(x_n)$$

for all sequences $\{x_n\}$ *converging weakly to* x. *Further* F *is w.l.s.c. if it is w.l.s.c. at all* $x \in X$.

(4) *The subset* $D(F) = \{x \in X : F(x) < \infty\}$ *of* X *is called the effective domain of* F.

(5) *The epigraph of* F *is defined by* $\text{epi}(F) = \{(x, c) \in X \times \mathbb{R} : F(x) \le c\}$.

Lemma 4.2. *A functional* $F : X \to (-\infty, \infty]$ *is l.s.c. if and only if its epigraph is closed.*

Proof. The lemma follows from the fact that epi$\,(F)$ is closed if and only if $(x_n, c_n) \in \text{epi}(F) \to (x, c)$ in $X \times \mathbb{R}$ implies $F(x) \le c$. $\square$

Lemma 4.3. *A functional* $F : X \to (-\infty, \infty]$ *is l.s.c. if and only if its level sets* $S_c = \{x \in X : F(x) \le c\}$ *are closed for all* $c \in \mathbb{R}$.

Proof. Assume that F is l.s.c. Let $c > 0$ and let $\{x_n\}$ be a sequence in S_c with limit x. Then $F(x) \le \liminf_{x \to \infty} F(x_n) \le c$ and hence $x \in S_c$ and S_c is closed. Conversely, assume that S_c is closed for all $c > 0$, and let $\{x_n\}$ be a sequence converging to x in X. Choose a subsequence $\{x_{n_k}\}$ with

$$\lim_{k \to \infty} F(x_{n_k}) = \liminf_{n \to \infty} F(x_n),$$

and suppose that $F(x) > \liminf_{n \to \infty} F(x_n)$. Then there exists $\bar{c} \in \mathbb{R}$ such that

$$\liminf_{n \to \infty} F(x_n) < \bar{c} < F(x),$$

and there exists an index m such that $F(x_{n_k}) < \bar{c}$ for all $k \ge m$. Since $S_{\bar{c}}$ is closed $x \in S_{\bar{c}}$, which is a contradiction to $\bar{c} < F(x)$. $\square$

Lemma 4.4. *Assume that all level sets of the functional* $F : X \to (-\infty, \infty]$ *are convex. Then* F *is l.s.c. if and only if it is w.l.s.c. on* X.

Proof. Assume that F is l.s.c. Suppose that $\{x_n\}$ converges weakly to $\bar{x}$ in X, and let $\{x_{n_k}\}$ be a subsequence such that $d = \liminf F(x_n) = \lim_{k\to\infty} F(x_{n_k})$. By Lemma 4.3 the sets $\{x : F(x) \le d + \epsilon\}$ are closed for every $\epsilon > 0$. By assumption they are also convex. Hence by Mazur's lemma they are also weakly (sequentially) closed. Hence $F(\bar{x}) \le d + \epsilon$ for every $\epsilon > 0$. Since ϵ is arbitrary, we have that $F(\bar{x}) \le d$ and F is w.l.s.c. The converse implication is obvious.

Note that for a convex function all associated level sets are convex, but not vice versa. □

Theorem 4.5. *Let F be a proper, l.s.c., convex functional on X. Then F is bounded below by an affine functional; i.e., there exist $x^* \in X^*$ and $c \in \mathbb{R}$ such that*

$$F(x) \ge \langle x^*, x\rangle_{X^*,X} + c \quad \textit{for all } x \in X.$$

Moreover, F is the pointwise supremum of a family of such continuous affine functionals.

Proof. Let $x_0 \in X$ and choose $\beta \in \mathbb{R}$ such that $F(x_0) > \beta$. Since $\text{epi}(F)$ is a closed convex subset of the product space $X \times \mathbb{R}$, it follows from the separation theorem for convex sets [EkTu] that there exists a closed hyperplane $H \subset X \times \mathbb{R}$ given by

$$H = \{(x, r) \in X \times \mathbb{R} : \langle x_0^*, x\rangle + ar = \alpha\} \quad \text{with } x_0^* \in X^*, \ a \in \mathbb{R}, \ \alpha \in \mathbb{R},$$

such that

$$\langle x_0^*, x_0\rangle + a\beta < \alpha < \langle x_0^*, x\rangle + ar \quad \text{for all } (x, r) \in \text{epi}(F).$$

Setting $x = x_0$ and $r = F(x_0)$, we have

$$\langle x_0^*, x_0\rangle + a\beta < \alpha < \langle x_0^*, x_0\rangle + aF(x_0)$$

and thus $a\,(F(x_0) - \beta) > 0$. If $F(x_0) < \infty$, then $a > 0$ and thus

$$\frac{\alpha}{a} - \frac{1}{a}\langle x_0^*, x\rangle \le r \quad \text{for all } (x, r) \in \text{epi}(F),$$

$$\beta < \frac{\alpha}{a} - \frac{1}{a}\langle x_0^*, x_0\rangle < F(x_0).$$

Hence, $b(x) = \frac{\alpha}{a} - \frac{1}{a}\langle x_0^*, x\rangle$ is a continuous affine function on X such that $b \le F$ and the first claim is established with $c = \frac{\alpha}{a}$ and $x^* = \frac{-x_0^*}{a}$. Moreover $\beta < b(x_0) < F(x_0)$. Therefore

$$F(x_0) = \sup \big\{b(x_0) = \langle x^*, x_0\rangle + c : \\ x^* \in X^*, c \in \mathbb{R}, b(x) \le F(x) \text{ for all } x \in X\big\}. \tag{4.2.1}$$

If $F(x_0) = \infty$, either $a > 0$ (thus we proceed as above) or $a = 0$. In the latter case $\alpha - \langle x_0^*, x_0\rangle > 0$ and $\alpha - \langle x_0^*, x\rangle < 0$ on $D(F)$. Since F is proper there exists an affine function $b(x) = \langle x^*, x\rangle + c$ such that $b \le F$. Thus

$$\langle x^*, x\rangle + c + \theta\,(\alpha - \langle x_0^*, x\rangle) < F(x)$$

for all x and $\theta > 0$. Choosing $\theta > 0$ large enough so that

$$\langle x^*, x\rangle + c + \theta\,(\alpha - \langle x_0^*, x\rangle) > \beta,$$

we have that $b(x) = \langle x^*, x\rangle + c + \theta\,(\alpha - \langle x_0^*, x\rangle)$ is a continuous affine function on X with $b \le F$ and $\beta < b(x_0)$. Therefore (4.2.1) holds at x_0 with $F(x_0) = \infty$ as well. □

Theorem 4.6. *If $F : X \to (-\infty, \infty]$ is convex and bounded above on an open set U, then F is continuous on U.*

Proof. We choose $M \in \mathbb{R}$ such that $F(x) \le M - 1$ for all $x \in U$. Let $\hat{x}$ be any element in U. Since U is open there exists a $\delta > 0$ such that the open ball $\{x \in X : |x - \hat{x}| < \delta\}$ is contained in U. For any $\epsilon \in (0, 1)$, let $\theta = \frac{\epsilon}{M - F(\hat{x})}$. Then for $x \in X$ satisfying $|x - \hat{x}| < \theta\,\delta$ we have

$$\left|\frac{x - \hat{x}}{\theta} + \hat{x} - \hat{x}\right| = \frac{|x - \hat{x}|}{\theta} < \delta.$$

Hence $\frac{x-\hat{x}}{\theta} + \hat{x} \in U$. By convexity of F

$$F(x) \le (1-\theta)F(\hat{x}) + \theta\, F\left(\frac{x - \hat{x}}{\theta} + \hat{x}\right) < (1-\theta)F(\hat{x}) + \theta\, M,$$

and thus

$$F(x) - F(\hat{x}) < \theta(M - F(\hat{x})) = \epsilon.$$

Similarly, $\frac{\hat{x}-x}{\theta} + \hat{x} \in U$ and

$$F(\hat{x}) \le \frac{\theta}{1+\theta}\, F\left(\frac{\hat{x} - x}{\theta} + \hat{x}\right) + \frac{1}{1+\theta}\, F(x) < \frac{\theta M}{1+\theta} + \frac{1}{1+\theta}\, F(x),$$

which implies

$$F(x) - F(\hat{x}) > -\theta(M - F(\hat{x})) = -\epsilon.$$

Therefore $|F(x) - F(\hat{x})| < \epsilon$ if $|x - \hat{x}| < \theta\delta$ and F is continuous in U. □

Theorem 4.7. *If $F\colon X \to (-\infty, \infty]$ is a proper, l.s.c., convex functional on X, and F is bounded above on a convex, open δ-neighborhood $\mathcal{U}$ of a bounded, convex set C, then F is Lipschitz continuous on C.*

Proof. By Theorem 4.5 and by assumption there exist constants M and m such that

$$m \le F(x) \le M \qquad \text{for all } x \in \mathcal{U}.$$

Let x and $\hat{x}$ be in C with $|x - \hat{x}| \le \frac{\delta}{M-m}$, $x \ne \hat{x}$, and set $\theta = \frac{2|x-\hat{x}|}{\theta}$. Without loss of generality we assume that $\frac{2}{M-m} < 1$ so that $\theta \in (0, 1)$. Then $y = \frac{x-\hat{x}}{\theta} + \hat{x} \in \mathcal{U}$ since $\hat{x} \in \mathcal{U}$ and $|y - \hat{x}| = \frac{x-\hat{x}|}{\theta} \le \frac{\delta}{2}$. Due to convexity of F we have

$$F(x) \le (1-\theta)F(\hat{x}) + \theta\left(F\left(\frac{x - \hat{x}}{\theta} + \hat{x}\right)\right) \le (1-\theta)F(\hat{x}) + \theta\, M$$

and hence

$$F(x) - F(\hat{x}) \le \theta(M - F(\hat{x})) \le \frac{2}{\delta}(M - m)|x - \hat{x}|.$$

Similarly, $\frac{\hat{x}-x}{\theta} + \hat{x} \in \mathcal{U}$ and

$$F(\hat{x}) \le \frac{\theta}{1+\theta} F\left(\frac{\hat{x}-x}{\theta} + \hat{x}\right) + \frac{1}{1+\theta} F(x) \le \frac{\theta M}{1+\theta} + \frac{1}{1+\theta} F(x),$$

which implies

$$-\theta(M - m) \le -\theta(M - F(\hat{x})) \le F(x) - F(\hat{x})$$

and therefore

$$|F(x) - F(\hat{x})| \le \frac{2}{\theta}(M - m)|x - \hat{x}| \text{ for } |x - \hat{x}| \le \frac{\delta}{M - m}.$$

Since C is bounded and convex, Lipschitz continuity for all $x, \hat{x} \in C$ follows. □

4.2.1 Conjugate and biconjugate functionals

Definition 4.8. *The functional* $F^* : X^* \to [-\infty, \infty]$ *defined by*

$$F^*(x^*) = \sup\{\langle x^*, x\rangle_{X^*,X} - F(x) : \ x \in X\}$$

is called the conjugate of F.

If F is bounded below by an affine functional $\langle x^*, \cdot\rangle - c$, then

$$F^*(x^*) = \sup_{x \in X}\{\langle x^*, x\rangle - F(x)\} \le \sup_{x \in X}\{\langle x^*, x\rangle - \langle x^*, x\rangle + c\} = c,$$

and hence F^* is not identically ∞. Note also that $D(F) = \emptyset$ implies that $F^* \equiv -\infty$ and conversely, if $F^*(x^*) = -\infty$ for some $x^* \in X$, then $D(F)$ is empty.

Example 4.9. If $F(x) = \frac{1}{p}|x|^p$, $x \in \mathbb{R}$, then $F^*(x^*) = \frac{1}{q}|x^*|^q$ for $1 < p < \infty$ and $\frac{1}{p} + \frac{1}{q} = 1$.

Example 4.10. If $F(x)$ is the indicator function of the closed unit ball of X, i.e., $F(x) = 0$ for $|x| \le 1$ and $F(x) = \infty$ otherwise, then $F^*(x^*) = |x^*|$. In fact, $F^*(x^*) = \sup\{\langle x^*, x\rangle : |x| \le 1\} = |x^*|$.

If F is a proper, l.s.c., convex functional on X, then by Theorem 4.5

$$F(x_0) = \sup\{\langle x^*, x_0\rangle - c : x^* \in X^*, c \in \mathbb{R}, \langle x^*, x\rangle - c \le F(x) \text{ for all } x \in X\}$$

for every $x_0 \in X$. For $x^* \in X$ define

$$c(x^*) = \inf\{c \in \mathbb{R} : c \ge \langle x^*, x\rangle - F(x) \text{ for all } x \in X\}$$

and observe that

$$F(x_0) = \sup_{x^* \in X^*} \{\langle x^*, x_0 \rangle - c(x^*)\}.$$

But $c(x^*) = \sup_{x \in X}\{\langle x^*, x \rangle - F(x)\} = F^*(x^*)$ and hence

$$F(x_0) = \sup_{x^* \in X^*} \{\langle x^*, x_0 \rangle - F^*(x^*)\}.$$

This suggests the following definition.

Definition 4.11. *For $F : X \to (-\infty, \infty]$ the biconjugate functional $F^{**} : X \to (-\infty, \infty]$ is defined by*

$$F^{**}(x) = \sup_{x^* \in X^*} \{\langle x^*, x \rangle - F^*(x^*)\}.$$

Above we proved the following result.

Theorem 4.12. *If F is a proper, l.s.c., convex functional, then $F = F^{**}$.*

It is simple to argue that $(F^*)^* = F^{**}$ if X is reflexive. Next we prove some important results on conjugate functionals.

Theorem 4.13. *For every functional $F : X \to (-\infty, \infty]$ the conjugate F^* is convex and l.s.c. on X^*.*

Proof. If F is identically ∞, then F^* is identically $-\infty$. Otherwise $D(F)$ is nonempty and $F(x^*) > -\infty$ for all $x^* \in X^*$, and

$$F^*(x^*) = \sup \{\langle x^*, x \rangle - F(x) : \quad x \in D(F)\}.$$

Thus F^* is the pointwise supremum of the family of continuous affine functionals $x^* \to \langle x^*, x \rangle - F(x)$ for $x \in D(F)$. This implies that F^* is convex. Moreover $\{x^* \in X^* : \langle x^*, x \rangle - F(x) \le c\}$ is closed for each $c \in \mathbb{R}$ and $x \in D(F)$. Consequently

$$\{x^* \in X^* : F^*(x^*) \le c\} = \bigcap_{x \in D(F)} \{x^* \in X^* : \langle x^*, x \rangle - F(x) \le c\}$$

is closed for each $c \in \mathbb{R}$. Hence F^* is l.s.c. by Lemma 4.3. $\square$

Theorem 4.14. *For every $F : X \to (-\infty, \infty]$ the biconjugate F^{**} is convex, l.s.c., and $F^{**} \le F$. Moreover for each $\bar{x} \in X$*

$$F^{**}(\bar{x}) = \sup \{\langle x^*, \bar{x} \rangle - c : \quad x^* \in X^*, \ c \in \mathbb{R}, \ \langle x^*, x \rangle - c \le F(x) \textit{ for all } x \in X\}. \quad (4.2.2)$$

Proof. If there is no continuous affine functional which is everywhere less than F, then $F^* \equiv \infty$ and hence $F^{**} \equiv -\infty$. In fact, if there exist $c \in \mathbb{R}$ and $x^* \in X^*$ with $c \ge F^*(x^*)$, then $\langle x^*, \cdot \rangle - c$ is everywhere less than F, which is impossible. The claims readily follow in this case.

Otherwise, assume that there exists a continuous affine functional everywhere less than F. Then $D(F^*)$ is nonempty and $F^{**}(x) > -\infty$ for all $x \in X$. If $F^*(x^*) = -\infty$ for some $x^* \in X^*$, then $F^{**} \equiv \infty$, $F \equiv \infty$, and the claims follow. In the remaining case $F^*(x^*)$ is finite for all x^* and we have

$$F^{**}(x) = \sup\{\langle x^*, x\rangle - F^*(x^*) : x^* \in X^*,\, F^*(x^*) \text{ finite}\}.$$

For every $x^* \in D(F^*)$ we have

$$\langle x^*, x\rangle - F^*(x^*) \le F(x) \text{ for all } x \in X,$$

hence $F^{**} \le F$ and

$$F^{**}(\bar{x}) \le \sup\{\langle x^*, \bar{x}\rangle - c :\; x^* \in X^*,\; c \in \mathbb{R},\; \langle x^*, x\rangle - c \le F(x) \text{ for all } x \in X\}.$$

Let x^* and c be such that $\langle x^*, x\rangle - c \le F(x)$ for all $x \in X$. Then

$$\langle x^*, x\rangle - c \le \langle x^*, x\rangle - F(x^*) \text{ for all } x \in X.$$

Therefore

$$\sup\{\langle x^*, \bar{x}\rangle - c :\; x^* \in X^*,\; c \in \mathbb{R},\; \langle x^*, x\rangle - c \le F(x) \text{ for all } x \in X\} \le F^{**}(\bar{x}),$$

and (4.2.2) follows. Moreover F^{**} is the pointwise supremum of a family of continuous affine functionals, and it follows as in the proof of Theorem 4.13 that F^* is convex and l.s.c. ☐

Theorem 4.15. *If $F : X \to (-\infty, \infty]$ is a convex functional which is finite and l.s.c. at x, then $F(x) = F^{**}(x)$.*

For the proof of this result we refer the reader to [EkTu, p. 104], for example.

Theorem 4.16. *For every $F : X \to (-\infty, \infty]$, we have $F^* = F^{***}$.*

Proof. Since $F^{**} \le F$ due to Theorem 4.14, $F^* \le F^{***}$ by the definition of conjugate functions. The definition of F^{**} implies that

$$\langle x^*, x\rangle - F^{**}(x) \le F(x^*) \tag{4.2.3}$$

if $F^{**}(x)$ and $F^*(x^*)$ are finite. If $F^*(x^*) = \infty$ or $F^{**}(x) = \infty$, inequality (4.2.3) holds as well. If $F^*(x^*) = -\infty$, we have that F and F^* are identically ∞. If $F^{**}(x) = -\infty$, then F^* is identically ∞, and (4.2.3) holds for all $(x, x^*) \in X \times X^*$. Thus,

$$F^{***}(x) = \sup_{x \in X}\{\langle x^*, x\rangle - F^{**}(x)\} \le F(x^*)$$

for all $x^* \in X^*$. Hence $F^{***} = F^*$. ☐

Theorem 4.17. *Let $F : X \to (-\infty, \infty]$ and assume that $\partial F(x) \ne \emptyset$ for some $x \in X$. Then $F(x) = F^{**}(x)$.*

Proof. Since $\partial F(x) \ne \emptyset$ there exists a continuous affine functional $\ell \le F$ with $\ell(x) = F(x)$. Due to Theorem 4.14 we have $\ell \le F^{**} \le F$. It follows that $\ell(x) = F^{**}(x) = F(x)$, as desired. ☐

4.2.2 Subdifferential

In this short section we summarize some important results on subdifferentials of functionals $F : X \to (-\infty, \infty]$. While F is not assumed to be convex, the applications that we have in mind are to convex functionals.

Definition 4.18. *Let $F : X \to (-\infty, \infty]$. The subdifferential of F at x is the (possibly empty) set*

$$\partial F(x) = \{x^* \in X^* : F(y) - F(x) \geq \langle x^*, y - x\rangle \textit{ for all } \; y \in X\}.$$

If $\partial F(x)$ is nonempty, then F is called subdifferentiable at x and the set of all points where F is subdifferentiable is denoted by $D(\partial F)$.

As a first observation, we note that $x_0 = \operatorname{argmin}_{x \in X} F(x)$ if and only if $0 \in \partial F(x_0)$. In fact, these two statements are equivalent to

$$F(x) \geq F(x_0) + (0, \; x - x_0) \quad \text{for all } x \in X.$$

Example 4.19. Let F be Gâteaux differentiable at x, i.e., there exists $w^* \in X^*$ such that

$$\lim_{t \to 0^+} \frac{F(x + t\,v) - F(x)}{t} = \langle w^*, v\rangle \quad \text{for all } \; v \in X,$$

and $w^* \in X^*$ is called the Gâteaux derivative of F at x. It is denoted by $F'(x)$. If in addition F is convex, then F is subdifferentiable at x and $\partial F(x) = \{F'(x)\}$. Indeed, for $v = y - x$

$$\frac{F(x + t\,(y - x)) - F(x)}{t} \leq F(y) - F(x), \qquad \text{where } 0 < t < 1.$$

As $t \to 0^+$ we obtain

$$\langle F'(x), y - x\rangle \leq F(y) - F(x) \quad \text{for all } \; y \in X,$$

and thus $F'(x) \in \partial F(x)$. On the other hand, if $w^* \in \partial F(x)$, we find for $y \in X$ and $t > 0$

$$\frac{F(x + t\,y) - F(x)}{t} \geq \langle w^*, y\rangle.$$

Taking the limit $t \to 0^+$, we obtain

$$\langle F'(x) - w^*, y\rangle \geq 0 \quad \text{for all } \; y \in X.$$

This implies that $w^* = F'(x)$.

Example 4.20. For $F(x) = \frac{1}{2}\,|x|^2$, $x \in X$, we will show that $\partial F(x) = \mathcal{F}(x)$, where $\mathcal{F} : X \to X^*$ denotes the duality mapping. In fact, if $x^* \in \mathcal{F}(x)$, then

$$\langle x^*, x - y\rangle = |x|^2 - \langle x^*, y\rangle \geq \frac{1}{2}\,(|x|^2 - |y|^2) \quad \text{for all } \; y \in X.$$

Thus $x^* \in \partial\varphi(x)$. Conversely, if $x^* \in \partial\varphi(x)$, then

$$\frac{1}{2}\,(|y|^2 - |x|^2) \geq \langle x^*, y - x\rangle \quad \text{for all} \quad y \in X. \tag{4.2.4}$$

We let $y = t\,x,\ 0 < t < 1$, and obtain

$$\frac{1+t}{2}\,|x|^2 \leq \langle x^*, x\rangle$$

and thus $|x|^2 \leq \langle x^*, x\rangle$. Similarly, if $t > 1$, then $|x|^2 \geq \langle x^*, x\rangle$ and therefore $|x|^2 = \langle x^*, x\rangle$ and $|x^*| \geq |x|$. On the other hand, setting $y = x + t\,u,\ t > 0$, in (4.2.4), we have

$$t\,\langle x^*, u\rangle \leq \frac{1}{2}\,(|x + t\,u|^2 - |x|^2) \leq t\,|u||x| + \frac{t^2}{2}\,|u|^2,$$

which implies $\langle x^*, u\rangle \leq |u||x|$. Hence $|x^*| \leq |x|$ and we obtain $|x|^2 = |x^*|^2 = \langle x^*, x\rangle$

Example 4.21. Let K be a closed convex subset of X and let ψ_K be the indicator function of K, i.e.,

$$\psi_K(x) = \begin{cases} 0 & \text{if } x \in K, \\ \infty & \text{otherwise.} \end{cases}$$

Obviously, ψ_K is convex and l.s.c. on X. By definition we have for $x \in K$

$$\partial\psi_K(x) = \{x^* \in X^* : \langle x^*, y - x\rangle \leq 0 \quad \text{for all} \quad y \in K\}$$

and $\partial\psi_K(x) = \emptyset$ if $x \notin X$. Thus $D(\psi_K) = D(\partial\psi_K) = K$ and $\partial\psi_K(x) = \{0\}$ for each interior point of K. Moreover, for $x \in K$, $\partial\psi_K(x)$ coincides with the definition of the normal cone to K at x.

Theorem 4.22. *Let $F : X \to (-\infty, \infty]$.*

(1) *We have*

$$x^* \in \partial F(\bar{x}) \textit{ if and only if } F(\bar{x}) + F^*(x^*) = \langle x^*, \bar{x}\rangle.$$

(2) *Assume that X is reflexive. Then $x^* \in \partial F(\bar{x})$ implies that $\bar{x} \in \partial F^*(x^*)$. If moreover F is convex and l.s.c., then $x^* \in \partial F(\bar{x})$ if and only if $\bar{x} \in \partial F^*(x^*)$.*

Proof. (1) Note that $x^* \in \partial F(\bar{x})$ if and only if

$$\langle x^*, x\rangle - F(x) \leq \langle x^*, \bar{x}\rangle - F(\bar{x}) \tag{4.2.5}$$

for all $x \in X$. By the definition of F^* this implies $F^*(x^*) = \langle x^*, \bar{x}\rangle - F(\bar{x})$. Conversely, if $F(x^*) + F(\bar{x}) = \langle x^*, \bar{x}\rangle$, then (4.2.5) holds for all $x \in X$.

(2) Since $F^{**} \leq F$ by Theorem 4.14, it follows from (1) that

$$F^{**}(\bar{x}) \leq \langle x^*, \bar{x}\rangle - F^*(x^*).$$

By the definition of F^{**},

$$F^{**}(\bar{x}) \geq \langle x^*, \bar{x}\rangle - F^*(x^*).$$

Hence

$$F^*(x^*) + F^{**}(\bar{x}) = \langle x^*, \bar{x}\rangle.$$

Since X is reflexive, $F^{**} = (F^*)^*$. Applying (1) to F^* we have $\bar{x} \in \partial F^*(x^*)$. If in addition F is convex and l.s.c., it follows from Theorem 4.12 that $F^{**} = F$. Thus if $\bar{x} \in \partial F^*(x^*)$, then

$$F(\bar{x}) + F^*(x^*) = F^*(x^*) + F^{**}(\bar{x}) = \langle x^*, \bar{x}\rangle$$

by applying (1) to F^*. Therefore $x^* \in \partial F(\bar{x})$ again by (1). □

Proposition 4.23. *For $F : X \to (-\infty, \infty]$ the set $\partial F(x)$ is closed and convex for every $x \in X$.*

Proof. If $F(x) = \infty$, then $\partial F(x) = \emptyset$. Henceforth let $F(x) < \infty$. For every $x^* \in X^*$ we have $F^*(x^*) \geq \langle x^*, x\rangle - F(x)$ and hence by Theorem 4.22

$$\partial F(x) = \{x^* \in X^* : F^*(x^*) - \langle x^*, x\rangle \leq -F(x)\}.$$

By Theorem 4.13 the functional $x^* \to F^*(x^*) - \langle x^*, x\rangle$ is convex and l.s.c. The claim follows from Lemma 4.3. □

Theorem 4.24. *If the convex function F is continuous at $\bar{x}$, then $\partial F(\bar{x})$ is not empty.*

Proof. Since F is continuous at x, there exists for every $\epsilon > 0$ an open neighborhood U_ϵ of $\bar{x}$ such that

$$F(x) \leq F(\bar{x}) + \epsilon, \quad x \in U_\epsilon.$$

Then $U_\epsilon \times (F(\bar{x})+\epsilon, \infty)$ is an open set in $X \times \mathbb{R}$, contained in epi F. Hence (epi $F)^o$, the relative interior of epi F, is nonempty. Since F is convex, epi F is convex and (epi $F)^o$ is convex. Note that $(\bar{x}, F(\bar{x}))$ is a boundary point of epi F. Hence by the Hahn–Banach separation theorem, there exists a closed hyperplane $S = \{(x, a) \in X \times \mathbb{R} : \langle x^*, x\rangle + \alpha\, a = \beta\}$ for nontrivial $(x^*, \alpha) \in X^* \times \mathbb{R}$ and $\beta \in \mathbb{R}$ such that

$$\begin{aligned} &\langle x^*, x\rangle + \alpha\, a > \beta \quad \text{for all } (x, a) \in (\text{epi } F)^o, \\ &\langle x^*, \bar{x}\rangle + \alpha\, F(\bar{x}) = \beta. \end{aligned} \tag{4.2.6}$$

Since $\overline{(\text{epi } F)^o} = \overline{\text{epi } F}$, every neighborhood of $(x, a) \in$ epi F contains an element of (epi $\varphi)^o$. Suppose $\langle x^*, x\rangle + \alpha\, a < \beta$. Then

$$\{(\tilde{x}, \tilde{a}) \in X \times \mathbb{R} : \langle x^*, \tilde{x}\rangle + \alpha\, \tilde{a} < \beta\}$$

is a neighborhood of (x, a) and contains an element of (epi $\varphi)^o$, which contradicts (4.2.6). Therefore

$$\langle x^*, x\rangle + \alpha\, a \geq \beta \quad \text{for all } \ (x, a) \in \text{epi } F. \tag{4.2.7}$$

Suppose $\alpha = 0$. For any $u \in U_\epsilon$ there is an $a \in \mathbb{R}$ such that $F(u) \leq a$. Then from (4.2.7)

$$\langle x^*, u\rangle = \langle x^*, u\rangle + \alpha\, a \geq \beta$$

and thus

$$\langle x^*, u - \bar{x}\rangle \geq 0 \quad \text{for all } u \in U_\epsilon.$$

Choose $\delta > 0$ such that $|u - \bar{x}| \leq \delta$ implies $u \in U$. For any nonzero element $x \in X$ let $t = \frac{\delta}{|x|}$. Then $|(tx + \bar{x}) - \bar{x}| = |tx| = \delta$ so that $tx + \bar{x} \in U_\epsilon$. Hence

$$\langle x^*, x\rangle = \langle x^*, (tx + \bar{x}) - \bar{x}\rangle / t \geq 0.$$

Similarly, $-t\, x + \bar{x} \in U_\epsilon$ and

$$\langle x^*, x\rangle = \langle x^*, (-tx + \bar{x}) - \bar{x}\rangle / (-t) \leq 0.$$

Thus, $\langle x^*, x\rangle$, $x^* = 0$, which is a contradiction. Therefore α is nonzero. From (4.2.6), (4.2.7) we have for $a > F(\bar{x})$ that $\alpha(a - F(\bar{x})) > 0$ and hence $\alpha > 0$. Employing (4.2.6), (4.2.7) again implies that

$$\left\langle -\frac{x^*}{\alpha}, x - \bar{x}\right\rangle + F(\bar{x}) \leq F(x)$$

for all $x \in X$ and therefore $-\frac{x^*}{\alpha} \in \partial F(\bar{x})$. □

4.3 Fenchel duality theory

In this section we discuss some elements of duality theory. We call

$$\inf_{x \in X} \ F(x) \tag{P}$$

the primal problem, where X is a real Banach space and $F : X \to (-\infty, \infty]$ is a proper l.s.c. convex function. We have the following result for the existence of a minimizer.

Theorem 4.25. *Let X be reflexive and let F be a l.s.c. proper convex functional defined on X satisfying*

$$\lim_{|x| \to \infty} \ F(x) = \infty. \tag{4.3.1}$$

Then there exists an $\bar{x} \in X$ such that

$$F(\bar{x}) = \inf\,\{F(x) :\ x \in X\}.$$

Proof. Let $\eta = \inf\{F(x) : x \in X\}$ and let $\{x_n\}$ be a minimizing sequence such that $\lim_{n\to\infty} F(x_n) = \eta$. Condition (4.3.1) implies that $\{x_n\}$ is bounded in X. Since X is reflexive, there exists a subsequence that converges weakly to some $\bar{x}$ in X and it follows from Lemma 4.4 that $F(\bar{x}) = \eta$. $\square$

We embed (P) into a family of perturbed problems

$$\inf_{x \in X} \; \Phi(x, y), \tag{P_y}$$

where $y \in Y$ is an embedding variable, Y is a Banach space, and $\Phi : X \times Y \to (-\infty, \infty]$ is a proper l.s.c. convex function with $\Phi(x, 0) = F(x)$. Thus $(P_0) = (P)$. For example, in terms of (4.1.1) we let

$$F(x) = f(x) + \varphi(\Lambda x) \tag{4.3.2}$$

and with $Y = H$

$$\Phi(x, y) = f(x) + \varphi(\Lambda x + y). \tag{4.3.3}$$

Definition 4.26. *The dual problem of (P) with respect to Φ is defined by*

$$\sup_{y^* \in Y^*} (-\Phi^*(0, y^*)). \tag{P^*}$$

The value function of (P_y) is defined by

$$h(y) = \inf_{x \in X} \; \Phi(x, y), \quad y \in Y.$$

In this section we analyze the relationship between the primal problem (P) and its dual (P^*). Throughout we assume that $h(y) > -\infty$ for all $y \in Y$.

Theorem 4.27. $\sup\,(P^*) \le \inf\,(P)$.

Proof. For any $(x, y) \in X \times Y$ and $(x^*, y^*) \in X^* \times Y^*$ we have

$$\langle x^*, x\rangle + \langle y^*, y\rangle - \Phi(x, y) \le \Phi^*(x^*, y^*).$$

Thus

$$0 = \langle 0, x\rangle + \langle y^*, 0\rangle \le F(x) + \Phi^*(0, y^*)$$

for all $x \in X$ and $y^* \in Y^*$. Therefore

$$\sup_{y^* \in Y^*} (-\Phi^*(0, y^*)) = \sup\,(P^*) \le \inf\,(P) = \inf_{x \in X} F(x). \qquad \square$$

Lemma 4.28. *h is convex.*

Proof. The proof is established by contradiction. Suppose there exist $y_1, y_2 \in Y$ and $\theta \in (0, 1)$ such that

$$\theta\, h(y_1) + (1 - \theta)\, h(y_2) < h(\theta\, y_1 + (1 - \theta)\, y_2).$$

Then there exist c and $\epsilon > 0$ such that

$$\theta\, h(y_1) + (1-\theta)\, h(y_2) < c - \epsilon < c < h(\theta\, y_1 + (1-\theta)\, y_2).$$

Set $a_1 = h(y_1) + \frac{\theta}{\epsilon}$ and

$$a_2 = \frac{c - \theta\, a_1}{1-\theta} = \frac{c - \epsilon - \theta\, h(y_1)}{1-\theta} > h(y_2).$$

By definition of h there exist $x_1, x_2 \in X$ such that

$$h(y_1) \le \Phi(x_1, y_1) \le a_1 \quad \text{and} \quad h(y_2) \le \Phi(x_2, y_2) \le a_2.$$

Thus

$$h(\theta\, y_1 + (1-\theta)\, y_2) \le \Phi(\theta\, x_1 + (1-\theta)\, x_2, \theta\, y_1 + (1-\theta)\, y_2)$$

$$\le \theta\, \Phi(x_1, y_1) + (1-\theta)\, \Phi(x_2, y_2) \le \theta\, a_1 + (1-\theta)\, a_2 = c,$$

which is a contradiction. Hence h is convex. $\square$

Lemma 4.29. *For all $y^* \in Y^*$, $h^*(y^*) = \Phi^*(0, y^*)$.*

Proof.

$$h^*(y^*) = \sup_{y \in Y} (\langle y^*, y\rangle - h(y)) = \sup_{y \in Y} (\langle y^*, y\rangle - \inf_{x \in X} \Phi(x, y))$$

$$= \sup_{y \in Y} \sup_{x \in X} (\langle y^*, y\rangle - \Phi(x, y)) = \sup_{y \in Y} \sup_{x \in X} (\langle 0, x\rangle + \langle y^*, y\rangle - \Phi(x, y))$$

$$= \sup_{(x,y) \in X \times Y} (\langle (0, y^*), (x, y)\rangle - \Phi(x, y)) = \Phi^*(0, y^*). \qquad \square$$

Theorem 4.30. *If h is l.s.c. at 0, then* $\inf\,(P) = \sup\,(P^*)$.

Proof. Since F is proper, $h(0) = \inf_{x \in X} F(x) < \infty$. Since h is convex by Lemma 4.28, it follows from Theorem 4.15 that $h(0) = h^{**}(0)$. Thus by Lemma 4.29

$$\sup\,(P^*) = \sup_{y^* \in Y^*} (-\Phi^*(0, y^*)) = \sup_{y^* \in Y^*} (\langle y^*, 0\rangle - h^*(y^*))$$

$$= h^{**}(0) = h(0) = \inf\,(P). \qquad \square$$

Theorem 4.31. *If h is subdifferentiable at* 0, *then* $\inf\,(P) = \sup\,(P^*)$ *and $\partial h(0)$ is the set of solutions of (P^*).*

Proof. By Lemma 4.29 we have that $\bar{y}^*$ solves (P^*) if and only if

$$-h^*(\bar{y}^*) = -\Phi^*(0, \bar{y}^*) = \sup_{y^* \in Y^*} (-\Phi^*(0, y^*))$$

$$= \sup_{y^* \in Y^*} (\langle y^*, 0\rangle - h^*(y^*)) = h^{**}(0).$$

By Theorem 4.22

$$h^{**}(0) + h^{***}(\bar{y}^*) = \langle \bar{y}^*, 0 \rangle = 0$$

if and only if $\bar{y}^* \in \partial h^{**}(0)$. Since $h^{***} = h^*$ by Theorem 4.16, we have $\bar{y}^* \in \partial h^{**}(y^*)$ if and only if $-h^*(\bar{y}^*) = h^{**}(0)$. Consequently $\bar{y}^*$ solves (P^*) if and only if $y^* \in \partial h^{**}(0)$. Since $\partial h(0)$ is not empty, $\partial h(0) = \partial h^{**}(0)$ by Theorem 4.17. Therefore $\partial h(0)$ is the set of all solutions of (P^*) and (P^*) has at least one solution. Let $y^* \in \partial h(0)$. Then

$$\langle y^*, x \rangle + h(0) \le h(x)$$

for all $x \in X$. If $\{x_n\}$ is a sequence in X such that $x_n \to 0$, then

$$\liminf_{n\to\infty} h(x_n) \ge \lim \langle y^*, x_n \rangle + h(0) = h(0)$$

and h is l.s.c. at 0. By Theorem 4.30 inf $(P) =$ sup (P^*). ☐

Corollary 4.32. *If there exists an $\bar{x} \in X$ such that $\Phi(\bar{x}, \cdot)$ is finite and continuous at 0, then h is continuous on an open neighborhood U of 0 and $h = h^{**}$. Moreover,*

$$\inf\ (P) = \sup\ (P^*)$$

and $\partial h(0)$ is the set of solutions of (P^*).

Proof. First show that h is continuous. Clearly, $\Phi(\bar{x}, \cdot)$ is bounded above on an open neighborhood U of 0. Since for all $y \in Y$

$$h(y) \le \Phi(\bar{x}, y),$$

h is bounded above on U. Since h is convex by Lemma 4.28 it is continuous by Theorem 4.6. Hence $h = h^{**}$ by Theorem 4.15. Moreover h is subdifferentiable at 0 by Theorem 4.24. The conclusion then follows from Theorem 4.31. ☐

Example 4.33. Consider the case (4.3.2)–(4.3.3), i.e.,

$$\Phi(x, y) = f(x) + \varphi(\Lambda x + y),$$

where $f : X \to (-\infty, \infty]$, $\varphi : Y \to (-\infty, \infty]$ are l.s.c. and convex and $\Lambda : X \to Y$ is a continuous linear operator. Let us calculate the conjugate of Φ:

$$\Phi^*(x^*, y^*) = \sup_{x\in X}\sup_{y\in Y}\{\langle x^*, x \rangle + \langle y^*, y \rangle - \Phi(x, y)\}$$

$$= \sup_{x\in X}\{\langle x^*, x \rangle - f(x) + \sup_{y\in Y}[\langle y^*, y \rangle - \varphi(\Lambda x + y)]\},$$

where

$$\sup_{y\in Y}[\langle y^*, y \rangle - \varphi(\Lambda x + y)] = \sup_{y\in Y}[\langle y^*, \Lambda x + y \rangle - \varphi(\Lambda x + y) - \langle y^*, \Lambda x \rangle]$$

$$= \sup_{z\in Y}[\langle y^*, z \rangle - \varphi(z)] - \langle y^*, \Lambda x \rangle = \varphi^*(y^*) - \langle y^*, \Lambda x \rangle.$$

Thus

$$\Phi^*(x^*, y^*) = \sup_{x\in X}\{\langle x^*, x\rangle - \langle y^*, \Lambda x\rangle - f(x) + \varphi^*(y^*)\}$$

$$= \sup_{x\in X}\{\langle x^* - \Lambda^* y^*, x\rangle - f(x) + \varphi^*(y^*)\} = f^*(x^* - \Lambda y^*) + \varphi^*(y^*).$$

Theorem 4.34. *For any $\bar{x} \in X$ and $\bar{y}^* \in Y^*$, the following statements are equivalent.*

(1) $\bar{x}$ solves (P), $\bar{y}^*$ solves (P^*), and $\min\,(P) = \max\,(P^*)$.

(2) $\Phi(\bar{x}, 0) + \Phi^*(0, \bar{y}^*) = 0$.

(3) $(0, \bar{y}^*) \in \partial\Phi(\bar{x}, 0)$.

Proof. Clearly (1) implies (2). If (2) holds, then

$$\Phi(\bar{x}, 0) = F(\bar{x}) \geq \inf\,(P) \geq \sup\,(P^*) \geq -\Phi^*(0, \bar{y}^*)$$

by Theorem 4.27. Thus

$$\Phi(\bar{x}, 0) = \min\,(P) = \max\,(P^*) = -\Phi^*(0, \bar{y}^*).$$

Therefore (2) implies (1). Since $\langle (0, \bar{y}^*), (\bar{x}, 0)\rangle = 0$, equivalence of (2) and (3) follows by Theorem 4.22. $\square$

Any solution y^* of (P^*) is called a Lagrange multiplier associated with Φ. For Example 4.33 the optimality condition implies

$$\begin{aligned} 0 &= \Phi(\bar{x}, 0) + \Phi^*(0, \bar{y}^*) = f(\bar{x}) + f^*(-\Lambda^*\bar{y}^*) + \varphi(\Lambda\bar{x}) + \varphi(\bar{y}^*) \\ &= [f(\bar{x}) + f^*(-\Lambda^*\bar{y}^*) - \langle -\Lambda^*\bar{y}^*, \bar{x}\rangle] + [\varphi(\Lambda\bar{x}) + \varphi^*(\bar{y}^*) - \langle \bar{y}^*, \Lambda\bar{x}\rangle]. \end{aligned} \tag{4.3.4}$$

Since each expression of (4.3.4) in square brackets is nonnegative, it follows that

$$f(\bar{x}) + f^*(-\Lambda^*\bar{y}^*) - \langle -\Lambda^*\bar{y}^*, \bar{x}\rangle = 0,$$

$$\varphi(\Lambda\bar{x}) + \varphi^*(\bar{y}^*) - \langle \bar{y}^*, \Lambda\bar{x}\rangle = 0.$$

By Theorem 4.22

$$\begin{aligned} -\Lambda^*\bar{y}^* &\in \partial f(\bar{x}), \\ \bar{y}^* &\in \partial\varphi(\Lambda\bar{x}). \end{aligned} \tag{4.3.5}$$

The functional $L : X \times Y^* \to (-\infty, \infty]$ defined by

$$-L(x, y^*) = \sup_{y\in Y}\,\{\langle y^*, y\rangle - \Phi(x, y)\} \tag{4.3.6}$$

is called the Lagrangian. Note that

$$\Phi^*(x^*, y^*) = \sup_{x\in X,\, y\in Y} \{\langle x^*, x\rangle + \langle y^*, y\rangle - \Phi(x, y)\}$$

$$= \sup_{x\in X} \langle x^*, x\rangle + \sup_{y\in Y} \{\langle y^*, y\rangle - \Phi(x, y)\}\} = \sup_{x\in X} (\langle x^*, x\rangle - L(x, y^*)).$$

Thus

$$-\Phi^*(0, y^*) = \inf_{x\in X} L(x, y^*) \tag{4.3.7}$$

and therefore the dual problem (P^*) is equivalent to

$$\sup_{y^*\in Y^*} \inf_{x\in X} L(x, y^*).$$

If Φ is a convex l.s.c. function that is finite at (x, y), then for the biconjugate of $\Phi_x : y \to \Phi(x, y)$ in y we have $\Phi_x(y)^{**} = \Phi(x, y)$ and

$$\Phi(x, y) = \Phi_x^{**}(x, y) = \sup_{y^*\in Y^*} \{\langle y^*, y\rangle - \Phi_x^*(y^*)\}$$

$$= \sup_{y^*\in Y^*} \{\langle y^*, y\rangle + L(x, y^*)\}.$$

Hence

$$\Phi(x, 0) = \sup_{y^*\in Y^*} L(x, y^*) \tag{4.3.8}$$

and the primal problem (P) is equivalently written as

$$\inf_{x\in X} \sup_{y^*\in Y} L(x, y^*).$$

Thus by means of the Lagrangian L, the problems (P) and (P^*) can be formulated as min-max problems, and by Theorem 4.27

$$\sup_{y^*} \inf_{x} L(x, y^*) \le \inf_{x} \sup_{y^*} L(x, y^*).$$

Theorem 4.35 (Saddle Point). *Assume that Φ is a convex l.s.c. function that is finite at $(\bar{x}, \bar{y}^*)$. Then the following are equivalent.*

(1) *$(\bar{x}, \bar{y}^*) \in X \times Y^*$ is a saddle point of L, i.e.,*

$$L(\bar{x}, y^*) \le L(\bar{x}, \bar{y}^*) \le L(x, \bar{y}^*) \quad \text{for all } x \in X,\ y^* \in Y^*. \tag{4.3.9}$$

(2) *$\bar{x}$ solves (P), $\bar{y}^*$ solves (P^*), and* $\min\,(P) = \max\,(P^*)$.

Proof. Suppose (1) holds. From (4.3.7) and (4.3.9)

$$L(\bar{x}, \bar{y}^*) = \inf_{x\in X} L(x, \bar{y}^*) = -\Phi^*(0, \bar{y}^*)$$

and from (4.3.8) and (4.3.9)

$$L(\bar{x}, \bar{y}^*) = \sup_{y^* \in Y^*} L(\bar{x}, y^*) = \Phi(\bar{x}, 0).$$

Thus, $\Phi^*(\bar{x}, 0) + \Phi(0, \bar{y}^*) = 0$ and (2) follows from Theorem 4.34.

Conversely, if (2) holds, then from (4.3.7) and (4.3.8)

$$-\Phi^*(0, \bar{y}^*) = \inf_{x \in X} L(x, \bar{y}^*) \le L(\bar{x}, \bar{y}^*),$$

$$\Phi(\bar{x}, 0) = \sup_{y^* \in Y^*} L(\bar{x}, y^*) \ge L(\bar{x}, \bar{y}^*).$$

Consequently $-\Phi^*(0, \bar{y}^*) = \Phi(\bar{x}, 0)$ by Theorem 4.34 and (4.3.9) holds. □

Theorem 4.35 implies that no duality gap between (P) and (P^*) is equivalent to the saddle point property of the pair $(\bar{x}, \bar{y}^*)$.

For Example 4.33 we have

$$L(x, y^*) = f(x) + \langle y^*, \Lambda x \rangle - \varphi(y^*). \tag{4.3.10}$$

If $(\bar{x}, \bar{y}^*)$ is a saddle point, then from (4.3.9)

$$\begin{aligned} -\Lambda^* \bar{y}^* &\in \partial f(\bar{x}), \\ \Lambda \bar{x} &\in \partial \varphi^*(\bar{x}). \end{aligned} \tag{4.3.11}$$

It follows from Theorem 4.22 that the second equation is equivalent to

$$\bar{y}^* \in \partial \varphi(\Lambda \bar{x})$$

and (4.3.11) is equivalent to (4.3.5), if X is reflexive. Thus the necessary optimality system for

$$\min_{x \in X} F(x) = f(x) + \varphi(\Lambda x)$$

is given by

$$\begin{aligned} -\Lambda^* \bar{y}^* &\in \partial f(\bar{x}), \\ \bar{y}^* &\in \partial \varphi^*(\Lambda \bar{x}). \end{aligned} \tag{4.3.12}$$

4.4 Generalized Yosida–Moreau approximation

Definition 4.36 (Monotone Operator). *Let A be a graph in $X \times X^*$.*

(1) *A is monotone if*

$$\langle y_1 - y_2, x_1 - x_2 \rangle \ge 0 \quad \textit{for all } x_1, x_2 \in D(A) \textit{ and } y_1 \in Ax_1,\ y_2 \in Ax_2$$

(2) *A is maximal monotone if any monotone extension of A coincides with A.*

Let φ be an l.s.c. convex function on X. For $x_1^* \in \partial\varphi(x_1)$ and $x_2^* \in \partial\varphi(x_2)$,

$$\varphi(x_1) - \varphi(x_2) \le \langle x_2^*, x_1 - x_2 \rangle,$$

$$\varphi(x_2) - \varphi(x_1) \le \langle x_1^*, x_2 - x_1 \rangle.$$

It follows that $\langle x_1^* - x_2^*, x_1 - x_2 \rangle \ge 0$. Hence $\partial\varphi$ is a monotone in $X \times X^*$. The following theorem characterizes maximal monotone operators by a range condition [ItKa].

Theorem 4.37 (Minty–Browder). *Assume that X and X^* are reflexive and strictly convex Banach spaces and let $F : X \to X^*$ denote the duality mapping. Then a monotone operator A is maximal monotone if and only if $R(\lambda F + A) = X^*$ for all $\lambda > 0$ (or, equivalently, for some $\lambda > 0$).*

Theorem 4.38 (Rockafeller). *Let X be a real Banach space. If φ is an l.s.c. proper convex functional on X, then $\partial\varphi$ is a maximal monotone operator from X into X^*.*

Proof. We prove the theorem for the case that X is reflexive . The general case is considered in [Roc2]. By Asplund's renorming theorem we can assume that after choosing an equivalent norm, X and X^* are strictly convex. Using the Minty–Browder theorem it suffices to prove that $R(F + \partial\varphi) = X^*$. For $x_0^* \in X^*$ we must show that the equation $x_0^* \in Fx + \partial\varphi(x)$ has at least a solution x_0. Note that Fx is single valued due to the fact that X^* is strictly convex. Define the proper convex functional $f : X \to (-\infty, \infty]$ by

$$f(x) = \frac{1}{2}\,|x|_X^2 + \varphi(x) - \langle x_0^*, x \rangle.$$

Since f is l.s.c. and $f(x) \to \infty$ as $|x| \to \infty$, there exists $x_0 \in D(f)$ such that $f(x_0) \le f(x)$ for all $x \in X$. The subdifferential of the mapping $x \to \frac{1}{2}|x|^2$ is given by the monotone operator F. Hence we find

$$\varphi(x) - \varphi(x_0) \ge \langle x_0^*, x - x_0 \rangle - \langle F(x), x - x_0 \rangle \text{ for all } x \in X.$$

Setting $x(t) = x_0 + t\,(u - x_0)$ with $u \in X$ and using convexity of φ we have

$$\varphi(u) - \varphi(x_0) \ge \langle x_0^*, u - x_0 \rangle - \langle F(x(t)), u - x_0 \rangle.$$

Taking the limit $t \to 0^+$ and using the fact that $x \to F(x)$ is continuous from X endowed with the norm topology to X^* endowed with the weak* topology, we obtain

$$\varphi(u) - \varphi(x_0) \ge \langle x_0^*, u - x_0 \rangle - \langle F(x_0), u - x_0 \rangle,$$

which implies that $x_0^* - F(x_0) \in \partial\varphi(x_0)$. $\square$

Throughout the remainder of this section let H denote a real Hilbert space which is identified with its dual H^*. Further let A denote a maximal monotone operator in $H \times H$. Recall that A is necessarily densely defined and closed. Moreover the resolvent

$J_\mu = (I + \mu A)^{-1}$, with $\mu > 0$, is a contraction defined on all of H; see [Bre, ItKa, Paz]. Moreover

$$|J_\mu x - x| \to 0 \quad \text{as } \mu \to 0^+ \text{ for each } x \in H. \tag{4.4.1}$$

The Yosida approximation A_μ of A is defined by

$$A_\mu x = \frac{1}{\mu}(x - J_\mu x).$$

The operator A_μ is single valued, monotone, everywhere defined, Lipschitz continuous with Lipschitz constant $\frac{1}{\mu}$ and $A_\mu x \in A J_\mu x$ for all $x \in H$.

Let φ be an l.s.c., proper, and convex function on H. Throughout the remainder of this chapter let A denote the maximal monotone operator $\partial\varphi$ on the Hilbert space $H = H^*$. For $x, \lambda \in H$ and $c > 0$ define the functional $\varphi_c(x, \lambda)$ by

$$\varphi_c(x, \lambda) = \inf_{u \in H} \left\{ \varphi(x - u) + (\lambda, u)_H + \frac{c}{2} |u|_H^2 \right\}. \tag{4.4.2}$$

Then $\varphi_c(x, \lambda)$ is a smooth approximation of φ in the following sense.

Theorem 4.39. *For x, $\lambda \in H$ the infimum in* (4.4.2) *is attained at a unique point*

$$u_c(x, \lambda) = x - J_{\frac{1}{c}}\left(x + \frac{\lambda}{c}\right).$$

$\varphi_c(x, \lambda)$ is convex, Lipschitz continuously Fréchet differentiable in x and

$$\varphi_c'(x, \lambda) = \lambda + c\, u_c(x, \lambda) = A_{\frac{1}{c}}\left(x + \frac{\lambda}{c}\right).$$

Moreover, $\lim_{c\to\infty} \varphi_c(x, \lambda) = \varphi(x)$ *and*

$$\varphi\left(J_{\frac{1}{c}}\left(x + \frac{\lambda}{c}\right)\right) - \frac{1}{2c}|\lambda|_H^2 \le \varphi_c(x, \lambda) \le \varphi(x) \quad \textit{for every } x,\ \lambda \in H.$$

Proof. First we observe that (4.4.2) is equivalent to

$$\varphi_c(x, \lambda) = \inf_{v \in H} \left\{ \varphi(v) + \frac{c}{2} \left| x + \frac{\lambda}{c} - v \right|^2 \right\} - \frac{1}{2c}|\lambda|^2, \tag{4.4.3}$$

where $v = x - u$. Note that $v \to \psi(v) = \varphi(v) + \frac{c}{2}|v - y|^2$, with $y = x + \frac{\lambda}{c}$, is convex and l.s.c. for every $y \in H$. Since φ is proper, $D(A) \neq \emptyset$, and thus for $x_0 \in D(A)$, and $x_0^* \in Ax_0$ we have

$$\varphi(x) \ge \varphi(x_0) + (x_0^*, x - x_0) \quad \text{for all } x \in H.$$

Thus, $\lim \psi(v) = \infty$ as $|v| \to \infty$. Hence there exists a unique minimizer $v_0 \in H$ of ψ. Let $\xi \in H$ and define $\eta = (1 - t)v_0 + t\xi$ for $0 < t < 1$. Since φ is convex we have

$$\varphi(\eta) - \varphi(v_0) \le t\,(\varphi(\xi) - \varphi(v_0)). \tag{4.4.4}$$

Moreover, since $\psi(v_0) \le \psi(\eta)$,

$$\begin{aligned}\varphi(\eta) - \varphi(v_0) &\ge \frac{c}{2}\,(|v_0 - y|^2 - |(1-t)v_0 + t\xi - y|^2)\\ &= tc\,(v_0 - \xi, v_0 - y) - \frac{t^2 c}{2}|v_0 - \xi|^2.\end{aligned} \tag{4.4.5}$$

From (4.4.4) and (4.4.5) we obtain, taking the limit $t \to 0^+$,

$$\varphi(\xi) - \varphi(v_0) \ge c\,(y - v_0, \xi - v_0).$$

Since $\xi \in H$ is arbitrary this implies that $y - v_0 \in \frac{1}{c}\,Av_0$. Thus $v_0 = J_{\frac{1}{c}}y$ and

$$u_c(x,\lambda) = x - v_0 = x - J_{\frac{1}{c}}\left(x + \frac{\lambda}{c}\right)$$

attains the minimum in (4.4.2). Note that this argument also implies that A is maximal.

For $x_1, x_2 \in X$ and $0 < t < 1$

$$\varphi_c((1-t)x_1 + tx_2, \lambda) = \psi((1-t)v_1 + tv_2) - \frac{1}{2c}|\lambda|^2,$$

where $y_i = x_i + \frac{\lambda}{c}$ and $v_i = J_{\frac{1}{c}}\,y_i$ for $i = 1, 2$. Hence the convexity of $x \to \varphi_c(x,\lambda)$ follows from the one of ψ.

Next, we show that $\partial\varphi_c(x,\lambda) = A_{\frac{1}{c}}(x + \frac{\lambda}{c})$. For $\hat{x} \in H$, let $\hat{y} = \hat{x} + \frac{\lambda}{c} \in H$ and $\hat{v} = J_{\frac{1}{c}}\,\hat{y}$. Then, we have

$$\varphi(\hat{v}) + \frac{c}{2}\,|\hat{v} - y|^2 \ge \varphi(v_0) + \frac{c}{2}\,|v_0 - y|^2$$

and

$$\varphi(v_0) + \frac{c}{2}\,|v_0 - \hat{y}|^2 \ge \varphi(\hat{v}) + \frac{c}{2}\,|\hat{v} - \hat{y}|^2.$$

Thus,

$$\frac{c}{2}\,(|v_0 - \hat{y}|^2 - |v_0 - y|^2) \ge \varphi_c(\hat{x},\lambda) - \varphi_c(x,\lambda) \ge \frac{c}{2}\,(|\hat{v} - \hat{y}|^2 - |\hat{v} - y|^2). \tag{4.4.6}$$

Since $|\hat{v} - v_0| \to 0$ as $|\hat{x} - x| \to 0$, it follows from (4.4.6) that

$$\frac{|\varphi_c(\hat{x},\lambda) - \varphi_c(x,\lambda) - \big(c(y - v_0), \hat{x} - x\big)|}{|\hat{x} - x|} \to 0$$

as $|\hat{x} - x| \to 0$. Hence $x \to \varphi_c(x,\lambda)$ is Fréchet differentiable with F-derivative

$$c(y - v_0) = \lambda + c\,u_c(x,\lambda) = A_{\frac{1}{c}}\left(x + \frac{\lambda}{c}\right). \quad \square$$

The dual representation of $\varphi_c(x, \lambda)$ is derived next. Define the functional $\Phi(v, y)$ on $H \times H$ by

$$\Phi(v, y) = \varphi(v) + \frac{c}{2}\,|v - (\hat{y} + y)|^2,$$

where we set $\hat{y} = x + \frac{\lambda}{c}$. Consider the family of primal problems

$$\inf_{v \in H} \;\Phi(v, y) \quad \text{for each } y \in H \tag{P_y}$$

and the dual problem

$$\sup_{y^* \in H} \{-\Phi^*(0, y^*)\}. \tag{P^*}$$

If $h(y)$ is the value functional of (P_y), i.e., $h(y) = \inf_{v \in H} \;\Phi(v, y)$, then from (4.4.3) we have

$$\varphi_c(x, \lambda) = h(0) - \frac{1}{2c}\,|\lambda|^2. \tag{4.4.7}$$

From the proof of Theorem 4.39 it follows that $h(y)$ is continuously Fréchet differentiable with $h'(0) = \varphi_c'(x, \lambda)$. It thus follows from Theorem 4.31 that $\inf\;(P_0) = \max\;(P^*)$ and $h'(0)$ is the solution of (P^*).

This leads to the following theorem.

Theorem 4.40. *For $x,\ \lambda \in H$*

$$\varphi_c(x, \lambda) = \sup_{y^* \in H} \left\{(x,\, y^*)_H - \varphi^*(y^*) - \frac{1}{2c}\,|y^* - \lambda|_H^2\right\}, \tag{4.4.8}$$

where the supremum is attained at a unique point $\lambda_c(x, \lambda)$ and we have

$$\lambda_c(x, \lambda) = \lambda + c\, u_c(x, \lambda) = \varphi_c'(x, \lambda) = A_{\frac{1}{c}}\left(x + \frac{\lambda}{c}\right). \tag{4.4.9}$$

Proof. For $(v^*, y^*) \in H \times H$ we have by definition

$$\begin{aligned}
\Phi^*(v^*, y^*) &= \sup_{v \in H}\,\sup_{y \in H} \left\{(v^*, v) + (y^*, y) - \varphi(v) - \frac{c}{2}\,|v - (\hat{y} + y)|^2\right\} \\
&= \sup_{v \in H} \left\{(v^*, v) + (y^*, v - \hat{y}) - \varphi(v) \right. \\
&\qquad \left. + \sup_{y \in H} \left[-\big(y^*, v - (\hat{y} + y)\big) - \frac{c}{2}\,|v - (\hat{y} + y)|^2\right]\right\} \\
&= \sup_{v \in H} \left\{(y^* + v^*, v) - \varphi(v) - (y^*, \hat{y}) + \frac{1}{2c}\,|y^*|^2\right\} \\
&= \varphi^*(y^* + v^*) - (y^*, \hat{y}) + \frac{1}{2c}\,|y^*|^2.
\end{aligned}$$

Hence,

$$h(0) = \sup_{y^* \in H} \{-\Phi^*(0, y^*)\} = \sup_{y^* \in H} \left\{ (y^*, \hat{y}) - \varphi^*(y^*) - \frac{1}{2c} |y^*|^2 \right\}$$

which implies (4.4.8) from (4.4.7) and since $\hat{y} = x + \frac{\lambda}{c}$. By Theorem 4.31 the maximum of $y^* \to \langle y^*, x\rangle - \varphi^*(y^*) - \frac{1}{2c} |y^* - \lambda|^2$ is attained at the unique point $h'(0) = \lambda_c(x, \lambda)$ that is given by (4.4.9). $\square$

The following theorem provides an equivalent characterization of $\lambda \in \partial\varphi(x)$. This results will be used in the following section to replace the differential inclusion, which relates the primal and adjoint variable by means of a nonlinear equation.

Theorem 4.41.

(1) *If* $\lambda \in \partial\varphi(x)$ *for* $x, \lambda \in H$, *then* $\lambda = \varphi_c'(x, \lambda)$ *for all* $c > 0$.

(2) *Conversely, if* $\lambda = \varphi_c'(x, \lambda)$ *for some* $c > 0$, *then* $\lambda \in \partial\varphi(x)$.

Proof. If $\lambda \in \partial\varphi(x)$, then by Theorems 4.39, 4.40, and 4.22

$$\varphi(x) \ge \varphi_c(x, \lambda) \ge \langle \lambda, x\rangle - \varphi^*(\lambda) = \varphi(x)$$

for every $c > 0$. Thus, $\lambda \in H$ attains the supremum in (4.4.8) and by Theorem 4.40 we have $\lambda = \varphi_c'(x, \lambda)$. Conversely, if $\lambda \in H$ satisfies $\lambda = \varphi_c'(x, \lambda)$ for some $c > 0$, then $u_c(x, \lambda) = 0$ by Theorem 4.40. Hence it follows from Theorem 4.39, (4.4.2), and Theorem 4.40 that

$$\varphi(x) = \varphi_c(x, \lambda) = \langle \lambda, x\rangle - \varphi^*(\lambda),$$

and thus $\lambda \in \partial\varphi(x)$ by Theorem 4.22. $\square$

4.5 Optimality systems

In this section we derive first order optimality systems for (4.1.1) based on Lagrange multipliers. In what follows we assume that $f : X \to \mathbb{R}$ and $\varphi : H \to (-\infty, \infty]$ are l.s.c., convex functions, that f is continuously differentiable, and that φ is proper. Further $\Lambda \in \mathcal{L}(X, H)$ and C is a closed convex set in X. In addition we require that

(A1) f, φ are bounded below by zero on K,

(A2) $\langle f'(x_1) - f'(x_2), x_1 - x_2\rangle_{X^*,X} \ge \sigma\, |x_1 - x_2|_X^2$

for some $\sigma > 0$ independent of $x_1, x_2 \in C$, and

(A3) $\varphi(\Lambda x_0) < \infty$ and φ is continuous at Λx_0 for some $x_0 \in C$.

As a consequence of (A2) we have

$$\begin{aligned} & f(x) - f(x_0) - \langle f'(x_0),\, x - x_0\rangle_{X^*,X} \\ &= \int_0^1 \langle f'(x_0 + t(x - x_0)) - f'(x_0),\, x - x_0\rangle_{X^*,X}\, dt \geq \frac{\sigma}{2}\, |x - x_0|_X^2 \end{aligned} \tag{4.5.1}$$

for all $x \in X$. Due to Theorem 4.24 there exists $y_0^* \in D(A(y_0)) = \partial\varphi(y_0)$ such that

$$\varphi(\Lambda x) - \varphi(\Lambda x_0) \geq (y_0^*,\, \Lambda x - \Lambda x_0)_H \quad \text{for } x_0 \in H. \tag{4.5.2}$$

Hence, $\lim f(x) + \varphi(\Lambda x) \to \infty$ as $|x|_X \to \infty$ and it follows from Theorem 4.25 and (A2) that there exists a unique minimizer $\bar{x} \in C$ of (4.1.1).

Theorem 4.42 (Optimality). *A necessary and sufficient condition for* $\bar{x} \in C$ *to be the minimizer of* (4.1.1) *is given by*

$$\langle f'(\bar{x}),\, x - \bar{x}\rangle_{X^*,X} + \varphi(\Lambda x) - \varphi(\Lambda \bar{x}) \geq 0 \quad \textit{for all } x \in C. \tag{4.5.3}$$

Proof. Assume that $\bar{x}$ is the minimizer of (4.1.1). Then for $x \in C$ and $0 < t < 1$ we have $\bar{x} + t(x - \bar{x}) \in C$ and

$$f(\bar{x} + t(x - \bar{x})) + \varphi(\Lambda(\bar{x} + t(x - \bar{x})) \geq f(\bar{x}) + \varphi(\Lambda\bar{x}).$$

Since

$$\varphi(\Lambda((1 - t)\bar{x} + tx)) - \varphi(\Lambda\bar{x}) \leq t\,(\varphi(\Lambda x) - \varphi(\Lambda\bar{x})),$$

we obtain

$$t^{-1}\,(f(\bar{x} + t(x - \bar{x})) - f(\bar{x})) + \varphi(\Lambda x) - \varphi(\Lambda\bar{x}) \geq 0.$$

Taking the limit $t \to 0^+$, we have (4.5.3) for all $x \in C$.

Conversely, assume that $\bar{x} \in C$ satisfies (4.5.3). Then, from (4.5.1),

$$\begin{aligned} & f(x) + \varphi(\Lambda x) - (f(\bar{x}) + \varphi(\Lambda\bar{x})) \\ &= f(x) - f(\bar{x}) - \langle f'(\bar{x}),\, x - \bar{x}\rangle + \langle f'(\bar{x}),\, x - \bar{x}\rangle + \varphi(\Lambda x) - \varphi(\Lambda\bar{x}) \\ &\geq \frac{\sigma}{2}\, |x - \bar{x}|_X^2 \end{aligned}$$

for all $x \in C$. Thus, $\bar{x}$ is a minimizer of (4.1.1). ☐

Next, φ is split from the optimality system in (4.5.3) by means of an appropriately defined Lagrange multiplier. For this purpose we consider the regularized minimization problems:

$$\min f(x) + \varphi_c(\Lambda x, \lambda) \quad \text{over } x \in C \tag{4.5.4}$$

for $c > 0$ and $\lambda \in H$. From Theorem 4.39 it follows that $x \to \varphi_c(\Lambda x, \lambda)$ is convex, continuously differentiable, and bounded below by $-\frac{1}{2c}|\lambda|_H^2$. Thus, for $\lambda \in H$, (4.5.4) has a unique solution $x_c \in C$ and $x_c \in C$ is the solution of (4.5.4) if and only if x_c satisfies

$$\langle f'(x_c),\, x - x_c\rangle + \big(\varphi_c'(\Lambda x_c, \lambda),\, \Lambda(x - x_c)\big)_H \geq 0 \quad \text{for all } x \in C. \tag{4.5.5}$$

It follows from Theorems 4.39 and 4.40 that

$$\varphi_c'(\Lambda x_c, \lambda) = A_{\frac{1}{c}}\left(\Lambda x_c + \frac{1}{c}\lambda\right) = \lambda_c \in H. \tag{4.5.6}$$

We have the following result.

Theorem 4.43 (Lagrange Multipliers). (1) *Suppose that x_c converges strongly to $\bar{x}$ in X as $c \to \infty$ and that $\{\lambda_c\}_{c\geq 1}$ has a weak cluster point. Then for each weak cluster point $\bar{\lambda} \in H$ of $\{\lambda_c\}_{c\geq 1}$*

$$\begin{aligned} &\bar{\lambda} \in \partial\varphi(\Lambda\bar{x}) \quad \text{and} \\ &\langle f'(\bar{x}),\, x - \bar{x}\rangle_{X^*,X} + \big(\bar{\lambda},\, \Lambda(x - \bar{x})\big)_H \geq 0 \text{ for all } x \in C. \end{aligned} \tag{4.5.7}$$

Conversely, if $(\bar{x}, \bar{\lambda}) \in C \times H$ satisfies (4.5.7), then $\bar{x}$ minimizes (4.1.1).

(2) *Assume that there exists $\tilde{\lambda}_c \in \partial\varphi(\Lambda x_c)$ for $c \geq 1$ such that $\{|\tilde{\lambda}_c|_H\}$, $c \geq 1$, is bounded. Then, x_c converges strongly to $\bar{x}$ in X as $c \to \infty$.*

Proof. To verify (1), assume that $x_c \to \bar{x}$ and let $\bar{\lambda}$ be a weak cluster point of λ_c in H. Then, from (4.5.5)–(4.5.6), we can conclude that $(\bar{x}, \bar{\lambda}) \in C \times H$ satisfies

$$\langle f'(\bar{x}),\, x - \bar{x}\rangle + \big(\bar{\lambda},\, \Lambda(x - \bar{x})\big)_H \geq 0$$

for all $x \in C$. It follows from Theorems 4.39 and 4.40 that

$$\begin{aligned} \big(\lambda_c, v_c\big)_H - \frac{1}{2c}|\lambda_c - \lambda|^2 &= \varphi_c(\Lambda x_c, \lambda) + \varphi^*(\lambda_c) \\ &\geq \varphi\left(J_{\frac{1}{c}}\left(v_c + \frac{\lambda}{c}\right)\right) - \frac{1}{2c}|\lambda|_H^2 + \varphi^*(\lambda_c). \end{aligned}$$

Since $J_{\frac{1}{c}}(v_c + \frac{\lambda}{c}) \to \bar{v} = \Lambda\bar{x}$ and φ and φ^* are l.s.c., letting $c \to \infty$, we obtain

$$\big(\bar{\lambda}, \bar{v}\big)_H \geq \varphi(\bar{v}) + \varphi^*(\bar{\lambda}),$$

which implies that $\bar{\lambda} \in \partial\varphi(\bar{v})$ by Theorem 4.22. Hence, $(\bar{x}, \bar{\lambda}) \in C \times H$ satisfies (4.5.7).

Conversely, suppose that $(\bar{x}, \bar{\lambda}) \in C \times H$ satisfies (4.5.7). Then $\varphi(\Lambda x) - \varphi(\Lambda\bar{x}) \geq \big(\bar{\lambda},\, \Lambda(x - \bar{x})\big)_H$ for all $x \in C$. Thus the inequality in (4.5.7) implies (4.5.3). It then follows from Theorem 4.42 that $\bar{x}$ minimizes (4.1.1).

For (2) we note that from (4.5.5) we have

$$\langle f'(x_c),\, x - x_c\rangle + \varphi_c(\Lambda x, \lambda) - \varphi_c(\Lambda x_c, \lambda) \geq 0 \quad \text{for all } x \in C.$$

Thus

$$\langle f'(x_c),\, \bar{x} - x_c\rangle + \varphi_c(\Lambda\bar{x}, \lambda) - \varphi_c(\Lambda x_c, \lambda) \geq 0.$$

Also, from (4.5.3)

$$\langle f'(\bar{x}),\, x_c - \bar{x}\rangle + \varphi(\Lambda x_c) - \varphi(\Lambda \bar{x}) \geq 0.$$

Adding these inequalities, we obtain

$$\begin{aligned}\langle f'(\bar{x}) - f'(x_c),\, \bar{x} - x_c\rangle + \varphi_c(\Lambda x_c, \lambda) - \varphi(\Lambda x_c)\\ \leq \varphi_c(\Lambda \bar{x}, \lambda) - \varphi(\Lambda \bar{x}) \leq 0.\end{aligned} \tag{4.5.8}$$

By (4.4.8)

$$\varphi(v_c) - \frac{1}{2c}\,|\tilde{\lambda}_c - \lambda|_H^2 = \langle v_c,\, \tilde{\lambda}_c\rangle_H - \varphi^*(\tilde{\lambda}_c) - \frac{1}{2c}\,|\tilde{\lambda}_c - \lambda|_H^2 \leq \varphi_c(v_c, \lambda), \tag{4.5.9}$$

where $v_c = \Lambda x_c$ and hence

$$\varphi(v_c) - \varphi_c(v_c, \lambda) \leq \frac{1}{2c}\,|\tilde{\lambda}_c - \lambda|_H^2.$$

Thus, from (4.5.8) and (A2)

$$\frac{\sigma}{2}\,|x_c - x^*|_X^2 \leq \frac{1}{2c}\,|\tilde{\lambda}_c - \lambda|_H^2$$

and since $\{|\tilde{\lambda}_c|_H\}_{c\geq 1}$ is bounded, $|x_c - x^*|_X \to 0$ as $c \to \infty$. □

The following lemma addresses the condition in Theorem 4.43 (2).

Lemma 4.44. (1) *If $D(\varphi) = H$ and bounded above, on bounded sets, then $\partial\varphi(\Lambda x_c)$ is nonempty and $|\partial\varphi(\Lambda x_c)|_H$ is uniformly bounded for $c \geq 1$.*

(2) *If $\varphi = \chi_K$ with K a closed convex set in H and $\Lambda x_c \in K$ for all $c > 0$, then $\tilde{\lambda}_c$ can be chosen to be 0 for all $c > 0$.*

Proof. (1) By definition of x_c and by Theorem 4.39

$$f(x_c) + \varphi_c(v_c, \lambda) \leq f(\bar{x}) + \varphi(\Lambda \bar{x})$$

and

$$f(x_c) + \varphi\left(J_{\frac{1}{c}}\left(v_c + \frac{\lambda}{c}\right)\right) \leq f(\bar{x}) + \varphi(\Lambda \bar{x}) + \frac{1}{2c}\,|\lambda|_H^2,$$

where $v_c = \Lambda x_c$. Since φ is bounded below by 0, there exists a constant $K_1 > 0$ independent of $c \geq 1$ such that

$$f(x_c) - f(\bar{x}) - \langle f'(\bar{x}),\, x_c - \bar{x}\rangle \leq K_1 + \|f'(\bar{x})\|\,|x_c - \bar{x}|_X.$$

By (4.5.1)

$$\frac{\sigma}{2}\,|x_c - \bar{x}|_X^2 \leq K_1 + \|f'(\bar{x})\|\,|x_c - \bar{x}|_X.$$

Thus there exists a constant M such that $|x_c - \bar{x}|_X \leq M$ for $c \geq 1$. Since φ is everywhere defined, convex, l.s.c., and assumed to be bounded above on bounded sets, it follows from

Theorem 4.7 that φ is Lipschitz continuous in the open set $B = \{v \in H : |v - \bar{v}| < (M+1)\,\|\Lambda\|\}$, where $\bar{v} = \Lambda\bar{x}$. Let L denote the Lipschitz constant. By Theorem 4.24 $\partial\varphi(v_c)$ is nonempty for $c \geq 1$. Let $\tilde{\lambda}_c \in \partial\varphi(v_c)$ for $c \geq 1$. Hence

$$L\,|v - v_c| \geq \varphi(v) - \varphi(v_c) \geq \left(\tilde{\lambda}_c,\, v - v_c\right)$$

for all $v \in B$. Since $v_c \in B$ we have $|\tilde{\lambda}_c| \leq L$, and the claim follows.

(2) In this case $\partial\varphi(v_c)$ is given by the normal cone $N_C(v_c) = \{w \in H : (w, v_c - v)_H \geq 0$ for all $v \in C\}$ and thus $0 \in \partial\varphi(v_c)$ for all $c > 0$. $\square$

Theorem 4.45 (Complementarity). *Assume that there exists a pair $(\bar{x}, \bar{\lambda}) \in C \times H$ that satisfies (4.5.7). Then the complementarity condition $\bar{\lambda} \in \partial\varphi(\Lambda\bar{x})$ can equivalently be expressed as*

$$\bar{\lambda} = \varphi_c'(\Lambda\bar{x}, \bar{\lambda}) \tag{4.5.10}$$

and $\bar{x}$ is the unique solution of

$$\min_{x \in C} f(x) + \varphi_c(\Lambda x, \bar{\lambda}) \tag{4.5.11}$$

for every $c > 0$.

Proof. The first claim follows directly from Theorem 4.41. From (4.5.5) we conclude that $\hat{x}$ is a minimizer of (4.5.11) if and only if $\hat{x} \in C$ satisfies

$$\langle f'(\hat{x}),\, x - \hat{x}\rangle_{X^*,X} + \left(\varphi_c'(\Lambda\hat{x}, \bar{\lambda}),\, \Lambda(x - \hat{x})\right)_H \geq 0 \quad \text{for all } x \in C.$$

From (4.5.7) and (4.5.10) it follows that $\bar{x} \in C$ satisfies this inequality as well and hence $\hat{x} = \bar{x}$. $\square$

Theorems 4.43 and 4.45 imply that if a pair $(\bar{x}, \bar{\lambda}) \in C \times H$ satisfies

$$\begin{cases} \langle f'(\bar{x}), x - \bar{x}\rangle + \left(\bar{\lambda},\, \Lambda(x - \bar{x})\right) \geq 0 & \text{for all } x \in C, \\ \bar{\lambda} = \varphi_c'(\Lambda\bar{x}, \bar{\lambda}) \end{cases} \tag{4.5.12}$$

for some $c > 0$, then $\bar{x}$ is the minimizer of (4.1.1). Conversely, if $\bar{x}$ is a minimizer of (4.1.1) and

$$\partial(\varphi \circ \Lambda + \psi_C)(\bar{x}) = \Lambda^*\partial\varphi(\Lambda\bar{x}) + \partial\psi_C(\bar{x}), \tag{4.5.13}$$

then there exists a $\bar{\lambda} \in H$ such that the pair $(\bar{x}, \bar{\lambda})$ satisfies (4.5.12) for all $c > 0$. Here ψ_C denotes the indicator function of the set C. In fact it follows from (4.5.7) that $-f'(\bar{x}) \in \partial(\varphi \circ \Lambda + \psi_C)(\bar{x})$ and by (4.5.13)

$$-f'(\bar{x}) \in \Lambda^*\partial\varphi(\Lambda\bar{x}) + \partial\psi_C(\bar{x}) = \Lambda^*\partial\varphi(\Lambda\bar{x}) + N_C(\bar{x}),$$

where $N_C(\bar{x}) = \{z \in X^* : \langle z,\, x - \bar{x}\rangle \leq 0$ for all $x \in C\}$. This implies that there exists some $\bar{\lambda} \in \partial\varphi(\Lambda\bar{x})$ such that (4.5.7) holds and also (4.5.12) for the pair $(\bar{x}, \bar{\lambda})$. Condition (4.5.13) holds, for example, if there exists $x \in \text{int}\,(C)$ and φ is continuous and finite at Λx (see, e.g., Propositions 12 and 13 in Section 3.2 of [EkTu]).

The following theorem discusses the equivalence between the existence of a Lagrange multiplier $\bar{\lambda}$ and uniform boundedness of λ_c.

Theorem 4.46. *Suppose Λ is compact and let $\lambda \in H$ be fixed. Then $\lambda_c = \varphi_c'(v_c, \lambda)$ is uniformly bounded in $c \geq 1$ if and only if there exists $\lambda^* \in \partial\varphi(\Lambda\bar{x})$ such that (4.5.7) holds. In either case there exists a subsequence $(x_{\hat{c}}, \lambda_{\hat{c}}) \in X \times H$ such that $x_{\hat{c}} \to \bar{x}$ strongly in X and $\lambda_{\hat{c}} \to \lambda^*$ weakly in H where $(\bar{x}, \bar{\lambda}) \in C \times H$ satisfies (4.5.7).*

Proof. Assume that $|\lambda_c|_H$ is uniformly bounded. From (4.5.5)

$$\langle f'(x_c) + \Lambda^*\lambda_c, x - x_c\rangle \geq 0 \quad \text{for all } x \in C,$$

where $\lambda_c = \varphi_c'(\Lambda x_c, \lambda)$. From (A2) it follows that

$$\sigma\, |x_c - \bar{x}|_X^2 \leq \langle f'(x_c) - f'(\bar{x}), x_c - \bar{x}\rangle \leq \langle f'(\bar{x}) - \Lambda^*\lambda_c, x_c - \bar{x}\rangle$$

$$\leq (\|\Lambda\|\,|\lambda_c| + \|f'(\bar{x})\|)\, |x_c - \bar{x}|,$$

which by assumption implies that $|x_c|_X$ is uniformly bounded. Since Λ is compact it follows that any weakly convergent sequence $\lambda_c \to \lambda^*$ in H satisfies $\Lambda^*\lambda_c \to \Lambda^*\bar{\lambda}$ strongly in X^*. Again, from (A2) we have

$$\sigma\, |x_c - x_{\hat{c}}|_X \leq \langle f'(x_c) - f'(x_{\hat{c}}), x_c - x_{\hat{c}}\rangle \leq |\Lambda^*\lambda_c - \Lambda^*\lambda_{\hat{c}}|_{X^*}$$

for any $c,\ \hat{c} > 0$. Hence $\{x_c\}$ is a Cauchy sequence in X and thus there exists $\bar{x} \in X$ such that $|x_c - \bar{x}|_X \to 0$. The rest of the proof for the "only if" part is identical to the one of the first part of Theorem 4.43. It will be shown in Theorem 4.49 that

$$\frac{\sigma}{2}\, |x_c - \bar{x}|_X^2 + \frac{1}{2c}\, |\lambda_c - \bar{\lambda}|_H^2 \leq \frac{1}{2c}\, |\lambda - \bar{\lambda}|_H^2$$

if (4.5.7) holds. This implies the "if" part. ☐

4.6 Augmented Lagrangian method

In this section we discuss the augmented Lagrangian method for (4.1.1). Throughout we assume that (A1)–(A3) of Section 4.5 hold and we set $L_c(x, \lambda) = f(x) + \phi_c(\Lambda x, \lambda)$.

Augmented Lagrangian Method.

Step 1: Choose a starting value $\lambda_1 \in H$ a positive number c and set $k = 1$.

Step 2: Given $\lambda_k \in H$ find $x_k \in C$ by

$$L_c(x_k, \lambda_k) = \min\ L_c(x, \lambda_k) \quad \text{over } x \in C.$$

Step 3: Update λ_k by $\lambda_{k+1} = \varphi_c'(\Lambda x_k, \lambda_k)$.

Step 4: If the convergence criterion is not satisfied, then set $k = k + 1$ and go to Step 2.

Before proving convergence of the augmented Lagrangian method we motivate the update of the Lagrange multiplier in Step 3. First, it follows from (4.5.12) that it is a fixed point iteration for the necessary optimality condition in the dual variable λ. To give the second motivation we define the dual functional $d_c : H \to \mathbb{R}$ by

$$d_c(\lambda) = \inf f(x) + \varphi_c(\Lambda x, \lambda) \quad \text{over } x \in C \tag{4.6.1}$$

for $\lambda \in H$. The update is then a steepest ascent step for $d_c(\lambda)$. In fact it will be shown in the following lemma that (4.6.1) attains the minimum at a unique minimizer $x(\lambda) \in C$ and that d_c is continuously Fréchet differentiable with F-derivative

$$d_c'(\lambda) = u(\Lambda x(\lambda), \lambda),$$

where $u_c(x, \lambda)$ is defined in Theorem 4.39. Thus the steepest ascent step is given by $\lambda_{k+1} = \lambda_k + c\, u(\Lambda x(\lambda_k), \lambda_k)$, which by Theorem 4.40 coincides with the update given in Step 3.

Lemma 4.47. *For $\lambda \in H$ and $c > 0$ the infimum in* (4.6.1) *is attained at a unique minimizer $x(\lambda) \in C$ and the mapping $\lambda \in H \to x(\lambda) \in X$ is Lipschitz continuous with Lipschitz constant σ^{-1}. Moreover, the dual functional d_c is continuously Fréchet differentiable with F-derivative*

$$d_c'(\lambda) = u(\Lambda x(\lambda), \lambda),$$

where $u_c(v, \lambda)$ is defined in Theorem 4.39.

Proof. The proof is given in three steps.

Step 1. Since $f(x) + \varphi_c(\Lambda x, \lambda)$ is convex and l.s.c. and since (4.5.1) holds, there exists a unique $x(\lambda)$ that attains its minimum over K. To establish Lipschitz continuity of $\lambda \to x(\lambda)$ we note that $x(\lambda)$ satisfies the necessary optimality condition

$$\langle f'(x(\lambda)), x - x(\lambda)\rangle + \big(\varphi_c'(\Lambda x(\lambda), \lambda),\ \Lambda(x - x(\lambda))\big) \geq 0 \quad \text{for all } x \in C. \tag{4.6.2}$$

Using this inequality at $\lambda,\ \mu \in H$, we obtain

$$\begin{aligned}&\langle f'(x(\mu)) - f'(x(\lambda)),\ x(\mu) - x(\lambda)\rangle\\ &\quad +\big(\varphi_c'(\Lambda x(\mu), \mu) - \varphi_c'(\Lambda x(\lambda), \lambda),\ \Lambda(x(\mu) - x(\lambda))\big) \leq 0.\end{aligned}$$

By (A2) and (4.5.6) this implies that

$$\sigma\, |x(\mu) - x(\lambda)|_X^2 + \left(A_{\frac{1}{c}}\left(\Lambda x(\mu) + \frac{\mu}{c}\right) - A_{\frac{1}{c}}\left(\Lambda x(\lambda) + \frac{\lambda}{c}\right),\ \Lambda(x(\mu) - x(\lambda))\right) \leq 0,$$

and thus

$$\begin{aligned}&\sigma\, |x(\mu) - x(\lambda)|_X^2 + \left(A_{\frac{1}{c}}\left(\Lambda x(\mu) + \frac{\mu}{c}\right) - A_{\frac{1}{c}}\left(\Lambda x(\lambda) + \frac{\mu}{c}\right),\ \Lambda(x(\mu) - x(\lambda))\right)\\ &\leq \left(A_{\frac{1}{c}}\left(\Lambda x(\lambda) + \frac{\lambda}{c}\right) - A_{\frac{1}{c}}\left(\Lambda x(\lambda) + \frac{\mu}{c}\right),\ \Lambda(x(\mu) - x(\lambda))\right).\end{aligned}$$

Since $A_{\frac{1}{c}}$ is monotone and Lipschitz continuous with Lipschitz constant c, this inequality yields

$$\sigma\, |x(\mu) - x(\lambda)|_X \le |\mu - \lambda|_H,$$

which shows the first assertion.

Step 2. We show that for every $v \in H$ the functional $\lambda \in H \to \varphi_c(v, \lambda)$ is Fréchet differentiable with F-derivative $\frac{\partial}{\partial\lambda}\varphi_c(v, \lambda)$ given by $u(v, \lambda)$. Since (4.4.2) can be equivalently written as (4.4.3), it follows from the proof of Theorem 4.39 that $\lambda \in H \to \varphi_c(v, \lambda)$ is Fréchet differentiable and

$$\frac{\partial}{\partial\lambda}\varphi_c(v, \lambda) = \frac{1}{c}\,(\lambda + c\,u(v, \lambda)) - \frac{\lambda}{c} = u(v, \lambda).$$

Step 3. To argue differentiability of $\lambda \to d_c(\lambda)$ note that it follows from (4.6.2) that for $\lambda,\ \mu \in H$

$$\begin{aligned} d_c(\mu) - d_c(\lambda) &= f(x(\mu)) - f(x(\lambda)) + \varphi_c(\Lambda x(\mu), \lambda) - \varphi_c(\Lambda x(\lambda), \lambda) \\ &\quad + \varphi_c(\Lambda x(\mu), \mu) - \varphi_c(\Lambda x(\mu), \lambda) \\ &\ge \langle f'(x(\lambda)),\ x(\mu) - x(\lambda)\rangle + \big(\varphi_c'(\Lambda x(\lambda), \lambda),\ \Lambda(x(\mu) - x(\lambda))\big) \qquad (4.6.3) \\ &\quad + \varphi_c(\Lambda x(\mu), \mu) - \varphi_c(\Lambda x(\mu), \lambda) \\ &\ge \varphi_c(\Lambda x(\mu), \mu) - \varphi_c(\Lambda x(\mu), \lambda). \end{aligned}$$

Similarly, we have

$$\begin{aligned} d_c(\mu) - d_c(\lambda) &= f(x(\mu)) - f(x(\lambda)) + \varphi_c(\Lambda x(\mu), \mu) - \varphi_c(\Lambda x(\lambda), \mu) \\ &\quad + \varphi_c(\Lambda x(\lambda), \mu) - \varphi_c(\Lambda x(\lambda), \lambda) \\ &\le -\langle f'(x(\mu)),\ x(\lambda) - x(\mu)\rangle + \langle \varphi_c'(\Lambda x(\mu), \mu),\ \Lambda(x(\lambda) - x(\mu))\rangle \qquad (4.6.4) \\ &\quad + \varphi_c(\Lambda x(\lambda), \mu) - \varphi_c(\Lambda x(\lambda), \lambda) \\ &\le \varphi_c(\Lambda x(\lambda), \mu) - \varphi_c(\Lambda x(\lambda), \lambda). \end{aligned}$$

It follows from Step 2 and Theorem 4.39 that

$$\begin{aligned} &|\varphi_c(\Lambda x(\mu), \mu) - \varphi_c(\Lambda x(\mu), \lambda) - (u(\Lambda x(\lambda)\lambda), \mu - \lambda)| \\ &\le \left| \int_0^1 (u(\Lambda x(\mu), \lambda + t\,(\mu - \lambda)) - u(\Lambda x(\lambda), \lambda), \mu - \lambda)\, dt \right| \\ &\le \frac{1}{c}\,|\mu - \lambda| \int_0^1 \left| A_{\frac{1}{c}}\left(\Lambda x(\mu) + \frac{1}{c}\,(\lambda + t(\mu - \lambda))\right) - A_{\frac{1}{c}}\left(\Lambda x(\lambda) + \frac{\lambda}{c}\right) + t(\lambda - \mu) \right| dt \\ &\le \frac{1}{c}\,|\mu - \lambda| \int_0^1 (c\|\Lambda\||x(\mu) - x(\lambda)| + 2t\,|\mu - \lambda|)\, dt \le \left(\frac{\|\Lambda\|}{\sigma} + \frac{1}{c}\right) |\mu - \lambda|^2. \end{aligned}$$

Hence (4.6.3)–(4.6.4) and Step 2 imply that $\lambda \in H \to d_c(\lambda)$ is Fréchet differentiable with F-derivative given by $u(\Lambda x(\lambda), \lambda)$, where $u(\Lambda x(\lambda), \lambda)$ is Lipschitz continuous in λ. □

The following theorem asserts convergence of the augmented Lagrangian method.

Theorem 4.48. *Assume that* (A1)–(A3) *hold and that there exists* $\bar{\lambda} \in \partial\varphi(\Lambda\bar{x})$ *such that* (4.5.7) *is satisfied. Then the sequence* (x_k, λ_k) *is well defined and satisfies*

$$\frac{\sigma}{2}\,|x_k - \bar{x}|_X^2 + \frac{1}{2c}\,|\lambda_{k+1} - \bar{\lambda}|_H^2 \le \frac{1}{2c}\,|\lambda_k - \bar{\lambda}|_H^2 \tag{4.6.5}$$

and

$$\sum_{k=1}^{\infty} \frac{\sigma}{2}\,|x_k - \bar{x}|_X^2 \le \frac{1}{2c}\,|\lambda_1 - \bar{\lambda}|_H^2, \tag{4.6.6}$$

which implies that $|x_k - \bar{x}|_X \to 0$ *as* $k \to \infty$.

Proof. It follows from Theorem 4.45 that $\bar{\lambda} = \varphi_c'(\bar{v}, \bar{\lambda})$, where $\bar{v} = \Lambda\bar{x}$. Next, we establish (4.6.5). From (4.5.5) and Step 3

$$\langle f'(x_k),\, \bar{x} - x_k\rangle + \big(\lambda_{k+1},\, \Lambda(\bar{x} - x_k)\big) \ge 0$$

and from (4.5.7)

$$\langle f'(\bar{x}),\, x_k - \bar{x}\rangle + \big(\bar{\lambda},\, \Lambda(x_k - \bar{x})\big) \ge 0.$$

Adding these two inequalities, we obtain

$$\langle f'(x_k) - f'(\bar{x}),\, x_k - \bar{x}\rangle + \big(\lambda_{k+1} - \bar{\lambda},\, \Lambda(x_k - \bar{x})\big) \le 0. \tag{4.6.7}$$

From Theorems 4.39 and 4.40

$$\lambda_{k+1} - \bar{\lambda} = A_{\frac{1}{c}}\left(v_k + \frac{\lambda_k}{c}\right) - A_{\frac{1}{c}}\left(\bar{v} + \frac{\bar{\lambda}}{c}\right),$$

where $v_k = \Lambda x_k$ and $\bar{v} = \Lambda\bar{x}$. From the definition of A_μ we have

$$\langle A_\mu v - A_\mu\hat{v},\, v - \hat{v}\rangle = \mu\,|A_\mu v - A_\mu\hat{v}|^2 + \langle A_\mu v - A_\mu\hat{v},\, J_\mu v - J_\mu\hat{v}\rangle \ge \mu\,|A_\mu v - A_\mu\hat{v}|^2$$

for $\mu > 0$ and $v,\ \hat{v} \in H$, since $A_\mu v \in AJ_\mu v$ and A is monotone. Thus,

$$(\lambda_{k+1} - \bar{\lambda},\, v_k - \bar{v}) = \left(\lambda_{k+1} - \bar{\lambda},\, v_k + \frac{\lambda_k}{c} - \left(\bar{v} + \frac{\bar{\lambda}}{c}\right)\right)$$

$$-\frac{1}{c}(\lambda_{k+1} - \bar{\lambda},\, \lambda_k - \bar{\lambda}) \ge \frac{1}{c}|\lambda_{k+1} - \bar{\lambda}|^2 - \frac{1}{c}(\lambda_{k+1} - \bar{\lambda},\, \lambda_k - \bar{\lambda})$$

$$\ge \frac{1}{2c}\,|\lambda_{k+1} - \bar{\lambda}|^2 - \frac{1}{2c}\,|\lambda_k - \bar{\lambda}|^2.$$

Hence, (4.6.5) follows from (A2) and (4.6.7). Summing up (4.6.5) with respect to k, we obtain (4.6.6). □

The duality (Uzawa) method is an iterative method for $\lambda_k \in H$. It is given by

$$\lambda_{k+1} = \varphi_c'(\Lambda x_k, \lambda_k), \tag{4.6.8}$$

where $x_k \in C$ solves

$$\langle f'(x_k) + \Lambda^* \lambda_k, x - x_k \rangle \geq 0 \quad \text{for all } x \in C. \tag{4.6.9}$$

It will be shown that the Uzawa method is conditionally convergent in the sense that there exists $0 < \underline{c} < \bar{c}$ such that it converges for $c \in [\underline{c}, \bar{c}]$. On the other hand, the augmented Lagrangian method can be written as

$$\lambda_{k+1} = \varphi_c'(\Lambda x_k, \lambda_k),$$

where $x_k \in C$ satisfies

$$\langle f'(x_k) + \Lambda^* \lambda_{k+1}, x - x_k \rangle \geq 0 \quad \text{for all } x \in C.$$

Note that the Uzawa method is explicit with respect to λ while the augmented Lagrangian method is implicit and converges unconditionally (see Theorem 4.48) with respect to c.

Theorem 4.49 (Uzawa Algorithm). *Assume that* (A1)–(A3) *hold and that there exists* $\bar{\lambda} \in \partial\varphi(\Lambda\bar{x})$ *such that* (4.5.7) *is satisfied. Then there exists* $\bar{c}$ *such that for the sequence* (x_k, λ_k) *generated by* (4.6.8)–(4.6.9) *we have* $|x_k - \bar{x}|_X \to 0$ *as* $k \to \infty$ *if* $0 < c \leq \bar{c}$.

Proof. As shown in the proof of Theorem 4.48 it follows from (4.5.7) and (4.6.9) that

$$\langle f'(x_k) - f'(\bar{x}),\ x_k - \bar{x} \rangle + \langle \lambda_k - \bar{\lambda},\ \Lambda(x_k - \bar{x}) \rangle \leq 0. \tag{4.6.10}$$

Since

$$\lambda_{k+1} - \bar{\lambda} = A_{\frac{1}{c}}\left(v_k + \frac{\lambda_k}{c}\right) - A_{\frac{1}{c}}\left(\bar{v} + \frac{\bar{\lambda}}{c}\right),$$

where $v_k = \Lambda x_k$, and since $\bar{v} = \Lambda\bar{x}$ and $A_{\frac{1}{c}}$ is Lipschitz continuous with Lipschitz constant c,

$$|\lambda_{k+1} - \bar{\lambda}|^2 \leq |\lambda_k - \bar{\lambda} + c\,(v_k - \bar{v})|^2$$

$$= |\lambda_k - \bar{\lambda}|^2 + 2c\langle \lambda_k - \bar{\lambda}, v_k - \bar{v} \rangle + c^2\,|v_k - \bar{v}|^2$$

$$\leq |\lambda_k - \bar{\lambda}|^2 - (2c\sigma - c^2\|\Lambda\|^2)|x_k - \bar{x}|^2,$$

where (4.6.10) was used for the last inequality. Choose $\bar{c} < 2\sigma(\|\Lambda\|)^{-1}$. Then $\beta = 2c\sigma - c^2\|\Lambda\|^2 > 0$ if $c \in (0, \bar{c}]$ and

$$\beta \sum_{k=0}^{\infty} |x_k - \bar{x}|^2 \leq |\lambda_0 - \bar{\lambda}|^2.$$

Consequently $|x_k - \bar{x}| \to 0$ as $k \to \infty$. $\square$

4.7 Applications

In this section we discuss applications of the results in Sections 4.5 and 4.6. In many cases the conjugate functional φ^* is given by

$$\varphi^*(v) = \psi_{K^*}(v),$$

where K^* is a closed convex set in H and ψ_S is the indicator function of a set S, i.e.,

$$\psi_S(x) = \begin{cases} 0 & \text{if } x \in S, \\ \infty & \text{if } x \notin S. \end{cases}$$

Then it follows from Theorem 4.40 that for $v, \lambda \in H$

$$\varphi_c(v, \lambda) = \sup_{y^* \in K^*} \left\{ -\frac{1}{2c}|y^* - (\lambda + c\,v)|_H^2 \right\} + \frac{1}{2c}(|\lambda + c\,v|_H^2 - |\lambda|_H^2). \tag{4.7.1}$$

Hence the supremum is attained at $\lambda_c(v, \lambda) = \text{Proj}_{K^*}(\lambda + c\,v)$, where $\text{Proj}_{K^*}(\phi)$ denotes the projection of $\phi \in H$ onto K^*. This implies that the update in Step 3 of the augmented Lagrangian algorithm is given by

$$\lambda_{k+1} = \text{Proj}_{K^*}(\lambda_k + c\,\Lambda x_k). \tag{4.7.2}$$

Example 4.50. For the equality constraint

$$\varphi(v) = \psi_K(v) \text{ with } K = \{0\}$$

and $\varphi^*(w) = 0$ for $w \in H$. Thus

$$\varphi_c(v, \lambda) = \frac{1}{2c}(|\lambda + c\,v|_H^2 - |\lambda|_H^2)$$

and

$$\lambda_{k+1} = \lambda_k + c\,\Lambda x_k,$$

which coincides with the first order augmented Lagrangian update for equality constraints discussed in Chapter 3.

Example 4.51. If $\varphi(v) = \psi_K(v)$, where K is a closed convex cone with vertex at the origin in H, then $\varphi^* = \psi_{K^+}$, where $K^+ = \{w \in H : \langle w, v\rangle_H \le 0 \text{ for all } v \in K\}$ is the dual cone of K. In particular, if $K = \{v \in L^2(\Omega) : v \le 0 \text{ a.e.}\}$, then $K^+ = \{w \in L^2(\Omega) : w \ge 0 \text{ a.e.}\}$. Thus for the inequality constraint in $H = L^2(\Omega)$ the update (4.7.2) becomes

$$\lambda_{k+1} = \max(0, \lambda_k + c\,\Lambda x_k),$$

where the max operation is defined pointwise in Ω. Here $L^2(\Omega)$ denotes the space of scalar-valued square integrable functions over a domain Ω in $\mathbb{R}^n$. For $K = \{v \in \mathbb{R}^n : v_i \le 0 \text{ for all } i\}$ the update (4.7.2) is given by $\lambda_{k+1} = \max(0, \lambda_k + c\Lambda x_k)$, where the maximum is taken coordinatewise. It coincides with the first order augmented Lagrangian update for finite rank inequality constraints in Chapter 3.

Example 4.52. If $\varphi(v) = |v|_H$, then $\varphi^* = \psi_B$, where B is the closed unit ball in H. Thus the update (4.7.2) becomes

$$\lambda_{k+1} = \frac{\lambda_k + c\,\Lambda_k x_k}{\max(1, |\lambda + c\,\Lambda x_k|)}.$$

Example 4.53. We consider the bilateral inequality constraint $\phi \le v \le \psi$ in $L^2(\Omega)$. We set

$$K = \left\{ v \in L^2(\Omega) : -\frac{\psi - \phi}{2} \le v - \frac{\phi + \psi}{2} \le \frac{\psi - \phi}{2} \right\}$$

and $\varphi = \psi_K$. It follows that

$$\varphi^*(w) = \left\langle w, \frac{\phi + \psi}{2} \right\rangle_{L^2} + \left\langle \frac{\psi - \phi}{2}, |w| \right\rangle_{L^2}.$$

From Theorem 4.40 we conclude that the expression $\langle x, w \rangle - \phi^*(w) - \frac{1}{2c}|w - \lambda|^2$ is maximized with respect to w at the element y^* satisfying

$$x - \frac{\phi + \psi}{2} - \frac{y^* - \lambda}{c} \in \frac{\psi - \phi}{2}\, \partial(|\cdot|)(y^*)$$

a.e. in Ω. Thus it follows from Theorem 4.40 that the complementarity condition (4.1.8) is given by

$$\bar{\lambda} = \max(0, \bar{\lambda} + c\,(\Lambda\bar{x} - \psi)) + \min(0, \bar{\lambda} + c\,(\Lambda\bar{x} - \phi)),$$

and the Lagrange multiplier update in Step 3 of the augmented Lagrangian method is

$$\lambda_{k+1} = \max(0, \lambda_k + c\,(\Lambda x_k - \psi)) + \min(0, \lambda_k + c\,(\Lambda x_k - \phi)),$$

where the max and min operations are defined pointwise a.e. in Ω.

Note that with obvious modifications, Λx in Sections 4.5 and 4.6 can be replaced by the affine function of the form $\Lambda x + a$ with $a \in H$.

4.7.1 Bingham flow

We consider the problem

$$\min \int_\Omega \left(\frac{\mu}{2}|\nabla u|^2 - \tilde{f}\,u \right) dx + g \int_\Omega |\nabla u|\, dx \quad \text{over } u \in H_0^1(\Omega), \tag{4.7.3}$$

where Ω is a bounded open set in $\mathbb{R}^2$ with Lipschitz boundary, g and μ are positive constants, and $\tilde{f} \in L^2(\Omega)$. For a discussion of (4.7.3) we refer the reader to [Glo, GLT] and the references therein. In the context of the general theory of Section 4.5 we choose

$$X = H_0^1(\Omega),\ H = L^2(\Omega) \times L^2(\Omega), \quad \text{and} \quad \Lambda = g\,\text{grad},$$

$K = X$, and define $f : X \to \mathbb{R}$ and $\varphi : H \to \mathbb{R}$ by

$$f(u) = \int_\Omega \left(\frac{\mu}{2} |\nabla u|^2 - \tilde{f}\, u \right) dx$$

and

$$\varphi(v_1, v_2) = \int_\Omega \sqrt{v_1^2 + v_2^2}\, dx.$$

Conditions (A1)–(A3) are clearly satisfied. Since $\overline{\text{dom}(\varphi)} = H$ it follows from Theorem 4.45 and Lemma 4.44 that there exists $\bar{\lambda}$ such that (4.5.7) holds. Moreover, it is not difficult to show that

$$\varphi^*(v) = \psi_{K^*}(v),$$

where K^* is given by

$$K^* = \{v \in H : |v(x)|_{\mathbb{R}^2} \leq 1 \text{ a.e. in } \Omega\}.$$

Hence it follows from (4.7.2) that Steps 2 and 3 in the augmented Lagrangian method are given by

$$-\mu\, \Delta u_k - g \operatorname{div} \lambda_{k+1} = \tilde{f}, \tag{4.7.4}$$

where

$$\lambda_{k+1} = \begin{cases} \lambda_k + c\, \nabla u_k & \text{on } A_k = \{x : |\lambda_k(x) + c\, \nabla u_k(x)|_{\mathbb{R}^2} \leq 1\}, \\ \dfrac{\lambda_k + c\, \nabla u_k}{|\lambda_k + c\, \nabla u_k|_{\mathbb{R}^2}} & \text{on } \Omega \setminus A_k. \end{cases} \tag{4.7.5}$$

4.7.2 Image restoration

For the image restoration problem introduced in (4.1.2) the analysis is similar to the one for the Bingham flow and Steps 2 and 3 in the augmented Lagrangian method are given by

$$-\mu\, \Delta u_k + \mathcal{K}^*(\mathcal{K} u_k - z) + g \operatorname{div} \lambda_{k+1} = 0, \quad \mu \frac{\partial u}{\partial n} + g\, \lambda_{k+1} = 0 \text{ on } \Gamma,$$

where

$$\lambda_{k+1} = \begin{cases} \lambda_k + c\, \nabla u_k & \text{on } A_k = \{x : |\lambda_k(x) + c\, \nabla u_k(x)|_{\mathbb{R}^2} \leq 1\}, \\ \dfrac{\lambda_k + c\, \nabla u_k}{|\lambda_k + c\, \nabla u_k|_{\mathbb{R}^2}} & \text{on } \Omega \setminus A_k \end{cases}$$

in the strong form. Here $\mathcal{K} : L^2(\Omega) \to L^2(\Omega)$ denotes the convolution operator defined by

$$(\mathcal{K} u)(x) = \int_\Omega k(x, s) u(s)\, ds.$$

4.7.3 Elastoplastic problem

We consider the problem

$$\min \int_\Omega \left(\frac{\beta}{2} |\nabla u|^2 - \tilde{f}\, u \right) dx \quad \text{over } u \in H_0^1(\Omega)$$
$$\text{subject to} \quad |\nabla u| \le 1 \ \text{ a.e. in } \Omega, \tag{4.7.6}$$

where β is a positive constant, Ω is a bounded domain in $\mathbb{R}^2$, and $\tilde{f} \in L^2(\Omega)$. In the context of the general theory we choose

$$X = H_0^1(\Omega), \ H = L^2(\Omega) \times L^2(\Omega), \quad \text{and} \quad \Lambda = \nabla,$$

$K = X$, and define $f : X \to \mathbb{R}$ and $\varphi : H \to \mathbb{R}$ by

$$f(u) = \int_\Omega \left(\frac{\beta}{2} |\nabla u|^2 - \tilde{f}\, u \right) dx \quad \text{and} \quad \varphi = \psi_{\hat{C}},$$

where $\hat{C}$ is the closed convex set defined by $C = \{v \in H : |v|_{\mathbb{R}^2} \le 1 \text{ a.e. in } \Omega\}$. Then

$$\varphi^*(w) = \int_\Omega |w|_{\mathbb{R}^2}\, dx.$$

The maximum of (4.4.8), i.e., $\langle \nabla u, w \rangle - \phi^*(w) - \frac{1}{2c}|w - \lambda|$ with respect to $w \in L^2(\Omega) \times L^2(\Omega)$, is attained at y^* such that

$$\lambda + c\,\nabla u = y^* + c\mu, \ \text{where } |\mu| \le 1 \text{ and } \mu \cdot y^* = |y^*|$$

a.e. in Ω. It thus follows from Theorem 4.40 that the Lagrange multiplier update is given by

$$\lambda^{k+1} = c \max\left(0, |\frac{\lambda^k + c\,\nabla u^k}{c}| - 1\right) \frac{\lambda^k + c\,\nabla u^k}{|\lambda^k + c\,\nabla u^k|}$$

a.e. in Ω. The existence of Lagrange multiplier $\bar{\lambda} \in L^\infty(\Omega)$ for the inequality constraint in (4.7.6) is shown in [Bre2] for $\tilde{f} = 1$. In general, existence is still an open problem.

4.7.4 Obstacle problem

We consider the problem

$$\min \int_\Omega \left(\frac{1}{2} |\nabla u|^2 - \tilde{f}\, u \right) dx \quad \text{over } u \in H_0^1(\Omega)$$
$$\text{subject to} \quad \phi \le u \le \psi \ \text{ a.e. in } \Omega. \tag{4.7.7}$$

In the context of the general theory we choose

$$X = H_0^1(\Omega), \ H = L^2(\Omega), \quad \text{and} \quad \Lambda = \text{the natural injection},$$

$C = X$, and define $f : X \to \mathbb{R}$ and $\varphi : H \to \mathbb{R}$ by

$$f(u) = \int_\Omega \left(\frac{1}{2} |\nabla u|^2 - \tilde{f}\, u \right) dx \quad \text{and} \quad \varphi(v) = \psi_K,$$

where K is the closed convex set defined by $K = \{v \in H : \phi \le v \le \psi \text{ a.e. in } \Omega\}$. For the one-sided constraint $u \le \psi$ (i.e., $\phi = -\infty$) it is shown in [IK4], for example, that there exists a unique $\bar{\lambda} \in \partial\varphi(\bar{u})$ such that (4.5.7) is satisfied provided that $\psi \in H^1(\Omega)$, $\psi|_\Gamma \ge 0$, and $\sup(0, \tilde{f} + \Delta\psi) \in L^2(\Omega)$. In fact, the optimal pair $(\bar{u}, \bar{\lambda}) \in (H^2 \cap H_0^1) \times L^2$ satisfies

$$\begin{aligned} -\Delta\bar{u} + \bar{\lambda} &= \tilde{f}, \\ \bar{\lambda} &= \max(0, \bar{\lambda} + c\,(\bar{u} - \psi)). \end{aligned} \tag{4.7.8}$$

In this case Steps 2 and 3 in the augmented Lagrangian method is given by

$$\begin{aligned} -\Delta u_k + \lambda_{k+1} &= \tilde{f}, \\ \lambda_{k+1} &= \max(0, \lambda_k + c\,(u_k - \psi)). \end{aligned}$$

For the two-sided constraint case we assume that $\phi, \psi \in H^1(\Omega)$ satisfy

$$\phi \le 0 \le \psi \quad \text{on } \Gamma, \max(0, \Delta\psi + \tilde{f}),\ \min(0, \Delta\phi + \tilde{f}) \in L^2(\Omega), \tag{4.7.9}$$

$$S_1 = \{x \in \Omega : \Delta\psi + \tilde{f} > 0\} \cap S_2 = \{x \in \Omega : \Delta\phi + \tilde{f} < 0\} \ = \ \emptyset, \tag{4.7.10}$$

and that there exists a $c_0 > 0$ such that

$$-\Delta(\psi - \phi) + c_0\,(\psi - \phi) \ge 0 \quad \text{a.e. in } \Omega. \tag{4.7.11}$$

Let $\hat{\lambda} \in H$ be defined by

$$\hat{\lambda}(x) = \begin{cases} \Delta\psi(x) + \tilde{f}(x), & x \in S_1, \\ \Delta\phi(x) + \tilde{f}(x), & x \in S_2, \\ 0 & \text{otherwise.} \end{cases} \tag{4.7.12}$$

Employing the regularization procedure of (4.5.4) we find the following theorem.

Theorem 4.54. *Assume that $\phi, \psi \in H^1(\Omega)$ satisfy* (4.7.9)–(4.7.12) *where $\hat{\lambda}$ is defined in* (4.7.12)*. Then* (4.5.5) *is given by*

$$-\Delta u_c + \lambda_c = \tilde{f}, \quad \lambda_c = \begin{cases} \hat{\lambda} + c\,(u_c - \psi) & \text{if } \hat{\lambda} + c\,(u_c - \psi) > 0, \\ \hat{\lambda} + c\,(u_c - \phi) & \text{if } \hat{\lambda} + c\,(u_c - \phi) < 0, \\ 0 & \text{otherwise}, \end{cases} \tag{4.7.13}$$

and $\phi \le u_c \le \psi$, $|\lambda_c| \le |\hat{\lambda}|$ a.e. in Ω for $c \ge c_0$. Moreover, as $c \to \infty$, $u_c \rightharpoonup u^*$ weakly in $H^2(\Omega)$ and $\lambda_c \rightharpoonup \bar{\lambda}$ weakly in $L^2(\Omega)$ where $\bar{u}$ is the solution of (4.7.7) *and $(\bar{u}, \bar{\lambda})$ satisfies the necessary and sufficient optimality condition*

$$-\Delta\bar{u} + \bar{\lambda} = \tilde{f}, \quad \bar{\lambda} = \begin{cases} \bar{\lambda} + c\,(\bar{u} - \psi) & \text{if } \bar{\lambda} + c\,(\bar{u} - \psi) > 0, \\ \bar{\lambda} + c\,(\bar{u} - \phi) & \text{if } \bar{\lambda} + c\,(\bar{u} - \phi) < 0, \\ 0 & \text{otherwise}. \end{cases}$$

Proof. From Theorem 4.40 it follows that

$$\lambda_c(u,\hat\lambda) = \begin{cases} \hat\lambda + c\,(u-\psi) & \text{if } \hat\lambda + c\,(u-\psi) > 0, \\ \hat\lambda + c\,(u-\phi) & \text{if } \hat\lambda + c\,(u-\phi) < 0, \\ 0 & \text{otherwise.} \end{cases} \tag{4.7.14}$$

We note that (4.5.4) has a unique solution $u_c \in H_0^1(\Omega)$. From (4.5.5)–(4.5.6) we deduce that u_c satisfies (4.7.13). Since $\hat\lambda \in L^2(\Omega)$ it follows that $\lambda_c(u_c, \hat\lambda) \in L^2(\Omega)$ and thus $u_c \in H^2(\Omega)$. Let $\eta = \sup(0, u_c - \psi)$. Then $\eta \in H_0^1(\Omega)$. Hence, we have

$$(\nabla(u_c - \psi), \nabla\eta) + (-(\Delta\psi + \tilde f) + \lambda_c, \eta) = 0.$$

For $x \in \Omega$ assume that $u_c(x) \ge \psi(x)$. If $\hat\lambda(x) > 0$, then $-(\Delta\psi + \tilde f) + \lambda_c = c\,(u_c - \psi) \ge 0$. If $\hat\lambda(x) = 0$, then $-(\Delta\psi + \tilde f) + \lambda_c \ge c\,(u_c - \psi) \ge 0$. If $\hat\lambda(x) < 0$, then $-(\Delta\psi + \tilde f) + \lambda_c \ge -\Delta(\psi - \phi) + c\,(\psi - \phi) \ge 0$ for $c \ge c_0$. Thus, we have $(-(\Delta\psi + \tilde f) + \lambda_c, \eta) \ge 0$ and $|\nabla\eta|^2 = 0$. This implies that $\eta = 0$ and $u_c \le \psi$ a.e. in Ω. Similarly, we can prove that $u_c \ge \phi$ a.e. in Ω by choosing the test function $\eta = \inf(0, u_c - \phi) \in H_0^1(\Omega)$. Moreover, it follows from (4.7.14) that $|\lambda_c| = |\lambda_c(u_c, \hat\lambda)| \le |\hat\lambda|$ a.e. in Ω and $|\lambda_c|_{L^2}$ is uniformly bounded. Thus, there exists a weakly convergent subsequence $(u_{\hat c}, \lambda_{\hat c}) \rightharpoonup (\bar u, \bar\lambda)$ in $H^2(\Omega) \times L^2(\Omega)$. Moreover the subsequence $u_{\hat c}$ converges strongly to $\bar u$ in $H_0^1(\Omega)$ and, as shown in the proof of Theorem 4.46,

$$-\Delta\bar u + \bar\lambda = \tilde f \quad \text{and} \quad \bar\lambda \in \partial\varphi(\bar u). \tag{4.7.15}$$

Hence, it follows from Theorem 4.43 that $\bar u$ minimizes (4.7.7). Since the solution to (4.7.7) is unique it follows from (4.7.15) that $\bar\lambda \in L^2(\Omega)$ is unique. The theorem now follows from Theorem 4.45.

From Theorem 4.54 it follows that Steps 2 and 3 in the augmented Lagrangian method for the two-sided constraint are given by

$$-\Delta u_k + \lambda_{k+1} = \tilde f, \qquad \lambda_{k+1} = \begin{cases} \lambda_k + c\,(u_k - \psi) & \text{if } \lambda_k + c\,(u_k - \psi) > 0, \\ \lambda_k + c\,(u_k - \phi) & \text{if } \lambda_k + c\,(u_k - \phi) < 0, \\ 0 & \text{otherwise.} \end{cases}$$

Note that the last equality can be equivalently expressed as

$$\lambda_{k+1} = \max(0, \lambda_k + c\,(u_k - \psi)) + \min(0, \lambda_k + c\,(u_k - \phi)). \qquad \square$$

4.7.5 Signorini problem

Consider the Signorini problem

$$\min \int_\Omega \frac{1}{2}(|\nabla u|^2 + |u|^2) - \tilde f\, u\, dx \quad \text{over } u \in H^1(\Omega)$$
$$\text{subject to } u \ge 0 \quad \text{on the boundary } \Gamma, \tag{4.7.16}$$

which is a simplified version of a contact problem arising in elasticity theory. In this case we choose

$$X = H^1(\Omega),\ H = L^2(\Gamma), \quad \text{and} \quad \Lambda = \text{the trace operator on boundary } \Gamma,$$

$C = X$, and define $f : X \to \mathbb{R}$ and $\varphi : H \to \mathbb{R}$ by

$$f(u) = \int_\Omega \frac{1}{2}\,(|\nabla u|^2 + |u|^2) - \tilde{f}\,u\,dx \quad \text{and} \quad \varphi(v) = \psi_K,$$

where K is the closed convex set defined by $K = \{v \in L^2(\Omega) : v \geq 0 \text{ a.e. in } \Gamma\}$. If $\tilde{f} \in L^2(\Omega)$, then the unique minimizer $\bar{u} \in H^2(\Omega)$ [Bre2] and it is shown in [Glo] that for $\bar{\lambda} = \frac{\partial}{\partial n}\bar{u}$, the pair $(\bar{u}, \bar{\lambda})$ satisfies (4.5.7). Note that range(Λ) is dense in H and Λ is compact. It thus follows from Theorems 4.43 and 4.46 that (4.5.7) has a unique solution and that $u_c \to \bar{u}$ strongly in X and $\lambda_c \to \bar{\lambda}$ weakly in H. Steps 2 and 3 in the augmented Lagrangian method are given by

$$-\Delta u_k + u_k = \tilde{f}, \quad \frac{\partial}{\partial n}u_k - \lambda_{k+1} = 0 \quad \text{on } \Gamma,$$

$$\lambda_{k+1} = \max(0, \lambda_k - c\,u_k) \quad \text{on } \Gamma.$$

4.7.6 Friction problem

Consider a simplified friction problem from elasticity theory

$$\min \int_\Omega \frac{1}{2}\,(|\nabla u|^2 + |u|^2) - \tilde{f}\,u\,dx + g\int_\Gamma |u|\,ds \quad \text{over } u \in H^1(\Omega), \tag{4.7.17}$$

where Ω is a bounded open domain of $\mathbb{R}^2$ with sufficiently smooth boundary Γ. In this case we choose

$$X = H^1(\Omega),\ H = L^2(\Gamma), \quad \text{and} \quad \Lambda = \text{the trace operator on boundary } \Gamma,$$

$C = X$, and define $f : X \to \mathbb{R}$ and $\varphi : H \to \mathbb{R}$ by

$$f(u) = \int_\Omega \frac{1}{2}\,(|\nabla u|^2 + |u|^2) - \tilde{f}\,u\,dx \quad \text{and} \quad \varphi(v) = g\int_\Gamma |v|\,ds.$$

Note that $\varphi = \psi_{K^*}$, where $K^* = \{v \in H : |v(x)| \leq 1 \text{ a.e. in } \Gamma\}$. Since $\mathrm{dom}(\varphi) = H$ it follows from Theorem 4.43 and Lemma 4.44 that there exists $\bar{\lambda}$ such that (4.5.7) holds. From (4.7.2) Steps 2 and 3 in the augmented Lagrangian method are given by

$$-\Delta u_k + u_k = \tilde{f}, \quad \frac{\partial}{\partial n}u_k + \lambda_{k+1} = 0 \quad \text{on } \Gamma,$$

$$\lambda_{k+1} = \begin{cases} \lambda_k + c\,u_k & \text{on } \Gamma_k = \{x \in \Gamma : |\lambda_k + c\,u_k| \leq 1\}, \\ \dfrac{\lambda_k + c\,u_k}{|\lambda_k + c\,u_k|} & \text{on } \Gamma \setminus \Gamma_k. \end{cases}$$

4.7.7 L^1-fitting

Consider the minimization problem

$$\min \int_\Omega |u - z|\, dx + \frac{\mu}{2} \int_\Omega |\nabla u|^2\, ds \quad \text{over } u \in H_0^1(\Omega) \tag{4.7.18}$$

for interpolation of noisy data $z \in L^2(\Omega)$ by minimizing the L^1-norm of the error $u - z$ over $u \in H_0^1(\Omega)$. Again $\mu > 0$ is fixed but should be adjusted to the statistics of noise. The analysis of this problem is very similar to that of a friction problem and Steps 2 and 3 in the augmented Lagrangian method are given by

$$-\mu\, \Delta u_k + \lambda_{k+1} = 0,$$

$$\lambda_{k+1} = \begin{cases} \lambda_k + c\,(u_k - z) & \text{on } \Omega_k = \{x : |\lambda_k(x) + c\,(u_k(x) - z)| \le 1\}, \\ \dfrac{\lambda_k + c\,(u_k - z)}{|\lambda_k + c\,(u_k - z)|} & \text{on } \Omega \setminus \Omega_k. \end{cases}$$

4.7.8 Control problem

Consider the optimal control problem

$$\begin{aligned} &\min \frac{1}{2} \int_0^T |x(t)|^2 + |u(t)|^2\, dt \quad \text{over } (x, u) \in L^2(0, T; \mathbb{R}^n) \times L^2(0, T; \mathbb{R}^m) \\ &\text{subject to } x(t) - x_0 - \int_0^t Ax(s) + Bu(s)\, ds = 0 \quad \text{and} \quad |u(t)| \le 1 \text{ a.e.,} \end{aligned} \tag{4.7.19}$$

where $A \in \mathbb{R}^{n \times n}$, $B \in \mathbb{R}^{n \times m}$, and $x_0 \in \mathbb{R}^n$ are fixed. In this case we formulate the problem as in the form (4.1.1) by choosing

$$X = L^2(0, T; \mathbb{R}^n) \times L^2(0, T; \mathbb{R}^m), \;\; H = L^2(0, T; \mathbb{R}^m), \quad \text{and} \quad \Lambda(x, u) = u \text{ for } (x, u) \in X,$$

and define $f : X \to \mathbb{R}$ and $\varphi : H \to$ by

$$f(x, u) = \frac{1}{2} \int_0^T (|x(t)|^2 + |u(t)|^2)\, dt \quad \text{and} \quad \varphi(v) = \psi_K(v),$$

where $K = \{v \in H : |v(t)| \le 1 \text{ a.e. in } (0, T)\}$. C is the closed affine space defined by

$$C = \left\{ (x, u) \in X : x(t) - x_0 - \int_0^t Ax(s) + Bu(s)\, ds = 0 \right\}.$$

It follows from Theorems 4.39 and 4.40 that

$$\lambda_c = c \left(v + \frac{\lambda}{c} - \frac{v + \frac{\lambda}{c}}{\max(1, |v + \frac{\lambda}{c}|)} \right) = c \max\left(0, \left|v + \frac{\lambda}{c}\right| - 1\right) \frac{\lambda + c\,v}{|\lambda + c\,v|} \tag{4.7.20}$$

for $v, \lambda \in H$. For $c > 0$ the regularized problem

$$f_c(x,u) = f(x,u) + \varphi_c(u,0) \quad \text{over } (x,u) \in C$$

has a unique solution $(x_c, u_c) \in C$. Define the Lagrangian

$$L(x,u,\mu) = f_c(x,u) + \int_0^T \left(x(t) - x_0 - \int_0^t Ax(s) + Bu(s)\,ds,\ \mu(t)\right)_{\mathbb{R}^n} dt.$$

Since the mapping $F : X \to L^2(0,T;\mathbb{R}^n)$ defined by

$$F(x,u) = x(t) - \int_0^t Ax(s) + Bu(s)\,ds, \qquad t \in (0,T),$$

is surjective it follows from the Lagrange multiplier theory that there exists a unique Lagrange multiplier $\mu_c \in L^2(0,T;\mathbb{R}^n)$ such that the Fréchet derivative of L with respect to (x,u) satisfies $L'(x_c,u_c,\mu_c)(h,v) = 0$ for all $(h,v) \in X$. Hence we obtain

$$\int_0^T \left[\left(h(t) - \int_0^t Ah(s)\,ds,\ \mu_c(t)\right) + (x(t),\, h(t))\right] dt = 0$$

for all $h \in L^2(0,T;\mathbb{R}^n)$ and

$$\int_0^T \left[\left(u(t) + c\,\max(0, |u_c| - 1)\frac{u_c}{|u_c|},\ v(t)\right) - \left(\int_0^t Bv(s)\,ds,\ \mu_c(t)\right)\right] dt = 0$$

for all $v \in L^2(0,T;\mathbb{R}^m)$. It is not difficult to show that if we define $p_c(t) = -\int_t^T \mu_c(s)\,ds$, then (x_c, u_c, p_c) satisfies

$$\begin{aligned} &\frac{d}{dt}p_c(t) + A^T p_c(t) + x_c(t) = 0, \qquad p_c(T) = 0, \\ &u_c(t) + c\,\max(0, |u_c(t)| - 1)\frac{u_c(t)}{|u_c(t)|} = -B^T p_c(t). \end{aligned} \tag{4.7.21}$$

Let $(\bar{x}, \bar{u})$ be a feasible point of (4.7.19); i.e., $(\tilde{x}, \tilde{u}) \in C$ and $\tilde{u} \in K$. Since $f_c(x_c, u_c) \le f_c(\tilde{x}, \tilde{u}) = f(\tilde{x}, \tilde{u})$ and $\varphi_c(u_c, 0) \ge 0$ it follows that $|(x_c, u_c)|_X$ is uniformly bounded in $c > 0$. Thus, there exists a subsequence of (x_c, u_c) that converges weakly to $(\bar{x}, \bar{u}) \in X$ as $c \to \infty$. We shall show that $(\bar{x}, \bar{u})$ is the solution of (4.7.19). First, since $F(x_c, u_c) = x_0$, x_c converges strongly to $\bar{x}$ in $L^2(0,T;\mathbb{R}^n)$ and weakly in $H^1(0,T;\mathbb{R}^n)$ and $(\bar{x}, \bar{u}) \in C$. From the first equation of (4.7.21) it follows that p_c converges strongly to $\bar{p}$ in $H^1(0,T;\mathbb{R}^n)$, where $\bar{p}$ satisfies

$$\frac{d}{dt}\bar{p}(t) + A^T\bar{p}(t) + \bar{x}(t) = 0, \qquad \bar{p}(T) = 0.$$

Next, from the second equation of (4.7.21) we have

$$|u_c(t)| + c\,\max(0, |u_c(t)| - 1) = |B^T p_c(t)|$$

and thus

$$|u_c(t)| = \begin{cases} |B^T p_c(t)| & \text{if } |B^T p_c(t)| \le 1, \\ 1 + \dfrac{|B^T p_c(t)| - 1}{c+1} & \text{if } |B^T p_c(t)| \ge 1, \end{cases} \quad t \in [0, T]. \tag{4.7.22}$$

Note that

$$u_c(t) = \frac{-B^T p_c(t)}{1 + \eta_c(t)|u_c(t)|^{-1}},$$
$$\text{where } \eta_c(t) = c \max(0, |u_c(t)| - 1) = \frac{c}{1+c} \max(0, |B^T p_c(t)| - 1). \tag{4.7.23}$$

Define the functions $\bar{\eta}, \hat{u} \in H^1(0, T)$ by

$$\bar{\eta}(t) = \max(0, |B^T \bar{p}(t)| - 1) \quad \text{and} \quad \hat{u}(t) = \frac{-B^T \bar{p}(t)}{1 + \bar{\eta}(t)} = \frac{-B^T \bar{p}(t)}{\max(1, |B^T \bar{p}(t)|)} \tag{4.7.24}$$

for $t \in [0, T]$. Then

$$|\hat{u}(t)| = \frac{|B^T \bar{p}(t)|}{\max(1, |B^T \bar{p}(t)|)}.$$

Since $B^T p_c \to B^T \bar{p}$ in $C(0, T)$ we have $|u_c| \to |\hat{u}|$ in $C(0, T)$ by (4.7.22) and it follows from (4.7.23)–(4.7.24) that $\eta_c \to \bar{\eta}$ and $u_c \to \hat{u}$ in $C(0, T)$. Since $u_c \to \bar{u}$ weakly in $L^2(0, T; \mathbb{R}^m)$ we have $\bar{u} = \hat{u}$. Therefore, we conclude that $u_c \to \bar{u}$ in $C(0, T)$ and if $\bar{\lambda}(t) = \bar{\eta}(t)\,\bar{u}(t)$, then $(\bar{x}, \bar{u}, \bar{p}, \bar{\lambda})$ satisfies $(\bar{x}, \bar{u}) \in C,\ \bar{u} \in C$ and

$$\begin{aligned} &\frac{d}{dt}\bar{p}(t) + A^T \bar{p}(t) + \bar{x} = 0, \quad \bar{p}(T) = 0, \\ &\bar{u}(t) + \bar{\lambda} = -B^T \bar{p}(t), \quad \bar{\lambda}(t) = \lambda_c(\bar{u}(t), \bar{\lambda}(t)) = \varphi_c'(\bar{u}, \bar{\lambda}), \end{aligned} \tag{4.7.25}$$

which is equivalent to (4.5.7). Now, from Theorem 4.43 we deduce that $(\bar{x}, \bar{u})$ is the solution to (4.7.19). The last equality in (4.7.25) can be verified by separately considering the cases $|\bar{u}(t)| < 1$ and $|\bar{u}(t)| = 1$. It follows from (4.7.20) that Steps 2 and 3 in the augmented Lagrangian method are given by

$$\frac{d}{dt}x_k(t) = Ax_k(t) + Bu_k(t), \quad x_k(0) = x_0,$$

$$\frac{d}{dt}p_k(t) + A^T p_k(t) + x_k = 0, \quad p_k(T) = 0,$$

$$u_k(t) + \lambda_{k+1}(t) = -B^T p_k(t), \quad \text{where } \lambda_{k+1} = c \max\left(0, \left|u_k + \frac{\lambda_k}{c}\right| - 1\right) \frac{\lambda_k + c\,u_k}{|\lambda_k + c\,u_k|}$$

for $t \in [0, T]$.

Chapter 5

Newton and SQP Methods

In this chapter we discuss the Newton method for the equality-constrained optimal control problem

$$\min J(y, u) \quad \text{subject to } e(y, u) = 0, \tag{P}$$

where $J : Y \times U \to \mathbb{R}$ and $e : Y \times U \to W$, and Y, U, and W are Hilbert spaces. The focus is on problems where for given u there exists a solution $y(u)$ to $e(y, u) = 0$, which is typical for optimal control problems. We shall refer to

$$\hat{J}(u) = J(y(u), u) \tag{5.0.1}$$

as the reduced cost functional. In the first section we give necessary and sufficient optimality conditions based on Taylor series expansion arguments. Section 5.2 is devoted to Newton's method to solve (P) and sufficient conditions for quadratic convergence are given. Section 5.3 contains a discussion of SQP (sequential quadratic programming) and reduced SQP techniques. We do not provide a convergence analysis here, since this can be obtained as a special case from the results on second order augmented Lagrangians in Chapter 6. The results are specialized to a class of optimal control problems for the Navier–Stokes equation in Section 5.4. Section 5.5 is devoted to the Newton method for weakly singular problems as introduced in Chapter 1.

5.1 Preliminaries

(I) Necessary Optimality Condition

The first objective is to characterize the derivative of $\hat{J}$ without recourse to the derivative of the state y with respect to the control u. This is in the same spirit as Theorem 1.17 but obtained under the less restrictive assumption of this chapter. We assume that

(C1) $(y, u) \in Y \times U$ satisfies $e(y, u) = 0$ and there exists a neighborhood $V(y) \times V(u)$ of (y, u) on which J and e are C^1 with Lipschitz continuous first derivatives such that for every $v \in V(u)$ there exists a unique $y(v) \in V(y)$ satisfying $e(y(v), v) = 0$.

Theorem 5.1. *Assume that the pair (y, u) satisfies $(C1)$ and that there exists $\lambda \in W^*$ such that*

(C2) $e_y(y,u)^*\lambda + J_y(y,u) = 0$, and

(C3) $\lim_{t\to 0^+} \frac{1}{t}|y(u+td) - y|_Y^2 = 0$ *for all* $d \in U$.

Then the Gâteaux derivative $\hat{J}'(u)$ *exists and*

$$\hat{J}'(u) = e_u(y,u)^*\lambda + J_u(y,u). \tag{5.1.1}$$

Proof. For $d \in U$ and t sufficiently small let $v = y(u+td) - y$. Then

$$\hat{J}(u+td) - \hat{J}(u) = \int_0^1 \big(J'(y+sv, u+std)(v,td) - J'(y,u)(v,td)\big)\, ds + J'(y,u)(v,td). \tag{5.1.2}$$

Similarly,

$$\begin{aligned} 0 &= \langle \lambda, e(v, u+td) - e(y,u)\rangle_{W^*,W} \\ &= \left\langle \lambda, \int_0^1 \big(e'(y+sv, u+std)(v,td) - e'(y,u)(v,td)\big)\, ds \right\rangle + \langle \lambda, e'(y,u)(v,td)\rangle, \end{aligned} \tag{5.1.3}$$

and hence by Lipschitz continuity of e' in $V(y) \times V(u)$ and (C3) we have

$$\lim_{t\to 0^+} \frac{1}{t}\langle \lambda, e_y(y,u)v\rangle = -\langle \lambda, e_u(y,u)d\rangle.$$

Since J' is Lipschitz continuous in $V(y) \times V(u)$, it follows from (5.1.2) and conditions (C2)–(C3) that

$$\begin{aligned} \lim_{t\to 0^+} \frac{\hat{J}(u+td) - \hat{J}(u)}{t} &= -\lim_{t\to 0^+} \frac{1}{t}\langle \lambda, e_y(y,u)v\rangle + J_u(y,u)d \\ &= (e_u(y,u)^*\lambda + J_u(y,u))d. \qquad \square \end{aligned}$$

Defining the Lagrangian $\mathcal{L} : Y \times U \times W^* \to \mathbb{R}$ by

$$\mathcal{L}(y,u,\lambda) = J(y,u) + \langle \lambda, e(y,u)\rangle,$$

(5.1.1) can be written as

$$\hat{J}'(u) = \mathcal{L}_u(y,u,\lambda).$$

If $e_y(y,u) : Y \to W$ is bijective, then (5.1.1) also follows from the implicit function theory. In fact, by the implicit function theorem there exists a neighborhood $V(y) \times V(u)$ of (y,u) such that $e(y,v) = 0$ has a unique solution $y(v) \in V(y)$ for $v \in V(u)$ and $y(\cdot) : V(u) \subset U \to Y$ is C^1 with

$$e_y(y,v)y'(v) + e_u(y,v) = 0, \quad v \in V(u).$$

Thus

$$\hat{J}'(u)v = J_y(y,u)y'(u)v + J_u(y,u)v = \langle e_u(y,u)^*\lambda, v\rangle + J_u(y,u)v.$$

If conditions (C1)–(C3) hold at a local minimizer (y^*, u^*) of (P), it follows from Theorem 5.1 that the first order necessary optimality condition for (P) is given by

$$\begin{cases} e_y(y^*,u^*)^*\lambda^* + J_y(y^*,u^*) = 0, \\ e_u(y^*,u^*)^*\lambda^* + J_u(y^*,u^*) = 0, \\ e(y^*,u^*) = 0, \end{cases} \tag{5.1.4}$$

or equivalently

$$\mathcal{L}'(y^*,u^*,\lambda^*) = 0, \quad e(y^*,u^*) = 0,$$

where, as in the previous chapter, prime with $\mathcal{L}$ denotes the derivative with respect to (y, u).

(II) Sufficient Optimality Condition

Assume that $(y^*, u^*, \lambda^*) \in Y_1 \times U \times W^*$ satisfies (C1) and that the necessary optimality condition (5.1.4) holds. For $(y,u) \in V(y^*) \times V(u^*)$ define the quadratic approximation to $J(y,u)$ and $\langle \lambda, e(y,u)\rangle$ by

$$E_1(y-y^*, u-u^*) = J(y,u) - J(y^*,u^*) - J_u(y^*,u^*)(u-u^*) - J_y(y^*,u^*)(y-y^*) \tag{5.1.5}$$

and

$$E_2(y-y^*, u-u^*) = \langle \lambda^*, e(y,u) - e(y^*,u^*) - e_y(y^*,u^*)(y-y^*) - e_u(y^*,u^*)(u-u^*)\rangle. \tag{5.1.6}$$

Since for $y = y(u)$ with $u \in V(u^*)$

$$E_2(y(u)-y^*, u-u^*) = -\langle \lambda^*, e_y(y^*,u^*)(y-y^*) + e_u(y^*,u^*)(u-u^*)\rangle,$$

we find by summing (5.1.5) and (5.1.6) with $y = y(u)$ and using (5.1.4)

$$\hat{J}(u) - \hat{J}(u^*) = J(y(u),u) - J(y^*,u^*) = E_1(y(u)-y^*, u-u^*) + E_2(y(u)-y^*, u-u^*).$$

Based on this identity, we have the following local sufficient optimality conditions.

- If $E(u) := E_1(y(u)-y^*, u-y^*) + E_2(y(u)-y^*, u-u^*) \geq 0$ for all $u \in V(u^*)$, then u^* is a local minimizer of (P). If $E(u) > 0$ for all $u \in V(u^*)$ with $u \neq u^*$, then u^* is a strict local minimizer.
- Assume that J is locally uniformly convex in the sense that for some $\alpha > 0$

$$E_1(y(u)-y^*, u-u^*) \geq \alpha(|y(u)-y^*|_Y^2 + |u-u^*|_U^2) \text{ for all } u \in V(u^*),$$

and let $\beta \geq 0$ be the size of the nonlinearity of e, i.e.,

$$|e(y,u) - e(y^*,u^*) - e_y(y^*,u^*)(y-y^*) - e_u(y^*,u^*)(u-u^*)| \leq \beta(|y-y^*|_Y^2 + |u-u^*|_U^2)$$

for all $(y, u) \in V(y^*) \times V(u^*)$. Then

$$E(u) \geq (\alpha - \beta|\lambda^*|_{W^*})\,(|y(u) - y^*|_Y^2 + |u - u^*|_U^2) \text{ for all } u \in V(u^*).$$

Thus, u^* is a strict local minimizer of (P) if $\alpha - \beta|\lambda^*|_{W^*} > 0$.

• Note that

$$E(u) = \mathcal{L}(x, \lambda^*) - \mathcal{L}(x^*, \lambda^*) - \mathcal{L}'(x^*, \lambda^*)(x - x^*),$$

where $x = (y, u)$, $x^* = (y^*, u^*)$, and $y^* = y(u^*)$. If $\mathcal{L}$ is C^2 with respect to $x = (y, u)$ in $V(y^*) \times V(u^*)$ with second derivative with respect to x denoted by $\mathcal{L}''$, we have

$$E(u) = \mathcal{L}''(x^*, \lambda^*)(x - x^*, x - x^*) + r(x - x^*), \tag{5.1.7}$$

where $|r(x - x^*)| = o(|x - x^*|^2)$. Thus, if there exists $\sigma > 0$ such that

$$\mathcal{L}''(x^*, \lambda^*)(\delta x, \delta x) \geq \sigma|\delta x|^2 \text{ for all } \delta x \in \ker\, e'(x^*), \tag{5.1.8}$$

and $e'(x^*) : Y \times U \to W$ is surjective, then there exist $\sigma_0 > 0$ and $\rho > 0$ such that

$$\hat{J}(u) - \hat{J}(u^*) = E(u) \geq \sigma_0(|y(u) - y(u^*))|^2 + |u - u^*|^2) \tag{5.1.9}$$

for all $y(u) \in V(y^*) \times V(u^*)$ with $|(y(u), u) - (y(u^*), u^*)| < \rho$. In fact, from Lemma 2.13, Chapter 2, with $S = \ker\, e'(x^*)$ there exist $\gamma > 0$, $\sigma_0 > 0$ such that

$$\begin{aligned} &\mathcal{L}''(x^*, \lambda^*)(x - x^*, x - x^*) \geq \sigma_0|x - x^*|^2 \\ &\text{for all } x \in V(y^*) \times V(u^*) \text{ with } |x_{\ker^\perp}| \leq \gamma|x_{\ker}|, \end{aligned} \tag{5.1.10}$$

where we decomposed $x - x^*$ as

$$x - x^* = x_{\ker} + x_{\ker^\perp} \in \ker\, e'(x^*) + (\ker\, e'(x^*))^\perp.$$

Since $0 = e(x) - e(x^*) = e'(x^*)(x - x^*) + \eta(x - x^*)$, where $|\eta(x - x^*)|_W = o(|x - x^*|)$, it follows that for $\delta \in (0, \frac{\gamma}{1+\gamma})$ there exists $\rho > 0$ such that

$$|x_{\ker^\perp}| \leq \delta|x - x^*| \quad \text{if } |x - x^*| < \rho.$$

Consequently

$$|x_{\ker^\perp}| \leq \frac{\delta}{1 - \delta}|x_{\ker}| < \gamma|x_{\ker}| \quad \text{if} \quad |x - x^*| < \rho,$$

and hence (5.1.9) follows from (5.1.7) and (5.1.10).

We refer the reader to [CaTr, RoTr] and the literature cited therein for detailed investigations on the topic of second order sufficient optimality. The aim of these results is to establish second order sufficient optimality conditions which are close to second order necessary conditions.

5.2 Newton method

In this section we describe the Newton method applied to the reduced form (5.0.1) of (P). That is, let $y(u)$ denote a solution to $e(y, u) = 0$. Then the constrained optimization problem is transformed to the unconstrained problem for u in U:

$$\min_{u \in U} \hat{J}(u) = J(y(u), u). \tag{5.2.1}$$

Let (y^*, u^*) denote a solution to (P), and assume that (C1) holds for (y^*, u^*) and that (C2), (C3) hold for all $(y(u), u) \in V(y^*) \times V(u^*)$. In addition it is assumed that J and e are C^2 in $V(y^*) \times V(u^*)$ with Lipschitz continuous second derivatives. From Theorem 5.1 the first derivative of $\hat{J}(u)$ is given by

$$\hat{J}'(u) = e_u(y, u)^* \lambda + J_u(y, u), \tag{5.2.2}$$

where $u \in V(u^*)$, $y = y(u)$, and $\lambda = \lambda(u)$ satisfy

$$e_y(y, u)^* \lambda = -J_y(y, u). \tag{5.2.3}$$

From (5.2.2) it follows that

$$\hat{J}'(u) = \mathcal{L}_u(y(u), u, \lambda(u)) \quad \text{for } u \in V(u^*). \tag{5.2.4}$$

We henceforth assume that for every $u \in V(u^*)$ and $y = y(u)$, $\lambda = \lambda(u)$,

$$\begin{pmatrix} \mathcal{L}_{yy}(y, u, \lambda) & e_y(y, u)^* \\ e_y(y, u) & 0 \end{pmatrix} \begin{pmatrix} \mu_1 \\ \mu_2 \end{pmatrix} + \begin{pmatrix} \mathcal{L}_{yu}(y, u, \lambda) \\ e_u(y, u) \end{pmatrix} = 0 \tag{5.2.5}$$

admits a solution $(\mu_1, \mu_2) \in \mathcal{L}(U, Y) \times \mathcal{L}(U, W^*)$. Note that (5.2.5) is an operator equation in $\mathcal{L}(U, Y^*) \times \mathcal{L}(U, W)$. It consists of the linearized primal equation for μ_1 and the adjoint operator equation with right-hand side $-e_y(y, u)^* \mu_2 - \mathcal{L}_{yu}(y, u, \lambda) \in \mathcal{L}(U, Y^*)$ for μ_2.

Using (5.24) and the adjoint operator in the form $\mathcal{L}_y(y, u, \lambda) = 0$ and $e(y, u) = 0$, we find

$$\hat{J}''(u) = \mathcal{L}_{yu}(y, u, \lambda)\mu_1 + e_u(y, u)^* \mu_2 + \mathcal{L}_{uu}(y, u, \lambda), \tag{5.2.6}$$

where (μ_1, μ_2) satisfy (5.2.5). In fact, since e and J are required to have Lipschitz continuous second derivatives in $V(y^*) \times V(u^*)$ we obtain the following relationships in W and Y^*:

$$\begin{aligned} &e_y(y, u)v + e_u(y, u)(td) = o(|t|), \\ &\mathcal{L}_{yy}(y, u, \lambda)v + e_y(y, u)^* w + \mathcal{L}_{yu}(y, u, \lambda)(td) = o(|t|), \end{aligned} \tag{5.2.7}$$

where $(y,u) \in V(y^*) \times V(u^*)$, $d \in U$, and $v = y(u+td) - y$, $w = \lambda(u+td) - \lambda$. By (5.2.4), (5.2.5), and (5.2.7) we find

$$\begin{aligned}\hat{J}'(u+td) - \hat{J}'(u) &= \mathcal{L}_{uy}(y,u,\lambda)v + \mathcal{L}_{uu}(y,u,\lambda)(td) + e_u(y,u)^* w + o(|t|)\\ &= -\mathcal{L}_{yy}(y,u,\lambda)(\mu_1, v) - \langle e_y(y,u)v, \mu_2\rangle_{W,\mathcal{L}(U,W^*)} + \mathcal{L}_{uu}(y,u,\lambda)(td)\\ &\qquad - \langle e_y(y,u)\mu_1, w\rangle_{\mathcal{L}(U,W^*),W} + o(|t|)\\ &= \langle e_y(y,u)\mu_1, w\rangle_{\mathcal{L}(U,W),W^*} + \mathcal{L}_{yu}(y,u,\lambda)(td,\mu_1) + \langle e_u^*(y,u)\mu_2, td\rangle_{\mathcal{L}(U,U^*),U}\\ &\qquad + \mathcal{L}_{uu}(y,u,\lambda)(td) - \langle e_y(y,u)\mu_1, w\rangle_{\mathcal{L}(U,W),W^*} + o(|t|).\end{aligned}$$

Dividing by t and letting $t \to 0$ we obtain (5.2.6).

From (5.2.5) we deduce

$$\mu_1 = -(e_y(y,u))^{-1} e_u(y,u),$$

$$\mu_2 = -(e_y(y,u)^*)^{-1}(\mathcal{L}_{yy}\mu_1 + \mathcal{L}_{yu}),$$

where $\mathcal{L}$ is evaluated at (y,u,λ). Hence the second derivative of $\hat{J}$ is given by

$$\begin{aligned}\hat{J}''(u) &= \mathcal{L}_{uy}\mu_1 - e_u(y,u)^*(e_y(y,u)^*)^{-1}(\mathcal{L}_{yy}\mu_1 + \mathcal{L}_{yu}) + \mathcal{L}_{uu}\\ &= T(y,u)^*\mathcal{L}''(y,u,\lambda)T(y,u),\end{aligned} \tag{5.2.8}$$

with

$$T(y,u) = \begin{pmatrix} -e_y(y,u)^{-1}e_u(y,u) \\ I \end{pmatrix},$$

where $T \in \mathcal{L}(U, Y \times U)$. After this preparation the Newton equation

$$\hat{J}''(u)\delta u + \hat{J}'(u) = 0 \tag{5.2.9}$$

can be expressed as

$$T(y,u)^*\mathcal{L}''(y,u,\lambda)\begin{pmatrix}\delta y\\ \delta u\end{pmatrix} + T(y,u)^*\begin{pmatrix} 0 \\ (e_u(y,u)^*\lambda + J_u(y,u)\end{pmatrix} = 0,$$

$$e_y(y,u)\delta y + e_u(y,u)\delta u = 0.$$

By the definition of $T(y,u)$ and the closed range theorem we have

$$\ker(T(y,u)^*) = \text{range}(T(y,u))^\perp = \ker(e'(y,u))^\perp = \text{range}(e'(y,u)^*)$$

provided that range$(e'(y,u))$, and hence range$(e'(y,u))^*$, is closed. As a consequence the Newton update can be expressed as

$$\begin{pmatrix} \mathcal{L}''(y,u,\lambda) & e'(y,u)^* \\ e'(y,u) & 0 \end{pmatrix}\begin{pmatrix}\delta y\\ \delta u \\ \delta\lambda\end{pmatrix} + \begin{pmatrix} 0 \\ e_u(y,u)^*\lambda + J_u(y,u) \\ 0\end{pmatrix} = 0, \tag{5.2.10}$$

where $(\delta y, \delta u, \delta\lambda) \in Y \times U \times W^*$ and the equality is understood in $Y^* \times U^* \times W$.

We now state the Newton iteration for (5.2.1).

Newton Method.

(i) Initialization: Choose $u_0 \in V(u^*)$, solve

$$e(y, u_0) = 0, \quad e_y(y, u_0)^*\lambda + J_y(y, u_0) = 0 \quad \text{for } (y_0, \lambda_0),$$

and set k = 0

(ii) Newton step: Solve for $(\delta y, \delta u, \delta\lambda) \in Y \times U \times W^*$

$$\begin{pmatrix} \mathcal{L}''(y_k, u_k, \lambda_k) & e'(y_k, u_k)^* \\ e'(y_k, u_k) & 0 \end{pmatrix} \left(\begin{pmatrix} \delta y \\ \delta u \\ \delta\lambda \end{pmatrix} \right) + \begin{pmatrix} 0 \\ e_u(y_k, u_k)^*\lambda_k + J_u(y_k, u_k) \\ 0 \end{pmatrix} = 0.$$

(iii) Update $u_{k+1} = u_k + \delta u$.

(iv) Feasibility step: Solve for $(y_{k+1}, \lambda_{k+1}) \in Y \times W^*$

$$e(y, u_{k+1}) = 0, \quad e_y(y_{k+1}, u_{k+1})^*\lambda + J_y(y, u_{k+1}) = 0.$$

(v) Stop, or set $k = k + 1$, and goto (ii).

The following theorem provides sufficient conditions for local quadratic convergence. Here we assume that $e_y(x^*) : Y \to W$ is a bijection at a local solution $x^* = (y^*, u^*)$ of (P). Then there corresponds a unique Lagrange multiplier λ^*.

Theorem 5.2. *Let $x^* = (y^*, u^*)$ be a local solution to (P) and let $V(y^*) \times V(u^*)$ denote a neighborhood in which J and e are C^2 with Lipschitz continuous second derivatives. Let us further assume that $e_y(x^*) : Y \to W$ is a bijection and that there exists $\kappa > 0$ such that*

$$\mathcal{L}''(x^*, \lambda^*)(h, h) \geq \kappa|h|^2 \textit{ for all } h \in \ker e'(x^*).$$

Then, if $|u_0 - u^*|$ is sufficiently small, $|(y_{k+1}, u_{k+1}, \lambda_{k+1}) - (y^*, u^*, \lambda^*)| \leq K|u_k - u^*|^2$ for a constant K independent of $k = 0, 1, \ldots$.

Proof. Since $e_y(x^*) : Y \to W$ is a homeomorphism and $x \to e_y(x)$ is continuous from $X \to \mathcal{L}(Y, X)$, there exist $r > 0$ and $\gamma > 0$ such that

$$\begin{cases} \|e_y(x)^{-1}\|_{\mathcal{L}(W,Y)} \leq \gamma & \text{for all } x \in B_r, \\ \|e_y(x)^{-*}\|_{\mathcal{L}(Y^*,W^*)} \leq \gamma & \text{for all } x \in B_r, \\ \langle T^*(x)\mathcal{L}''(x, \lambda)T(x)u, u\rangle \geq \frac{\kappa}{2}|u|^2 & \text{for all } x \in B_r, |\lambda - \lambda^*| < r, \end{cases} \tag{5.2.11}$$

where $B_r = \{(y, u) \in Y \times U : |y - y^*| < r, |u - u^*| < r\}$. These estimates imply that, possibly after decreasing $r > 0$, there exists some $\hat{K} > 0$ such that

$$|y(u) - y^*|_Y \leq \hat{K}|u - u^*|_U \text{ and } |\lambda(u) - \lambda^*|_{W^*} \leq \hat{K}|u - u^*|_U \text{ for all } |u - u^*| < r,$$

where $e_y(y(u),u)^*\lambda(u) = -J_y(y(u),u)$. Thus for $\rho = \min(r, \frac{r}{\hat{K}})$ the inequalities in (5.2.11) hold with $x = (y(u),u)$ and $\lambda = \lambda(u)$, provided that $|u - u^*| < \rho$. Let $S(u) = T(y(u),u)^*\mathcal{L}''(y(u),u,\lambda(u))T(y(u),u)$. Then there exists $\mu > 0$ such that

$$|S(u) - S(u^*)| \le \mu|u - u^*| \quad \text{for } |u - u^*| < \rho.$$

Let $|u_0 - u^*| < \min(\frac{\kappa}{\mu}, \rho)$ and, proceeding by induction, assume that $|u_k - u^*| \le |u_0 - u^*|$. Then

$$\begin{aligned} S(u_k)(u_{k+1} - u^*) &= S(u_k)(u_k - u^*) - \hat{J}'(u_k) + \hat{J}'(u^*) \\ &= \int_0^1 (S(u_k + s(u^* - u_k)) - S(u_k))(u^* - u_k)\, ds \end{aligned} \tag{5.2.12}$$

and hence

$$|u_{k+1} - u^*| \le \frac{2}{\kappa}\frac{\mu}{2}|u_k - u^*|^2 \le \frac{\mu}{\kappa}|u_0 - u^*|\,|u_0 - u^*| \le |u_0 - u^*| < \rho.$$

It further follows that $|y(u_{k+1}) - y^*|_Y \le \frac{\hat{K}\mu}{\kappa}|u_k - u^*|^2$ and $|y(u_{k+1}) - y^*|_Y \le \hat{K}|u_{k+1} - u^*| < \hat{K}\rho \le r$. In a similar manner we have $|\lambda(u_{k+1}) - \lambda^*|_{W^*} \le \frac{\hat{K}\mu}{\kappa}|u_k - u^*|^2$ and $|\lambda(u_{k+1}) - \lambda^*|_{W^*} < r$. The claim follows from these estimates. ☐

Remark 5.2.1. If $|\lambda^*|_{W^*}$ is small, then it is suggested to approximate the Hessian of $\mathcal{L}''(y,u,\lambda)$ by

$$\mathcal{L}''(y,u,\lambda) \sim J''(y,u)$$

and the reduced Hessian $T^*\mathcal{L}''T$ by

$$\hat{J}''(u) \sim T(y,u)^*J''(y,u)T(y,u).$$

Under the assumptions of Theorem 5.2, $T^*J''T$ is positive definite in a neighborhood of the solution, provided that λ^* is sufficiently small. Moreover, if $\lambda^* = 0$, then the Newton method converges superlinearly to (y^*, u^*), if it converges. Indeed we can follow the proof of Theorem 5.2 and replace (5.2.12) by

$$\begin{aligned} S_0(u_k)(u_{k+1} - u^*) &= S(u_k)(u_k - u^*) - \hat{J}'(u_k) + \hat{J}'(u^*) \\ &\quad - T^*(y(u_k),u_k)(e''(x_k)^*(\lambda_k - \lambda^*))T(y(u_k),u_k)(u_k - u^*), \end{aligned}$$

where $S_0(u) = T^*(y(u),u)\,J''(y(u),u)\,T(y(u),u)$.

Remark 5.2.2. Note that the reduced Hessian $S = T^*\mathcal{L}''T$ is a Schur complement of the linear system (5.2.10). That is, if we eliminate δy and $\delta\lambda$ by

$$\delta y = -e_y^{-1}e_u\delta u, \quad \delta\lambda = -(e_y^*)^{-1}(\mathcal{L}_{yy}\delta y + \mathcal{L}_{yu}\delta u),$$

we obtain (5.2.9) in the form

$$S\delta u + \mathcal{L}_u(y,u,\lambda) = 0. \tag{5.2.13}$$

Remark 5.2.3. If (5.2.13) is solved by an iterative method based on Krylov subspaces, then this uses Sr for given $r \in U$ and requires that we perform the forward solution

$$\widehat{\delta y} = -e_y^{-1}(y,u)e_u(y,u)r,$$

the adjoint solution

$$\widehat{\delta\lambda} = -(e_y(y,u)^*)^{-1}(\mathcal{L}_{yy}\widehat{\delta y} + \mathcal{L}_{yu}r),$$

and evaluation of the expression

$$Sr = \mathcal{L}_{uy}\widehat{\delta y} + \mathcal{L}_{uu}r + e_u(y,u)^*\widehat{\delta\lambda}.$$

In applications where the discretization of U is much smaller than that of $Y \times U \times W^*$ this procedure may offer a significant reduction in storage and execution time over solving the saddle point problem (5.2.10).

5.3 SQP and reduced SQP methods

In this section we describe the SQP (sequential quadratic programing) and reduced SQP methods without entering into technical details. Throughout this section we set $x = (y,u) \in X = Y \times U$. Let x^* denote a local solution to

$$\begin{cases} \min J(x) \quad \text{over } x \in X \\ \text{subject to } e(x) = 0. \end{cases} \tag{5.3.1}$$

Further let

$$\mathcal{L}(x,\lambda) = J(x) + \langle \lambda, e(x) \rangle$$

denote the Lagrangian and let λ^* be a Lagrange multiplier at the solution x^*. Then a necessary first order optimality condition is given by

$$\mathcal{L}'(y^*,u^*,\lambda^*) = 0, \quad e(x^*) = 0, \tag{5.3.2}$$

and a second order sufficient optimality condition by

$$\mathcal{L}''(x^*,\lambda^*)(h,h) \ge \sigma |h|_X^2 \text{ for all } h \in \ker e'(x^*), \tag{5.3.3}$$

where $\sigma > 0$. Differently from the Newton method described in the previous section, in the SQP method both y and u are considered as independent variables related by the equality constraint $e(y,u) = 0$ which is realized by a Lagrangian term. The SQP method then consists essentially in a Newton method applied to the necessary optimality condition (5.3.2) to iteratively solve for (y^*,u^*,λ^*). This results in determining updates from the linear system

$$\begin{pmatrix} \mathcal{L}''(y_k,u_k,\lambda_k) & e'(y_k,u_k)^* \\ e'(y_k,u_k) & 0 \end{pmatrix} \left(\begin{pmatrix} \delta y \\ \delta u \\ \delta\lambda \end{pmatrix} \right) + \left(\begin{matrix} \begin{pmatrix} e_y(y_k,u_k)^*\lambda_k + J_y(y_k,u_k) \\ e_u(y_k,u_k)^*\lambda_k + J_u(y_k,u_k) \end{pmatrix} \\ e(y_k,u_k) \end{matrix} \right) = 0. \tag{5.3.4}$$

If the values for y_k and λ_k are obtained by means of a feasibility step as in (iv) of the Newton algorithm, then the first and the last components on the right-hand side of (5.3.4) are 0 and we arrive at the Newton algorithm. This will be further investigated in Section 5.5 for weakly singular problems. Note that for affine constraints the iterates, except possibly the initialization, are feasible by construction, and hence $e(x_k) = 0$.

Let us note that given an iterate $x_k = (y_{k+1}, u_{k+1})$ near x^* the SQP step for (5.3.1) is also obtained as the necessary optimality to the quadratic subproblem

$$\begin{cases} \min \langle \mathcal{L}'(x_k, \lambda_k), h \rangle + \dfrac{1}{2}\, \mathcal{L}''(x_k, \lambda_k)(h, h) \\ \text{subject to } \; e'(x_k)h + e(x_k) = 0, \;\; h \in X, \end{cases} \tag{5.3.5}$$

for $h_k = (\delta y_k, \delta u_k)$ and setting $x_{k+1} = x_k + h_k$.

If (5.3.1) admits a solution (x^*, λ^*) and J and e are C^2 in a neighborhood of x^* with Lipschitz continuous second derivatives, and if further $e'(x^*)$ is surjective and there exists $\kappa > 0$ such that

$$\mathcal{L}''(x^*, \lambda^*)(h, h) \geq \kappa |h|_X^2 \quad \text{for all } h \in \ker e'(x^*),$$

then

$$|(x_{k+1}, \lambda_{k+1}) - (x^*, \lambda^*)|^2 \leq K |(x_k, \lambda_k) - (x^*, \lambda^*)|$$

for a constant K independent of k, provided that $|(x_0, \lambda_0) - (x^*, \lambda^*)|$ is sufficiently small. Thus the SQP method is locally quadratically convergent. The proof is quite similar to the one of Theorem 6.4.

Assume next that there exists a null-space representation of $e'(x)$ for all x in a neighborhood of x^*; i.e., there exist a Hilbert space H and an isomorphism $T(x)$ from H onto $\ker e'(x)$, where $\ker e'(x^*)$ is endowed with the norm of X. Then (5.3.2) and (5.3.3) can be expressed as

$$T(x^*)^* J'(x^*) = 0 \tag{5.3.6}$$

and

$$T(x^*)^* \mathcal{L}''(x^*, \lambda^*) T(x^*) \geq \sigma\, I, \tag{5.3.7}$$

respectively.

Referring to (5.3.5), let $q_k \in \ker e'(x_k)^\perp \subset X$ satisfy $e'(x_k)q_k + e(x_k) = 0$. Then $h_k \in X$ can be expressed as

$$h_k = q_k + T(x_k)w.$$

Thus, (5.3.5) is reduced to

$$\begin{aligned} &\min \langle \mathcal{L}'(x_k, \lambda_k), T(x_k)w \rangle + \langle T(x_k)w, \mathcal{L}''(x_k, \lambda_k) q_k \rangle \\ &+ \frac{1}{2} \, \langle T(x_k)w, \mathcal{L}''(x_k, \lambda_k) T(x_k) w \rangle, \end{aligned} \tag{5.3.8}$$

and the solution h_k to (5.3.5) can be expressed as

$$h_k = q_k + T(x_k)w_k, \tag{5.3.9}$$

where w_k is a solution to the unconstrained problem (5.3.8) in H, given by

$$T^*(x_k)\mathcal{L}''(x_k, \lambda_k)T(x_k)w_k = -T^*(x_k)(J'(x_k) - \mathcal{L}''(x_k, \lambda_k)R_k(x_k)e(x_k)). \tag{5.3.10}$$

Therefore the full SQP step is decomposed into a minimization step in H, a restoration step to the linearized equation

$$e'(x_k)q + e(x_k) = 0,$$

and an update of the Lagrange multiplier according to

$$e'(x_k)^* \delta\lambda = -e'(x_k)^*\lambda_k - J_y(x_k) - \mathcal{L}''(x_k, \lambda_k)h_k.$$

If $e'(x_k) \in \mathcal{L}(X, W)$ admits a right inverse $R(x_k) \in \mathcal{L}(W, X)$ satisfying $e'(x_k)R(x_k) = I_W$, then $q_k = -R(x_k)e(x_k)$ in (5.3.9) and

$$h_k = -R_k e(x_k) - T_k\big(T_k^*\mathcal{L}''(x_k, \lambda_k)T_k\big)^{-1}T_k^*(J'(x_k) - \mathcal{L}''(x_k, \lambda_k)R_k e(x_k)\big), \tag{5.3.11}$$

where $T_k = T(x_k)$ and $R_k = R(x_k)$ and we used that $T^*w = 0$ for $w \in$ range $e'(x)^*$. Note that a right inverse to $e'(x)$ for x in a neighborhood of x^* exists if $e'(x^*)$ is surjective and $x \to e'(x)$ is continuous.

An alternative to deriving the update w_k is given by differentiating

$$T(x)^*J'(x) = T(x)^*\mathcal{L}'(x, \lambda)$$

with respect to x, evaluating at $x = x^*$, $\lambda = \lambda^*$, and using (5.3.2):

$$\frac{d}{dx}(T(x^*)^*J'(x^*)) = T(x^*)^*\mathcal{L}''(x^*, \lambda^*).$$

This representation holds in general only at the solution x^*. But if we use its structure for an update of the form $h_k = q_k + T(x_k)w_k$ in a Newton step to (5.3.6), we arrive at

$$T_k^*\mathcal{L}''(x_k, \lambda_k)h_k = T_k^*\mathcal{L}''(x_k, \lambda_k)(T_k w_k - R(x_k)e(x_k)) = -T_k^*J'(x_k),$$

which results in an update as in (5.3.10) above.

In the *reduced SQP* approach the term $\mathcal{L}''(x_k, \lambda_k)R_k e(x_k)$ is deleted from the expression in (5.3.11). Recall that for Newton's method this term vanishes since $e(x_k) = 0$ at each iteration level. This results in

$$T_k^*\mathcal{L}''(x_k, \lambda_k)T(x_k)w_k = -T_k^*J'(x_k). \tag{5.3.12}$$

This equation for the update of the control coincides with the Schur complement form of the Newton update given in (5.2.13), since

$$\mathcal{L}_u(x_k, \lambda_k) = e_u^*(x_k)\lambda_k + J_u(x_k) = T_k^*J'(x_k),$$

where we used the fact that in the Newton method $e_y^*(x_k)\lambda_k + J_y(x_k) = 0$. The updates for the state and the adjoint state differ, however, for the Newton and the reduced SQP methods.

A second distinguishing feature for reduced SQP methods is that often the reduced Hessians $T_k^*\mathcal{L}''(x_k, \lambda_k)T_k$ are approximated by invertible operators $B_k \in \mathcal{L}(H)$ suggested by secant methods. Thus a reduced SQP step has the form $x_{k+1} = x_k + h_k^{\text{RED}}$, where

$$h_k^{\text{RED}} = -R_k\, e(x_k) - T_k\, B_k^{-1}\, T_k^*\, J'(x_k). \tag{5.3.13}$$

For the reduced SQP method the step h_k^{RED} in general depends on the specific choice of the null-space representation and the right inverse. In the full SQP method the step h_k in (5.3.11) is invariant with respect to these choices. A third distinguishing feature of reduced SQP methods is the choice of the Lagrange multiplier, which is required to specify the update of B_k. The λ-update is typically not derived from the first equation in (5.3.4). From the first order condition we have

$$\langle J'(x^*), R(x^*)h\rangle + \langle \lambda^*, e'(x^*)R(x^*)h\rangle = \langle J'(x^*), R(x^*)h\rangle + \langle \lambda^*, h\rangle = 0 \quad \text{for all } h \in W.$$

This suggests the λ-update

$$\lambda^+ = -R(x)^*\, J'(x). \tag{5.3.14}$$

Other choices for the Lagrange multiplier update are possible. For convergence proofs of reduced SQP methods this update is required to be locally Lipschitz continuous; see [Kup, KuSa].

In the case of problem (P), a vector $(\delta y, \delta u) \in X$ lies in the null-space of $e'(x)$ if

$$e_y(x)\delta y + e_u(x)\delta u = 0.$$

Assuming that $e_y(x)\colon Y \to W$ is invertible, we have

$$\ker e'(x) = \{(\delta y, \delta u)\colon \delta y = -e_y(x)^{-1}e_u(x)\,\delta u\}.$$

This suggests choosing $H = U$ and using the following definitions for the null-space representation and the right inverse:

$$T(x) = (-e_y(x)^{-1}e_u(x), I), \quad R(x) = (e_y(x)^{-1}, 0). \tag{5.3.15}$$

In this case a *reduced SQP step* can be decomposed as follows:

(i) solve $B_k\, \delta u = -T(x_k)^* J'(x_k)$,

(ii) solve $e_y(x_k)\delta y = -e_u'(x_k)\delta u - e(x_k)$,

(iii) set $x_{k+1} = x_k + (\delta y, \delta u)$,

(iv) update λ.

The update of the Lagrange multiplier λ is needed for the approximation B_k to the reduced Hessian $T^*(x_k)\mathcal{L}''(x_k, \lambda_k)T(x_k)$. A popular choice is given by the BFGS-update formula

$$B_{k+1} = B_k + \frac{1}{\langle z_k, \delta u_k\rangle}\langle z_k, \cdot\rangle z_k - \frac{1}{\langle \delta u_k, B_k\delta u_k\rangle}\langle B_k\delta u_k, \cdot\rangle B_k\delta u_k,$$

where

$$z_k = T^*(x_k)\mathcal{L}'\big(x_k + T(x_k)\delta u_k, \lambda_k\big) - J'(x_k).$$

Each SQP step requires at least three linear systems solves, one in W^* for the evaluation of $T(x_k)^*J'(x_k)$, one in U for δu, and another one in Y for δy. The update of the BFGS formula requires one additional system solve. The typical convergence behavior that can be proved for reduced SQP methods with BFGS update is local two-step superlinear convergence [Kup, KuSa, KSa],

$$\lim_{k\to\infty} \frac{|x_{k+1} - x^*|}{|x_{k-1} - x^*|} = 0$$

provided that

$$B_0 - T(x^*)^*\mathcal{L}''(x^*, \lambda^*)T(x^*) \quad \text{is compact.} \tag{5.3.16}$$

Here B_0 denotes the initialization to the reduced Hessian approximation, and (x^*, λ^*) is a solution to (5.1.4). If e is C^1 at x^* and $e_y(x^*)$ is continuously invertible and compact, then (5.3.16) holds for the choice $B_0 = \mathcal{L}_{uu}(x^*, \lambda^*)$. In [HiK2], condition (5.3.16) is analyzed for a class of optimal control problems related to the stationary Navier–Stokes equations. We recall that condition (5.3.16) is related to an analogous condition in the context of solving nonlinear equations $F(x) = 0$ in infinite-dimensional Hilbert spaces by means of secant methods, in particular the Broyden method. The initialization of the Jacobian must be a compact perturbation of the linearization of F at the solution to ascertain local Q-superlinear convergence [Grie, Sa].

If we deal with finite-dimensional problems with e a mapping from $\mathbb{R}^N$ into $\mathbb{R}^M$, with $M < N$, then a common null-space representation is provided by the QR decomposition. Let $H = \mathbb{R}^{N-M}$ and

$$e'(x)^T = Q(x)R(x) = (Q_1(x)\ Q_2(x))\begin{pmatrix} R_1(x) \\ 0 \end{pmatrix},$$

where $Q(x)$ is orthogonal, $Q_1(x) \in \mathbb{R}^{N\times M}$, $Q_2(x) \in \mathbb{R}^{N\times(N-M)}$, and $R_1(x) \in \mathbb{R}^{M\times M}$ is an upper triangular matrix. Then the null-space representation and the right inverse to $e'(x)$ are given by

$$T(x) = Q_2(x) \in \mathbb{R}^{N\times(N-M)} \quad \text{and} \quad R(x) = Q_1(x)(R_1^T(x))^{-1} \in \mathbb{R}^{N\times M}.$$

In the finite-dimensional setting one often works with the "least squares solution" for the λ-update, i.e., with the solution to

$$\min_\lambda |J'(x) + e'(x)^*\lambda|,$$

which results in $\lambda(x)^+ = -R_1^{-1}(x)Q_1^T(x)J'(x)$.

To derive another interesting feature of a variant of a reduced SQP method we recapture the derivation from above for (P) with T and R chosen as in (5.3.15), as

$$\begin{cases} T(x)^*\mathcal{L}''(x,\lambda)T(x)w = -T(x)^*\mathcal{L}'(x,\lambda), \\ h = (\delta y, \delta u) = q + T(x)w, \ \text{where } q - R(x)e(x) = 0, \\ \lambda + \delta\lambda = -R(x)^*J'(x). \end{cases} \tag{5.3.17}$$

In particular, the term $\mathcal{L}''(x_k,\lambda_k)R_k e(x_k)$ is again deleted from the expression in (5.3.10) and the Lagrange multiplier update is chosen according to (5.3.14). However, differently from (5.3.13), the reduced Hessian is not approximated in (5.3.17). Here we do not indicate the iteration index k, and we assume that $e_y(x) : Y \to W$ is bijective and that (5.3.7) holds.

Then $(q, w, \delta\lambda)$ is a solution to (5.3.17) if and only if $(\delta y, \delta u, \delta\lambda)$ is a solution to the system

$$\begin{pmatrix} 0 & 0 & e_y(x)^* \\ 0 & T(x)^*\mathcal{L}''(x,\lambda)T(x) & e_u(x)^* \\ e_y(x) & e_u(x) & 0 \end{pmatrix} \begin{pmatrix} \delta y \\ \delta u \\ \delta\lambda \end{pmatrix} = -\begin{pmatrix} \mathcal{L}_y(x,\lambda) \\ \mathcal{L}_u(x,\lambda) \\ e(x) \end{pmatrix}, \tag{5.3.18}$$

with $w = \delta u$, $q = -R(x)e(x)$, and $(\delta y, \delta u) = q + T(x)\delta u$. In fact, $R^*(x)$ is a left inverse to $e'(x)^*$, and from the first equation in (5.3.18) we have

$$\delta\lambda = -(e_y(x)^*)^{-1}(J_y(x) + e_y^*(x)\lambda) = -R(x)^*J'(x) + \lambda.$$

From the second equation

$$T(x)^*\mathcal{L}''(x,\lambda)T(x)\delta u = -\mathcal{L}_u(x,\lambda) - e_u(x)^*\delta\lambda$$
$$= -\mathcal{L}_u(x,\lambda) + e_u(x)^*e_y(x)^{-*}\mathcal{L}_y(x,\lambda) = -T(x)^*\mathcal{L}'(x,\lambda) = -T(x)^*J'(x).$$

The third equation is equivalent to

$$(\delta y, \delta u) = -R(x)e(x) + T(x)\delta u.$$

System (5.3.18) should be compared to (5.3.4). We note that the reduced SQP step is equivalent to a block tridiagonal system, where the "$\mathcal{L}_{uu}$" element in the system matrix is replaced by the reduced Hessian while the elements corresponding to "$\mathcal{L}_{yy}$" and "$\mathcal{L}_{yu}$" are zero. The system matrix of (5.3.18) can be used advantageously as preconditioning for iteratively solving (5.3.4). In fact, let us denote the system matrices in (5.3.4) and (5.3.18) by $\mathcal{S}$ and $\mathcal{S}_{red}$, respectively, and consider the iteration

$$\delta z_{n+1} = \delta z_n - \mathcal{S}_{red}^{-1}\left(\mathcal{S}\,\delta z_k + \text{col}\,(\mathcal{L}_y(x,\lambda), \mathcal{L}_u(x,\lambda), e(x))\right). \tag{5.3.19}$$

In [IKSG] it was proved that the iteration matrix $I - \mathcal{S}_{red}^{-1}\mathcal{S}$ is nilpotent of degree 3. Hence the iteration (5.3.19) converges in 3 steps to the solution of (5.3.4) with $(x_k, \lambda_k) = (x, \lambda)$.

5.4 Optimal control of the Navier–Stokes equations

In this section we consider the optimal control problem

$$\begin{cases} \min J(y,u) = \int_0^T \big(\ell(y(t)) + h(u(t))\big)dt \\ \text{over } u \in U = L^2(0,T;\tilde{U}) \text{ subject to} \\ \frac{d}{dt}y(t) = A_0 y(t) + F(y(t)) + Bu(t) \text{ for } t \in (0,T], \\ y(0) = y_0, \end{cases} \tag{5.4.1}$$

where A_0 is a densely defined, self-adjoint operator in a Hilbert space H which satisfies $(-A_0\phi,\phi)_H \ge \alpha|\phi|_H^2$ for some $\alpha > 0$ independent of $\phi \in \mathrm{dom}A_0$. We set $V = \mathrm{dom}((-A_0)^{\frac{1}{2}})$ endowed with $|\phi|_V = |(-A_0)^{\frac{1}{2}}\phi|_H$ as norm. The V-coercive form $(v,w) \to ((-A_0)^{\frac{1}{2}}v, (-A_0)^{\frac{1}{2}}w)$ also defines an operator $A \in \mathcal{L}(V,V^*)$ satisfying $\langle -Av, w\rangle_{V^*,V} = ((-A_0)^{\frac{1}{2}}v, (-A_0)^{\frac{1}{2}}w)_H$ for all $v, w \in W$. Since A and A_0 coincide on dom A_0, and dom A_0 is dense in V, we shall not distinguish between A and A_0 as operators in $\mathcal{L}(V,V^*)$. In (5.4.1), moreover, $\tilde{U}$ is the Hilbert space of controls, $B \in \mathcal{L}(\tilde{U},V^*)$, and $y_0 \in H$. We assume that $h \in C^1(\tilde{U},\mathbb{R})$ with Lipschitz continuous first derivative and $\ell \in C^1(V,\mathbb{R})$, with ℓ' Lipschitz continuous on bounded subsets of H. The nonlinearity F is supposed to satisfy

$$\text{(H1)} \quad \begin{cases} F: V \to V^* \text{ is continuously differentiable and there exists} \\ \text{a constant } c > 0 \text{ such that for every } \varepsilon > 0 \\ \langle F(y) - F(\bar{y}), y - \bar{y}\rangle_{V^*,V} \le \varepsilon|y-\bar{y}|_V^2 + \frac{c}{\varepsilon}(|\bar{y}|_V^2 + |y|_H^2|y|_V^2)|y-\bar{y}|_H^2, \\ \langle F'(\bar{y})y, y\rangle_{V^*,V} \le \varepsilon|y|_V^2 + \frac{c}{\varepsilon}|y|_H^2|\bar{y}|_V^2(1+|\bar{y}|_H^2), \\ |(F'(y) - F'(\bar{y}))v|_{V^*} \le c|y-\bar{y}|_H^{\frac{1}{2}}|y-\bar{y}|_V^{\frac{1}{2}}\,|v|_H^{\frac{1}{2}}\,|v|_V^{\frac{1}{2}} \\ \qquad \text{for all } y, \bar{y}, \text{ and } v \text{ in } V. \end{cases}$$

$$\text{(H2)} \quad \begin{cases} \text{For any } u \in L^2(0,T;\tilde{U}) \text{ there exists a unique weak} \\ \text{solution } y = y(u) \text{ in} \\ W(0,T) = L^2(0,T;V) \cap H^1(0,T;V^*) \text{ satisfying} \\ \langle \frac{d}{dt}y(t), \psi\rangle_{V^*,V} = \langle A_0 y(t) + F(y(t)) + Bu(t), \psi\rangle_{V^*,V} \\ \text{for all } \psi \in V, \text{ and } y(0) = y_0. \\ \text{Moreover } \{y(u) : |u|_{L^2(0,T;\tilde{U})} \le r\} \text{ is bounded in} \\ W(0,T) \text{ for each } r > 0. \end{cases}$$

With these preliminaries the dynamical system in (5.4.1) can be considered with values in V^*. The conditions on A_0 and F are motivated by the two-dimensional incompressible

Navier–Stokes equations with distributed control given by

$$\begin{cases} \frac{d}{dt}y + (y\cdot\nabla)y + \nabla p = \nu\nabla y + Bu \text{ in } (0,T]\times\Omega, \\ \nabla y = 0 \quad \text{in } (0,T]\times\Omega, \\ y(0,\cdot) = y_0 \quad \text{in } \Omega, \end{cases} \tag{5.4.2}$$

where Ω is a bounded domain with Lipschitz continuous boundary $\partial\Omega$, $y = y(t,x) \in \mathbb{R}^2$ is the the velocity field, $p = p(t,x) \in \mathbb{R}$ is the pressure, and ν is the normalized viscosity. Let

$$V = \{\phi \in H_0^1(\Omega)^2 : \nabla\cdot\phi = 0\},\ \ H = \{\phi \in L^2(\Omega)^2 : \nabla\cdot\phi = 0,\ n\cdot\phi = 0 \text{ on } \delta\Omega\},$$

where n is the outer unit normal vector to $\partial\Omega$, let Δ denote the Laplacian in H, and let PF denote the orthogonal projection of $L^2(\Omega)^2$ onto the closed subspace H. Then $A_0 = \nu P\Delta$ is the Stokes operator in H. It is a self-adjoint operator with domain $\mathrm{dom}((-A_0)^{\frac{1}{2}}) = V$ and

$$-(A_0\phi, \psi)_H = \nu(\nabla\phi, \nabla\psi)_{L^2} \text{ for all } \phi \in \mathrm{dom}(A_0),\ \psi \in V.$$

If $\partial\Omega$ is sufficiently smooth, e.g., $\partial\Omega$ is C^2, then $\mathrm{dom}(A_0) = H^2(\Omega)^2 \cap V$. The nonlinearity in (5.4.2) satisfies the following properties:

$$\int_\Omega (u\cdot\nabla v)\, w\, dx = \int_\Omega (u\cdot\nabla w)\, v\, dx, \tag{5.4.3}$$

$$\int_\Omega (u\cdot\nabla v)\, w\, dx \le c|u|_H^{\frac{1}{2}}\, |u|_V^{\frac{1}{2}}\, |v|_V\, |w|_H^{\frac{1}{2}}\, |w|_V^{\frac{1}{2}} \tag{5.4.4}$$

for a constant c independent of $u, v, w \in V$; see, e.g., [Te]. In terms of the general formulation (5.4.1) the nonlinearity $F : V \to V^*$ is given by

$$(F(\phi), v)_{V^*,V} = -\int_\Omega (\phi\cdot\nabla)\phi\, v\, dx.$$

It is well defined due to (5.4.4). Moreover (H1) follows from (5.4.3) and (5.4.4). From the theory of variational solutions to the Navier–Stokes equations it follows that there exists a constant C such that for all $y_0 \in H$ and $u \in L^2(0,T;\tilde{U})$ there exists a unique solution $y \in W(0,T)$ such that

$$|y|_{C(0,T;H)} + |y|_{W(0,T)} \le C(|y_0|_H + |u|_{L^2(0,T;\tilde{U})} + |y_0|_H^2 + |u|^2_{L^2(0,T;\tilde{U})}),$$

where $|y|_{W(0,T)} = |y|_{L^2(0,T;V)} + |\frac{d}{dt}y|_{L^2(0,T;V^*)}$ and we recall that $C([0,T];H)$ is embedded continuously into $W(0,T)$; see, e.g., [LiMa, Tem]. Assuming the existence of a local solution to (5.4.1) we shall present in the following subsections first and second order optimality conditions, as well as the steps which are necessary to realize the Newton algorithm.

Control and especially optimal control has received a considerable amount of attention in the literature. Here we only mention a few [Be, FGH, Gu, HiK2] and refer the reader to further references given there.

5.4.1 Necessary optimality condition

Let $x^* = (y^*, u^*)$ denote a local solution to (5.4.1). We shall derive a first order optimality condition by verifying the assumptions of Theorem 5.1 and applying (5.1.4). In the context of Section 5.1 we set

$$\begin{aligned} &Y = W(0,T) = L^2(0,T;V) \cap H^1(0,T;V^*), \\ &W = L^2(0,T;V^*) \times H, \; U = L^2(0,T;\tilde{U}), \\ &J(y,u) = \textstyle\int_0^T \big(\ell(y(t)) + h(u(t))\big)\, dt, \\ &e(y,u) = (y_t - A_0 y - F(y) - Bu, y(0)). \end{aligned}$$

To verify (C1)–(C3) with $(y,u) = (y^*, u^*)$, let $V(y^*) \times V(u^*)$ denote a bounded neighborhood of (y^*, u^*). Let $V(y^*)$ be chosen such that $y(u) \in V(y^*)$ for every $u \in V(u^*)$. This is possible by (H2). Since $Y = W(0,T)$ is embedded continuously into $C([0,T];H)$, the continuity assumptions for ℓ and h imply the continuity requirements J in (C1). Note that $e'(y,u) : Y \times U \to W$ is given by

$$e'(y,u)(\delta y, \delta u) = ((\delta y)_t - A_0 \delta y - F'(y)\delta y - B\delta u, \delta y(0)),$$

and global Lipschitz continuity of $(y,u) \to e'(y,u)$ from $V(y^*) \times V(u^*) \subset Y \times U$ to $\mathcal{L}(Y \times U, W)$ follows from the last condition in (H1). Solvability of $e(y,u) = 0$ with respect to y for given u follows from (H2) and hence (C1) holds. Since, by (H1), the bounded bilinear form $a(t;\cdot,\cdot) : V \times V \to \mathbb{R}$ defined by

$$t \to a(t;\phi,\psi) = -\langle A_0\phi, \psi\rangle_{V^*,V} - \langle F'(y^*(t))\phi, \psi\rangle_{V^*,V}$$

satisfies

$$a(t;\phi,\phi) \ge |\phi|_V^2 - \varepsilon|\phi|_V^2 - \frac{c}{\varepsilon}|\phi|_V^2 |y^*(t)|_V^2 (1 + |y^*|^2_{C(0,T;H)}),$$

there exists a constant $\bar{c} > 0$ such that

$$a(t;\phi,\phi) \ge \frac{1}{2}|\phi|_V^2 - \bar{c}|y^*(t)|_V^2 |\phi|_H^2 \text{ for } t \in (0,T),$$

where $|y^*|_V^2 \in L^1(0,T)$. Consequently, the adjoint equation

$$\begin{cases} -\frac{d}{dt}p^*(t) = A_0 p^*(t) + F'(y^*(t))^* p^*(t) + \ell'(y^*(t)), \\ p^*(T) = 0 \end{cases} \tag{5.4.5}$$

admits a unique solution $p^* \in W(0,T)$, with

$$|p^*(t)|_H^2 + \int_t^T |p^*(s)|_V^2 ds \le e^{C\int_0^T |y^*(s)|_V^2 ds} \int_t^T |\ell'(y^*(s))|_{V^*}^2 ds \tag{5.4.6}$$

for a constant C independent of $t \in [0,T]$. In fact,

$$-\frac{1}{2}\frac{d}{dt}|p^*(t)|_H^2 + |p^*(t)|_V^2 \le |\langle F'(y^*(t))p^*(t), p^*(t)\rangle_{V^*,V}| + |\ell'(y^*(t))|_{V^*}|p^*(t)|_V.$$

Hence by (H1) there exists a constant C such that

$$-\frac{d}{dt}|p^*(t)|_H^2 - C|y^*(t)|_V^2|p^*(t)|_H^2 + |p^*(t)|_V^2 \le |\ell'(y^*(t))|_{V^*}^2.$$

Multiplying by $\exp(-\int_t^T \bar{\rho}(s)ds)$, where $\bar{\rho}(s) = C|y^*(s)|_V^2$, we find

$$|p^*(t)|_H^2 + \int_t^T |p^*(s)|_V^2 \exp\int_t^s \bar{\rho}(\tau)d\tau\, ds \le \int_t^T |\ell'(y^*(s))|_{V^*}^2 \exp\int_t^s \bar{\rho}(\tau)d\tau\, ds$$

and (5.4.6) follows. This implies (C2) with $\lambda^* = (p^*, p^*(0))$. For any $y \in W(0,T)$ we have

$$\frac{d}{dt}|y(t)|_H^2 = 2\left\langle \frac{d}{dt}y(t), y(t)\right\rangle_{V^*,V} \quad \text{for } t \in (0,T). \tag{5.4.7}$$

Let $u \in V(u^*)$ and denote by $y = y(u) \in V(y^*)$ the solution to the dynamical system in (5.4.1). From (5.4.7) we have

$$\begin{aligned}&\frac{1}{2}\frac{d}{dt}|y(t) - y^*(t)|_H^2 + |y(t) - y^*(t)|_V^2 \\ &= \langle F(y(t)) - F(y^*(t)), y(t) - y^*(t)\rangle_{V^*,V} + \langle Bu(t) - Bu^*(t), y(t) - y^*(t)\rangle_{V^*,V}.\end{aligned}$$

By (H2) the set $\{y(u) : u \in V(u^*)\}$ is bounded in $W(0,T)$. Hence by (H1) there exists a constant C independent of $u \in V(u^*)$ and $t \in [0,T]$ such that

$$\frac{d}{dt}|y(t) - y^*(t)|_H^2 + |y(t) - y^*(t)|_V^2 \le C(\rho(t)|y(t) - y^*(t)|_H^2 + |u(t) - u^*(t)|_{\tilde{U}}^2),$$

where $\rho(t) = |y(t)|_V^2 + |y^*(t)|_V^2$. By Gronwall's inequality this implies that

$$|y(t) - y^*(t)|_H^2 + \int_0^t |y(t) - y^*(t)|_V^2 ds \le C\exp\left(C\int_0^T \rho(\tau)d\tau\right)\int_0^t |u(s) - u^*(s)|^2 ds,$$

where $\int_0^T \rho(\tau)d\tau$ is bounded independent of $u \in V(u^*)$. Utilizing the equations satisfied by $y(u)$ and $y(u^*)$ it follows that

$$|y(u) - y(u^*)|_{W(0,T)} \le \hat{C}|u - u^*|_{L^2(0,T;\tilde{U})} \tag{5.4.8}$$

for a constant $\hat{C}$ independent of $u \in V(u^*)$. Hence (C3) follows and Theorem 5.1 implies the optimality system

$$\begin{aligned}&\frac{d}{dt}y^*(t) = A_0 y^*(t) + F(y^*(t)) + Bu^*(t), \quad y(0) = y_0,\\ &-\frac{d}{dt}p^*(t) = A_0(y^*(t)) + F'(y^*(t))^* p^*(t) + \ell'(y^*(t)), \quad p^*(T) = 0,\\ &B^* p^*(t) + h'(u^*(t)) = 0.\end{aligned}$$

5.4.2 Sufficient optimality condition

We assume J to be of the form

$$J(y,u) = \frac{1}{2}\int_0^T (Q(y(t)-\bar{y}(t)), y(t)-\bar{y}(t))_H dt + \frac{\beta}{2}\int_0^T |u(t)|^2_{\tilde{U}}\, dt,$$

where Q is symmetric and positive semidefinite on H, $\bar{y} \in L^2(0,T;H)$, and $\beta > 0$. We shall verify the requirements in (II) of Section 5.1. By (5.4.8) it follows that

$$E_1(y(u)-y^*, u-u^*) \geq \alpha(|y(u)-y^*|^2_{W(0,T)} + |u-u^*|^2_{L^2(0,T;\tilde{U})}) \quad \text{for all } u \in V(u^*). \tag{5.4.9}$$

Moreover,

$$\begin{aligned}
&|e(y(u),u) - e(y^*,u^*) - e'(y^*,u^*)(y(u)-y^*, u-u^*)|_{L^2(0,T;V^*)\times H} \\
&= |F(y(u)) - F(y^*) - F'(y^*)(y(u)-y^*)|_{L^2(0,T,V^*)} \\
&= \left(\int_0^T |F(y) - F(y^*) - F'(y^*)(y(u)-y^*)|^2_{V^*}\right)^{\frac{1}{2}} \\
&\leq c\left(\int_0^T \left(\int_0^1 s|y(t)-y^*(t)|_H\, |y(t)-y^*(t)|_V ds\right)^2 dt\right)^{\frac{1}{2}} \\
&\leq \frac{c}{\sqrt{2}} |y-y^*|_{C(0,T;H)} |y-y^*|_{L^2(0,T;V)},
\end{aligned}$$

and therefore

$$\begin{aligned}
&|e(y(u),u) - e(y^*,u^*) - e'(y^*,u^*)(y(u)-y^*)|_{L^2(0,T;V^*)\times H} \\
&\leq \bar{C}|y(u)-y^*|^2_{W(0,T)}
\end{aligned} \tag{5.4.10}$$

for a constant $\bar{C}$ independent of $u \in V(u^*)$. From (5.4.6)

$$|p^*|_{L^2(0,T;V)} \leq \exp\left(\frac{C}{2}|y^*|_{L^2(0,T;V)}\right)|Q(y^*-\bar{y})|_{L^2(0,T;V^*)}. \tag{5.4.11}$$

Combining (5.4.9)–(5.4.11) implies that if $|Q(y^*-\bar{y})|_{L^2(O,T,V^*)}$ is sufficiently small, then (y^*, u^*) is a strict local minimum.

5.4.3 Newton's method for (5.4.1)

Here we specify the steps necessary to carry out one Newton iteration for the optimal control problem related to the Navier–Stokes equation. Let $u \in L^2(0,T;\tilde{U})$ and let $y = y(u) \in W(0,T)$ denote the associated velocity field. For $\delta u \in L^2(0,T;\tilde{U})$ let δy, λ, and μ in $W(0,T)$ denote the solutions to the sensitivity equation

$$\frac{d}{dt}\delta y = A_0\delta y + F'(y)\delta y + B\delta u, \quad \delta y(0) = 0, \tag{5.4.12}$$

the adjoint equation

$$-\frac{d}{dt}\lambda = A_0\lambda + F'(y)^*\lambda + \ell'(y), \quad \lambda(T) = 0, \tag{5.4.13}$$

and the adjoint equation arising in the Hessian

$$-\frac{d}{dt}\mu = A_0\mu + F'(y)^*\mu + (F''(y)^*\lambda)\delta y + \ell''(y)\delta y, \ \mu(T) = 0. \tag{5.4.14}$$

Then the operators characterizing the Newton step

$$\hat{J}''(u)\,\delta u = -\hat{J}'(u) \tag{5.4.15}$$

can be expressed by

$$\hat{J}'(u) = h'(u) + B^*\lambda$$

and

$$\hat{J}''(u)\delta u = h''(u)\delta u + B^*\mu.$$

Let us also note that

$$\begin{aligned} -\langle F'(y)\delta y, \psi\rangle_{V^*,V} &= b(y, \delta y, \psi) + b(\delta y, y, \psi), \\ -\langle F'(y)^*\lambda, \psi\rangle &= b(y, \psi, \lambda) + b(\psi, y, \lambda), \\ -\langle (F''(y)^*\lambda)\,\delta y, \psi\rangle &= b(\delta y, \psi, \lambda) + b(\psi, \delta y, \lambda) \end{aligned}$$

for all $\psi \in V$, where $b(u, v, w) = \int_\Omega (u \cdot \nabla) v \; w \, dx$. The evaluation of the gradient $\hat{J}'(u)$ requires one forward in time solve for $y(u)$ and one adjoint solve for λ. Each evaluation of the Hessian necessitates one forward solve of the sensitivity equation for δy and one solve of the second adjoint equation (5.4.14) for μ.

If the approximation as in Remark 5.2.1 is used, then this results in not including the term $(F''(y)^*\lambda)\delta y$ in the second adjoint equation and each evaluation requires us to solve the linear time-varying Hamiltonian system

$$\begin{aligned} \frac{d}{dt}y &= A(t)\,\delta y + B\,\delta u, \quad \delta y(0) = 0, \\ -\frac{d}{dt}\mu &= A(t)^*\mu + \ell''(y)\delta y, \quad \mu(T) = 0, \end{aligned}$$

where $A(t) = A_0 + F'(y(t))$.

5.5 Newton method for the weakly singular case

In this section we discuss the Newton method for the weakly singular case as introduced in Section 1.5, and we utilize the notation given there. In particular $J : Y \times U \to \mathbb{R}$, $e : Y_1 \times U \to W$, with $e_y(y, u)$ considered as an operator in Y. Assuming that $(y, u, \lambda) \in Y_1 \times U \times \operatorname{dom}(e_y(y, u)^*)$, the SQP step for $(\delta y, \delta u, \delta\lambda)$ is given by

$$\begin{pmatrix} \mathcal{L}''(y, u, \lambda) & e'(y, u)^* \\ e'(y, u) & 0 \end{pmatrix} \begin{pmatrix} \delta y \\ \delta u \\ \delta\lambda \end{pmatrix} = - \begin{pmatrix} e_y(y, u)^*\lambda + J_y(y, u) \\ e_u(y, u)^*\lambda + J_u(y, u) \\ e(y, u) \end{pmatrix}.$$

The updates $(y+\delta y, u+\delta u, \lambda+\delta\lambda)$ will not necessarily remain in $Y_1 \times U \times \mathrm{dom}(e_y(y,u)^*)$ since it is not assumed that e' is surjective from $Y_1 \times U$ to W. However, the feasibility steps consisting in solving the primal equation

$$e(y, u^+) = 0 \tag{5.5.1}$$

for $y = y^+$ and the adjoint equation

$$e_y(y^+, u^+)^*\lambda + J_y(y^+, u^+) = 0 \tag{5.5.2}$$

for the dual variable $\lambda = \lambda^+$ will guarantee that $(y^+, \lambda^+) \in Y_1 \times Z_1$ holds. Here $u^+ = u+\delta u$ and $Z_1 \subset \mathrm{dom}\,(e_y(y^+, u^+)^*)$ denotes a Banach space densely embedded into W^*, with $\lambda^* \in Z_1$. Since Y_1 and Z_1 are contained in Y and W^*, the feasibility steps (5.5.1)–(5.5.2) can also be considered as smoothing steps. Thus we obtain the Newton method for the singular case.

Algorithm

- Initialization: Choose $u_0 \in V(u^*)$, solve

$$e(y, u_0) = 0, \quad e_y(y_0, u_0)^*\lambda + J_y(y_0, u_0) = 0 \text{ for } (y_0, \lambda_0) \in Y_1 \times Z_1,$$

  and set $k = 0$.

- Newton step: Solve for $(\delta y, \delta u, \delta\lambda) \in Y \times U \times W^*$

$$\begin{pmatrix} \mathcal{L}''(y_k, u_k, \lambda_k) & e'(y_k, u_k)^* \\ e'(y_k, u_k) & 0 \end{pmatrix} \begin{pmatrix} \delta y \\ \delta u \\ \delta\lambda \end{pmatrix} = - \begin{pmatrix} 0 \\ e_u(y_k, u_k)^*\lambda_k + J_u(y_k, u_k) \\ 0 \end{pmatrix}.$$

- Newton update: $u_{k+1} = u_k + \delta u$.

- Feasibility step: Solve for $(y_{k+1}, \lambda_{k+1}) \in Y_1 \times Z_1$:

$$e(y, u_{k+1}) = 0, \quad e_y(y_{k+1}, u_{k+1})^*\lambda + J_y(y_{k+1}, u_{k+1}) = 0.$$

- Stop, or set $k = k + 1$, and goto the Newton step.

Remark 5.5.1. The algorithm is essentially the Newton method of Section 5.2. Because of the feasible step, the first and the last components on the right-hand side of the SQP step are zero, and Newton's method and the SQP method coincide. Let us point out that the SQP iteration may not be well defined without the feasibility step since the updates (y^+, λ^+) may only be in $Y \times W^*$, while e and $e_y'(y^+, u^+)^*$ are not necessarily well defined on $Y \times U$ and W^*.

We next specify the assumptions which justify the above derivation and under which well-posedness and convergence of the algorithm can be proved. Thus let $(y^*, u^*, \lambda^*) \in Y_1 \times U \times Z_1$ be a solution to (5.1.4), or equivalently to (1.5.3), and let

$$V(y^*) \times V(u^*) \times V(\lambda^*) \subset Y_1 \times U \times Z_1$$

be a convex bounded neighborhood of the solution triple (y^*, u^*, λ^*).

(H5) (a) For every $u \in V(u^*)$ there exists a unique solution $y = y(u) \in V(y^*)$ of $e(y, u) = 0$. Moreover, there exists $M > 0$ such that $|y(u) - y^*|_{Y_1} \leq M|u - u^*|_U$.

(b) For every $(y, u) \in V(y^*) \times V(u^*)$ there exists a unique solution $\lambda = \lambda(y, u) \in V(\lambda^*)$ of $e_y(y, u)^*\lambda + J_y(y, u) = 0$ and $|\lambda(y, u) - \lambda^*|_{Z_1} \leq M|(y, u) - (y^*, u^*)|_{Y_1 \times U}$.

(H6) J is twice continuously Fréchet differentiable on $Y \times U$ with the second derivative locally Lipschitz continuous.

(H7) The operator $e : V(y^*) \times V(u^*) \subset Y_1 \times U \to W$ is Fréchet differentiable with Lipschitz continuous Fréchet derivative $e'(y, u) \in \mathcal{L}(Y_1 \times U, W)$. Moreover, for each $(y, u) \in V(y^*) \times V(u^*)$ the operator $e'(y, u)$ with domain in $Y \times U$ has closed range.

(H8) For every $\lambda \in V(\lambda^*)$ the mapping $(y, u) \to \langle \lambda, e(y, u)\rangle_{W^*,W}$ from $V(y^*) \times V(u^*)$ to $\mathbb{R}$ is twice Fréchet differentiable and the mapping $(y, u, \lambda) \to \langle \lambda, e''(y, u)(\cdot, \cdot)\rangle_{W^*,W}$ from $V(y^*) \times V(u^*) \times V(\lambda^*) \subset Y_1 \times U \times Z_1 \to \mathcal{L}(Y_1 \times U, Y^* \times U^*)$ is Lipschitz continuous. Moreover, for each $(y, u, \lambda) \in Y_1 \times U \times Z_1$, the bilinear form $\langle \lambda, e''(y, u)(\cdot, \cdot)\rangle_{W^*,W}$ can be extended as a continuous bilinear form on $(Y \times U)^2$.

(H9) For every $(y, u) \in V(y^*) \times V(u^*) \subset Y_1 \times U$ the operator $e'(y, u)$ can be extended as continuous linear operator from $Y \times U$ to W, and the mapping $(y, u) \to e'(y, u)$ from $V(y^*) \times V(u^*) \subset Y_1 \times U \to \mathcal{L}(Y \times U, W)$ is continuous and $(y, u, \lambda) \to \langle \lambda, e''(y, u)(\cdot, \cdot)\rangle$ from $V(y^*) \times V(u^*) \times V(\lambda^*) \subset Y_1 \times U \times Z_1 \to \mathcal{L}(Y \times U, Y^* \times U^*)$ is continuous.

(H10) There exists $\kappa > 0$ such that

$$\langle \mathcal{L}''(y^*, u^*, \lambda^*)v, v\rangle_{Y^* \times U^*, Y \times U} \geq \kappa\, |v|^2_{Y \times U}$$

for all $v \in \ker\, e'(y^*, u^*) \subset Y \times U$.

(H11) $e'(y^*, u^*) \in \mathcal{L}(Y \times U, W)$ is surjective.

Condition (H5) requires well-posedness of the primal and the adjoint equation in Y_1, respectively, Z_1. The adjoint equations arise from linearization of e at elements of $Y_1 \times U$. Condition (H6) requires smoothness of J. In (H7) and (H8) the necessary regularity requirements for e as mapping on $Y_1 \times U$ and in $Y \times U$ are specified. From (H5) it follows that the initialization as well as the feasibility step are well defined provided that $u_k \in V(u^*)$. As a consequence the derivatives of J and e that are required for defining the Newton step are taken at elements $(y_k, u_k, \lambda_k) \in Y_1 \times U \times Z_1$.

For $x = (y, u, \lambda) \in V(y^*) \times V(u^*) \times V(\lambda^*)$ let $A(x)$ denote the operator

$$A(x) = \begin{pmatrix} \mathcal{L}''(y, u, \lambda) & e'(y, u)^* \\ e'(y, u) & 0 \end{pmatrix}.$$

Conditions (H6) and (H9) guarantee that the operator $A(x) \in \mathcal{L}(Y \times U \times W^*, Y^* \times U^* \times W)$ for $x \in V(y^*) \times V(u^*) \times V(\lambda^*)$. Conditions (H6), (H9)–(H11) imply that

(H12) there exist a neighborhood $V(x^*) \subset V(y^*) \times V(u^*) \times V(\lambda^*)$ and M_1, such that for every $x = (y, u, \lambda) \in V(x^*)$ and $\delta w \in Y^* \times U^* \times W$

$$A(x)\delta x = \delta w$$

admits a unique solution $\delta x \in Y \times U \times W^*$ satisfying

$$|\delta x|_{Y \times U \times W^*} \leq M_1 \, |\delta w|_{Y^* \times U^* \times W}.$$

Theorem 5.3. *If* (H5)–(H8) *and* (H12) *hold at a solution* $(y^*, u^*, \lambda^*) \in Y_1 \times U \times Z_1$ *of* (5.1.4) *and* $|u_0 - u^*|_U$ *is sufficiently small, then the iterates of the algorithm are well defined and they satisfy*

$$|(y_{k+1}, u_{k+1}, \lambda_{k+1}) - (y^*, u^*, \lambda^*)|_{Y_1 \times U \times Z_1} \leq K \, |u_k - u^*|_U^2 \tag{5.5.3}$$

for a constant K independent of k.

Proof. Well-posedness of the algorithm in a neighborhood of (y^*, u^*, λ^*) follows from (H5) and (H12). To prove convergence let us denote by x^* the triple (y^*, u^*, λ^*), and similarly $\delta x = (\delta y, \delta u, \delta \lambda)$ and $x_k = (y_k, u_k, \lambda_k)$. Without loss of generality we assume that $\min(M, M_1) \geq 1$. The Newton step of the algorithm can be expressed as

$$A(x_k)\delta x = -F(x_k),$$

with $F : Y_1 \times U \times Z_1 \to Y^* \times U^* \times W$ defined by

$$F(y, u, \lambda) = -(e_y(y, u)^*\lambda + J_y(y, u), e_u(y, u)^*\lambda + J_u(y, u), e(y, u)).$$

Due to the smoothing step the first and third coordinates of F are 0 at x_k. By (H5) there exists $M > 0$ such that

$$|y_k - y^*|_{Y_1} \leq M \, |u_k - u^*|_U \quad \text{if } u_k \in V(u^*) \tag{5.5.4}$$

and

$$|\lambda_k - \lambda^*|_{Z_1} \leq M \, |(y_k, u_k) - (y^*, u^*)|_{Y_1 \times U} \quad \text{if } (y_k, u_k) \in V(y^*) \times V(u^*), \tag{5.5.5}$$

where (y_k, λ_k) are determined by the feasibility step. Let us assume that $x_k \in V(x^*)$. Then it follows from (H6)–(H8) that, possibly after increasing M, it can be chosen such that for every $x_k \in V(y^*) \times V(u^*) \times V(\lambda^*)$

$$|F(x^*) - F(x_k) - F'(x_k)(x^* - x_k)|_{Y^* \times U^* \times W}$$

$$= \int_0^1 |(F'(x_k + s(x^* - x_k)) - F'(x_k))(x^* - x_k)|_{Y^* \times U^* \times W} \, ds$$

$$= \int_0^1 |(A(x_k + s(x^* - x_k)) - A(x_k))(x^* - x_k)|_{Y^* \times U^* \times W} \, ds \leq \frac{M}{2} |x^* - x_k|_{Y_1 \times U \times Z_1}^2.$$

Moreover,

$$|A(x_k)(x_k+\delta x - x^*)| = |F(x^*) - F(x_k) - F'(x_k)(x^* - x_k)| \le \frac{M}{2}\,|x_k - x^*|^2_{Y_1\times U\times Z_1}.$$

Consequently, by (H12)

$$|u_{k+1} - u^*|_U \le |x_{k+1} - x^*|_{Y\times U\times W^*} \le \frac{MM_1}{2}|x_k - x^*|^2_{Y_1\times U\times Z_1}, \tag{5.5.6}$$

provided that $x_k \in V(x^*)$. The proof will be completed by an induction argument with respect to k. Let r be such that $2M^5M_1 r < 1$ and that $|x - x^*| < 2M^2 r$ implies $x \in V(x^*)$. Assume that $|u_0 - u^*| \le r$. Then $|y_0 - y^*|_{Y_1} \le Mr$ by (5.5.4) and $|\lambda_0 - \lambda^*|_{Z_1} \le \sqrt{2}M^2 r$ by (5.5.5). It follows that

$$|x_0 - x^*|_{Y_1\times U\times Z_1} \le 2M^2|u_0 - u^*|^2_U \le 2M^2 r$$

and hence $x_0 \in V(x^*)$. Let $|x_k - x^*|_{Y_1\times U\times Z_1} \le 2M^2 r$. Then from (5.5.6)

$$|u_{k+1} - u^*|_U = 2M^5M_1r^2 \le r. \tag{5.5.7}$$

Consequently (5.5.4)–(5.5.6) are applicable and imply

$$|x_{k+1} - x^*|_{Y_1\times U\times Z_1} \le 4M^2|u_{k+1} - u^*|^2_U \le M^3M_1|x_k - x^*|^2_{Y_1\times U\times Z_1}. \tag{5.5.8}$$

It follows that

$$|x_{k+1} - x^*|_{Y_1\times U\times Z_1} \le 4M^7M_1|u_k - u^*|^2_U,$$

which implies (5.5.3) with $K = 4M^7M_1$. From (5.5.7)–(5.5.8) finally $|x_{k+1} - x^*|_{Y_1\times U\times Z_1} \le 2M^2 r$. ☐

Let us return now to some of the examples of Chapter 1, Section 1.5, and discuss the applicability of conditions (H5)–(H11).

Example 1.19 revisited. Condition (H5)(a) is a direct consequence of Lemma 1.14. Condition (H5)(b) corresponds to

$$(\nabla\lambda, \nabla\phi) + (e^y\lambda, \phi) + (y - z, \phi) = 0 \text{ for all } \phi \in Y, \tag{5.5.9}$$

given $y \in Y_1$. Let $Z_1 = Y_1 = Y \cap L^\infty(\Omega)$. It follows from [Tr, Chapter 2.3] and the proof of Lemma 1.14 that there exists a unique solution $\lambda = \lambda(y) \in Z_1$ to (5.5.9). Moreover, if $y \in V(y^*)$ and $w = \lambda(y) - \lambda(y^*)$, then $w \in Z_1$ satisfies

$$(\nabla w, \nabla\phi) + (e^{y^*}w, \phi) + ((e^y - e^{y^*})\lambda, \phi) + (y - y^*, \phi) = 0 \text{ for all } \phi \in Y.$$

From the proof of Lemma 4.1 it follows that there exists $M > 0$ such that

$$|\lambda(y) - \lambda(y^*)|_{Z_1} \le M\,|y - y^*|_{Y_1} \text{ for all } y \in V(y^*)$$

and thus (H5)(b) is satisfied. It is simple to argue the validity of (H6)–(H9). Note that $e'(y^*, u^*)$ is surjective from $Y \times U$ to W and thus (H11) is satisfied. As for (H10), this condition is equivalent to the requirement that

$$|\delta y|^2_{L^2(\Omega)} + (\lambda^* e^{y^*}, (\delta y)^2) + \beta\,|\delta u|^2_U \ge \kappa(|\delta y|^2_Y + |\delta u|^2_U)$$

for all $(\delta y, \delta u) \in Y \times U$ satisfying

$$(\nabla \delta y, \nabla \phi) + (e^{y^*} \delta y, \phi) - (\delta u, \phi) = 0 \text{ for all } \phi \in Y, \tag{5.5.10}$$

where λ^* is the solution to (5.3.11) with $y = y^*$. Then there exists $\bar{k} > 0$ such that

$$|\delta y|_Y^2 \leq \bar{k} |\delta u|^2$$

for all $(\delta y, \delta u)$ satisfying (5.5.10). It follows that (H10) holds if $|(\lambda^*)^-|_{L^\infty}$ is sufficiently small. This is the case, for example, in the case of small residue problems, in the sense that $|y^* - z|_{L^2(\Omega)}^2$ is small enough. If $z \geq y^*$, then the weak maximum principle [Tr] is applied to (5.5.9) and gives $\lambda^* \geq 0$.

With quite analogous arguments it can be shown that (H5)–(H11) also hold for Example 1.20 of Chapter 1.

Example 1.22 revisited. The constraint and adjoint equations are given by

$$(\nabla y, \nabla \phi) - (u\, y, \nabla \phi) = (f, \phi) \text{ for all } \phi \in H_0^1(\Omega)$$

and

$$(\nabla \lambda, \nabla \phi) + (u\, \lambda, \nabla \phi) + (y - z, \phi) = 0 \text{ for all } \phi \in Y = H_0^1(\Omega), \tag{5.5.11}$$

where $\operatorname{div} u = 0$, $f \in L^2(\Omega)$, and $z \in L^2(\Omega)$. As already discussed in Examples 1.15 and 1.22 they admit unique solutions in $Y_1 = H_0^1(\Omega) \cap L^\infty(\Omega)$ for $u \in U = L_n^2(\Omega)$. Let $w_1 = y(u) - y(u^*)$ and $w_2 = \lambda(y, u) - \lambda(y^*, u^*)$. Then

$$(\nabla w_1, \nabla \phi) - (u\, w_1, \nabla \phi) = ((u - u^*) y^*, \nabla \phi)$$

and

$$(\nabla w_2, \nabla \phi) + (u\, w_2, \nabla \phi) = -((u - u^*)\, \lambda^*, \nabla \phi) - (y - y^*, \phi) \text{ for all } \phi \in Y.$$

From (1.4.22) it follows that (H5) holds if y^* and λ^* are in $W^{1,\infty}$. Conditions (H6)–(H8) are easily verified. Here we address only the closed range property of $e'(y, u)$, with $(y, u) \in Y_1 \times U$. For $u \in C_n^\infty(\Omega)$ with $\operatorname{div} u = 0$ surjectivity follows from the Lax–Milgram lemma, and a density argument asserts surjectivity for every $u \in U$. We turn to (H12) and assume that $u \in C_n^\infty(\Omega)$, $\operatorname{div} u = 0$ first. Then $e'(y, u) \in \mathcal{L}(V \times U, H^{-1}(\Omega))$ and $e'(y, u)(\delta y, \delta u) = 0$ can be expressed as

$$(\nabla \delta y, \nabla \phi) - (\delta u\; y, \nabla \phi) - (u\, \delta y, \nabla \phi) = 0 \text{ for all } \phi \in Y.$$

Hence

$$|\delta y|_Y \leq |y|_{L^\infty} |\delta u|_U \tag{5.5.12}$$

for all $(\delta y, \delta u) \in \ker e'(y, u)$. Henceforth we assume that

$$\beta - 2|y^*|_{L^\infty} |\lambda^*|_{L^\infty} > 0.$$

Then there exists a neighborhood $V(y^*) \times V(\lambda^*)$ of (y^*, λ^*) and $\kappa > 0$ such that

$$\beta - 2|y|_{L^\infty(\Omega)}|\lambda|_{L^\infty} \geq \kappa \tag{5.5.13}$$

for all $(y, \lambda) \in V(y^*) \times V(\lambda^*)$. For $(\delta y, \delta u) \in Y \times U$ and $(y, \lambda) \in V(y^*) \times V(\lambda^*)$ we have by (5.5.12) and (5.5.13)

$$\begin{aligned} &\mathcal{L}''(y, u, \lambda)((\delta y, \delta u), (\delta y, \delta u)) = |\delta y|^2_{L^2(\Omega)} + \beta|\delta u|^2_U - 2(\delta u \nabla \delta y, \lambda) \\ &\geq \beta|\delta u|^2 - 2|\lambda|_{L^\infty}|\nabla \delta y|_Y|\delta u|_U \geq \kappa|\delta u|^2_U. \end{aligned}$$

This estimate, together with (5.5.12), implies that $\mathcal{L}''(y, u, \lambda)$ is coercive on $\ker e'(y, u)$ and hence $A(x)\delta x = \delta w$ admits a unique solution in $Y \times U \times Y$ for every δw. To estimate δx in terms of δw, we note that the last equation in the system $A(x)\delta x = \delta w$ implies that

$$|\delta y|_Y \leq |y|_{L^\infty}|\delta u|_U + |w_3|_{H^{-1}}. \tag{5.5.14}$$

Similarly, we find from the first equation in the system that

$$|\delta\lambda|^2_Y \leq |\delta y||\delta\lambda| + |\lambda|_{L^\infty}|\delta\lambda|_Y|\delta u|_U + |w_1|_{H^{-1}}|\delta\lambda|_Y$$

and consequently

$$|\delta\lambda|_Y \leq c|\delta y| + |\lambda|_{L^\infty(\Omega)}|\delta u|_U + |w_1|_{H^{-1}}, \tag{5.5.15}$$

where c is the embedding constant of $H^1_0(\Omega)$ into $L^2(\Omega)$. Moreover, using (5.5.14) we find

$$\begin{aligned} &\mathcal{L}''(y, u, \lambda)((\delta y, \delta u), (\delta y, \delta u)) \geq |\delta y|^2_{L^2} + \beta|\delta u|^2_U - 2|\lambda|_{L^\infty}|\delta y|_Y|\delta u|_U \\ &\geq |\delta y|^2_{L^2} + \beta|\delta u|^2_U - 2|\lambda|_{L^\infty}|y|_{L^\infty(\Omega)}|\delta u|^2_U - 2|\lambda|_{L^\infty}|w_3|_{H^{-1}}|\delta u|_U \\ &\geq |\delta y|^2 + \kappa|\delta u|^2_U - 2|\lambda|_{L^\infty}|w_3|_{H^{-1}}|\delta u|_U. \end{aligned} \tag{5.5.16}$$

From (5.5.15), (5.5.16) and utilizing $A(x)\delta x = \delta w$ we obtain

$$|\delta y|^2 + \kappa|\delta u|^2 \leq (2|\lambda|_{L^\infty}|\delta u|_U + |\delta\lambda|_Y)|w_3|_{H^{-1}} + |w_1|_{H^{-1}}|\delta y|_Y + |w_2|_U|\delta u|_U. \tag{5.5.17}$$

Inequalities (5.5.14), (5.5.15), and (5.5.17) imply the existence of a constant M_1 such that

$$|\delta x|_{Y \times U \times Y} \leq M_1|\delta w|_{H^{-1} \times U \times H^{-1}} \tag{5.5.18}$$

for all $(y, \lambda) \in V(y^*) \times V(\lambda^*)$ and every $b \in C^\infty_n(\Omega)$, with $\operatorname{div} b = 0$. A density argument with respect to u implies that $A(x)\delta x = \delta w$ admits a solution for all $x \in V(y^*) \times U \times V(\lambda^*)$ and that (5.5.18) holds for all such x.

Chapter 6

Augmented Lagrangian-SQP Methods

6.1 Generalities

This chapter is devoted to second order augmented Lagrangian methods for optimization problems with equality constraints of the type

$$\begin{cases} \min f(x) \text{ over } x \in X \\ \text{subject to } e(x) = 0, \end{cases} \tag{6.1.1}$$

and to problems with equality constraints as well as additional constraints

$$\begin{cases} \min f(x) \text{ over } x \in C \\ \text{subject to } e(x) = 0, \end{cases} \tag{6.1.2}$$

where $f : X \to \mathbb{R}$, $e : X \to W$, with X and W real Hilbert spaces and C a closed convex set in X. We shall show that for equality constraints it is possible to replace the first order Lagrangian update that was developed in Chapter 4 by a second order one. It will become evident that second order augmented Lagrangian methods are closely related to SQP methods. For equality-constrained problems the SQP method also coincides with the Newton method applied to the first order optimality conditions. Just like the Newton method, the SQP method and second order augmented Lagrangian methods are convergent with second order convergence rate if the initialization is sufficiently close to a local solution of (6.1.1) and if appropriate additional regularity conditions are satisfied. In case the initialization is not in the region of attraction, globalization techniques such as line searches or trust region techniques may be necessary. We shall not focus on such methods within this monograph. However, the penalty term, which, together with the Lagrangian term, characterizes the augmented Lagrangian method, also has a globalization effect. This will become evident from the analytical as well as the numerical results of this chapter. Let us stress that for the concepts that are analyzed in this chapter we do not advocate choosing c large.

As in Chapter 3, throughout this chapter we identify the dual of the space W with itself, and we consider $e'(x)^*$ as an operator from W to X. In Section 6.2 we present the second order augmented Lagrangian method for (6.1.1). Problems with additional constraints as in (6.1.2) are considered in Section 6.3. Applications to optimal control problems will be given in Section 6.4. Section 6.5 is devoted to short discussions of miscellaneous topics including reduced SQP methods and mesh independence. In Section 6.6 we give a short description of related literature.

6.2 Equality-constrained problems

In this section we consider

$$\begin{cases} \min f(x) \text{ over } x \in X \\ \text{subject to } e(x) = 0, \end{cases} \tag{6.2.1}$$

$f : X \to \mathbb{R}$, $e : X \to W$, with X and W real Hilbert spaces. Let x^* be a local solution of (6.2.1). As before derivatives as well as partial derivatives with respect to the variable x will be denoted by primes. We shall not distinguish by notation between the functional f' in the dual X^* of X and its Riesz representation in X. As in Chapter 4 we shall identify the topological duals of W and Z with themselves.

It is assumed throughout that

$$\begin{cases} f \text{ and } e \text{ are twice continuously Fréchet differentiable} \\ \text{with Lipschitz continuous second derivatives} \\ \text{in a convex neighborhood of } V(x^*) \text{ of } x^* \end{cases} \tag{6.2.2}$$

and

$$e'(x^*) \text{ is surjective.} \tag{6.2.3}$$

The Lagrangian functional associated with (6.2.1) is denoted by $\mathcal{L} : X \times W \to \mathbb{R}$ and it is given by

$$\mathcal{L}(x, \lambda) = f(x) + \big(\lambda, e(x)\big)_W.$$

With (6.2.3) holding there exists a Lagrange multiplier $\lambda^* \in W$ such that the following first order necessary optimality condition is satisfied:

$$\mathcal{L}'(x^*, \lambda^*) = 0, \quad e(x^*) = 0. \tag{6.2.4}$$

We shall also make use of the following second order sufficient optimality condition:

$$\begin{cases} \text{there exists } \kappa > 0 \text{ such that} \\ \mathcal{L}''(x^*, \lambda^*)(h, h) \geq \kappa\, |h|_X^2 \quad \text{for all } \ h \in \ker e'(x^*). \end{cases} \tag{6.2.5}$$

Here $\mathcal{L}''(x^*, \lambda^*)$ denotes the bilinear form characterizing the second Fréchet derivative of $\mathcal{L}$ with respect to x at (x^*, λ^*). For any $c > 0$ the augmented Lagrangian functional $\mathcal{L}_c : X \times W \to \mathbb{R}$ is defined by

$$\mathcal{L}_c(x, \lambda) = f(x) + \big(\lambda, e(x)\big)_W + \frac{c}{2}\, |e(x)|_W^2 .$$

We note that the necessary optimality condition implies

$$\mathcal{L}_c'(x^*, \lambda^*) = 0 \quad e(x^*) = 0 \ \text{ for all } \ c \geq 0. \tag{6.2.6}$$

Lemma 6.1. *Let* (6.2.3) *and* (6.2.5) *hold. Then there exists a neighborhood* $V(x^*, \lambda^*)$ *of* (x^*, λ^*), $\bar{c} > 0$ *and* $\bar{\sigma} > 0$ *such that*

$$\mathcal{L}_c''(x, \lambda)(h, h) \geq \bar{\sigma}\, |h|_X^2 \quad \textit{for all } h \in X, \ (x, \lambda) \in V(x^*, \lambda^*), \quad \textit{and} \ \ c \geq \bar{c}.$$

Proof. Corollary 3.2 and conditions (6.2.3) and (6.2.5) imply the existence of $\bar{\sigma} > 0$ and $\bar{c} > 0$ such that

$$\mathcal{L}_c''(x^*, \lambda^*)(h, h) \geq 2\bar{\sigma}\, |h|_X^2 \quad \text{for all } \ h \in X \ \text{ and } \ c \geq \bar{c}.$$

Due to continuity of $(x, \lambda) \to \mathcal{L}_c''(x, \lambda)$ the conclusion of the lemma follows. ☐

Lemma 6.1 implies in particular that $x \to \mathcal{L}_c(x, \lambda^*)$ can be bounded from below by a quadratic function. This fact is referred to as augmentability of (6.2.1) at (x^*, λ^*).

Lemma 6.2. *Let* (6.2.3) *and* (6.2.5) *hold. Then there exist* $\bar{\sigma} > 0, \bar{c} > 0$, *and a neighborhood* $\tilde{V}(x^*)$ *of* x^* *such that*

$$\mathcal{L}_c(x, \lambda^*) \geq \mathcal{L}_c(x^*, \lambda^*) + \bar{\sigma}\, \left|x - x^*\right|_X^2 \ \textit{for all } x \in \tilde{V}(x^*) \textit{ and } c \geq \bar{c}. \tag{6.2.7}$$

Proof. Due to Taylor's theorem, Lemma 6.1, and (6.2.6) we find for $x \in V(x^*)$

$$\mathcal{L}_c(x, \lambda^*) = \mathcal{L}_c(x^*, \lambda^*) + \frac{1}{2}\mathcal{L}_c''(x^*, \lambda^*)(x - x^*, x - x^*) + o(|x - x^*|_X^2)$$

$$\geq \mathcal{L}_c(x^*, \lambda^*) + \frac{\bar{\sigma}}{2}\, \left|x - x^*\right|_X^2 + o(\left|x - x^*\right|_X^2).$$

The claim follows from this estimate. ☐

Without loss of generality we may assume that the neighborhoods $V(x^*)$ and $\tilde{V}(x^*)$ of (6.2.2) and Lemma 6.2 coincide and that $V(x^*, \lambda^*)$ of Lemma 6.1 equals $V(x^*) \times V(\lambda^*)$. Due to (6.2.2) we can further assume that $e'(x)$ is surjective for all $x \in V(x^*)$.

To iteratively determine (x^*, λ^*) one can apply Newton's method to (6.2.6). Given a current iterate (x, λ) the next iterate $(\hat{x}, \hat{\lambda})$ is the solution to

$$\begin{pmatrix} \mathcal{L}_c''(x, \lambda) & e'(x)^* \\ e'(x) & 0 \end{pmatrix} \begin{pmatrix} \hat{x} - x \\ \hat{\lambda} - \lambda \end{pmatrix} = -\begin{pmatrix} \mathcal{L}_c'(x, \lambda) \\ e(x) \end{pmatrix}. \tag{6.2.8}$$

Alternatively, (6.2.8) can be used to define the λ-update only, whereas the x-update is calculated by a different technique. We shall demonstrate next that (6.2.8) can be solved for $\hat{\lambda}$ without recourse to $\hat{x}$.

For $(x, \lambda) \in V(x^*, \lambda^*)$ and $c \geq \bar{c}$ we define $B(x, \lambda) \in \mathcal{L}(W)$ by

$$B(x, \lambda) = e'(x)\mathcal{L}_c''(x, \lambda)^{-1}e'(x)^*. \tag{6.2.9}$$

Here $\mathcal{L}(W)$ denotes the space of bounded linear operators from W into itself. Note that $B(x,\lambda)$ is invertible. In fact, there exists a constant $k>0$ such that

$$\begin{aligned}\big(B(x,y)y,y\big)_W &= \big(\mathcal{L}_c''(x,\lambda)^{-1}e'(x)^*y, e'(x)^*y\big)_X\\ &\ge k\,\big|e'(x)^*y\big|_X^2\end{aligned}$$

for all $y\in W$. Since $e'(x)^*$ is injective and has closed range, there exists $\hat{k}$ such that

$$\big|e'(x)^*y\big|_X^2 \ge \hat{k}\,|y|_W^2 \text{ for all } y\in W,$$

and by the Lax–Milgram theorem continuous invertibility of $B(x,\lambda)$ follows, provided that $(x,\lambda)\in V(x^*,\lambda^*)$ and $c\ge\bar{c}$. Premultiplying the first equation in (6.2.8) by $e'(x)\mathcal{L}_c''(x,\lambda)^{-1}$ we obtain

$$\begin{cases}\hat{\lambda}=\lambda+B(x,\lambda)^{-1}\left[e(x)-e'(x)\mathcal{L}_c''(x,\lambda)^{-1}\mathcal{L}_c'(x,\lambda)\right],\\ \hat{x}=x-\mathcal{L}_c''(x,\lambda)^{-1}\mathcal{L}_c'(x,\hat{\lambda}).\end{cases} \tag{6.2.10}$$

Whenever the dependence of $(\hat{x},\hat{\lambda})$ on (x,λ,c) is important, $(\hat{x}(x,\lambda,c),\hat{\lambda}(x,\lambda,c))$ will be written in place of $(\hat{x},\hat{\lambda})$. If, for fixed λ, $x=x(\lambda)$ is chosen as a local solution to

$$\min \mathcal{L}_c(x,\lambda) \text{ subject to } x\in X, \tag{6.2.11}$$

then $\mathcal{L}_c'(x(\lambda),\lambda)=0$ and

$$\hat{\lambda}(x(\lambda),\lambda,c)=\lambda+B(x(\lambda),\lambda)^{-1}e(x(\lambda)). \tag{6.2.12}$$

We point out that (6.2.12) can be interpreted as a second order update to the Lagrange variable. To acknowledge this, let d_c denote the dual functional associated with $\mathcal{L}_c$, i.e.,

$$d_c(\lambda)=\min \mathcal{L}_c(x,\lambda) \text{ subject to } \{x : \big|x-x^*\big|\le\epsilon\}$$

for some $\epsilon>0$. Then the first and second derivatives of d_c with respect to λ satisfy

$$\nabla_\lambda d_c(\lambda)=e(x(\lambda))$$

and

$$\nabla_\lambda^2 d_c(\lambda)=-B(x(\lambda),\lambda),$$

and (6.2.12) presents a Newton step for maximizing d_c.

Returning to (6.2.8) we note that its blockwise solution given in (6.2.10) requires setting up and inverting $\mathcal{L}_c''(x,\lambda)$. Following an argument due to Bertsekas [Be] we next argue that $\mathcal{L}_c''(x,\lambda)$ can be avoided during the iteration. One requires only that $\mathcal{L}_0''(x,\lambda)=\mathcal{L}''(x,\lambda)$. In fact, we find

$$\mathcal{L}_c'(x,\lambda)=\mathcal{L}_0'(x,\lambda+ce(x))$$

and

$$\mathcal{L}_c''(x,\lambda)=\mathcal{L}_0''(x,\lambda+ce(x))+c\big(e'(x)(\cdot),e'(x)(\cdot)\big)_W.$$

Consequently (6.2.8) can be expressed as

$$\begin{pmatrix} \mathcal{L}_0''(x, \lambda + ce(x)) + ce'(x)^*e'(x) & e'(x)^* \\ e'(x) & 0 \end{pmatrix} \begin{pmatrix} \hat{x} - x \\ \hat{\lambda} - \lambda \end{pmatrix}$$
$$= -\begin{pmatrix} \mathcal{L}_0'(x, \lambda + ce(x)) \\ e(x) \end{pmatrix}. \tag{6.2.13}$$

Using the second equation $e'(x)(\hat{x} - x) = -e(x)$ in the first equation of (6.2.13) we arrive at

$$\mathcal{L}_0''(x, \lambda + ce(x))(\hat{x} - x) - ce'(x)^*e(x) + e'(x^*)(\hat{\lambda} - \lambda) = -\mathcal{L}_0'(x, \lambda + ce(x)),$$

and hence (6.2.8) is equivalent to

$$\begin{pmatrix} \mathcal{L}_0''(x, \lambda + ce(x)) & e'(x)^* \\ e'(x) & 0 \end{pmatrix} \begin{pmatrix} \hat{x} - x \\ \hat{\lambda} - (\lambda + ce(x)) \end{pmatrix}$$
$$= -\begin{pmatrix} \mathcal{L}_0'(x, \lambda + ce(x)) \\ e(x) \end{pmatrix}. \tag{6.2.14}$$

Solving (6.2.8) is thus equivalent to

(i) carrying out the first order multiplier iteration

$$\tilde{\lambda} = \lambda + ce(x),$$

(ii) solving

$$\begin{pmatrix} \mathcal{L}_0''(x, \tilde{\lambda}) & e'(x)^* \\ e'(x) & 0 \end{pmatrix} \begin{pmatrix} \hat{x} - x \\ \hat{\lambda} - \tilde{\lambda} \end{pmatrix} = -\begin{pmatrix} \mathcal{L}_0'(x, \tilde{\lambda}) \\ e(x) \end{pmatrix}$$

for $(\hat{x}, \hat{\lambda})$.

It will be convenient to introduce the matrix of operators

$$M(x, \lambda) = \begin{pmatrix} \mathcal{L}_0''(x, \lambda) & e'(x)^* \\ e'(x) & 0 \end{pmatrix}.$$

With (6.2.3) and (6.2.5) holding there exist a constant $\kappa > 0$ and a neighborhood $U(x^*, \lambda^*) \subset V(x^*, \lambda^*)$ of (x^*, λ^*) such that

$$\| M^{-1}(x, \lambda) \|_{\mathcal{L}(X \times W)} \le \kappa \text{ for all } (x, \lambda) \in U(x^*, \lambda^*). \tag{6.2.15}$$

Lemma 6.3. *Assume that* (6.2.3) *and* (6.2.5) *hold. Then there exists a constant* $K > 0$ *such that for any* $(x, \lambda) \in U(x^*, \lambda^*)$ *the solution* $(\hat{x}, \hat{\lambda})$ *of*

$$M(x, \lambda) \begin{pmatrix} \hat{x} - x \\ \hat{\lambda} - \lambda \end{pmatrix} = -\begin{pmatrix} \mathcal{L}'(x, \lambda) \\ e(x) \end{pmatrix}$$

satisfies

$$\left|(\hat{x}, \hat{\lambda}) - (x^*, \lambda^*)\right|_{X \times W} \le K \left|(x, \lambda) - (x^*, \lambda^*)\right|^2_{X \times W}. \tag{6.2.16}$$

Proof. Note that

$$M(x,\lambda)\begin{pmatrix}\hat{x}-x\\ \hat{\lambda}-\lambda\end{pmatrix} = -\begin{pmatrix}\mathcal{L}'(x,\lambda)-\mathcal{L}'(x^*,\lambda^*)\\ e(x)-e(x^*)\end{pmatrix}$$

and consequently

$$M(x,\lambda)\begin{pmatrix}\hat{x}-x^*\\ \hat{\lambda}-\lambda^*\end{pmatrix} = -\begin{pmatrix}\mathcal{L}'(x,\lambda)-\mathcal{L}'(x^*,\lambda^*)\\ e(x)-e(x^*)\end{pmatrix} + M(x,\lambda)\begin{pmatrix}x-x^*\\ \lambda-\lambda^*\end{pmatrix}.$$

This equality further implies

$$M(x,\lambda)\begin{pmatrix}\hat{x}-x^*\\ \hat{\lambda}-\lambda^*\end{pmatrix} = \int_0^1 \Big[M(x,\lambda) - M(tx+(1-t)x^*, t\lambda+(1-t)\lambda^*)\Big]\begin{pmatrix}x-x^*\\ \lambda-\lambda^*\end{pmatrix}dt.$$

The regularity properties of f and e imply that $(x,\lambda)\to M(x,\lambda)$ is Lipschitz continuous on $U(x^*,\lambda^*)$ for a Lipschitz constant $\gamma>0$. Thus we obtain

$$\left|M(x,\lambda)\begin{pmatrix}\hat{x}-x^*\\ \hat{\lambda}-\lambda^*\end{pmatrix}\right|_{X\times W} \le \frac{\gamma}{2}\,|(x,\lambda)-(x^*,\lambda^*)|^2_{X\times W},$$

and by (6.2.15)

$$\left|(\hat{x},\hat{\lambda})-(x^*,\lambda^*)\right|_{X\times W} \le \frac{\gamma\kappa}{2}\,|(x,\lambda)-(x^*,\lambda^*)|^2_{X\times W},$$

which implies the claim. □

We now describe three algorithms and analyze their convergence. They are all based on (6.2.14) and differ only in the choice of x. Recall that if x is a solution to (6.2.11), then (6.2.12) and hence (6.2.14) provide a second order update to the Lagrange multiplier. Solving (6.2.11) implies extra computational cost. In the results which follow we show that as a consequence a larger region of attraction with respect to the initial condition and an improved rate of convergence factor is obtained, compared to methods which solve (6.2.11) only approximately or skip it all together. As a first choice x in (6.2.14) is determined by solving

$$\min \mathcal{L}_c(x,\lambda_n) \text{ subject to } x\in\overline{V(x^*)}.$$

The second choice is to take x only as an appropriate suboptimal solution to this optimization problem, and the third choice is to simply choose x as the solution $\hat{x}$ of the previous iteration of (6.2.14).

Algorithm 6.1.

(i) Choose $\lambda_0\in W$, $c\in(\bar{c},\infty)$ and set $\sigma=c-\bar{c}$, $n=0$.

(ii) Determine $\tilde{x}$ as a solution of

$$\min \mathcal{L}_c(x,\lambda_n) \text{ subject to } x\in\overline{V(x^*)}. \qquad (P_{aux})$$

(iii) Set $\tilde{\lambda} = \lambda_n + \sigma e(\tilde{x})$.

(iv) Solve for $(\hat{x}, \hat{\lambda})$:

$$M(\tilde{x}, \tilde{\lambda}) \begin{pmatrix} \hat{x} - \tilde{x} \\ \hat{\lambda} - \tilde{\lambda} \end{pmatrix} = - \begin{pmatrix} \mathcal{L}_0'(\tilde{x}, \tilde{\lambda}) \\ e(\tilde{x}) \end{pmatrix}.$$

(v) Set $\lambda_{n+1} = \hat{\lambda}$, $n = n + 1$, and goto (ii).

The existence of a solution to (P_{aux}) is guaranteed if, for example, the following conditions on f and e hold:

$$\begin{cases} f : X \to \mathbb{R} \text{ is weakly lower semicontinuous,} \\ e : X \to W \text{ maps weakly convergent sequences} \\ \text{to weakly convergent sequences.} \end{cases} \tag{6.2.17}$$

Under the conditions of Theorem 6.4 below it follows that the solutions $\tilde{x}$ of (P_{aux}) satisfy $\tilde{x} \in V(x^*)$.

Theorem 6.4. *If* (6.2.3), (6.2.5), *and* (6.2.17) *hold and* $\frac{1}{c-\bar{c}}\, |\lambda_0 - \lambda^*|^2$ *is sufficiently small, then Algorithm* 6.1 *is well defined and*

$$\left|(x_{n+1}, \lambda_{n+1}) - (x^*, \lambda^*)\right|_{X \times W} \le \frac{\hat{K}}{c - \bar{c}} \left|\lambda_n - \lambda^*\right|_W^2 \tag{6.2.18}$$

for a constant $\hat{K}$ *independent of* c *and* $n = 0, 1, \ldots$.

Proof. Let $\hat{\eta}$ be the largest radius for a ball centered at (x^*, λ^*) and contained in $U(x^*, \lambda^*)$, and let γ be a Lipschitz constant for f' and e' on $U(x^*, \lambda^*)$. Further let $E = e'(x^*)$ and note that $(EE^*)^{-1}E \in \mathcal{L}(X)$ as a consequence of (6.2.3). We define

$$\bar{M} = 1 + 2(\bar{c}\gamma)^2 + 8\gamma^2(1 + \mu)^2\, \|(EE^*)^{-1}E\|^2,$$

where $\mu = \left(1 + \frac{\bar{c}\gamma}{\sqrt{2\sigma\bar{\sigma}}}\right) |\lambda_0 - \lambda^*|_W + |\lambda^*|_W$, and we put

$$\eta = \min\left(\hat{\eta}\sqrt{\frac{2\sigma\bar{\sigma}}{\bar{M}}}, \frac{2\sigma\bar{\sigma}}{K\bar{M}}\right), \tag{6.2.19}$$

where K is given in Lemma 6.3. Let us assume that

$$\left|\lambda_0 - \lambda^*\right|_W < \eta.$$

The proof will be given by induction on n. The case $n = 0$ follows from the general arguments given below. For the induction step we assume that

$$\left|\lambda_i - \lambda^*\right|_W \le \left|\lambda_{i-1} - \lambda^*\right|_W \text{ for } i = 1, \ldots, n. \tag{6.2.20}$$

Using (ii) of Algorithm 6.1 we have

$$\begin{aligned} f(x^*) \geq \mathcal{L}_c(\tilde{x}, \lambda_n) &= f(\tilde{x}) + \big(\lambda^*, e(\tilde{x})\big)_W + \big(\lambda_n - \lambda^*, e(\tilde{x})\big)_W \\ &\quad + \frac{1}{2}\bar{c}\,|e(\tilde{x})|_W^2 + \frac{1}{2}(c-\bar{c})\,|e(\tilde{x})|_W^2 \\ &\geq \mathcal{L}_{\bar{c}}(\tilde{x}, \lambda^*) + \big(\lambda_n - \lambda^*, e(\tilde{x})\big)_W + \frac{\sigma}{2}\,|e(\tilde{x})|_W^2 \\ &= \mathcal{L}_{\bar{c}}(\tilde{x}, \lambda^*) + \frac{1}{2\sigma}\Big(\,|\lambda_n + \sigma e(\tilde{x}) - \lambda^*|_W^2 - |\lambda_n - \lambda^*|_W^2\,\Big). \end{aligned}$$

In view of Lemma 6.2 and the fact that $\tilde{x}$ is chosen in $V(x^*)(\subset \tilde{V}(x^*))$ we find

$$\bar{\sigma}\,\left|\tilde{x} - x^*\right|_X^2 + \frac{1}{2\sigma}\left|\tilde{\lambda} - \lambda^*\right|_W^2 \leq \frac{1}{2\sigma}\,|\lambda_n - \lambda^*|_W^2. \tag{6.2.21}$$

This implies in particular that $|\tilde{x} - x^*| \leq \sqrt{\frac{\bar{M}}{2\sigma\bar{\sigma}}}\,|\lambda_0 - \lambda^*| < \hat{\eta}$ and hence $\tilde{x} \in V(x^*)$. The necessary optimality conditions for (6.2.1) and (P_{aux}) with $\tilde{x} \in V(x^*)$ are given by

$$f'(x^*) + E^*\lambda^* = 0$$

and

$$f'(\tilde{x}) + e'(\tilde{x})^*(\lambda_n + ce(\tilde{x})) = 0.$$

Subtracting these two equations and defining $\bar{\lambda} = \lambda_n + ce(\tilde{x})$ give

$$E^*(\lambda^* - \bar{\lambda}) = f'(\tilde{x}) - f'(x^*) + (e'(\tilde{x})^* - e'(x^*)^*)(\bar{\lambda})$$

and consequently

$$\lambda^* - \bar{\lambda} = (EE^*)^{-1}E\left[f'(\tilde{x}) - f'(x^*) + (e'(\tilde{x})^* - e'(x^*)^*)(\bar{\lambda})\right].$$

We obtain

$$|\lambda^* - \bar{\lambda}|_W \leq 2\gamma(1 + |\bar{\lambda}|_W)\|(EE^*)^{-1}E\|\,\left|\tilde{x} - x^*\right|_X. \tag{6.2.22}$$

To estimate $\bar{\lambda}$ observe that due to (6.2.20) and (6.2.21)

$$\begin{aligned} \left|\bar{\lambda}\right|_W &\leq \left|\bar{\lambda} - \tilde{\lambda}\right|_W + \left|\tilde{\lambda}\right|_W = \bar{c}\,\left|e(\tilde{x}) - e(x^*)\right|_W + \left|\lambda^*\right|_W + \left|\tilde{\lambda} - \lambda^*\right|_W \\ &\leq \bar{c}\gamma\,\left|\tilde{x} - x^*\right|_W + \left|\lambda^*\right|_W + \left|\lambda_n - \lambda^*\right|_W \leq \left(1 + \frac{\bar{c}\gamma}{\sqrt{2\sigma\bar{\sigma}}}\right)\left|\lambda_n - \lambda^*\right|_W \\ &\quad + \left|\lambda^*\right|_W \leq \mu. \end{aligned} \tag{6.2.23}$$

Using (6.2.20)–(6.2.23) we find

$$\begin{aligned} &|(\tilde{x}, \tilde{\lambda}) - (x^*, \lambda^*)|_{X\times W}^2 \leq |\tilde{x} - x^*|_X^2 + 2|\tilde{\lambda} - \bar{\lambda}|_W^2 + 2|\bar{\lambda} - \lambda^*|_W^2 \\ &= |\tilde{x} - x^*|_X^2 + 2(\bar{c}\gamma)^2|\tilde{x} - x^*|_X^2 + 8\gamma^2(1+\mu)^2\|(EE^*)^{-1}E\|^2\,|\tilde{x} - x^*|_X^2 \\ &= \frac{\bar{M}}{2\sigma\bar{\sigma}}\,|\lambda_n - \lambda^*|_X^2 < \frac{\bar{M}}{2\sigma\bar{\sigma}}\eta^2 \leq \hat{\eta}^2. \end{aligned}$$

This implies that Lemma 6.3 is applicable with $(x, \lambda) = (\tilde{x}, \tilde{\lambda})$. We find

$$\left|(x_{n+1}, \lambda_{n+1}) - (x^*, \lambda^*)\right|_{X\times W} \le \frac{K\bar{M}}{2\sigma\bar{\sigma}} \left|\lambda_n - \lambda^*\right|_X^2, \tag{6.2.24}$$

and (6.2.18) is proved with $\hat{K} = \frac{K\bar{M}}{2\bar{\sigma}}$. From (6.2.20), (6.2.24), and the definition of η we also obtain

$$\left|(x_{n+1}, \lambda_{n+1}) - (x^*, \lambda^*)\right|_{X\times W} \le \left|\lambda_n - \lambda^*\right|_W < \eta.$$

This implies (6.2.20) for $i = n + 1$ and the proof is finished. □

Algorithm 6.2. This coincides with Algorithm 6.1 except for (i) and (ii), which are replaced by

(i)′ Choose $\lambda_0 \in W$, $c \in (\bar{c}, \infty)$, and $\sigma \in (0, c - \bar{c}]$ and set $n = 0$.

(ii)′ Determine $\tilde{x} \in V(x^*)$ such that

$$\mathcal{L}_c(\tilde{x}, \lambda_n) \le \mathcal{L}_c(x^*, \lambda_n) = f(x^*).$$

Theorem 6.5. *Let* (6.2.3) *and* (6.2.5) *hold. If* $|\lambda_0 - \lambda^*|_W$ *is sufficiently small, then Algorithm* 6.2 *is well defined and*

$$\left|(x_{n+1}, \lambda_{n+1}) - (x^*, \lambda^*)\right|_{X\times W} \le K\left(1 + \frac{1}{\sigma\bar{\sigma}}\right)\left|\lambda_n - \lambda^*\right|_W^2 \tag{6.2.25}$$

for all $n = 0, 1, \ldots$. *Here* x_{n+1} *stands for* $\hat{x}$ *of step* (iv) *of Algorithm* 6.1 *and* K, *independent of* c, *is given in* (6.2.16).

Proof. Let $\hat{\eta}$ be defined as in the proof of Theorem 6.4. We define

$$\eta = \min\left\{\frac{\hat{\eta}}{\sqrt{a}}, \frac{1}{Ka}\right\}, \tag{6.2.26}$$

where $a = 1 + \frac{1}{\sigma\bar{\sigma}}$. The proof is based on an induction argument. If

$$\left|\lambda_0 - \lambda^*\right| < \eta,$$

then the first iterate of Algorithm 6.2 is well defined,

$$\left|(x_1, \lambda_1) - (x^*, \lambda^*)\right|_{X\times W} < \eta$$

and (6.2.25) holds with $n = 0$. This will follow from the general arguments given below. Assuming that $|(x_n, \lambda_n) - (x^*, \lambda^*)|_{X\times W} < \eta$, we show that Algorithm 6.2 is well defined for $n + 1$, that

$$\left|(x_{n+1}, \lambda_{n+1}) - (x^*, \lambda^*)\right|_{X\times W} < \eta,$$

and that (6.2.25) holds. As in the proof of Theorem 6.4 one argues that (6.2.21) holds and consequently

$$\left|(\tilde{x}, \tilde{\lambda}) - (x^*, \lambda^*)\right|^2_{X \times W} \le \left(1 + \frac{1}{2\sigma\bar{\sigma}}\right) |\lambda_n - \lambda^*|^2_W . \tag{6.2.27}$$

This implies that $|(\tilde{x}, \tilde{\lambda}^*) - (x^*, \lambda^*)|_{X\times W} < \hat{\eta}$ and hence Lemma 6.3 is applicable with $(x, \lambda) = (\tilde{x}, \tilde{\lambda})$ and (iv) of Algorithm 6.2 is well defined. Combining (6.2.16) with (6.2.27) we find

$$\left|(x_{n+1}, \lambda_{n+1}) - (x^*, \lambda^*)\right|_{X \times W} \le K \left(1 + \frac{1}{2\sigma\bar{\sigma}}\right) |\lambda_n - \lambda^*|^2_W ,$$

which is (6.2.25). By the definition of η we further obtain

$$\left|(x_{n+1}, \lambda_{n+1}) - (x^*, \lambda^*)\right|_{X \times W} \le \left|\lambda_n - \lambda^*\right|_W < \eta. \qquad \square$$

Remark 6.2.1. In the proof of Theorem 6.4 as well as in that of Theorem 6.5 we utilize (6.2.7) from Lemma 6.2. Conditions (6.2.3) and (6.2.5) are sufficient conditions for (6.2.7) to hold for $c \ge \bar{c} > 0$. If (6.2.7) can be shown to hold for all $c \ge 0$, then $\bar{c}$ can be chosen equal to 0 and $\sigma = c$ is admissible in (i′) of Algorithm 6.2.

In the third algorithm we delete the second step of Algorithms 6.1 and 6.2 and directly iterate (6.2.14).

Algorithm 6.3.

(i) Choose $(x_0, \lambda_0) \in X \times W$, $c \ge 0$ and put $n = 0$.

(ii) Set $\tilde{\lambda} = \lambda_n + c e(x_n)$.

(iii) Solve for $(\hat{x}, \hat{\lambda})$:

$$M(x_n, \tilde{\lambda}) \begin{pmatrix} \hat{x} - x_n \\ \hat{\lambda} - \tilde{\lambda} \end{pmatrix} = - \begin{pmatrix} \mathcal{L}_0'(x_n, \tilde{\lambda}) \\ e(x_n) \end{pmatrix} .$$

(iv) Set $(x_{n+1}, \lambda_{n+1}) = (\hat{x}, \hat{\lambda})$, $n = n + 1$, and goto (ii).

Theorem 6.6. *Let* (6.2.3) *and* (6.2.5) *hold. If* $\max(1, c)\, |(x_0, \lambda_0) - (x^*, \lambda^*)|_{X\times W}$ *is sufficiently small, then Algorithm* 6.3 *is well defined and*

$$\left|(x_{n+1}, \lambda_{n+1}) - (x^*, \lambda^*)\right|_{X \times W} \le \tilde{K} \left|(x_n, \lambda_n) - (x^*, \lambda^*)\right|^2_{X \times W}$$

for a constant $\tilde{K}$ *independent of* $n = 0, 1, 2, \ldots,$ *but depending on* c.

Proof. Let $\hat{\eta}$ and γ be defined as in the proof of Theorem 6.4. We introduce

$$\eta = \min \left\{ \frac{\hat{\eta}}{\sqrt{a}}, \frac{1}{Ka} \right\}, \tag{6.2.28}$$

where $a = \max\left(2, 1 + 2c^2\gamma^2\right)$ and K is defined in Lemma 6.3.

Let us assume that

$$\left|(x_0, \lambda_0) - (x^*, \lambda^*)\right|_{X\times W} < \eta.$$

Again we proceed by induction and the case $n = 0$ follows from the general arguments given below. Let us assume that $|(x_n, \lambda_n) - (x^*, \lambda^*)|_{X\times W} < \eta$. Then

$$\begin{aligned}\left|(x_n, \tilde{\lambda}) - (x^*, \lambda^*)\right|^2_{X\times W} &\le \left|x_n - x^*\right|^2_X + 2c^2\left|e(x_n) - e(x^*)\right|^2_W + 2\left|\lambda_n - \lambda^*\right|^2_W \\ &\le \left|x_n - x^*\right|^2_X + 2c^2\gamma^2\left|x_n - x^*\right|^2_X + 2\left|\lambda_n - \lambda^*\right|^2_W \\ &\le a\left|(x_n, \lambda_n) - (x^*, \lambda^*)\right|^2_{X\times W} < \hat{\eta}^2,\end{aligned}$$

and thus Lemma 6.3 is applicable with $(x, \lambda) = (x_n, \tilde{\lambda})$. ☐

Remark 6.2.2. (i) If c is set equal to 0 in Algorithm 6.3, then we obtain the well-known SQP algorithm for the equality-constrained problem (6.2.1). It is well known to have a second order convergence rate which also follows from Theorem 6.6 since $\tilde{K}$ is finite for $c = 0$.

(ii) Theorem 6.6 suggests that in case (P_{aux}) is completely skipped in the second order augmented Lagrangian update the penalty parameter may have a negative effect on the region of attraction and on the convergence rate estimate. Our numerical experience, without additional globalization techniques, indicates that moderate values of c do not impede the behavior of the algorithm when compared to $c = 0$, which results in the SQP algorithm. Choosing $c > 0$ may actually enlarge the region of attraction when compared to $c = 0$. For parameter estimation problems $c > 0$ is useful, because in this way Algorithm 6.3 becomes a hybrid algorithm combining the output least squares and the equation error formulations [IK9].

6.3 Partial elimination of constraints

Here we consider

$$\begin{cases} \min f(x) \text{ subject to} \\ \\ e(x) = 0,\ g(x) \le 0, \text{ and } \ell(x) \in K, \end{cases} \tag{6.3.1}$$

where $f\colon X \to \mathbb{R}$, $e\colon X \to W$, $g\colon X \to \mathbb{R}^m$, $\ell\colon X \to Z$, where X, W, and Z are real Hilbert spaces, K is a closed convex cone in Z, and ℓ is an affine constraint. In the remainder of this chapter we identify the dual of Z with itself. Let x^* be a local solution of (6.3.1) and assume that

$$\begin{cases} f, e, \text{ and } g \text{ are twice continuously Fréchet differentiable} \\ \text{with second derivatives Lipschitz continuous} \\ \text{in a neighborhood of } x^*. \end{cases} \tag{6.3.2}$$

The objective of this section is to approximate (6.3.1) by a sequence of problems with quadratic cost and affine constraints. For this purpose we utilize the Lagrangian $\mathcal{L}: X \times W \times \mathbb{R}^m \times Z \to \mathbb{R}$ associated with (6.3.1) given by

$$\mathcal{L}(x, \lambda, \mu, \eta) = f(x) + \big(\lambda, e(x)\big)_W + \big(\mu, g(x)\big)_{\mathbb{R}^m} + \big(\eta, \ell(x)\big)_Z.$$

We shall require the following assumption (cf. Chapter 3):

$$x^* \text{ is a regular point, i.e.,} \tag{6.3.3}$$

$$0 \in \operatorname{int}\left\{ \begin{pmatrix} e'(x^*) \\ g'(x^*) \\ L \end{pmatrix} X + \begin{pmatrix} 0 \\ \mathbb{R}^m_+ \\ -K \end{pmatrix} + \begin{pmatrix} 0 \\ g(x^*) \\ \ell(x^*) \end{pmatrix} \right\},$$

where the interior is taken in $W \times \mathbb{R}^m \times Z$. With (6.3.3) holding there exists a Lagrange multiplier $(\lambda^*, \mu^*, \eta^*) \in W \times \mathbb{R}^{m,+} \times K^+$ such that

$$0 \in \begin{cases} \mathcal{L}'(x^*, \lambda^*, \mu^*, \eta^*), \\ -e(x^*), \\ -g(x^*) + \partial\, \psi_{\mathbb{R}^{m,+}}(\mu^*), \\ -\ell(x^*) + \partial\, \psi_{K^+}(\eta^*). \end{cases}$$

As in Section 3.1, we assume that the coordinates of the inequalities $g(x) \le 0$ are arranged so that $\mu^* = (\mu^{*,+}, \mu^{*,0}, \mu^{*,-})$ and $g = (g^+, g^0, g^-)$ with $g^+: X \to \mathbb{R}^{m_1},\ g^0: \mathbb{R}^{m_2},\ g^-: X \to \mathbb{R}^{m_3},\ m = m_1 + m_2 + m_3$, and

$$\begin{aligned} g^+(x^*) = 0,\ \mu^{*,+} > 0, \\ g^0(x^*) = 0,\ \mu^{*,0} = 0, \\ g^-(x^*) < 0,\ \mu^{*,-} = 0. \end{aligned}$$

We further define $G_+ = g^+(x^*)'$, $G_0 = g^0(x^*)'$,

$$E_+ = \begin{pmatrix} E \\ G_+ \end{pmatrix} : X \to W \times \mathbb{R}^{m_1},$$

and for $z \in Z$ we define the operator $\mathcal{E}(z): X \times \mathbb{R} \to (W \times \mathbb{R}^{m_1}) \times \mathbb{R}^{m_2} \times Z$ by

$$\mathcal{E}(z) = \begin{pmatrix} E_+ & 0 \\ G_0 & 0 \\ L & z \end{pmatrix}.$$

The following additional assumptions are required:

$$\text{there exists } \kappa > 0 \text{ such that } \mathcal{L}''(x^*, \lambda^*, \mu^*)\,(x, x) \ge \kappa |x|_X^2 \text{ for all } x \in \ker(E_+) \tag{6.3.4}$$

and

$$\mathcal{E}(\ell(x^*)) \text{ is surjective.} \tag{6.3.5}$$

The SQP method for (6.2.1) with elimination of the equality and finite rank inequality constraints is given next. We set

$$\mathcal{L}(x, \lambda, \mu) = f(x) + \big(\lambda, e(x)\big)_W + \big(\mu, g(x)\big)_{\mathbb{R}^m}.$$

Algorithm 6.4.

(i) Choose $(x_0, \lambda_0, \mu_0) \in X \times W \times \mathbb{R}^m$ and set $n = 0$.

(ii) Solve for $(x_{n+1}, \lambda_{n+1}, \mu_{n+1})$:

$$\begin{cases} \min \frac{1}{2}\mathcal{L}''(x_n, \lambda_n, \mu_n)(x - x_n, x - x_n) + f'(x_n)(x - x_n), \\ e(x_n) + e'(x_n)(x - x_n) = 0, \\ g(x_n) + g'(x_n)(x - x_n) \le 0, \ \ell(x_n) + L(x - x_n) \in K, \end{cases} \tag{P_{aux}}$$

where $(\lambda_{n+1}, \mu_{n+1})$ are the Lagrange multipliers associated to the equality and inequality constraints.

(iii) Set $n = n + 1$ and goto (ii).

Since (6.3.3)–(6.3.5) imply (H1), $(\widetilde{H}2)$, (H3) of Chapter 2, there exist by Corollary 2.18 neighborhoods $\hat{U}(x^*, \lambda^*, \mu^*)$ and $U(x^*)$ such that the auxiliary problem (P_{aux}) of Algorithm 6.4 admits a unique local solution x_{n+1} in $U(x^*)$ provided that $(x_n, \lambda_n, \mu_n) \in \hat{U}(x^*, \lambda^*, \mu^*) = \hat{U}(x^*) \times \hat{U}(\lambda^*, \mu^*)$. To obtain a convergence rate estimate for x_n, Lagrange multipliers to the constraints in (P_{aux}) are introduced. Since the regular point condition is stable with respect to perturbations in x^*, we can assume that $\hat{U}(x^*)$ is chosen sufficiently small such that for $x_n \in \hat{U}(x^*)$ there exist $(\lambda_{n+1}, \mu_{n+1}, \eta_{n+1}) \in W \times \mathbb{R}^m \times Z$ such that

$$0 \in \mathcal{G}(x_n, \lambda_n, \mu_n)(x_{n+1}, \lambda_{n+1}, \mu_{n+1}, \eta_{n+1}) + \begin{pmatrix} 0 \\ 0 \\ \partial\, \psi_{\mathbb{R}^{m,+}}(\mu_{n+1}) \\ \partial_{K^+}(\eta_{n+1}) \end{pmatrix}, \tag{6.3.6}$$

where

$$\mathcal{G}(x_n, \lambda_n, \mu_n)(x, \lambda, \mu, \eta) = \begin{pmatrix} \mathcal{L}''(x_n, \lambda_n, \mu_n)(x - x_n) + f'(x_n) + e'(x_n)^*\lambda + g'(x_n)^*\mu + L^*\eta \\ -e(x_n) - e'(x_n)(x - x_n) \\ -g(x_n) - g'(x_n)(x - x_n) \\ -\ell(x_n) - L(x - x_n) \end{pmatrix}.$$

Theorem 6.7. *Assume that* (6.3.2)–(6.3.5) *are satisfied at* (x^*, λ^*, μ^*) *and that* $|(x_0, \lambda_0, \mu_0) - (x^*, \lambda^*, \mu^*)|$ *is sufficiently small. Then there exists* $\bar{K} > 0$ *such that*

$$|(x_{n+1}, \lambda_{n+1}, \mu_{n+1}, \eta_{n+1}) - (x^*, \lambda^*, \mu^*, \eta^*)| \le \bar{K}|(x_n, \lambda_n, \mu_n) - (x^*, \lambda^*, \mu^*)|^2$$

for all $n = 1, 2, \ldots$.

Proof. The optimality system for (6.3.1) can be expressed by

$$0 \in \mathcal{G}(x^*, \lambda^*, \mu^*)(x^*, \lambda^*, \mu^*, \eta^*) + \begin{pmatrix} 0 \\ 0 \\ \partial\, \psi_{\mathbb{R}^{m,+}}(\mu^*) \\ \partial\, \psi_{K^+}(\eta^*) \end{pmatrix},$$

or equivalently

$$0 \in \begin{pmatrix} a_n \\ b_n \\ c_n \\ d_n \end{pmatrix} + \mathcal{G}(x_n, \lambda_n, \mu_n)(x^*, \lambda^*, \mu^*, \eta^*) + \begin{pmatrix} 0 \\ 0 \\ \partial\,\psi_{\mathbb{R}^{m,+}}(\mu^*) \\ \partial_k^+(\eta^*) \end{pmatrix}, \tag{6.3.7}$$

where $(a_n, b_n, c_n, d_n) = (a(x_n, \lambda_n, \mu_n), b(x_n), c(x_n), d(x_n))$ and

$$\begin{aligned} a(x, \lambda, \mu) &= f'(x^*) - f'(x) + (e'(x^*)^* - e'(x)^*)\lambda^* \\ &\quad + (g'(x^*)^* - g'(x)^*)\mu^* - \mathcal{L}''(x, \lambda, \mu)(x^* - x), \\ b(x) &= e(x) + e'(x)(x^* - x) - e(x^*), \\ c(x) &= g(x) + g'(x)(x^* - x) - g(x^*), \\ d(x) &= 0. \end{aligned}$$

Without loss of generality we may assume that the first and second derivatives of f, e, and g are Lipschitz continuous in $\hat{U}(x^*)$. It follows that there exists $\tilde{L}$ such that

$$\begin{aligned} (|a(x, \lambda)|^2 + |b(x)|^2 + |c(x)|^2)^{1/2} &\le \tilde{L}(|x - x^*|^2 + |\lambda - \lambda^*|^2 + |\mu - \mu^*|^2) \\ \text{for all } (x, \lambda, \mu) &\in \hat{U}(x^*) \times \hat{U}(\lambda^*, \mu^*). \end{aligned} \tag{6.3.8}$$

Let $\tilde{K}$ be determined from Corollary 2.18 and let $B((x^*, \lambda^*, \mu^*), r)$ denote a ball in $\hat{U}(x^*) \times \hat{U}(\lambda^*, \mu^*)$ with center (x^*, λ^*, μ^*) and radius r, where $r\,\tilde{K}\,\tilde{L} < 1$.

Proceeding by induction, assume that $(x_n, \lambda_n, \mu_n) \in B((x^*, \lambda^*, \mu^*), r)$. Then from Corollary 2.18, with $(\bar{x}_1, \bar{\lambda}_1, \bar{\mu}_1) = (\bar{x}_2, \bar{\lambda}_2, \bar{\mu}_2) = (x_n, \lambda_n, \mu_n)$, and (6.3.6) – (6.3.8) we find

$$\begin{aligned} &|(x_{n+1}, \lambda_{n+1}, \mu_{n+1}, \eta_{n+1}) - (x^*, \lambda^*, \mu^*, \eta^*)| \\ &\le \tilde{K}|(a_n, b_n, c_n)|_{X \times W \times \mathbb{R}^m} \le \tilde{K}\,\tilde{L}(|x_n - x^*|^2 + |\lambda_n - \lambda^*|^2 + |\mu_n - \mu^*|^2) \\ &\le (\tilde{K}\,\tilde{L}\,r)r < r. \end{aligned}$$

This estimate implies that $(x_{n+1}, \lambda_{n+1}, \mu_{n+1}) \in B((x^*, \lambda^*, \mu^*), r)$, as well as the desired local quadratic convergence of Algorithm 6.4. □

Remark 6.3.1. Let $\mathcal{L}(x, \lambda) = f(x) + \big(\lambda, e(x)\big)_W$ and consider Algorithm 6.4 with $\mathcal{L}''(x_n, \lambda_n, \mu_n)$ replaced by $\mathcal{L}''(x_n, \lambda_n)$; i.e., only the equality constraints are eliminated. If (6.3.4) is satisfied with $\mathcal{L}''(x^*, \lambda^*, \mu^*)$ replaced by $\mathcal{L}''(x^*, \lambda^*)$, then Theorem 6.7 holds for the resulting algorithm with the triple (x, λ, μ) replaced by (x, λ).

Next we consider

$$\begin{cases} \min f(x) \text{ subject to} \\ e(x) = 0, \; x \in C, \end{cases} \tag{6.3.9}$$

where f and e are as in (6.3.1) and C is a closed convex set in X. Let x^* denote a local solution of (6.3.9) and assume that

$$\begin{cases} f \text{ and } e \text{ are twice continuously Fréchet differentiable,} \\ f'' \text{ and } e'' \text{ are Lipschitz continuous in a neighborhood of } x^*, \end{cases} \tag{6.3.10}$$

$$0 \in \text{ int } e'(x^*)(C - x^*), \tag{6.3.11}$$

and

$$\begin{aligned} &\text{there exists a constant } \kappa > 0 \text{ such that } \mathcal{L}''(x^*, \lambda^*)(x, x) \geq \kappa |x|_X^2 \\ &\text{for all } x \in \ker e'(x^*). \end{aligned} \tag{6.3.12}$$

To solve (6.3.9) we consider the SQP algorithm with elimination of the equality constraint.

Algorithm 6.5.

(i) Choose $(x_0, \lambda_0) \in X \times W^*$ and set $n = 0$.

(ii) Solve for (x_{n+1}, λ_{n+1}):

$$\begin{cases} \min \frac{1}{2}\mathcal{L}''(x_n, \lambda_n)(x - x_n, x - x_n) + f'(x_n)(x - x_n), \\ \\ e(x_n) + e'(x_n)(x - x_n) = 0, \ x \in C, \end{cases} \tag{P_{aux}}$$

where λ_{n+1} is the Lagrange multiplier associated to the equality constraint.

(iii) Set $n = n + 1$ and goto (ii).

The optimality condition for (P_{aux}) in Algorithm 6.5 is given by

$$0 \in \mathcal{G}(x_n, \lambda_n)(x_{n+1}, \lambda_{n+1}) + \begin{pmatrix} \partial\psi_C(x_{n+1}) \\ 0 \end{pmatrix}, \tag{6.3.13}$$

where

$$\mathcal{G}(x_n, \lambda_n)(x, \lambda) = \begin{pmatrix} \mathcal{L}''(x_n, \lambda_n)(x - x_n) + f'(x_n) \\ -e(x_n) - e'(x_n)(x - x_n) \end{pmatrix}.$$

Due to (6.3.11) and (6.3.12) there exists a neighborhood $U(x^*, \lambda^*)$ of (x^*, λ^*) such that (6.3.13) admits a solution (x_{n+1}, λ_{n+1}) and x_{n+1} is the solution of (P_{aux}) of Algorithm 6.5, provided that $(x_n, \lambda_n) \in U(x^*, \lambda^*)$. We also require the following condition:

$$\begin{cases} \text{There exist neighborhoods } \hat{U}(x^*) \times \hat{U}(\lambda^*) \text{ of } (x^*, \lambda^*) \text{ and} \\ V \text{ of the origin in } X \times W \text{ and a constant } \tilde{K} \text{ such that} \\ \text{for all } q_1, q_2 \in V \text{ and } (\bar{x}, \bar{\lambda}) \in \hat{U}(x^*) \times \hat{U}(\lambda^*) \text{ there exists} \\ \text{a unique solution} \\ (x, \lambda) = (x(\bar{x}, \bar{\lambda}, q), \lambda(\bar{x}, \bar{\lambda}, q_1), \lambda(\bar{x}, \bar{\lambda}, q_1)) \in \hat{U}(x^*) \times \hat{U}(\lambda^*) \text{ of} \\ 0 \in q_1 + \mathcal{G}(\bar{x}, \bar{\lambda})(x, \lambda) + \partial\psi_C(x) \text{ and} \\ |(x(\bar{x}, \bar{\lambda}, q_1), \lambda(\bar{x}, \bar{\lambda}, q_1)) - (x(\bar{x}, \bar{\lambda}, q_2), \lambda(\bar{x}, \bar{\lambda}, q_2))| \leq \tilde{K}|q_1 - q_2|. \end{cases} \tag{6.3.14}$$

Remark 6.3.2. From the discussion in the first part of this section it follows that (6.3.14) holds for problem (6.3.1) with $C = \{x : g(x) \le 0,\ \ell(x) \in K\}$, provided that g is convex. Condition (6.3.14) was analyzed for optimal control problems in several papers; see, for instance, [AlMa, Tro].

Theorem 6.8. *Suppose that* (6.3.10)–(6.3.13) *hold at a local solution* x^* *of* (6.3.9). *If* $|(x_0, \lambda_0) - (x^*, \lambda^*)|$ *is sufficiently small, then the iterates* (x_n, λ_n) *converge quadratically to* (x^*, λ^*).

Proof. The optimality system for (6.3.9) can be expressed as

$$0 \in \mathcal{G}(x^*, \lambda^*)(x^*, \lambda^*) + \begin{pmatrix} \partial\psi_C(x^*) \\ 0 \end{pmatrix},$$

or equivalently

$$0 \in \begin{pmatrix} a_n \\ b_n \end{pmatrix} + \mathcal{G}(x_n, \lambda_n)(x^*, \lambda^*) + \begin{pmatrix} \partial\psi_C(x^*) \\ 0 \end{pmatrix}, \tag{6.3.15}$$

where

$$\begin{aligned} a_n &= f'(x^*) - f'(x_n) + (e'(x_n)^* - e'(x^*)^*)\lambda^* - \mathcal{L}''(x_n, \lambda_n)(x^* - x_n), \\ b_n &= e(x_n) + e'(x_n)(x^* - x_n) - e(x^*). \end{aligned}$$

In view of (6.3.10) we may assume that $\hat{U}(x^*)$ is chosen sufficiently small such that f'' and e'' are Lipschitz continuous on $\hat{U}(x^*)$. Consequently there exists $\tilde{L}$ such that

$$|(a_n, b_n)| \le \tilde{L}(|x_n - x^*|^2 + |\lambda_n - \lambda^*|^2), \text{ provided that } x_n \in \hat{U}(x^*). \tag{6.3.16}$$

Let $B((x^*, \lambda^*), r)$ denote a ball with radius $r < (\tilde{K}\tilde{L})^{-1}$ and center (x^*, λ^*) contained in $U(x^*, \lambda^*)$ and in $\hat{U}(x^*) \times \hat{U}(\lambda^*)$, and assume that $(x_n, \lambda_n) \in B((x^*, \lambda^*), r)$. Then by (6.3.13)–(6.3.16) we have

$$|(x_{n+1}, \lambda_{n+1}) - (x^*, \lambda^*)| \le \tilde{K}\tilde{L}\ |(x_n, \lambda_n) - (x^*, \lambda^*)|^2 < |(x_n, \lambda_n) - (x^*, \lambda^*)| < r.$$

The proof now follows by an induction argument. □

In Section 6.2 the combination of the second order method (6.2.8) with first order augmented Lagrangian updates was analyzed for equality-constrained problems. This approach can also be taken for problems with inequality constraints. We present the analogue of Theorem 6.4. Let

$$\mathcal{L}_c(x, \lambda, \mu) = f(x) + \big(\lambda, e(x)\big)_W + \big(\mu, \hat{g}(x, \mu, c)\big)_{\mathbb{R}^m} + \frac{c}{2}\,|e(x)|_W^2 + \frac{c}{2}\,|\hat{g}(x, \mu, c)|_{\mathbb{R}^m}^2,$$

where $\hat{g}(x, \mu, c) = \max(g(x), -\frac{\mu}{c})$, as defined in Chapter 3. Below $\tilde{c} \ge \bar{c}$ denote the constants and B_δ the closed ball of radius δ around x^* that were utilized in Corollary 3.7.

Algorithm 6.6.

(i) Choose $(\lambda_0, \mu_0) \in W \times \mathbb{R}^m$, $c \in [\bar{c}, \infty)$ and set $n = 0$.

(ii) Determine $\tilde{x}$ as the solution to

$$\min \mathcal{L}_c(x, \lambda_n, \mu_n) \text{ subject to } x \in B_\delta,\ \ell(x) \in K. \tag{P^1_{aux}}$$

(iii) Update $\tilde{\lambda} = \lambda_n + (c - \bar{c})e(\tilde{x})$, $\tilde{\mu} = \mu_n + (c - \bar{c})\hat{g}(\tilde{x}, \mu_n, c)$.

(iv) Solve (P_{aux}) of Algorithm 6.4 with $(x_n, \lambda_n, \mu_n) = (\tilde{x}, \tilde{\lambda}, \tilde{\mu})$ for $(x_{n+1}, \lambda_{n+1}, \eta_{n+1})$.

(v) Set $n = n + 1$ and goto (ii).

Theorem 6.9. *Assume that* (3.4.7), (3.4.9) *of Chapter 3 and* (6.3.2)–(6.3.4) *of this chapter hold at* (x^*, λ^*, μ^*). *If* $\frac{1}{c-\bar{c}}(|\lambda_0 - \lambda^*|^2 + |\mu_0 - \mu^*|^2)$ *is sufficiently small, then Algorithm 6.6 is well defined and*

$$\begin{aligned}|(x_{n+1}, \lambda_{n+1}, \mu_{n+1}, \eta_{n+1}) - (x^*, \lambda^*, \mu^*, \eta^*)|_{X \times W \times \mathbb{R}^m \times Z}\\ \leq \frac{\hat{K}}{c - \bar{c}}(|\lambda_n - \lambda^*|^2_W + |\mu_n - \mu^*|^2_{\mathbb{R}^m})\end{aligned}$$

for a constant $\hat{K}$ *independent of* $n = 0, 1, 2, \ldots$.

Proof. The assumptions guarantee that Corollary 3.7, Proposition 3.9, and Theorem 3.10 of Chapter 3 and Theorem 6.7 of the present chapter are applicable. The proof is now similar to that of Theorem 6.4, with Theorem 6.7 replacing Lemma 6.3 to estimate the step of (P_{aux}). Let $\hat{\eta}$ be the radius of the largest ball U centered at (x^*, λ^*, μ^*) such that Theorem 6.7 is applicable and that the x-coordinates of elements in U are also in B_δ. Let

$$\eta = \min\left(\hat{\eta}\,\kappa\,\sqrt{c - \bar{c}},\ \frac{\kappa^2(c - \bar{c})}{\bar{K}}\right), \quad \text{where } \kappa^2 = \frac{2(c - \bar{c})}{1 + 2K^2},$$

with K defined in Theorem 3.10 and $\bar{K}$ in Theorem 6.7,

$$|(\lambda_0, \mu_0) - (\lambda^*, \mu^*)|_{W \times \mathbb{R}^m} < \eta.$$

We proceed by induction with respect to n. The case $n = 0$ is simple and we assume that

$$|(\lambda_i, \mu_i) - (\lambda^*, \mu^*)| \leq |(\lambda_{i-1}, \mu_{i-1}) - (\lambda^*, \mu^*)| \text{ for } i = 1, \ldots, n. \tag{6.3.17}$$

From Theorems 3.8 and 3.10 we have

$$\begin{aligned}|(\tilde{x} - x^*, \tilde{\lambda} - \lambda^*, \tilde{\mu} - \mu^*)| &\leq \frac{1}{\kappa\sqrt{c - \bar{c}}}|(\lambda_n, \mu_n) - (\lambda^*, \mu^*)|\\ &< \frac{\eta}{\kappa\sqrt{c - \bar{c}}} \leq \hat{\eta}.\end{aligned} \tag{6.3.18}$$

This estimate implies that $\tilde{x} \in \text{int } B_\delta$ and that Theorem 6.7 with (x_n, λ_n, μ_n) replaced by $(\tilde{x}, \tilde{\lambda}, \tilde{\mu})$ is applicable. It implies

$$|(x_{n+1}, \lambda_{n+1}, \mu_{n+1}, \eta_{n+1}) - (x^*, \lambda^*, \mu^*, \eta^*)| \le \bar{K}|(\tilde{x}, \tilde{\lambda}, \tilde{\mu}) - (x^*, \lambda^*, \mu^*)|^2, \quad (6.3.19)$$

and combined with (6.3.18)

$$\begin{aligned} |(x_{n+1}, \lambda_{n+1}, \mu_{n+1}, \eta_{n+1}) - (x^*, \lambda^*, \mu^*, \eta^*)| &\le \frac{\bar{K}}{\kappa^2(c-\bar{c})}|(\lambda_n, \mu_n) - (\lambda^*, \mu^*)|^2 \\ &\le |(\lambda_n, \mu_n) - (\lambda^*, \mu^*)|^2. \end{aligned}$$

This implies (6.3.17) with $i = n+1$ as well as the claim of the theorem. □

6.4 Applications

6.4.1 An introductory example

Consider the optimal control problem

$$\begin{cases} \min \frac{1}{2}\int_Q |y-z|^2 dxdt + \frac{\beta}{2}|u|_U^2, \\ y_t = \Delta y + g(y) + Bu \ \text{ in } \ Q, \\ y = 0 \ \text{ on } \ \Sigma, \\ y(0,\cdot) = \varphi \ \text{ on } \ \Omega, \end{cases} \quad (6.4.1)$$

where $Q = (0,T) \times \Omega$ and $\Sigma = (0,T) \times \partial\Omega$. Define $e : X = Y \times U \to W$ by

$$e(y,u) = y_t - \Delta y - g(y) - Bu.$$

Here $B \in \mathcal{L}(U,Y)$, where U is the Hilbert space of controls, and the choice for Y can be

$$Y = \{y \in L^2(H_0^1(\Omega)) \colon y_t \in L^2(H^{-1}(\Omega))\}$$

or

$$Y = L^2(H_0^1(\Omega) \cap H^2(\Omega)) \cap W^{1,2}(L^2(\Omega)),$$

for example. Here $L^2(H_0^1(\Omega))$ is an abbreviation for $L^2(0,T;H_0^1(\Omega))$. The former choice corresponds to variational formulations, the latter to strong formulations of the partial differential equation and both require regularity assumptions for g. Matching choices for W are

$$W = L^2(H^{-1}(\Omega)) \ \text{ and } \ W = L^2(L^2(\Omega)).$$

We proceed with $Y = \{y \in L^2(H_0^1(\Omega)) \colon y_t \in L^2(H^{-1}(\Omega))\}$. The Lagrangian is given by

$$\begin{aligned} \mathcal{L}(y,u,\lambda) &= J(y,u) + \langle \lambda, e(y,u)\rangle_{L^2(H^{-1})} \\ &= J(y,u) + \int_0^T \langle \lambda, (-\Delta)^{-1}(y_t - \Delta y - g(y) - Bu)\rangle_{H^{-1},H_0^1} dt, \end{aligned}$$

where Δ denotes the Laplacian with Dirichlet boundary conditions. The augmented Lagrangian SQP step as described below (6.2.14) is given by

$$\tilde{\lambda} = \lambda + c(y_t - \Delta y - g(y) - Bu)$$

and

$$\begin{pmatrix} I + g''(y)\Delta^{-1}\tilde{\lambda} & 0 & (\frac{\partial}{\partial t} + \Delta + g'(y))\Delta^{-1} \\ 0 & \beta I & B^*\Delta^{-1} \\ (-\Delta)^{-1}(\frac{\partial}{\partial t} - \Delta - g'(y)) & \Delta^{-1}B & 0 \end{pmatrix} \begin{pmatrix} \delta y \\ \delta u \\ \delta\lambda \end{pmatrix}$$
$$= - \begin{pmatrix} y - z - ((\frac{\partial}{\partial t} + \Delta + g'(y))(-\Delta)^{-1}\tilde{\lambda} \\ \beta u + B^*\Delta^{-1}\tilde{\lambda} \\ (-\Delta)^{-1}(y_t - \Delta y - g(y) - Bu) \end{pmatrix},$$

with $\delta y(0,\cdot) = \delta\lambda(T,\cdot) = 0$. Inspection of the previous system (e.g., for the sake of symmetrization) suggests introducing the new variable $\Lambda = -\Delta^{-1}\lambda$.

This results in an update for Λ given by

$$\tilde{\Lambda} = \Lambda + c(-\Delta)^{-1}(y_t - \Delta y - g(y) - Bu), \tag{6.4.2}$$

and the transformed system has the form

$$\begin{pmatrix} I - g''(y)\tilde{\Lambda} & 0 & -(\frac{\partial}{\partial t} + \Delta + g'(y)) \\ 0 & \beta I & -B^* \\ (\frac{\partial}{\partial t} - \Delta - g'(y)) & -B & 0 \end{pmatrix} \begin{pmatrix} \delta y \\ \delta u \\ \delta\Lambda \end{pmatrix}$$
$$= - \begin{pmatrix} y - z - (\frac{\partial}{\partial t} + \Delta + g'(y))\tilde{\Lambda} \\ \beta u - B^*\tilde{\Lambda} \\ y_t - \Delta y - g(y) - Bu \end{pmatrix}. \tag{6.4.3}$$

Let us also point out that if we had immediately discretized (6.4.1), then the differences between topologies tend to get lost and $(-\Delta)^{-1}$ in (6.4.2) may have been forgotten. On the discretized level the effect of $(-\Delta)^{-1}$ can be restored by preconditioning.

Let us now return to the system (6.4.3). Due to its size it will—after discretization—not be solved by a direct method but rather by iteration based on conjugate gradients. The question of choice for preconditioners arises. The following two choices for block preconditioners were successful in our tests [KaK]:

$$\begin{pmatrix} 0 & 0 & P^* \\ 0 & \beta I & 0 \\ P & 0 & 0 \end{pmatrix}^{-1} = \begin{pmatrix} 0 & 0 & P^{-1} \\ 0 & \beta^{-1}I & 0 \\ P^{-*} & 0 & 0 \end{pmatrix} \tag{6.4.4}$$

and

$$\begin{pmatrix} I & 0 & P^* \\ 0 & \beta I & 0 \\ P & 0 & 0 \end{pmatrix}^{-1} = \begin{pmatrix} 0 & 0 & P^{-1} \\ 0 & \beta^{-1}I & 0 \\ P^{-*} & 0 & P^{-*}P^{-1}, \end{pmatrix}, \tag{6.4.5}$$

where $P = \partial_t - \Delta$. Note that (6.4.4) requires 2, whereas (6.4.5) needs 4, parabolic solves per iteration. Numerically, it appears that (6.4.4) is preferable to (6.4.5). For further investigations on preconditioning of SQP-like systems we refer the reader to [BaSa]. This may still be extended by the results of Battermann and Sachs.

6.4.2 A class of nonlinear elliptic optimal control problems

The general framework of the previous section will be applied to optimal control problems governed by partial differential equations of the type

$$\begin{cases} -\Delta y + g(y) = \hat{f} & \text{in } \Omega, \\ \frac{\partial y}{\partial n} = u & \text{on } \Gamma_1, \\ \frac{\partial y}{\partial n} = u_f & \text{on } \Gamma_2, \end{cases} \tag{6.4.6}$$

where $\hat{f} \in L^2(\Omega)$ and $u_f \in L^2(\Gamma_2)$ are fixed and $u \in L^2(\Gamma_1)$ will be the control variable. Here Ω is a bounded domain in $\mathbb{R}^2$ with $C^{1,1}$ boundary or Ω is convex. For the case of higher-dimensional domains we refer the reader to [IK10]. The boundary $\partial\Omega$ is assumed to consist of two disjoint sets Γ_1, Γ_2, each of which is connected, or possibly consisting of finitely many connected components, with $\Gamma = \partial\Omega = \Gamma_1 \cup \Gamma_2$, and Γ_2 possibly empty. Further it is assumed that $g \in C^2(\mathbb{R})$ and that $g(H^1(\Omega)) \subset L^{1+\varepsilon}(\Omega)$ for some $\varepsilon > 0$. Equation (6.4.6) is understood in the variational sense, i.e.,

$$(\nabla y, \nabla\varphi)_\Omega + (g(y), \varphi)_\Omega = (\hat{f}, \varphi)_\Omega + (\tilde{u}, \varphi)_\Gamma \tag{6.4.7}$$

for all $\varphi \in H^1(\Omega)$, where

$$\tilde{u} = \begin{cases} u & \text{on } \Gamma_1, \\ u_f & \text{on } \Gamma_2, \end{cases}$$

$(\cdot,\cdot)_\Gamma$ denotes the L^2-inner product on Γ, and $(\cdot,\cdot)_\Omega$ stands for duality pairing between functions in $L^p(\Omega)$ and $L^q(\Omega)$ with $p^{-1} + q^{-1} = 1$ for appropriate choices of p. We recall that $H^1(\Omega)$ is continuously embedded into $L^p(\Omega)$ for every $p \geq 1$ if $n = 2$.

In (6.4.7) we should more precisely write $\langle \tilde{u}, \tau_\Gamma \varphi\rangle_\Gamma$ instead of $\langle \tilde{u}, \varphi\rangle_\Gamma$, with τ_Γ the zero order trace operator on Γ. However, we shall frequently suppress this notation. We refer to y as a solution of (6.4.6) if (6.4.7) holds. The optimal control problem is given by

$$\begin{cases} \min \frac{1}{2}|Cy - y_d|_Z^2 + \frac{\alpha}{2}|u|_{L^2(\Gamma_1)}^2 \\ \\ \text{subject to } (y, u) \in H^1(\Omega) \times L^2(\Gamma_1) \text{ a solution of (6.4.6).} \end{cases} \tag{6.4.8}$$

Here C is a bounded linear (observation) operator from $H^1(\Omega)$ to a Hilbert space Z, and $y_d \in Z$ and $\alpha > 0$ are fixed.

To express (6.4.8) in the form (6.2.1) of Section 6.2 we introduce

$$\tilde{e} : H^1(\Omega) \times L^2(\Gamma_1) \to H^1(\Omega)^*$$

with

$$\langle \tilde{e}(y, u), \varphi\rangle_{(H^1)^*, H^1} = (\nabla y, \nabla\varphi)_\Omega + (g(y) - \hat{f}, \varphi)_\Omega - (\tilde{u}, \varphi)_\Gamma$$

and

$$e : H^1(\Omega) \times L^2(\Gamma) \to H^1(\Omega)$$

by

$$e = \mathcal{N}\tilde{e},$$

where $\mathcal{N} : H^1(\Omega)^* \to H^1(\Omega)$ is the Neumann solution operator associated with

$$(\nabla v, \nabla\varphi)_\Omega + (v, \varphi) = (h, \varphi)_\Omega \text{ for all } \varphi \in H^1(\Omega),$$

where $h \in H^1(\Omega)^*$. In the context of Section 6.2 we set

$$X = H^1(\Omega) \times L^2(\Gamma_1), \quad Y = H^1(\Omega),$$

with $x = (y, u) \in X$, and

$$f(x) = f(y, u) = \frac{1}{2}|Cy - y_d|_Z^2 + \frac{\alpha}{2}|u|^2_{L^2(\Gamma_1)}.$$

We assume that (6.4.8) admits a solution $x^* = (y^*, u^*)$. The regularity requirements (6.2.2) of Section 6.2 are clearly met by the mapping g. Those for e are implied by

(h0) $y \to g(y)$ is twice continuously differentiable from $H^1(\Omega)$ to $L^{1+\varepsilon}(\Omega)$ for some $\varepsilon > 0$ with Lipschitz continuous second derivative in a neighborhood of y^*.

We shall also require the following hypothesis:

(h1) $g'(y^*) \in L^{2+\varepsilon}(\Omega)$ for some $\varepsilon > 0$.

With (h1) holding, $g'(y^*)\varphi \in L^2(\Omega)$ for every $\varphi \in H^1(\Omega)$. It is simple to argue that

$$\ker e'(x^*) = \{(v, h) : \left(\nabla v, \nabla\varphi\right)_\Omega + \left(g'(y^*)v, \varphi\right)_\Omega = \left(h, \varphi\right)_{\Gamma_1} \text{ for all } \varphi \in H^1(\Omega)\},$$

i.e., $(v, h) \in \ker e'(x^*)$ if and only if (v, h) is a variational solution of

$$\begin{cases} -\Delta v + g'(y^*)v = 0 & \text{in } \Omega, \\ \frac{\partial v}{\partial n} = h & \text{on } \Gamma_1, \\ \frac{\partial v}{\partial n} = 0 & \text{on } \Gamma_2. \end{cases} \tag{6.4.9}$$

If (6.2.3) is satisfied, i.e., if $e'(x^*)$ is surjective, then there exists a unique Lagrange multiplier $\lambda^* \in H^1(\Omega)$ associated to x^* such that

$$e'(x^*)^*\lambda^* + (\mathcal{N}C^*(Cy^* - y_d), \alpha u^*) = 0 \text{ in } H^1(\Omega) \times L^2(\Gamma_1), \tag{6.4.10}$$

where $e'(x^*)^* : H^1(\Omega) \to H^1(\Omega) \times L^2(\Gamma_1)$ denotes the adjoint of $e'(x^*)$ and $C^* : Z \to H^1(\Omega)^*$ stands for the adjoint of $C : H^1(\Omega) \to Z$ with Z a pivot space. More precisely we have the following proposition.

Proposition 6.10 (Necessary Condition). *If x^* is a solution to* (6.4.8)*, $e'(x^*)$ is surjective, and* (h1) *holds, then λ^* is a variational solution of*

$$\begin{cases} -\Delta\lambda^* + g'(y^*)\lambda^* = -C^*(Cy^* - y_d) \text{ in } \Omega, \\ \dfrac{\partial \lambda^*}{\partial n} = 0 \text{ on } \Gamma, \end{cases} \tag{6.4.11}$$

i.e., $(\nabla\lambda^*, \nabla\varphi) + (g'(y^*)\lambda^*, \varphi) + (Cy^* - y_d, C\varphi)_Z = 0$ *for all* $\varphi \in H^1(\Omega)$ *and*

$$\tau_{\Gamma_1}\lambda^* = \alpha u^* \text{ on } \Gamma_1. \tag{6.4.12}$$

Proof. The Lagrangian associated with (6.4.8) can be expressed by

$$\begin{aligned} \mathcal{L}(y, u, \lambda) = &\frac{1}{2}|Cy - y_d|_Z^2 + \frac{\alpha}{2}|u|_{L^2(\Gamma_1)}^2 + \left(\nabla\lambda, \nabla y\right)_\Omega \\ &+\left(\lambda, g(y) - \hat{f}\right)_\Omega - \left(\lambda, \tilde{u}\right)_\Gamma. \end{aligned}$$

For every $(v, h) \in H^1(\Omega) \times L^2(\Gamma_1)$ we find

$$\mathcal{L}_y(y^*, u^*, \lambda^*)(v) = \left(Cy^* - y_d, Cv\right)_Z + \left(\nabla\lambda^*, \nabla v\right)_\Omega + \left(\lambda^*, g'(y^*)v\right)_\Omega$$

and

$$\mathcal{L}_u(y^*, u^*, \lambda^*)(h) = \alpha\left(u^*, h\right)_{\Gamma_1} - \left(\lambda^*, h\right)_{\Gamma_1}.$$

Thus the claim follows. □

We now aim for a priori estimates for λ^*. We define $B : H^1(\Omega) \to H^1(\Omega)^*$ as the differential operator given by the left-hand side of (6.4.11); i.e., $Bv = \varphi$ is characterized as the solution to

$$\left(\nabla v, \nabla\psi\right)_\Omega + \left(g'(y^*)v, \psi\right)_\Omega = \langle\varphi, \psi\rangle_{(H^1)^*, H^1} \text{ for all } \psi \in H^1(\Omega).$$

We shall use the following hypothesis:

(h2) 0 is not an eigenvalue of B.

Note that (h2) holds, for example, if

$$g'(y^*) \ge \underline{\beta} \quad \text{a.e. on } \Omega$$

for some $\underline{\beta} > 0$. With (h2) holding, B is an isomorphism from $H^1(\Omega)$ onto $H^1(\Omega)^*$. Moreover, (h2) implies surjectivity of $e'(x^*)$.

Lemma 6.11. *Let the conditions of Proposition* 6.10 *hold.*

(i) *There exists a constant $K(x^*)$ such that*

$$|\lambda^*|_{H^1} \le K(x^*)|(\mathcal{N}C^*(Cy^* - y_d), \alpha u^*)|_X.$$

(ii) *If moreover* (h2) *is satisfied and* $C^*(Cy^* - y_d) \in L^2(\Omega)$, *then there exists a constant* $K(y^*)$ *such that*

$$|\lambda^*|_{H^2} \le K(y^*)|C^*(Cy^* - y_d)|_{L^2(\Omega)}.$$

Proof. Due to surjectivity of $e'(x^*)$ we have $(e'(x^*)e'(x^*)^*)^{-1} \in \mathcal{L}(H^1(\Omega))$ and thus (i) follows from (6.4.10). Let us turn to (ii). Due to (h2) and (6.4.11) there exists a constant K_{y^*} such that

$$|\lambda^*|_{H^1} \le K_{y^*}|C^*(Cy^* - y_d)|_{(H^1)^*}. \tag{6.4.13}$$

To obtain the desired $H^2(\Omega)$ estimate for λ^* one applies the well-known H^2 a priori estimate for Neumann problems to

$$-\Delta\lambda^* + \lambda^* = w \text{ in } \Omega,$$
$$\frac{\partial \lambda^*}{\partial n} = 0 \text{ in } \partial\Omega$$

with $w = \lambda^* - g'(y^*)\lambda^* - C^*(Cy^* - y_d)$. This gives

$$|\lambda^*|_{H^2} \le K\left(|\lambda^*|_{L^2} + |g'(y^*)\lambda^*|_{L^2} + |C^*(Cy^* - y_d)|_{L^2}\right)$$

for a constant K (depending on Ω but independent of y^*). Since

$$|g'(y^*)\lambda^*|_{L^2} \le |g'(y^*)|_{L^{1+\epsilon}}|\lambda^*|_{H^1},$$

the desired result follows from (6.4.13). $\square$

To calculate the second Fréchet derivative we shall use

(h3) $g''(y^*) \in L^{1+\varepsilon}(\Omega)$ for some $\varepsilon > 0$.

Proposition 6.12. *Let* (6.2.3), (h1), *and* (h3) *hold. Then*

$$\mathcal{L}''(y^*, u^*, \lambda^*)((v, h), (v, h)) = |Cv|_Z^2 + \alpha|h|^2_{L^2(\Gamma_1)} + \left(\lambda^*, g''(y^*)v^2\right)_\Omega \tag{6.4.14}$$

for all $(v, h) \in X$.

Proof. It suffices to observe that by Sobolev's embedding theorem there exists a constant K_e such that

$$|\left(\lambda^*, g''(y^*)v^2\right)_\Omega| \le K_e|\lambda^*|_{H^1}|g''(y^*)|_{L^{1+\epsilon}}|v|^2_{H^1} \tag{6.4.15}$$

for all $v \in H^1(\Omega)$. $\square$

We turn now to an analysis of the second order sufficient optimality condition (6.2.5) for the optimal control problem (6.4.8). In view of (6.4.14) the crucial term is given by $\left(\lambda^*, g''(y^*)v^2\right)_\Omega$. Two types of results will be given. The first will rely on $|\left(\lambda^*, g''(y^*)v^2\right)_\Omega|$

being sufficiently small. This can be achieved by guaranteeing that λ^* or, in view of (6.4.11), $Cy^* - y_d$ is small. We refer to this case as small residual problems. The second class of assumptions rests on guaranteeing that $\lambda^* g''(y^*) \geq 0$ a.e. on Ω.

In the statement of the following theorem we use K_e from (6.4.15) and $K(x^*)$, $K(y^*)$ from Lemma 6.11. Further $\|B^{-1}\|$ denotes the norm of B^{-1} as an operator from $H^1(\Omega)^*$ to $H^1(\Omega)$.

Theorem 6.13. *Let* (6.2.3), (h1)–(h3) *hold.*

(i) *If* $Z = H^1(\Omega)$, $C = \text{id}$, *and*

$$K_e K(x^*)|(y^* - y_d, \alpha g^*)|_X |g''(y^*)|_{L^q} < 1, \tag{6.4.16}$$

then the second order sufficient optimality condition (6.2.5) *holds.*

(ii) *If* $Z = L^2(\Omega)$, *C is the injection of* $H^1(\Omega)$ *into* $L^2(\Omega)$, *and*

$$\tilde{k}_e K(y^*)|y^* - y_d|_{L^2(\Omega)}|g''(y^*)|_{L^\infty(\Omega)} \leq 1, \tag{6.4.17}$$

where $\tilde{k}_e$ *is the embedding constant of* $H^2(\Omega)$ *into* $L^\infty(\Omega)$, *then* (6.2.5) *is satisfied.*

(iii) *If*

$$2\|B^{-1}\|\|\tau_{\Gamma_1}\|K_{y^*}|C^*(Cy^* - y_d)|_{(H^1)^*} < \alpha, \tag{6.4.18}$$

where $\|\tau_{\Gamma_1}\|$ *is the norm of the trace operator from* $H^1(\Omega)$ *onto* $L^2(\Gamma_1)$ *and* K_{y^*} *is defined in* (6.4.13), *then* (6.2.5) *is satisfied.*

Proof. (i) By (6.4.14) and (6.4.15) we have for every $(v, h) \in X$

$$\begin{aligned}
&\mathcal{L}''(y^*, u^*, \lambda^*)((v,h),(v,h)) \\
&\geq |v|^2_{H^1(\Omega)} + \alpha|h|^2_{L^2(\Gamma_1)} - K_e|\lambda^*|_{H^1(\Omega)}|g''(y^*)|_{L^{1+\epsilon}(\Omega)}|v|^2_{H^1(\Omega)} \\
&\geq \left(1 - K_e K(x^*)|(y^* - y_d, \alpha u^*)|_X |g''(y^*)|_{L^{1+\epsilon}(\Omega)}\right)|v|^2_{H^1} + \alpha|h|^2_{L^2(\Gamma_1)},
\end{aligned}$$

where in the last estimate we used Lemma 6.11 (i). The claim now follows from (6.4.16). We observe that in this case $\mathcal{L}''(y^*, u^*, \lambda^*)$ is positive definite on all of X, not only on $\ker e'(x^*)$.

(ii) By (6.4.9) and (h2) we obtain

$$|v|_{H^1(\Omega)} \leq \|B^{-1}\|\|\tau_{\Gamma_1}\||h|_{L^2(\Gamma_1)} \text{ for all } (v,h) \in \ker e'(x^*). \tag{6.4.19}$$

Here $\|\tau_{\Gamma_1}\|$ denotes the norm of the trace operator from $H^1(\Omega)$ onto $L^2(\Gamma_1)$. Hence by Lemma 6.11 (ii) and (6.4.19) we find for every $(v, h) \in \ker e'(x^*)$

$$\begin{aligned}
&\mathcal{L}''(v^*, u^*, \lambda^*)((v,h),(v,h)) = |v|^2_{L^2(\Omega)} + \alpha|h|^2_{L^2(\Gamma_1)} - \left(\lambda^*, g''(y^*)v^2\right)_{L^2(\Omega)} \\
&\geq |v|^2_{L^2(\Omega)} + \alpha|h|^2_{L^2(\Gamma_1)} - |\lambda^*|_{L^\infty(\Omega)}|g''(y^*)|_{L^\infty(\Omega)}|v|^2_{L^2(\Omega)} \\
&\geq \left[1 - \tilde{k}_e K(y^*)|g''(y^*)|_{L^\infty(\Omega)}|y^* - y_d|_{L^2(\Omega)}\right]|v|^2_{L^2(\Omega)} \\
&\quad + \frac{\alpha}{2}\left(|h|^2_{L^2(\Gamma_1)} + \frac{1}{\|B^{-1}\|^2\|\tau_{\Gamma_1}\|^2}|v|^2_{H^1(\Omega)}\right).
\end{aligned}$$

Due to (6.4.17) the expression in brackets is nonnegative and the result follows.

(iii) In this case C can be a boundary observation operator, for example. As in (i) we find

$$\begin{aligned}\mathcal{L}''(y^*, u^*, \lambda^*)((v,h),(v,h)) &\geq |Cv|_Z^2 + \alpha|h|^2_{L^2(\Gamma_1)} - K_e|\lambda^*|_{H^1(\Omega)}|g''|_{L^{1+\epsilon}(\Omega)}|v|^2_{H^1(\Omega)} \\ &\geq |Cv|_Z^2 + \alpha|h|^2_{L^2(\Gamma_1)} - K_{y^*}|C^*(Cy^* - y_d)|_{(H^1)^*}|g''|_{L^\epsilon(\Omega)}|v|^2_{H^1(\Omega)},\end{aligned}$$

where (6.4.17) was used. This estimate and (6.4.19) imply that for $(v,h) \in \ker e'(x^*)$

$$\begin{aligned}\mathcal{L}''(y^*, u^*, \lambda^*)((v,h),(v,h)) &\geq |Cv|_Z^2 + \frac{\alpha}{2}|h|^2_{L^2(\Gamma_1)} \\ &+ \left[\frac{\alpha}{2\|B^{-1}\|\|\tau_{\Gamma_1}\|} - K_{y^*}|C^*(Cy^* - y_d)|_{(H^1)^*}|g''|_{L^{1+\epsilon}(\Omega)}\right]|v|^2_{H^1(\Omega)}.\end{aligned}$$

The desired result follows from (6.4.18). □

In view of (6.4.16)–(6.4.18) and the fact that $|y^* - y_d|$ is decreasing with $\alpha \to 0^+$, the question arises, whether decreasing α results in the fact that the second order sufficient optimality condition. This is nontrivial since the term $|y^* - y_d|$ in (6.4.16)–(6.4.18) is multiplied by factors which depend on x^* and hence on α itself. We refer the reader to [IK10] for a treatment of this issue.

In Theorem 6.13 the second order optimality condition was guaranteed by small residue conditions. Alternatively one can proceed by assuming that

(h4) $\lambda^* g''(y^*) \geq 0$ a.e. on Ω.

Theorem 6.14. *Assume that* (6.2.3), (h1), (h3), *and* (h4) *hold and that*

(a) $Z = H^1(\Omega)$ *and* $C = \mathrm{id}$,

or

(b) *(h2) is satisfied.*

Then (6.2.5) *holds.*

Proof. By Proposition 6.12 and (h4) we find for all $(v,h) \in X$

$$\mathcal{L}''(y^*, u^*, \lambda^*)((v,h),(v,h)) \geq |Cv|_Z^2 + \alpha|h|^2_{L^2(\Gamma_1)}.$$

In the case that (a) holds the conclusion is obvious and $\mathcal{L}''(v^*, u^*, \lambda^*)$ is positive not only on $\ker e'(x^*)$ but on all of X. In case (b) we use (6.4.19) to conclude that

$$\mathcal{L}''(y^*, u^*, \lambda^*)((v,h),(v,h)) \geq |Cv|_Z^2 + \frac{\alpha}{2}\left(|h|^2_{L^2(\Gamma_1)} + \frac{1}{\|B^{-1}\|^2\|\tau_{\Gamma_1}\|^2}|v|^2_{H^1(\Omega)}\right)$$

for all $(v,h) \in \ker e'(x^*)$. □

Next we give a sufficient condition for (h4).

Theorem 6.15. *Let* (6.2.3), (h1) *hold and assume that*

(i) $\langle C^*(y_d - Cy^*), \psi\rangle_{(H^1)^*,H^1} \geq 0$ *for all* $\psi \in H^1(\Omega)$ *with* $\psi \geq 0$, *a.e.,*

(ii) $g'(y^*) \geq 0$ *a.e. on* Ω, *and*

(iii) $g''(y^*) \geq 0$ *a.e. on* Ω.

Then (h4) *holds. The conclusion remains correct if the inequalities in* (i) *and* (iii) *are reversed.*

Proof. Set $\varphi = \inf(0, \lambda^*) \in H^1(\Omega)$ in (6.4.11). Then we have

$$\int_\Omega |\nabla\varphi|^2 dx + (g'(y^*)\varphi, \varphi) + \langle c^*(Cy^* - y_d), \varphi\rangle_{(H^1)^* \times H^1} = 0.$$

Since $g'(y^*) \geq 0$ it follows from (i) that $|\nabla\varphi|^2_{L^2(\Omega)} = 0$ and $\lambda^* \geq 0$. Together with (iii) we find $\lambda^* g''(y^*) \geq 0$ a.e. on Ω. If the inequalities in (i) and (iii) are reversed, we take $\varphi = \sup(0, \lambda^*)$. $\square$

Example 6.16. We consider

$$\begin{aligned} -\Delta y + y^3 - y &= \hat{f} \text{ in } \Omega, \\ \frac{\partial y}{\partial n} &= u \text{ on } \Gamma \end{aligned} \tag{6.4.20}$$

and the associated optimal control problem

$$\begin{cases} \min \frac{1}{2}\int_\Omega |y - y_d|^2 dx + \frac{\alpha}{2}\int_\Gamma u^2 dx \\ \text{subject to } (y, u) \in H^1(\Omega) \times L^2(\Gamma) \text{ a solution of (6.4.20).} \end{cases} \tag{6.4.21}$$

In the context of the problem (6.4.8) we set $\Gamma_1 = \Gamma = \partial\Omega$, $\Gamma_2 = \emptyset$, and

$$Z = L^2(\Omega), \quad C : H^1(\Omega) \to L^2(\Omega) \text{ canonical injection}, \quad g(t) = t^3 - t.$$

Equation (6.4.20) represents a simplified Ginzburg–Landau model for superconductivity with y denoting the wave function, which is valid in the absence of internal magnetic fields (see [Tin, Chapters 1, 4]). Both (6.4.20) and $-\Delta y + y^3 + y = \hat{h}$ are of interest in this context, but here we concentrate on (6.4.20) since it has three equilibria, ± 1 and 0, of which ± 1 are stable and 0 is unstable.

Proposition 6.17. *Problem* (6.4.21) *admits a solution* $(y^*, u^*) \in H^1(\Omega) \times L^1(\Gamma)$.

Proof. We first argue that the set of admissible pairs $(y, u) \in H^1(\Omega) \times L^2(\Gamma)$ for (6.4.21) is not empty. For this purpose we consider

$$\begin{aligned} -\Delta y + y^3 - y &= \hat{f} \quad \text{in } \Omega, \\ y &= 0 \quad \text{on } \partial\Omega. \end{aligned} \tag{6.4.22}$$

Let $T: L^6(\Omega) \to L^6(\Omega)$ be the operator defined by

$$T(y) = (-\Delta)^{-1}(\hat{h} + y - y^3),$$

where Δ denotes the Laplacian with Dirichlet boundary conditions. Clearly T is completely continuous and $(I - T)y = 0$ implies that y is a solution to (6.4.22) with $y \in H^2(\Omega) \cap H_0^1(\Omega)$. By a variant of Schauder's fixed point theorem due to Schäfer [Dei, p. 61], either $(I - t\,T)y = 0$ admits a solution for every $t \in [0, 1]$ or $(I - t\,T)y = 0$ admits an unbounded set of solutions for some $\hat{t} \in (0, 1)$. Assume that the latter is the case and let $y = y(\hat{t})$ satisfy $(I - \hat{t}\,T)y = 0$. Then $-(\hat{t})^{-1}\Delta\, y + y^3 - y = \hat{f}$, and hence

$$\begin{aligned} \hat{t}^{-1} \int_\Omega |\nabla y|^2 dx + \int_\Omega y^4 dx &= \int_\Omega y^2 dx + \int \hat{f}\, y\, dx \\ &\le \int_{|y(x)|>2} y^2 dx + \int_{|y(x)|\le 2} y^2 dx + \int_\Omega |y|\,|\hat{f}| dx. \end{aligned} \tag{6.4.23}$$

This implies that

$$\hat{t}^{-1} \int_\Omega |\nabla y|^2 dx \;\le\; 4\,|\Omega| \;+ \int_\Omega |y|\,|\hat{f}| dx.$$

Since $y \in H_0^1(\Omega)$, Poincaré's inequality implies the existence of a constant $\hat{C}$ independent of $y = y(\hat{t})$ such that

$$|y(\hat{t})|_{H_0^1} \;\le\; \hat{C}(|\hat{f}|_{L^2} \;+\; |\Omega|).$$

Consequently the set of solutions $y(\hat{t})$ to $(I - \hat{t}\,T)y = 0$ is necessarily bounded in $L^6(\Omega)$, and (6.4.22) admits a solution $y \in H^2(\Omega) \cap H_0^1(\Omega)$. Setting $u = \frac{\partial y}{\partial u} \in H^{1/2}(\Gamma) \subset L^2(\Gamma)$ we find the existence of an admissible pair for (6.4.21).

Let $(y_n, u_n) \in H^1(\Omega) \times L^2(\Gamma)$ be a minimizing sequence for (6.4.21). Due to the choice of the cost functional the sequence $\{u_n\}_{n=1}^\infty$ is bounded in $L^2(\Gamma)$. Proceeding as in (6.4.23) it is simple to argue that $\{y_n\}_{n=1}^\infty$ is bounded in $H^1(\Omega)$. Standard weak convergence arguments then imply that every weak subsequential limit of $\{(y_n, u_n)\}$ is a solution to (6.4.21). $\square$

Proposition 6.18. *For every $(y, u) \in X = H^1(\Omega) \times L^2(\Gamma)$ the linearization $e'(y, u)\colon X \to X$ is surjective.*

Proof. We follow [GHS]. Since $e = \mathcal{N}\,\tilde{e}$ with $\mathcal{N}$ defined below (6.4.8) it suffices to show that $E\colon = \tilde{e}(y, u)\colon X \to H^1(\Omega)^*$ is surjective, where E is characterized by

$$\langle E(v, h), w\rangle_{H^{1,*},H^1} = \left(\nabla v, \nabla w\right)_\Omega + \left(q\, v, w\right)_\Omega - \left(h, \tau_\Gamma\, w\right)_\Gamma,$$

with $q = 3y^2 - 1 \in H^1(\Omega)$. Consider the operator $E_0\colon H^1(\Omega) \to H^1(\Omega)^*$ defined by

$$\langle E_0(v), w\rangle_{H^{1,*},H^1} = \left(\nabla v, \nabla w\right)_\Omega + \left(q\, v, w\right)_\Omega,$$

and observe that $E_0 = -\Delta + I + (q-1)I$; i.e., E_0 is a compact perturbation by the operator $(q-1)I \in \mathcal{L}(H^1(\Omega), H^1(\Omega)^*)$ of the homeomorphism $(-\Delta + I) \in \mathcal{L}(H^1(\Omega), H^1(\Omega)^*)$.

By the Fredholm alternative either E_0, and hence E, is surjective or there exists a finite-dimensional kernel of E_0 spanned by the basis $\{v_i\}_{i=1}^N$ in $H^1(\Omega)$ and $E_0 v = z$, for $z \in H^1(\Omega)^*$, is solvable if and only if $\langle z, v_i\rangle_{H^{1,*},H^1} = 0$ for all $i = 1, \dots, N$. In the latter case v_i, $i = 1, \dots, N$, are nontrivial on Γ by Lemma 6.19 below. Without loss of generality we may assume that $(v_i, v_j)_\Gamma = \delta_{ij}$ for $i, j = 1, \dots, N$. For $z \in H^1(\Omega)^*$ we define $\hat{z} \in H^1(\Omega)^*$ by $\langle \hat{z}, w\rangle_{H^{1,*},H^1} = \langle z, w\rangle_{H^{1,*},H^1} + (h, w)_\Gamma$, where $h = -\sum_{i=1}^N \langle z, v_i\rangle_{H^{1,*},H^1} \tau_\Gamma v_i \in L^2(\Gamma)$. Then $\langle \hat{z}, v_i\rangle_{H^{1,*},H^1} = 0$ for all $i = 1, \dots, N$ and hence there exists $v \in H^1(\Omega)$ such that $E_0 v = \hat{z}$ or equivalently

$$(\nabla v, \nabla w)_\Omega + (q\, v, w)_\Omega = \langle z, w\rangle_{H^{1,*},H^1} \text{ for all } v \in H^1(\Omega).$$

Consequently E is surjective. □

Lemma 6.19. *If for $q \in L^2(\Omega)$ the function $v \in H^1(\Omega)$ satisfies*

$$(\nabla v, \nabla w)_\Omega + (q\, v, w)_\Omega = 0 \text{ for all } v \in H^1(\Omega) \tag{6.4.24}$$

and $v = 0$ on Γ, then $v = 0$ in Ω.

Proof. Let $\tilde{\Omega}$ be a bounded domain with smooth boundary strictly containing Ω. Define

$$\tilde{v} = \begin{cases} v & \text{in } \Omega, \\ 0 & \text{in } \tilde{\Omega} \setminus \Omega, \end{cases}$$

and let $\tilde{q}$ denote the extension by 0 of q. Since $v = 0$ on $\Gamma = \partial\Omega$ we have $\tilde{v} \in H^1(\Omega)$ and $(\nabla \tilde{v}, \nabla w)_{\tilde{\Omega}} + (\tilde{q}\, \tilde{v}, w)_{\tilde{\Omega}} = 0$ for all $w \in H_0^1(\tilde{\Omega})$ by (6.4.24). This, together with the fact that $\tilde{v} = 0$ on an open nonempty subset of $\tilde{\Omega}$, implies that $\tilde{v} = 0$ in $\tilde{\Omega}$ and $v = 0$ in Ω; see [Geo]. □

Let us discuss the validity of the conditions (hi) for the present problem. Sobolev's embedding theorem and

$$g'(t) = 3t^2 - 1 \quad \text{and} \quad g''(t) = 6t$$

imply that (h1), (h3), and (h4) hold. The location of the equilibria of g suggests that for $h = 0$, $y_d \geq 1$ implies $1 \leq y^* \leq y_d$ and similarly that $y_d \leq -1$ implies $y_d \leq y^* \leq -1$. This was confirmed in numerical experiments [IK10]. In these cases $g'(y^*) \geq \underline{\beta} > 0$ and (i)–(iii) of Theorem 6.15 hold.

The numerical results in [IK10] are based on Algorithm 6.3. While it does not contain a strategy on the choice of c, it was observed that $c > 0$ and of moderate size is superior (with respect to region of attraction for convergence) to $c = 0$, which corresponds to the SQP method without line search. For example, for the choice $y_d = 1$ and the initialization $y_0 = -2$, the iterates y_n have to pass through the stable equilibrium -1 and the unstable equilibrium 0 to reach the desired state y_d. This can be accomplished with Algorithm 6.3 with $c > 0$, without globalization strategy, but not with $c = 0$. Similar comments apply to the case of Neumann boundary controls.

Example 6.20. This is the singular system

$$
\begin{aligned}
-\Delta y - y^3 &= h \text{ in } \Omega, \\
\frac{\partial y}{\partial n} &= u \text{ on } \Gamma,
\end{aligned}
\tag{6.4.25}
$$

and the associated optimal control problem

$$
\begin{cases}
\min \frac{1}{2}|y - y_d|^2_{H^1(\Omega)} + \frac{\alpha}{2} \int_\Gamma u^2 ds \\
\\
\text{subject to } (y, u) \in H^1(\Omega) \times L^2(\Gamma_1) \text{ a solution of (6.4.25)},
\end{cases}
\tag{6.4.26}
$$

where $y_d \in H^1(\Omega)$. If (6.4.25) admits at least one feasible pair (y, u), then it is simple to argue that (6.4.26) has a solution $x^* = (y^*, u^*)$. We refer the reader to [Lio2, Chapter 3] for existence results in the case that the cost functional is of the form $|y - y_d|^r_{L^r(\Omega)} + \alpha |g|^2_{L^2(\Gamma)}$ for appropriately chosen $r > 2$. The existence of a Lagrange multiplier is assured in the same manner as in Example 6.6. Clearly (h1) and (h3) are satisfied. For $y_d = \text{const} \geq \frac{1}{2}$ we observed numerically that $0 \leq y^* \leq y_d$, $\lambda^* < 0$, which in view of (h4) and Theorem 6.14 explains the second order convergence rate that is observed numerically.

6.5 Approximation and mesh-independence

This section is devoted to a brief description of some aspects concerning approximation of Algorithm 6.3. For this purpose let h denote a discretization parameter tending to zero and, for each h, let X_h and W_h be finite-dimensional subspaces of X and W, respectively. In this section we set $Z = X \times W$ and $Z_h = X_h \times W_h$. The finite-dimensional spaces X_h and W_h are endowed with the inner products and norms induced by X and W. We introduce surjective restriction operators $R_h^X \in \mathcal{L}(X, X_h)$ and $R_h^W \in \mathcal{L}(W, W_h)$ and define $R_h = (R_h^X, R_h^W)$. For each h, let $f_h : X_h \to \mathbb{R}$ and $e_h : X_h \to W_h$ denote discretizations of the cost functional f and the constraint e. Together with the infinite-dimensional problem (6.2.1) we consider the finite-dimensional discrete problems

$$
\begin{cases}
\min f_h(x_h) \text{ over } x \in X_h \\
\text{subject to } e_h(x) = 0.
\end{cases}
\tag{6.5.1}
$$

The Lagrangians for (6.5.1) are given by

$$
\mathcal{L}_h(x_h, \lambda_h) = f_h(x_h) + \big(e_h(x_h), \lambda_h\big)_W,
$$

and the approximations of (6.2.6) are

$$
\mathcal{L}'_h(x_h, \lambda_h + c e_h(x_h)) = 0, \quad e_h(x_h) = 0,
\tag{6.5.2}
$$

which are first order necessary optimality conditions for (6.5.1).

We require the following assumptions.

Condition 6.5.1. The approximations (Z_h, R_h) of the space Z are uniformly bounded; i.e., there exists a constant $c_R > 0$ independent of h satisfying

$$\|R_h\|_{\mathcal{L}(Z,Z_h)} \le c_R\,.$$

Condition 6.5.2. For every h problem (6.5.1) has a local solution x_h^*. The mappings f_h and e_h are twice continuously Fréchet differentiable in neighborhoods $\tilde{V}_h^*$ of x_h^*, and the operators e_h are Lipschitz continuous on $\tilde{V}_h^*$ with a uniform Lipschitz constant $\xi_e > 0$ independent of h.

Condition 6.5.3. For every h there exists a nonempty open and convex set $\hat{V}_h^* \subseteq \tilde{V}_h^*$ such that a uniform Babuška–Brezzi condition is satisfied on $\hat{V}_h^*$; i.e., there exists a constant $\beta > 0$ independent of h satisfying

$$\inf_{w_h \in W_h} \sup_{q_h \in X_h} \frac{\left(e_h'(x_h)q_h, w_h\right)_W}{\|q_h\|_X \, \|w_h\|_W} \ge \beta \qquad \text{for all } x_h \in \hat{V}_h^*\,.$$

The Babuška–Brezzi condition implies that the operators $e_h'(x_h)$ are surjective on $\hat{V}_h^*$. Hence, if Condition 6.5.3 holds, there exists for every $h > 0$ a Lagrange multiplier $\lambda_h^* \in W_h$ such that (x_h^*, λ_h^*) solves (6.5.2).

Condition 6.5.4. There exist a scalar $r > 0$ independent of h and a neighborhood $V(\lambda_h^*)$ of λ_h^* for every h such that for $V_h^* = \hat{V}_h^* \times V(\lambda_h^*)$

(i) $B\left((x_h^* \lambda_h^*); r\right) \subseteq V_h^*$ and

(ii) a uniform second order sufficient optimality condition is satisfied on V_h^*; i.e., there exists a constant $\bar{\kappa} > 0$ such that for all $(x_h, \lambda_h) \in V_h^*$

$$\mathcal{L}_h''(x_h, \lambda_h)(q_h)^2 \ge \bar{\kappa}\, \|q_h\|_X^2$$

for all $q_h \in \ker e_h'(x_h)$.

We define

$$F(x,\lambda) = \begin{pmatrix} \mathcal{L}'(x,\lambda) \\ e(x) \end{pmatrix}.$$

For every h and for all $(x_h, \lambda_h) \in V_h^*$ we introduce approximations of F and M by

$$F_h(x_h,\lambda_h) = \begin{pmatrix} \mathcal{L}_h'(x_h,\lambda_h) \\ e_h(x_h) \end{pmatrix},$$

$$M_h(x_h,\lambda_h) = \begin{pmatrix} \mathcal{L}_h''(x_h,\lambda_h) & e_h'(x_h)^\star \\ e_h'(x_h) & 0 \end{pmatrix}.$$

By Conditions 6.5.3 and 6.5.4 there exists a bound η which may depend on β and κ^* but is independent of h so that

$$\|M_h^{-1}(x_h,\lambda_h)\|_{\mathcal{L}(Z_h)} \le \eta \tag{6.5.3}$$

for all $(x_h, \lambda_h) \in V_h^*$. We require the following consistency properties of F_h and M_h, where $V(x^*, \lambda^*) = V(x^*) \times V(\lambda^*)$ denotes the neighborhood of the solution x^* of (6.2.1) and the associated Lagrange multiplier λ^* defined above (6.2.8).

Condition 6.5.5. For each h we have $R_h(V(x^*, \lambda^*)) \subseteq V_h^*$, and the approximations of the operators F and M are consistent on $V(x^*)$ and $V(x^*, \lambda^*)$, respectively, i.e.,

$$\lim_{h \to 0} \|F_h(R_h(x, \lambda)) - R_h F(x, \lambda)\|_Z = 0$$

and

$$\lim_{h \to 0} \left\| M_h(R_h(x, \lambda)) R_h \begin{pmatrix} q \\ w \end{pmatrix} - R_h M(x, \lambda) \begin{pmatrix} q \\ w \end{pmatrix} \right\|_Z = 0$$

for all $(q, w) \in Z$ and $(x, \lambda) \in V(x^*, \lambda^*)$. Moreover, for every h let the operator M_h be Lipschitz continuous on V_h^* with a uniform Lipschitz constant $\xi_M > 0$ independent of h.

Now we consider the following approximation of Algorithm 6.3.

Algorithm 6.7.

(i) Choose $(x_h^0, \lambda_h^0) \in V_h^*$, $c \geq 0$ and put $n = 0$.

(ii) Set $\tilde{\lambda}_h^n = \lambda_h^n + c\, e_h(x_h^n)$.

(iii) Solve for $(\hat{x}_h, \hat{\lambda}_h)$:

$$M_h(x_h^n, \tilde{\lambda}_h^n) \begin{pmatrix} \hat{x}_h - x_h^n \\ \hat{\lambda}_h - \tilde{\lambda}_h^n \end{pmatrix} = - \begin{pmatrix} \mathcal{L}_h'(x_h^n, \tilde{\lambda}_h^n) \\ e_h(x_h^n) \end{pmatrix}.$$

(iv) Set $(x_h^{n+1}, \lambda_h^{n+1}) = (\hat{x}_h, \hat{\lambda}_h)$, $n = n + 1$, and goto (ii).

Theorem 6.21. *Let $c\, \|(x_h^0, \lambda_h^0) - (x_h^*, \lambda_h^*)\|_Z$ be sufficiently small for all h and let* (6.2.3), (6.2.5), *and Conditions* 6.5.2–6.5.4 *hold. Then we have the following:*

(a) *Algorithm* 6.7 *is well defined and*

$$\|(x_h^{n+1}, \lambda_h^{n+1}) - (x_h^*, \lambda_h^*)\|_Z \leq C\, \|(x_h^n, \lambda_h^n) - (x_h^*, \lambda_h^*)\|_Z^2,$$

where $C = \frac{1}{2}\, \eta\, \xi_M\, \max(2, 1 + 2c^2 \xi_e^2)$ is independent of h.

(b) *Let (x^0, λ^0) be a startup value of Algorithm* 6.3 *such that $c\, \|(x^0, \lambda^0) - (x^*, \lambda^*)\|_Z$ is sufficiently small and assume that*

$$\lim_{h \to 0} \left\| (x_h^0, \lambda_h^0) - R_h(x^0, \lambda^0) \right\|_Z = 0. \tag{6.5.4}$$

If in addition Conditions 6.5.1 *and* 6.5.5 *are satisfied, we obtain*

$$\lim_{h \to 0} \left\| (x_h^n, \lambda_h^n) - R_h(x^n, \lambda^n) \right\|_Z = 0$$

for all n, where (x_h^n, λ_h^n) and (x^n, λ^n) are the nth iterates of the finite- and infinite-dimensional methods, respectively.

Corollary 6.22. *Under the hypotheses of Theorem* 6.21

$$\lim_{h \to 0} \|(x_h^*, \lambda_h^*) - R_h(x^*, \lambda^*)\|_Z = 0$$

holds.

Corollary 6.23. *Let the hypotheses of Theorem* 6.21 *hold and let* $\{(Z_{h(n)}, R_{h(n)})\}$ *be a sequence of approximations with* $\lim_{n\to\infty} h(n) = 0$. *Then we have*

$$\lim_{n \to \infty} \|(x_{h(n)}^n, \lambda_{h(n)}^n) - R_{h(n)}(x^*, \lambda^*)\|_Z = 0.$$

We turn to a brief account of mesh-independence. This is an important feature of iterative approximation schemes for infinite-dimensional problems. It asserts that the number of iterations to reach a specified approximation quality $\varepsilon > 0$ is independent of the mesh size. We require the following notation:

$$n(\varepsilon) = \min \left\{ n_0 | \text{ for } n \geq n_0 : \left\| F\left(x^n, \lambda^n + ce(x^n)\right) \right\|_Z < \varepsilon \right\},$$
$$n_h(\varepsilon) = \min \left\{ n_0 | \text{ for } n \geq n_0 : \left\| F_h\left(x_h^n, \lambda_h^n + ce_h(x_h^n)\right) \right\|_Z < \varepsilon \right\}.$$

We point out that both $n(\varepsilon)$ and $n_h(\varepsilon)$ depend on the startup values (x^0, λ^0) and (x_h^0, λ_h^0) of the infinite- and finite-dimensional methods.

Theorem 6.24. *Under the assumptions of Theorem* 6.21 *there exists for each* $\varepsilon > 0$ *a constant* $h_\varepsilon > 0$ *such that*

$$n(\varepsilon) - 1 \leq n_h(\varepsilon) \leq n(\varepsilon) \qquad \textit{for } h \in (0, h_\varepsilon].$$

The proofs of the results of this section as well as numerical examples which demonstrate that mesh-independence is also observed numerically can be found in [KVo1, Vol].

6.6 Comments

Second order augmented Lagrangian methods as discussed in Section 6.2 were treated in [Be] for finite-dimensional and in [IK9, IK10] for infinite-dimensional problems. The history for SQP methods is a long one; see [Han, Pow1] and [Sto, StTa], for example, and the references therein for the analysis of finite-dimensional problems. Infinite-dimensional problems are considered in [Alt5], for example. Mesh-independence of augmented Lagrangian-SQP methods was analyzed in [Vol]. This paper also contains many references on mesh-independence for SQP and Newton methods. Methods which replace equality- and inequality-constrained optimization problems by a linear quadratic approximation with respect to the equality constraints and the cost functional, while leaving the equality constraints as explicit constraints, are frequently referred to as Lagrange–Newton methods in the literature. We quote [Alt4, AlMa, Don, Tro] and the references therein. Reduced SQP methods in infinite-dimensional spaces were analyzed in [JaSa, KSa, Kup] among others. For optimization problems with equality and simple inequality constraints the SQP method can advantageously be combined with projection methods; see [Hei], for example. We have not considered the topic of approximating the Hessian by secant updates. In part this is due to the fact that in the context of optimal control of partial differential equations the structure of the nonlinearities is such that the continuous problem may allow a rather straightforward characterization of the gradient and Hessian, provided, of course, that it is sufficiently regular. Secant methods for SQP methods are considered in, e.g., [Sto, KuSa, Kup, StTa].

Chapter 7

The Primal-Dual Active Set Method

This chapter is devoted to the primal-dual active set strategy for variational problems with simple constraints. This is an efficient method for solving the optimality systems arising from quadratic programming problems with unilateral or bilateral affine constraints and it is equally well applicable to certain complementarity problems. The algorithm and some of its basic properties are described in Section 7.1. In the ensuing sections sufficient conditions for its convergence with arbitrary initialization and without globalization are presented for a variety of different classes of problems. Sections 7.2 and 7.3 are devoted to the finite-dimensional case where the system matrix is an M-matrix or has a cone-preserving property, respectively. Operators which are of diagonally dominant type are considered in Sections 7.4 and 7.5 for unilateral, respectively, bilateral problems. In Section 7.6 nonlinear optimal control problems with control constraints are investigated.

7.1 Introduction and basic properties

Here we investigate the primal-dual active method. Let us consider the quadratic programming problem

$$\begin{cases} \min_{x\in X} \dfrac{1}{2}\left(Ax, x\right)_X - \left(a, x\right)_X \\ \\ \text{subject to } Ex = b, \quad Gx \le \psi, \end{cases} \tag{7.1.1}$$

where $A \in \mathcal{L}(X)$ is a self-adjoint operator in the real Hilbert space X, $E \in \mathcal{L}(X, W)$, $G \in \mathcal{L}(X, Z)$, with W a real Hilbert space, and $Z = \mathbb{R}^n$ or $Z = L^2(\Omega)$, with Ω a domain in $\mathbb{R}^d$, endowed with the usual Hilbert space structure and the natural ordering. We assume that (7.1.1) admits a unique solution denoted by x. If x is a regular point in the sense of Definition 1.5 (with $C = X$ and $g(x) = (Ex - b, Gx - \psi)$), then there exists a Lagrange multiplier $(\lambda, \mu) \in W \times Z$ such that

$$\begin{aligned} &Ax + E^*\lambda + G^*\mu = a, \\ &Ex = b, \\ &\mu = \max(0, \mu + c\,(Gx - \psi)), \end{aligned} \tag{7.1.2}$$

where $c > 0$ is a fixed constant and max is interpreted pointwise a.e. in Ω if $Z = L^2(\Omega)$ and coordinatewise if $Z = \mathbb{R}^n$. The third equation in (7.1.2) constitutes the complementarity condition associated with the inequality constraint in (7.1.1), as discussed in Examples 4.51 and 4.53. Note that the auxiliary problems in the augmented Lagrangian-SQP methods of Chapter 6 (see for instance (ii) of Algorithm 6.4 or (ii) of Algorithm 6.6) take the form of (7.1.1).

The primal-dual active set method that will be analyzed in this chapter is an efficient technique for solving (7.1.2). While (7.1.2) is derived from (7.1.1) with A self-adjoint, this assumption is not essential in the remainder of this chapter, and we therefore drop it unless it is explicitly specified.

We next give two examples of constrained optimal control problems which are special cases of (7.1.1).

Example 7.1. Let $X = L^2(\hat{\Omega})$, where $\hat{\Omega} \subset \Omega$ is the control domain and consider the optimal control problem

$$\begin{cases} \min\limits_{u \in X} \dfrac{1}{2} \displaystyle\int_\Omega |y - y_d|^2\, dx + \frac{\alpha}{2} |u|_X^2 \\ \\ \text{subject to} \quad -\Delta y = Bu, \quad y = 0 \text{ on } \partial\Omega \quad \text{and} \quad u \le \psi, \end{cases}$$

where $\alpha > 0$, $y_d \in L^2(\Omega)$, $\psi \in L^2(\hat{\Omega})$, and $B \in \mathcal{L}(X, L^2(\Omega))$ is the extension-by-zero operator $\hat{\Omega}$ to Ω. This problem can be formulated as (7.1.1) without equality constraint ($E = 0$) by setting

$$A = \alpha I + B^*(-\Delta)^{-2}B, \quad a = B^*(-\Delta)^{-1}\bar{y}, \quad \text{and} \quad G = I,$$

where Δ denotes the Laplacian with homogeneous Dirichlet boundary conditions.

Example 7.2. Here we consider optimal control of the heat equation with Neumann boundary control and set, $X = L^2(0, T; L^2(\hat{\Gamma}))$, where $\hat{\Gamma}$ is a subset of the boundary $\partial\Omega$ of Ω:

$$\begin{cases} \min\limits_{u \in X} \dfrac{1}{2} \displaystyle\int_0^T \int_\Omega |y - y_d|^2\, dx\, dt + \frac{\alpha}{2} |u|_U^2 \\ \\ \text{subject to} \quad \frac{d}{dt} y = \Delta y, \quad y(0, \cdot) = y_0 \quad y = 0 \text{ on } \partial\Omega \setminus \hat{\Gamma}, \\ \\ \qquad\qquad \nu \cdot \nabla y(t) = Bu(t) \text{ in } \hat{\Gamma}, \quad u(t) \le \psi, \end{cases}$$

where $\alpha > 0$, $y_d \in L^2(0, T; L^2(\Omega))$, $y_0 \in L^2(\Omega)$, $\hat{\Gamma}$ is a subset of the boundary $\partial\Omega$, ν is the outer normal to $\hat{\Gamma}$, and B is the extension-by-zero operator $\hat{\Gamma}$ to $\partial\Omega$. For $u \in X$ there exists a unique solution $y = y(u) \in Y = L^2(0,T;H^1(\Omega)) \cap H^1(0,T;L^2(\Omega))$ to the initial boundary value problem arising as the equality constraints. Let $T \in \mathcal{L}(U,Y)$ denote the solution operator given by $y = T(u)$. Setting

$$A = \alpha I + T^*T, \quad a = T^* y_d, \quad \text{and} \quad G = I,$$

this optimal control problem is equivalent to (7.1.1).

In these examples A is bounded on X. More specifically it is a compact perturbation of a multiple of the identity. There are important problems which are not covered by (7.1.1). For the obstacle problem, considered in Section 4.7.4, the choice $X = L^2(\Omega)$ results in the unbounded operator $A = -\Delta$, and hence it is not covered by (7.1.1). If X is chosen as $H_0^1(\Omega)$ and G as the inclusion of $H_0^1(\Omega)$ into $L^2(\Omega)$, then the solution does not satisfy the regular point condition. While special techniques allow one to guarantee the existence of a Lagrange multiplier μ associated with the constraint $y \le \psi$ in $L^2(\Omega)$, the natural space for the Lagrange multiplier is $H^{-1}(\Omega)$. This suggests combining the primal-dual active set method with a regularization method, which will be discussed in a more general context in Section 8.6. Discretization of the obstacle problem by central finite differences leads to a system matrix A on $X = \mathbb{R}^n$ that is symmetric positive definite and is an M-matrix. This important class of finite-dimensional variational problems will be discussed in Section 7.2.

Another important class of problems are state-constrained optimal control problem, for example,

$$\min \frac{1}{2}\int_\Omega |y - \bar{y}|^2\, dx + \frac{\alpha}{2}|u|^2_{L^2(\Omega)}$$

subject to

$$-\Delta y = u, \quad y = 0 \text{ on } \partial\Omega, \quad y \le \psi$$

for $y \in H_0^1(\Omega)$ and $u \in L^2(\Omega)$. This problem can be formulated as (7.1.1) with $X = (H^2(\Omega) \cap H_0^1(\Omega)) \times L^2(\Omega)$ and G the inclusion of X into $L^2(\Omega)$. The solution, however, will not satisfy a regular point condition, and the Lagrange multiplier associated with the state constraint $y \le \psi$ is only a measure in general. This class of problems will be discussed in Section 8.6.

Let us return to (7.1.2). If the active set

$$\mathcal{A} = \{\mu + c(Gx - \psi) > 0\}$$

is known, then the linear system reduces to

$$Ax + E^*\lambda + G^*\mu = a,$$

$$Ex = b,$$

$$Gx = \psi \text{ in } \mathcal{A} \quad \text{and} \quad \mu = 0 \text{ in } \mathcal{A}^c.$$

Here and below we frequently use the notation $\{f > 0\}$ to stand for $\{x : f(x) > 0\}$ if $f \in L^2(\Omega)$ and $\{i : f_i > 0\}$ if $f \in \mathbb{R}^n$. The active set at the solution, however, is unknown.

The primal-dual active set method uses the complementarity condition

$$\mu = \max(0, \mu + c\,(Gx - \psi))$$

as a prediction strategy. Based on the current primal-dual pair (x, μ) the updates for the active and inactive sets are determined by

$$\mathcal{I} = \{\mu + c(Gx - \psi) \le 0\} \quad \text{and} \quad \mathcal{A} = \{\mu + c(Gx - \psi) > 0\}.$$

This leads to the following Newton-like method.

Primal-Dual Active Set Method.

(i) Initialize x^0, μ^0. Set $k = 0$.

(ii) Set $\mathcal{I}_k = \{\mu^k + c(Gx^k - \psi) \le 0\}$, $\mathcal{A}_k = \{\mu^k + c(Gx^k - \psi) > 0\}$.

(iii) Solve for $(x^{k+1}, \lambda^{k+1}, \mu^{k+1})$:

$$\begin{aligned} &Ax^{k+1} + E^*\lambda^{k+1} + G^*\mu^{k+1} = a, \\ &Ex^{k+1} = b, \\ &Gx^{k+1} = \psi \text{ in } \mathcal{A}_k \quad \text{and} \quad \mu^{k+1} = 0 \text{ in } \mathcal{I}_k. \end{aligned}$$

(iv) Stop, or set $k = k + 1$, and return to (ii).

It will be shown in Section 8.4 that the above algorithm can be interpreted as a semismooth Newton method for solving (7.1.2). This will allow the local convergence analysis of the algorithm. In this chapter we concentrate on its global convergence, i.e., convergence for arbitrary initializations and without the necessity for a line search. In case A is positive definite, a sufficient condition for the existence of solutions to the auxiliary systems of step (iii) of the algorithm is given by surjectivity of E and surjectivity $G : N(E) \to Z$. In fact in this case (iii) is the necessary optimality condition for

$$\begin{cases} \min_{x \in X} \dfrac{1}{2}(Ax, x)_X - (a, x)_X \\[2ex] \text{subject to } Ex = b, \quad Gx = \psi \text{ on } \mathcal{A}_k. \end{cases}$$

In the remainder of this chapter we focus on the reduced form of (7.1.2) given by

$$Ax + \mu = a, \qquad \mu = \max(0, \mu + c\,(x - \psi)), \tag{7.1.3}$$

where $A \in \mathcal{L}(Z)$.

We now derive sufficient conditions which allow us to transform (7.1.2) into (7.1.3). In a first step we assume that

$$G \text{ is surjective, } \operatorname{range}(G^*) \subset \ker E, \ \ E\bar{x} = b \text{ for some } \bar{x} \in (\ker E)^{\perp}. \tag{7.1.4}$$

Note that (7.1.4) implies that $G : N(E) \to Z$ is surjective. If not, then there exists a nonzero $z \in Z$ such that $(z, Gx)_Z = (G^*z, x)_X = 0$ for all $x \in \ker E$. If we let $x = G^*z$, then $|x|^2 = 0$ and $z = 0$, since G^* is injective. Let P_E denote the orthogonal projection in X onto $\ker E$. Then (7.1.2) is equivalent to

$$\mathcal{A}\hat{x} + G^*\mu = P_E(a - A\bar{x}), \quad \mu = \max(0, \mu + c\,(G\hat{x} - (\psi - G\bar{x}),$$

$$E^*\lambda = (I - P_E)(a - A(P_E\hat{x} + \bar{x})),$$

with $\mathcal{A} = P_E A P_E$ and $x = \hat{x} + \bar{x} \in \ker E + (\ker E)^\perp$. The first of the above equations is equivalent to the system

$$\begin{aligned}&(I - P_G)\mathcal{A}((I - P_G)\,\hat{x} + P_G\,\hat{x}) + G^*\mu = (I - P_G)P_E(a - A\,\bar{x}),\\ &P_G\,\mathcal{A}((I - P_G)\,\hat{x} + P_G\,\hat{x}) = P_G\,P_E\,(a - A\,\bar{x}),\end{aligned} \tag{7.1.5}$$

where $P_G = I - G^*(G\,G^*)^{-1}G$ is the orthogonal projection in $\ker E \subset X$ onto $\ker G$. Since G^* is injective the first equation in (7.1.5) is equivalent to

$$(G\,G^*)^{-1}G\,\mathcal{A}(G^*(G\,G^*)^{-1}y + x_2) + \mu = (G\,G^*)^{-1}G\,P_E(a - A\,\bar{x}),$$

where $y = G\,x_1$ for $x_1 \in \ker E \cap (\ker G)^\perp$, $x_2 \in \ker G$, and $\hat{x} = x_1 + x_2$. Let

$$A_{11} = (GG^*)^{-1}G\mathcal{A}G^*(GG^*)^{-1}, \qquad A_{12} = (GG^*)^{-1}G\mathcal{A}P_G,\; A_{22} = P_G\mathcal{A}P_G$$

and

$$a_1 = (GG^*)^{-1}GP_E(a - A\bar{x}), \qquad a_2 = P_G P_E(a - A\bar{x}).$$

Then (7.1.5) is equivalent to the following equation in $Z \times (\ker E \cap \ker G)$:

$$\begin{pmatrix} A_{11} & A_{12} \\ A_{21} & A_{22} \end{pmatrix}\begin{pmatrix} y \\ x_2 \end{pmatrix} + \begin{pmatrix} \mu \\ 0 \end{pmatrix} = \begin{pmatrix} a_1 \\ a_2 \end{pmatrix}. \tag{7.1.6}$$

Equation (7.1.6) together with

$$\mu = \max(0, \mu + c(y - (\psi - G\bar{x}))) \tag{7.1.7}$$

and $E^*\lambda = (I - P_E)(a - A(\hat{x} + \bar{x}))$ are equivalent to (7.1.2) where $x = \hat{x} + \bar{x}$, with $\hat{x} = x_1 + x_2 \in \ker E$, $x_2 \in \ker E \cap \ker G$, $x_1 \in \ker E \cap (\ker G)^\perp$, $y = Gx_1$.

Note that the system matrix in (7.1.6) is positive definite if A restricted to $\ker E$ is positive definite.

Let us now further assume that

$$A_{22} \text{ is nonsingular.} \tag{7.1.8}$$

Then (7.1.6), (7.1.7) are equivalent to

$$(A_{11} - A_{12}A_{22}^{-1}A_{12}^*)y + \mu = a_1 - A_{12}A_{22}^{-1}a_2$$

and (7.1.7), which is of the desired form (7.1.3). In the finite-dimensional case, (7.1.3) admits a unique solution for every $a \in \mathbb{R}^n$ if and only if A is a P-matrix; see [BePl, Theorem 10.2.15.]. Recall that A is called a P-matrix if all its principal minors are positive. In view of the fact that the reduction of (7.1.6) to (7.1.3) was achieved by taking the Schur complement with respect to A_{22} it is also worthwhile to recall that the Schur complement of a P-matrix (resp., M-matrix) is again a P-matrix (resp., M-matrix); see [BePl, p. 292].

For further reference it will be convenient to specify the primal-dual active set algorithm for the reduced system (7.1.3).

Primal-Dual Active Set Method for Reduced System.

(i) Initialize x^0, μ^0. Set $k = 0$.

(ii) Set $\mathcal{I}_k = \{\mu^k + c(x^k - \psi) \le 0\}$, $\mathcal{A}_k = \{\mu^k + c(x^k - \psi) > 0\}$.

(iii) Solve for (x^{k+1}, μ^{k+1}):

$$Ax^{k+1} + \mu^{k+1} = a,$$

$$x^{k+1} = \psi \text{ in } \mathcal{A}_k \quad \text{and} \quad \mu^{k+1} = 0 \text{ in } \mathcal{I}_k.$$

(iv) Stop, or set $k = k + 1$, and return to (ii).

We use (7.1.3) rather than (7.1.6) for the convergence analysis for the reason of avoiding additional notation. All convergence results that follow equally well apply to (7.1.6) where it is understood that the coordinates corresponding to the variable x_2 are treated as inactive throughout the algorithm and the corresponding Lagrange multipliers are set and updated by 0.

In the following subsections convergence will be proved under various different conditions. These conditions will imply also the existence of a solution to the subsystems in step (iii) of the algorithm as well as the existence of a unique solution to (7.1.3).

For $\tilde{\Omega} \subset \{1, \ldots, n\}$, respectively, $\tilde{\Omega} \subset \Omega$, let $R_{\tilde{\Omega}}$ denote the restriction operator to $\tilde{\Omega}$. Then $R^*_{\tilde{\Omega}}$ is the extension-by-zero operator to $\tilde{\Omega}^c$. For any $\mathcal{A}$ let $\mathcal{I}$ be its complement in $\mathbb{R}^n$, respectively, Ω, and denote

$$A_{\mathcal{A}} = R_{\mathcal{A}} A R^*_{\mathcal{A}}, \quad A_{\mathcal{A},\mathcal{I}} = R_{\mathcal{A}} A R^*_{\mathcal{I}},$$

and analogously for $A_{\mathcal{I}}$ and $A_{\mathcal{I},\mathcal{A}}$, and

$$\delta x_{\mathcal{A}} = R_{\mathcal{A}}(x^{k+1} - x^k), \quad \delta x_{\mathcal{I}} = R_{\mathcal{I}}(x^{k+1} - x^k),$$

and analogously for $\delta\mu_{\mathcal{A}}$ and $\delta\mu_{\mathcal{I}}$. From (iii) above we have

$$\begin{aligned} A_{\mathcal{A}_k}\delta x_{\mathcal{A}_k} + A_{\mathcal{A}_k,\mathcal{I}_k}\delta x_{\mathcal{I}_k} + \delta\mu_{\mathcal{A}_k} &= 0, \\ A_{\mathcal{I}_k}\delta x_{\mathcal{I}_k} + A_{\mathcal{I}_k,\mathcal{A}_k}\delta x_{\mathcal{A}_k} - \mu^k_{\mathcal{I}_k} &= 0. \end{aligned} \tag{7.1.9}$$

The following properties for $k = 1, 2, \ldots$ follow from steps (ii) and (iii):

$$\mu^k(x^k - \psi) = 0, \quad \mu^k + (x^k - \psi) > 0 \text{ on } \mathcal{A}_k,$$
$$x^k - \psi \geq 0, \; \mu^k \geq 0 \text{ on } \mathcal{A}_k, \quad x^k \leq \psi, \; \mu^k \leq 0 \text{ on } \mathcal{I}_k, \tag{7.1.10}$$
$$\delta x_{\mathcal{A}_k} \leq 0, \quad \delta \mu_{\mathcal{I}_k} \geq 0,$$

all statements holding coordinatewise if $Z = \mathbb{R}^n$ and pointwise a.e. for $Z = L^2(\Omega)$.

Remark 7.1.1. From (7.1.10) it follows that $\mathcal{A}_k = \mathcal{A}_{k+1}$ implies that the solution is found, i.e., $(x_k, \mu_k) = (x^*, \mu^*)$. In numerical practice it was observed that $\mathcal{A}_k = \mathcal{A}_{k+1}$ can be used as a stopping criterion; see [HiIK, IK20, IK22].

Remark 7.1.2. The primal-dual active set strategy can be interpreted as a prediction strategy which, on the basis of (x^k, μ^k), predicts the true active and inactive sets for (7.1.3), i.e., the sets

$$\mathcal{A}^* = \{\mu^* + c(x^* - \psi) > 0\} \quad \text{and} \quad \mathcal{I}^* = (\mathcal{A}^*)^c.$$

To further pursue this point we assume that the systems in (7.1.3) and step (iii) of the algorithm admit solutions, and we define the following partitioning of the index set at iteration level k:

$$\mathcal{I}_G = \mathcal{I}_k \cap \mathcal{I}^*, \quad \mathcal{I}_B = \mathcal{I}_k \cap \mathcal{A}^*, \quad \mathcal{A}_G = \mathcal{A}_k \cap \mathcal{A}^*, \quad \mathcal{A}_B = \mathcal{A}_k \cap \mathcal{I}^*.$$

The sets $\mathcal{I}_G$, $\mathcal{A}_G$ give a *good* prediction and the sets $\mathcal{I}_B$, $\mathcal{A}_B$ give a *bad* prediction. Let us denote $\Delta x = x^{k+1} - x^*$, $\Delta\mu = \mu^{k+1} - \mu^*$, and we denote by $G((x^k, \mu^k))$ the system matrix for step (iii) of the algorithm:

$$G(x_k, \mu_k) = \begin{pmatrix} A_{\mathcal{I}_k} & A_{\mathcal{I}_k\mathcal{A}_k} & I_{\mathcal{I}_k} & 0 \\ A_{\mathcal{A}_k\mathcal{I}_k} & A_{\mathcal{A}_k} & 0 & I_{\mathcal{A}_k} \\ 0 & 0 & I_{\mathcal{I}_k} & 0 \\ 0 & -cI_{\mathcal{A}_k} & 0 & 0 \end{pmatrix}.$$

Then we have the identity

$$G(x^k, \mu^k) \begin{pmatrix} \Delta x_{\mathcal{I}_k} \\ \Delta x_{\mathcal{A}_k} \\ \Delta \mu_{\mathcal{I}_k} \\ \Delta \mu_{\mathcal{A}_k} \end{pmatrix} = -\text{col}\left(0_{\mathcal{I}_k}, 0_{\mathcal{A}_k}, 0_{\mathcal{I}_G}, \mu^*_{\mathcal{I}_B}, 0_{\mathcal{A}_G}, c(\psi - x^*)_{\mathcal{A}_B}\right). \tag{7.1.11}$$

Here we assumed that the components of the equation $\mu - \max\{0, \mu + c(x - \psi)\} = 0$ are ordered as $(\mathcal{I}_G, \mathcal{I}_B, \mathcal{A}_G, \mathcal{A}_B)$. Since $x^k \geq \psi$ on $\mathcal{A}_k$ and $\mu^k \leq 0$ on $\mathcal{I}_k$, we have

$$|\psi - x^*|_{\mathcal{A}_B} \leq |x^k - x^*|_{\mathcal{A}_B} \quad \text{and} \quad |\mu^*|_{\mathcal{I}_B} \leq |\mu^k - \mu^*|_{\mathcal{I}_B}. \tag{7.1.12}$$

By the definition

$$\Delta y_{\mathcal{A}_G} = 0, \quad \Delta x_{\mathcal{A}_B} = (\psi - x^*)_{\mathcal{A}_B}, \quad \Delta \mu_{\mathcal{I}_G} = 0, \quad \Delta \mu_{\mathcal{I}_B} = -\mu^*_{\mathcal{I}_B}. \tag{7.1.13}$$

On the basis of (7.1.11)–(7.1.13) we can draw the following conclusions.

(i) If $x^k \to x^*$, then, in the finite-dimensional case, there exists an index $\bar{k}$ such that $\mathcal{I}_B = \mathcal{A}_B = \emptyset$ for all $k \geq \bar{k}$. Consequently the convergence occurs in finitely many steps.

(ii) By (7.1.11)–(7.1.12) there exists a constant $\kappa \geq 1$ independent of k such that

$$|\Delta x| + |\Delta \mu| \leq \kappa \left(|(x^k - x^*)_{\mathcal{A}_B}| + |(\mu^k - \mu^*)_{\mathcal{I}_B}| \right).$$

Thus if the incorrectly predicted sets are small in the sense that

$$|(x^k - x^*)_{\mathcal{A}_B}| + |(\mu^k - \mu^*)_{\mathcal{I}_{B'}}| \leq \frac{1}{2\kappa - 1} (|(x^k - x^*)_{\mathcal{A}_{B,c}}| + |(\mu^k - \mu^*)_{\mathcal{I}_{B,c}}|),$$

where $\mathcal{A}_{B,c}$, $\mathcal{I}_{B,c}$ denote the complement of the indices $\mathcal{A}_B$, $\mathcal{I}_B$, respectively, then

$$|x^{k+1} - x^*| + |\mu^{k+1} - \mu^*| \leq \frac{1}{2} (|x^k - y^*| + |\mu^k - \mu^*|),$$

and convergence follows.

(iii) If $x^* < \psi$ and $\mu^0 + c\,(x^0 - \psi) \leq 0$ (e.g., $y^0 = \psi$, $\mu^0 = 0$), then the algorithm converges in one step. In fact, in this case $\mathcal{A}_B = \mathcal{I}_B = \emptyset$.

7.2 Monotone class

In this section we assume that $Z = \mathbb{R}^n$ and that A is an M-matrix, i.e., it is nonsingular, its nondiagonal elements are nonnegative, and $A^{-1} \geq 0$. Then there exists a unique solution (x^*, μ^*) to (7.1.3).

Example 7.3. Consider the discretized obstacle problem on the square $(0, 1) \times (0, 1)$. Let $h = \frac{1}{N}$ and let $u_{i,j}$ denote the approximation of the solution at the nodes (ih, jh), $0 \leq i,\ j \leq N$. Using the central difference approximation results in a finite-dimensional variational inequality satisfying

$$\frac{4u_{i,j} - u_{i+1,j} - u_{i-1,j} - u_{i,j+1} - u_{i,j-1}}{h^2} + \mu_{i,j} = f_{i,j},$$

$$\mu_{i,j} = \max(0, \mu_{i,j} + c\,(u_{i,j} - \psi_{i,j})$$

for $1 \leq i,\ j \leq N - 1$, and $u_{0,j} = u_{N,j} = u_{i,0} = u_{i,N} = 0$. The resulting matrix A in $\mathbb{R}^{(N-1)^2}$ is an M-matrix.

The following theorem asserts the convergence of the iterates of the primal-dual active set method for (7.1.3).

Theorem 7.4. *Assume that A is an M-matrix. Then $x_k \to x^*$ for arbitrary initial data. Moreover $x^* \leq x^{k+1} \leq x^k$ for all $k \geq 1$, $x^k \leq \psi$ for all $k \geq 2$, and there exists k_0 such that $\mu_k \geq 0$ for all $k \geq k_0$.*

Proof. Since A is an M-matrix we have $A_{\mathcal{I}}^{-1} \geq 0$ and $A_{\mathcal{I}}^{-1} A_{\mathcal{I},\mathcal{A}} \leq 0$ for every index partition of $\{1, \ldots, n\}$ into $\mathcal{I}$ and $\mathcal{A}$. Since $\delta x_{\mathcal{I}_k} = -A_{\mathcal{I}_k}^{-1} A_{\mathcal{I}_k \mathcal{A}_k} \delta x_{\mathcal{A}_k} - A_{\mathcal{I}_k}^{-1} \delta \mu_{\mathcal{I}_k}$ by (7.1.9) it follows that $\delta x_{\mathcal{I}_k} \le 0$. Together with $\delta x_{\mathcal{A}_k} \le 0$, which follows from the third equation in (7.1.10), this implies that $x^{k+1} \le x^k$ for $k \ge 1$. Next we show that x^k is feasible for $k \ge 2$. Due to monotonicity of x^k with respect to k it suffices to show this for $k = 2$. For i such that $(x^1 - \psi)_i > 0$ we have $\mu_i^1 = 0$ by (7.1.10) and hence $\mu_i^1 + c\,(x^1 - \psi)_i > 0$ and $i \in \mathcal{A}_1$. Since $x^2 = \psi$ on $\mathcal{A}_1$ and $x^2 \le x^1$ it follows that $x^2 \le \psi$.

To verify that $x^* \le x^k$ for $k \ge 1$, note that

$$\begin{aligned} a_{\mathcal{I}_{k-1}} &= \mu^*_{\mathcal{I}_{k-1}} + A_{\mathcal{I}_{k-1}} x^*_{\mathcal{I}_{k-1}} + A_{\mathcal{I}_{k-1}\mathcal{A}_{k-1}} x^*_{\mathcal{A}_{k-1}} \\ &= A_{\mathcal{I}_{k-1}} x^k_{\mathcal{I}_{k-1}} + A_{\mathcal{I}_{k-1}\mathcal{A}_{k-1}} \psi_{\mathcal{A}_{k-1}}, \end{aligned}$$

and consequently

$$A_{\mathcal{I}_{k-1}} \left(x^k_{\mathcal{I}_{k-1}} - x^*_{\mathcal{I}_{k-1}} \right) = \mu^*_{\mathcal{I}_{k-1}} + A_{\mathcal{I}_{k-1}\mathcal{A}_{k-1}} \left(x^*_{\mathcal{A}_{k-1}} - \psi_{\mathcal{A}_{k-1}} \right).$$

Since $\mu^*_{\mathcal{I}_{k-1}} \ge 0$ and $x^*_{\mathcal{A}_{k-1}} \le \psi_{\mathcal{A}_{k-1}}$, the M-matrix properties of A imply that $x^k_{\mathcal{I}_{k-1}} \ge x^*_{\mathcal{I}_{k-1}}$ and consequently $x^k \ge x^*$ for all $k \ge 1$.

Turning to the feasibility of μ^k assume that for a pair of indices $(\bar{k}, i)$, $\bar{k} \ge 1$, we have $\mu_i^{\bar{k}} < 0$. Then necessarily $i \in \mathcal{A}_{\bar{k}-1}$, $x_i^{\bar{k}} = \psi_i$, and $\mu_i^{\bar{k}} + c(x_i^{\bar{k}} - \psi_i) < 0$. It follows that $i \in \mathcal{I}_{\bar{k}}$, $\mu_i^{\bar{k}+1} = 0$, and $\mu_i^{\bar{k}+1} + c(x_i^{\bar{k}+1} - \psi_i) \le 0$, since $x_i^{k+1} \le \psi_i$, $k \ge 1$. Consequently $i \in \mathcal{I}_{\bar{k}+1}$ and by induction $i \in \mathcal{I}_k$ for all $k \ge \bar{k}+1$. Thus, whenever a coordinate of μ^k becomes negative at iteration $\bar{k}$, it is zero from iteration $\bar{k}+1$ onwards, and the corresponding primal coordinate is feasible. Due to finite-dimensionality of $\mathbb{R}^n$ it follows that there exists k_o such that $\mu^k \ge 0$ for all $k \ge k_o$.

Monotonicity of x^k and $x^* \le x^k \le \psi$ for $k \ge 2$ imply the existence of $\bar{x}$ such that $\lim x^k = \bar{x} \le \psi$. Since $\mu^k = Ax^k + a \ge 0$ for all $k \ge k_o$, there exists $\bar{\mu}$ such that $\lim \mu^k = \bar{\mu} \ge 0$. Together with the complementarity property $\bar{\mu}(\bar{x} - \psi)$, which is a consequence of the first equation in (7.1.10), it follows that $(\bar{x}, \bar{\mu}) = (x^*, \mu^*)$. □

7.3 Cone sum preserving class

For rectangular matrices $B \in \mathbb{R}^{n \times m}$ we denote by $\|\cdot\|_1$ the subordinate matrix norm when both $\mathbb{R}^n$ and $\mathbb{R}^m$ are endowed with the 1-norms. Moreover, B_+ denotes the $n \times m$ matrix containing the positive parts of the elements of B. Recall that a square matrix is called a P-matrix if all its principle minors are positive.

Theorem 7.5. *If A is a P-matrix and for every partitioning of the index set $\{1, \ldots, n\}$ into disjoint subsets $\mathcal{I}$ and $\mathcal{A}$ we have $\|(A_{\mathcal{I}}^{-1} A_{\mathcal{I}\mathcal{A}})_+\|_1 < 1$ and $\sum_{i \in \mathcal{I}} (A_{\mathcal{I}}^{-1} x_{\mathcal{I}})_i > 0$ for $x_{\mathcal{I}} \ge 0$ with $x_{\mathcal{I}} \ne 0$, then $\lim_{k \to \infty} x^k = x^*$.*

The third condition in the above theorem motivates the terminology cone sum preserving. If A is an M-matrix, then the conditions of Theorem 7.5 are satisfied. The proof will reveal that $\mathcal{M}(x^k) = \sum_{i=1}^n x_i^k$ is a merit function.

Proof. From (7.1.9) and the fact that $x^{k+1} = \psi$ on $\mathcal{A}_k$ we have for $k = 1, 2, \ldots$

$$(x^{k+1} - x^k)_{\mathcal{I}_k} = A_{\mathcal{I}_k}^{-1} A_{\mathcal{I}_k \mathcal{A}_k} (x^k - \psi)_{\mathcal{A}_k} + A_{\mathcal{I}_k}^{-1} \mu^k_{\mathcal{I}_k}$$

and upon summation over the inactive indices

$$\sum_{i\in\mathcal{I}_k}(x_i^{k+1}-x_i^k)=\sum_{i\in\mathcal{I}_k}\left(A_{\mathcal{I}_k}^{-1}A_{\mathcal{I}_k\mathcal{A}_k}(x^k-\psi)_{\mathcal{A}_k}\right)_i+\sum_{i\in\mathcal{I}_k}(A_{\mathcal{I}_k}^{-1}\mu_{\mathcal{I}_k}^k)_i. \tag{7.3.1}$$

Using again that $x^{k+1}=\psi$ on $\mathcal{A}_k$, this implies that

$$\sum_{i=1}^n(x_i^{k+1}-x_i^k)=-\sum_{i\in\mathcal{A}_k}(x_i^k-\psi_i)+\sum_{i\in\mathcal{I}_k}(A_{\mathcal{I}_k}^{-1}A_{\mathcal{I}_k\mathcal{A}_k}(x^k-\psi)_{\mathcal{A}_k})_i+\sum_{i\in\mathcal{I}_k}(A_{\mathcal{I}_k}^{-1}\mu_{\mathcal{I}_k}^k)_i. \tag{7.3.2}$$

Since $x_{\mathcal{A}_k}^k\geq\psi_{\mathcal{A}_k}$ it follows that

$$\sum_{i=1}^n(x_i^{k+1}-x_i^k)\leq(\|(A_{\mathcal{I}_k}^{-1}A_{\mathcal{I}_k\mathcal{A}_k})_+\|_1-1)\,|x^k-\psi|_{1,\mathcal{A}_k}+\sum_{i\in\mathcal{I}_k}(A_{\mathcal{I}_k}^{-1}\mu_{\mathcal{I}_k}^k)_i<0 \tag{7.3.3}$$

unless x^k is the solution to (7.1.3). In fact, if $|x^k-\psi|_{1,\mathcal{A}_k}$, then $x^k\leq\psi$ on $\mathcal{A}_k$ and $\mu^k\geq0$ on $\mathcal{A}_k$. If moreover $\mu_{\mathcal{I}_k}=0$, then $\mu^k\geq0$ and $x^k\leq\psi$ on Ω. Together with the first equation in (7.1.10), this implies that $\{(x^k,\mu^k)\}$ satisfies the complementarity conditions. It also satisfies $Ax^k+\mu^k=a$ and hence (x^k,μ^k) is a solution to (7.1.3). Consequently

$$x^k\to\mathcal{M}(x^k)=\sum_{i=1}^n x_i^k$$

acts as a merit function for the algorithm. Since there are only finitely many possible choices for active/inactive sets, there exists an iteration index $\bar{k}$ such that $\mathcal{I}_{\bar{k}}=\mathcal{I}_{\bar{k}+1}$. In this case $(x^{\bar{k}+1},\mu^{\bar{k}+1})$ is a solution to (7.1.3). In fact, in view of (iii) of the algorithm it suffices to show that $x^{\bar{k}+1}$ and $\mu^{\bar{k}+1}$ are feasible. This follows from the fact that due to $\mathcal{I}_{\bar{k}}=\mathcal{I}_{\bar{k}+1}$ we have $c(x_i^{\bar{k}+1}-\psi_i)=\mu_i^{\bar{k}+1}+c(x_i^{\bar{k}+1}-\psi_i)\leq0$ for $i\in\mathcal{I}_{\bar{k}}$ and $\mu_i^{\bar{k}+1}+c(x_i^{\bar{k}+1}-\psi_i)=\mu_i^{\bar{k}+1}>0$ for $i\in\mathcal{A}_{\bar{k}}$. From (7.1.10) we deduce $\mu^{\bar{k}+1}(x^{\bar{k}+1}-\psi)=0$, and hence the complementarity conditions hold and the algorithm converges in finitely many steps. ☐

A perturbation result. We now discuss the primal-dual active set strategy for the case where the matrix A can be expressed as an additive perturbation of an M-matrix.

Theorem 7.6. *Assume that $A=M+K$ with M an M-matrix and K an $n\times n$ matrix. If $\|K\|_1$ is sufficiently small, then the primal-dual active set algorithm is well defined and $\lim_{k\to\infty}(x^k,\mu^k)=(x^*,\mu^*)$, where (x^*,μ^*) is a solution to* (7.1.3). *If $y^TAy>0$ for $y\neq0$, then the solution to* (7.1.3) *is unique.*

Proof. As a consequence of the assumption that M is an M-matrix all principal submatrices of M are M-matrices as well [BePl]. Let $\mathcal{S}$ denote the set of all subsets of $\{1,\dots,n\}$ and $\mathcal{A}$ its complement. Define

$$\rho=\sup_{\mathcal{I}\in\mathcal{S}}\|M_{\mathcal{I}}^{-1}K_{\mathcal{I}}\|_1\text{ and }\sigma=\sup_{\mathcal{I}\in\mathcal{S}}\|B_{\mathcal{I}\mathcal{A}}(K)\|_1, \tag{7.3.4}$$

where $B_{\mathcal{I}\mathcal{A}}(K) = M_{\mathcal{I}}^{-1}K_{\mathcal{I}\mathcal{A}} - M_{\mathcal{I}}^{-1}K_{\mathcal{I}}(M+K)_{\mathcal{I}}^{-1}A_{\mathcal{I}\mathcal{A}}$. Assume that K is chosen such that $\rho < \frac{1}{2}$ and $\sigma < 1$. For every subset $\mathcal{I} \in \mathcal{S}$ the inverse of $A_{\mathcal{I}}$ exists and can be expressed as

$$A_{\mathcal{I}}^{-1} = \big(I_{\mathcal{I}} + \sum_{i=1}^{\infty} (-M_{\mathcal{I}}^{-1}K_{\mathcal{I}})^i\big)\, M_{\mathcal{I}}^{-1}.$$

Consequently the algorithm is well defined. Proceeding as in the proof of Theorem 7.5 we arrive at

$$\sum_{i=1}^{n}(x_i^{k+1} - x_i^k) = -\sum_{i\in\mathcal{A}_k}(x_i^k - \psi_i) + \sum_{i\in\mathcal{I}_k}\left(A_{\mathcal{I}_k}^{-1}A_{\mathcal{I}_k\mathcal{A}_k}(x^k-\psi)_{\mathcal{A}_k}\right)_i + \sum_{i\in\mathcal{I}_k}(A_{\mathcal{I}_k}^{-1}\mu_{\mathcal{I}_k}^k)_i,$$

where $\mu_i^k \le 0$ for $i \in \mathcal{I}_k$ and $x_i^k \ge \psi_i$ for $i \in \mathcal{A}_k$. Below we drop the index k with $\mathcal{I}_k$ and $\mathcal{A}_k$. Note that $A_{\mathcal{I}}^{-1}A_{\mathcal{I}\mathcal{A}} \le M_{\mathcal{I}}^{-1}K_{\mathcal{I}\mathcal{A}} - M_{\mathcal{I}}^{-1}K_{\mathcal{I}}(M+K)_{\mathcal{I}}^{-1}A_{\mathcal{I}\mathcal{A}} = B_{\mathcal{I}\mathcal{A}}(K)$. Here we used $(M+K)_{\mathcal{I}}^{-1} - M_{\mathcal{I}}^{-1} = -M_{\mathcal{I}}^{-1}K_{\mathcal{I}}(M+K)_{\mathcal{I}}^{-1}$ and $M_{\mathcal{I}}^{-1}M_{\mathcal{I}\mathcal{A}} \le 0$. This implies

$$\sum_{i=1}^{n}(x_i^{k+1} - x_i^k) \le -\sum_{i\in\mathcal{A}}(x_i^k - \psi_i) + \sum_{i\in\mathcal{I}}\left(B_{\mathcal{I}\mathcal{A}}(K)(x^k-\psi)_{\mathcal{A}}\right)_i + \sum_{i\in\mathcal{I}}(A_{\mathcal{I}}^{-1}\mu_{\mathcal{I}}^k)_i. \tag{7.3.5}$$

We estimate

$$\sum_{i\in\mathcal{I}}(A_{\mathcal{I}}^{-1}\mu_{\mathcal{I}}^k)_i = \sum_{i\in\mathcal{I}}\left(M_{\mathcal{I}}^{-1}\mu_{\mathcal{I}}^k + \sum_{j=1}^{\infty}(-M_{\mathcal{I}}^{-1}K_{\mathcal{I}})^j M_{\mathcal{I}}^{-1}\mu_{\mathcal{I}}^k\right)_i$$

$$\le -|M_{\mathcal{I}}^{-1}\mu_{\mathcal{I}}^k|_1 + \sum_{j=1}^{\infty}\rho^i|M_{\mathcal{I}}^{-1}\mu_{\mathcal{I}}^k|_1$$

$$= (\alpha-1)|M_{\mathcal{I}}^{-1}\mu_{\mathcal{I}}^k|_1 + \left(\frac{1}{1-\rho} - (\alpha+1)\right)|M_{\mathcal{I}}^{-1}\mu_{\mathcal{I}}^k|_1 = (\alpha-1)|M_{\mathcal{I}}^{-1}\mu_{\mathcal{I}}^k|_1,$$

where we set $\alpha = \frac{\rho}{1-\rho} \in (0,1)$ by (7.3.4). This estimate, together with (7.3.4) and (7.3.5), implies that

$$\sum_{i=1}^{n}(x_i^{k+1} - x_i^k) \le (\sigma-1)|x_{\mathcal{A}_k}^k - \psi_{\mathcal{A}_k}|_1 + (\alpha-1)|M_{\mathcal{I}_k}^{-1}\mu_{\mathcal{I}_k}^k|_1.$$

Now it can be verified in the same manner as in the proof of Theorem 7.5 that $x^k \to \mathcal{M}(x^k) = \sum_{i=1}^n x_i^k$ is a merit function for the algorithm and convergence of (x^k, μ^k) to a solution (x^*, μ^*) follows. If there are two solutions to (7.1.3), then their difference y satisfies $y^t A y \le 0$ and hence $y = 0$ and uniqueness follows. □

Observe that the M-matrix property is not stable under arbitrarily small perturbations since off-diagonal elements may become positive. Theorem 7.6 guarantees that convergence of the primal-dual active set strategy for arbitrary initial data is preserved for sufficiently small perturbations K of an M-matrix. Therefore, Theorem 7.6 is also of interest in connection with numerical implementations of the primal-dual active set algorithm.

7.4 Diagonally dominated class

We again consider the reduced problem (7.1.3),

$$Ax + \mu = a, \quad \mu = \max(0, \mu + c\,(x - \psi)), \tag{7.4.1}$$

but differently from Sections 7.2 and 7.3 we admit also the infinite-dimensional case. Sufficient conditions related to diagonal dominance of A will be given which imply that

$$\mathcal{M}(x^{k+1}, \mu^{k+1}) = \max\left(\beta \int_{\mathcal{I}_k} |(x^{k+1} - \psi)^+|\, dx, \int_{\mathcal{A}_k} |(\mu^{k+1})^-|\, dx \right) \tag{7.4.2}$$

with $\beta > 0$ acts as a merit functional for the primal-dual algorithm. Here we set $\phi^+ = \max(\phi, 0)$ and $\phi^- = -\min(\phi, 0)$. The natural norm associated to this merit functional is the $L^1(\Omega)$-norm and consequently we assume that

$$A \in \mathcal{L}(L^1(\Omega)), \; a \in L^1(\Omega), \text{ and } \psi \in L^1(\Omega). \tag{7.4.3}$$

The analysis of this section can also be used to obtain convergence in the $L^p(\Omega)$-norm for any $p \in (1, \infty)$ if the norms in the integrands of $\mathcal{M}$ are replaced by $|\cdot|^p$-norms and the $L^1(\Omega)$-norms below are replaced by $L^p(\Omega)$-norms as well.

The results also apply for $Z = \mathbb{R}^n$. In this case the integrals in (7.4.2) are replaced by sums over the respective index sets.

We assume that there exist constants $\rho_i, i = 1, \dots, 5$, such that for all partitions $\mathcal{A}$ and $\mathcal{I}$ of Ω and for all $\phi_{\mathcal{A}} \geq 0$ in $L^2(\mathcal{A})$ and $\phi_{\mathcal{I}} \geq 0$ in $L^2(\mathcal{I})$

$$\begin{aligned} &|[A_{\mathcal{I}}^{-1} \phi_{\mathcal{I}}]^-| \leq \rho_1\, |\phi_{\mathcal{I}}|, \\ &|[A_{\mathcal{I}}^{-1} A_{\mathcal{I}\mathcal{A}} \phi_{\mathcal{A}}]^+| \leq \rho_2\, |\phi_{\mathcal{A}}| \end{aligned} \tag{7.4.4}$$

and

$$\begin{aligned} &|[A_{\mathcal{A}} \phi_{\mathcal{A}}]^-| \leq \rho_3\, |\phi_{\mathcal{A}}|, \\ &|[A_{\mathcal{A}\mathcal{I}} A_{\mathcal{I}}^{-1} \phi_{\mathcal{I}}]^-| \leq \rho_4\, |\phi_{\mathcal{I}}|, \\ &|[A_{\mathcal{A}\mathcal{I}} A_{\mathcal{I}}^{-1} A_{\mathcal{I}\mathcal{A}} \phi_{\mathcal{A}}]^+| \leq \rho_5\, |\phi_{\mathcal{A}}|. \end{aligned} \tag{7.4.5}$$

Here $|\cdot|$ denotes the $L^1(\Omega)$-norm. Assumption (7.4.4) requires in particular the existence of $A_{\mathcal{I}}^{-1}$. By a Schur complement argument with respect to the sets $\mathcal{I}_k$ and $\mathcal{A}_k$ this implies existence of a solution to the linear systems in step (iii) of the algorithm for every k.

Theorem 7.7. *If* (7.4.3), (7.4.4), (7.4.5) *hold and* $\rho = \max(\beta\,\rho_1 + \rho_2, \frac{\rho_3}{\beta} + \rho_4 + \frac{\rho_5}{\beta}) < 1$, *then* $\mathcal{M}$ *is a merit function for the primal-dual algorithm of the reduced system and* $\lim_{k\to\infty}(x^k, \mu^k) = (x^*, \mu^*)$ *in* $L^1(\Omega) \times L^1(\Omega)$, *with* (x^*, μ^*) *a solution to* (7.4.1).

Proof. For every $k \geq 1$ we have $(x^{k+1} - \psi)^+ \leq (x^{k+1} - x^k)^+$ on $\mathcal{I}_k$ and $(\mu^{k+1})^- = (\delta\mu)^-$ on $\mathcal{A}_k$. Therefore

$$\mathcal{M}(x^{k+1}, \mu^{k+1}) \leq \max\left(\beta \int_{\mathcal{I}_k} (\delta x_{\mathcal{I}_k})^+, \int_{\mathcal{A}_k} (\delta \mu_{\mathcal{A}_k})^- \right). \tag{7.4.6}$$

From (7.1.9) we deduce that

$$\delta x_{\mathcal{I}_k} = -A_{\mathcal{I}_k}^{-1}(-\mu_{\mathcal{I}_k}^k) + A_{\mathcal{I}_k}^{-1} A_{\mathcal{I}_k \mathcal{A}_k}(-\delta x_{\mathcal{A}_k}),$$

with $\mu_{\mathcal{I}_k}^k \le 0$ and $\delta x_{\mathcal{A}_k} \le 0$. By (7.4.4) therefore

$$\begin{aligned} |(\delta x_{\mathcal{I}_k})^+| &\le \rho_1 |\mu_{\mathcal{I}_k}^k| + \rho_2 |\delta x_{\mathcal{A}_k}| \\ &= \rho_1 \int_{\mathcal{I}_k \cap \mathcal{A}_{k-1}} |(\mu_{\mathcal{I}_k}^k)^-| + \rho_2 \int_{\mathcal{A}_k \cap \mathcal{I}_{k-1}} (x_k - \psi)^+ \\ &\le \left(\rho_1 + \frac{\rho_2}{\beta}\right) \mathcal{M}(x^k, \mu^k). \end{aligned} \tag{7.4.7}$$

Similarly, by (7.1.9)

$$\delta \mu_{\mathcal{A}_k} = A_{\mathcal{A}_k}(-\delta x_{\mathcal{A}_k}) + A_{\mathcal{A}_k \mathcal{I}_k} A_{\mathcal{I}_k}^{-1}(-\mu_{\mathcal{I}_k}^k) - A_{\mathcal{A}_k \mathcal{I}_k} A_{\mathcal{I}_k}^{-1} A_{\mathcal{I}_k \mathcal{A}_k}(-\delta x_{\mathcal{A}_k}).$$

Since $\delta x_{\mathcal{A}_k} \le 0$ and $\mu_{\mathcal{I}_k}^k \le 0$, we find by (7.4.5)

$$|(\delta \mu_{\mathcal{A}_k})^-| \le \rho_3 |\delta x_{\mathcal{A}_k}| + \rho_4 |\mu_{\mathcal{I}_k}^k| + \rho_5 |\delta x_{\mathcal{A}_k}| \le \left(\frac{\rho_3 + \rho_5}{\beta} + \rho_4\right) \mathcal{M}(x^k, \nu^k), \tag{7.4.8}$$

and therefore

$$\mathcal{M}(x^{k+1}, \mu^{k+1}) \le \max\left(\beta\rho_1 + \rho_2, \frac{\rho_3 + \rho_5}{\beta} + \rho_4\right) \mathcal{M}(x^k, \mu^k) = \rho\, \mathcal{M}(x^k, \mu^k).$$

Thus, if $\rho < 1$, then $\mathcal{M}$ is a merit functional. Furthermore $\mathcal{M}(x^{k+1}, \mu^{k+1}) \le \rho^k M(x^1, \mu^1)$. Together with (7.4.7), (7.4.8), and (7.1.9) it follows that (x^k, μ^k) is a Cauchy sequence. Hence there exists (x^*, μ^*) such that $\lim_{k\to\infty}(x^k, \mu^k) = (x^*, \mu^*)$ and $Ax^* + \mu^* = a$, $\mu^*(x^* - \psi) = 0$ a.e. in Ω. Since $(x^k - \psi)^+ \to (x^* - \psi)^+$ as $k \to \infty$ and $\lim_{k\to\infty} \int_\Omega (x^{k+1} - \psi)^+ = 0$, it follows that $x^* \le \psi$. Similarly, one argues that $\mu^* \ge 0$. Thus (x^*, μ^*) is a solution to (7.4.1). □

Concerning the uniqueness of the solution to (7.4.1), assume that $A \in \mathcal{L}(L^2(\Omega))$ and that $(Ay, y)_{L^2(\Omega)} > 0$ for all $y \ne 0$. Assume further that (x^*, μ^*) and $(\hat{x}, \hat{\mu})$ are solutions to (7.4.1) with $\hat{x} - x^* \in L^2(\Omega)$. Then $(\hat{x} - x^*, A(\hat{x} - x^*))_{L^2(\Omega)} \le 0$ and therefore $\hat{x} - x^* = 0$.

Remark 7.4.1. In the finite-dimensional case the integrals in the definition of $\mathcal{M}$ must be replaced by sums over the active/inactive index sets. If A is an M-matrix, then $\rho_1 = \rho_2 = \rho_5 = 0$ and $\rho < 1$ if $\frac{\rho_3}{\beta} + \rho_4 < 1$. This is the case if A is diagonally dominant in the sense that $\rho_4 < 1$ and β is chosen sufficiently large. If these conditions are met, then $\rho < 1$ is stable under perturbations of A.

Remark 7.4.2. Consider the infinite-dimensional case with $A = \alpha I + K$, where $\alpha > 0$, $K \in \mathcal{L}(L^1(\Omega))$, and $K\phi \ge 0$ for all $\phi \ge 0$. This is the case for the operators in Examples 7.1 and 7.2, as can be argued by using the maximum principle. Let $\|K\|$ denote the norm of

K in $K \in \mathcal{L}(L^1(\Omega))$. For $\|K\| < \alpha$ and any $\mathcal{I} \subset \Omega$ we have $A_{\mathcal{I}}^{-1} = \frac{1}{\alpha} I_{\mathcal{I}} - \frac{1}{\alpha} K_{\mathcal{I}} A_{\mathcal{I}}^{-1}$ and hence $\rho_1 \le \frac{\|K\|}{\alpha(\alpha - \|K\|)}$. Moreover $\rho_3 = 0$. The conditions involving ρ_2, ρ_4, and ρ_5 are satisfied with $\rho_2 = \frac{\|K\|}{\alpha - \|K\|}$, $\rho_4 = \frac{\|K\|^2}{\alpha(\alpha - \|K\|)}$, and $\rho_5 = \frac{\|K\|^2}{\alpha - \|K\|}$.

7.5 Bilateral constraints, diagonally dominated class

Consider the quadratic programming with the bilateral constraints

$$\min \frac{1}{2}(Ax, x)_X - (a, x)_X$$

subject to

$$Ex = b, \quad \varphi \le Gx \le \psi$$

with conditions on A, E, and G as in (7.1.1). From Example 4.53 of Section 4.7, we recall that the necessary optimality condition for this problem is given by

$$\begin{aligned} & Ax + E^*\lambda + G^*\mu = a, \\ & Ex = b, \\ & \mu = \max(0, \mu + c(Gx - \psi)) + \min(0, \mu + c(Gx - \varphi)), \end{aligned} \tag{7.5.1}$$

where max as well as min are interpreted as pointwise a.e. operations if $Z = L^2(\Omega)$ and coordinatewise for $Z = \mathbb{R}^n$.

Primal-Dual Active Set Algorithm.

(i) Initialize x^0, μ^0. Set $k = 0$.

(ii) Given (x^k, μ^k), set
$\mathcal{A}_k^+ = \{\mu^k + c(Gx^k - \psi) > 0\}$,
$\mathcal{I}_k = \{\mu^k + c(Gx^k - \psi) \le 0 \le \mu^k + c(Gx^k - \varphi)\}$,
$\mathcal{A}_k^- = \{\mu^k + c(Gx^k - \varphi) < 0\}$.

(iii) Solve for $(x^{k+1}, \lambda^{k+1}, \mu^{k+1})$:

$$\begin{aligned} & Ax^{k+1} + E^*\lambda^{k+1} + G^*\mu^{k+1} = a, \\ & Ex^{k+1} = b, \\ & Gx^{k+1} = \psi \text{ in } \mathcal{A}_k^+, \quad \mu^{k+1} = 0 \text{ in } \mathcal{I}_k, \text{ and } Gx^{k+1} = \varphi \text{ in } \mathcal{A}_k^-. \end{aligned}$$

(iv) Stop, or set $k = k + 1$, and return to (ii).

We assume the existence of a solution to the auxiliary systems in step (iii). In case A is positive definite a sufficient condition for existence is given by surjectivity of E and surjectivity of $G : N(E) \to Z$. As in the unilateral case we shall consider a transformed version of (7.5.1). If (7.1.4) and (7.1.8) are satisfied, then (7.5.1) can be transformed into the equivalent system

$$Ax + \mu = a, \quad \mu = \max(0, \mu + c\,(x - \psi)) + \min(0, \mu + c\,(x - \varphi)), \tag{7.5.2}$$

where A is a bounded operator on Z. The algorithm for this reduced system is obtained from the above algorithm by replacing G by I and deleting the terms involving E and E^*. As in the unilateral case, if one does not carry out the reduction step from (7.1.6) to (7.5.2), then the coordinates corresponding to x_2 are treated as inactive ones in the algorithm. We henceforth concentrate on the infinite-dimensional case and give sufficient conditions for

$$\begin{aligned}\mathcal{M}(x^{k+1}, \mu^{k+1}) = \max\Bigg(&\int_{\mathcal{I}_k} ((x^{k+1} - \psi)^+ + (x^{k+1} - \varphi)^-)\,dx, \\ &\int_{\mathcal{A}_k^+} (\mu^{k+1})^-\,dx + \int_{\mathcal{A}_k^-} (\mu^{k+1})^+\,dx\Bigg)\end{aligned} \tag{7.5.3}$$

to act as a merit function for the algorithm applied to the reduced system. In the finite-dimensional case the integrals must be replaced by sums over the respective index sets. We note that (iii) with $G = I$ implies the complementarity property

$$(x^k - \psi)(x^k - \varphi)\mu^k = 0 \text{ a.e. in } \Omega. \tag{7.5.4}$$

As in the previous section the merit function involves L^1-norms and accordingly we aim for convergence in $L^1(\Omega)$. We henceforth assume that

$$A \in \mathcal{L}(L^1(\Omega)), \;\; a \in L^1(\Omega), \;\; \psi \text{ and } \varphi \in L^1(\Omega). \tag{7.5.5}$$

Below $\|\cdot\|$ denotes the norm of operators in $\mathcal{L}(L^1(\Omega))$. The following conditions will be used: There exist constants $\rho_i, i = 1, \ldots, 5$, such that for arbitrary partitions $\mathcal{A} \cup \mathcal{I} = \Omega$ we have

$$\begin{aligned}&\|A_{\mathcal{I}}^{-1}\| \le \rho_1, \\ &\|A_{\mathcal{I}}^{-1} A_{\mathcal{I}\mathcal{A}}\| \le \rho_2\end{aligned} \tag{7.5.6}$$

and

$$\begin{aligned}&\|A_{\mathcal{A}_k} - c\,I\| \le \rho_3, \\ &\|A_{\mathcal{A}\mathcal{I}} A_{\mathcal{I}}^{-1}\| \le \rho_4, \\ &\|A_{\mathcal{A}\mathcal{I}} A_{\mathcal{I}}^{-1} A_{\mathcal{I}\mathcal{A}}\| \le \rho_5.\end{aligned} \tag{7.5.7}$$

We further set $\rho = 2\max(\,\max(\rho_1, \rho_2, \frac{\rho_2}{c}), \max(\rho_3 + \rho_5, \rho_4), \frac{\rho_3+\rho_5}{c}\,)$.

Theorem 7.8. *If* (7.5.5), (7.5.6), (7.5.7) *hold and* $\rho < 1$, *then* $\mathcal{M}$ *is a merit function for the primal-dual algorithm of the reduced system* (7.5.2) *and* $\lim_{k\to\infty}(x^k, \mu^k) = (x^*, \mu^*)$ *in* $L^1(\Omega) \times L^1(\Omega)$, *with* (x^*, μ^*) *a solution to* (7.5.2).

Proof. For $\delta x = x^{k+1} - x^k$ and $\delta\mu = \mu^{k+1} - \mu^k$ we have

$$\begin{aligned} &A_{\mathcal{A}_k^+}\delta x_{\mathcal{A}_k^+} + A_{\mathcal{A}_k^+\mathcal{I}_k}\delta x_{\mathcal{I}_k} + A_{\mathcal{A}_k^+\mathcal{A}_k^-}\delta x_{\mathcal{A}_k^-} + \delta\mu_{\mathcal{A}_k^+} = 0,\\ &A_{\mathcal{I}_k}\delta x_{\mathcal{I}_k} + A_{\mathcal{I}_k\mathcal{A}_k^+}\delta x_{\mathcal{A}_k^+} + A_{\mathcal{I}_k\mathcal{A}_k^-}\delta x_{\mathcal{A}_k^-} - \mu^k_{\mathcal{I}_k} = 0,\\ &A_{\mathcal{A}_k^-}\delta x_{\mathcal{A}_k^-} + A_{\mathcal{A}_k^-\mathcal{I}_k}\delta x_{\mathcal{I}_k} + A_{\mathcal{A}_k^-\mathcal{A}_k^+}\delta x_{\mathcal{A}_k^+} + \delta\mu_{\mathcal{A}_k^-} = 0 \end{aligned} \tag{7.5.8}$$

with

$$\mu^k_{\mathcal{A}_k^+} \begin{cases} > 0 & \text{on} \quad \mathcal{A}_{k-1}^+ \cap \mathcal{A}_k^+,\\ = 0 & \text{on} \quad \mathcal{I}_{k-1} \cap \mathcal{A}_k^+,\\ > c\,(\psi - \varphi) & \text{on} \quad \mathcal{A}_{k-1}^- \cap \mathcal{A}_k^+, \end{cases} \tag{7.5.9}$$

$$\mu^k_{\mathcal{I}_k} \in \begin{cases} [c(\varphi - \psi), 0) & \text{on} \quad \mathcal{A}_{k-1}^+ \cap \mathcal{I}_k,\\ = 0 & \text{on} \quad \mathcal{I}_{k-1} \cap \mathcal{I}_k,\\ (0, c(\psi - \varphi)] & \text{on} \quad \mathcal{A}_{k-1}^- \cap \mathcal{I}_k, \end{cases} \tag{7.5.10}$$

$$\mu^k_{\mathcal{A}_k^-} \begin{cases} < c\,(\varphi - \psi) & \text{on} \quad \mathcal{A}_{k-1}^+ \cap \mathcal{A}_k^-,\\ = 0 & \text{on} \quad \mathcal{I}_{k-1} \cap \mathcal{A}_k^-,\\ < 0 & \text{on} \quad \mathcal{A}_{k-1}^- \cap \mathcal{A}_k^-, \end{cases} \tag{7.5.11}$$

$$\delta x_{\mathcal{A}_k^+} \begin{cases} = 0 & \text{on} \quad \mathcal{A}_{k-1}^+ \cap \mathcal{A}_k^+,\\ < 0 & \text{on} \quad \mathcal{I}_{k-1} \cap \mathcal{A}_k^+,\\ = \psi - \varphi < \frac{\mu^k}{c} & \text{on} \quad \mathcal{A}_{k-1}^- \cap \mathcal{A}_k^+, \end{cases} \tag{7.5.12}$$

$$\delta x_{\mathcal{A}_k^-} \begin{cases} = \varphi - \psi > \frac{\mu^k}{c} & \text{on} \quad \mathcal{A}_{k-1}^+ \cap \mathcal{A}_k^-,\\ > 0 & \text{on} \quad \mathcal{I}_{k-1} \cap \mathcal{A}_k^-,\\ = 0 & \text{on} \quad \mathcal{A}_{k-1}^- \cap \mathcal{A}_k^-. \end{cases} \tag{7.5.13}$$

From (7.5.4)

$$(x^{k+1}_{\mathcal{I}_k} - \psi_{\mathcal{I}_k})^+ \le (\delta x_{\mathcal{I}_k})^+ \text{ and } (x^{k+1}_{\mathcal{I}_k} - \varphi_{\mathcal{I}_k})^- \le (\delta x_{\mathcal{I}_k})^-. \tag{7.5.14}$$

This implies that

$$\mathcal{M}(x^{k+1}, \mu^{k+1}) \le \max\left(\int_{\mathcal{I}_k} |\delta x_{\mathcal{I}_k}|,\ \int_{\mathcal{A}_k^+} (\mu^{k+1})^- + \int_{\mathcal{A}_k^-} (\mu^{k+1})^+\right). \tag{7.5.15}$$

From (7.5.8), (7.5.6), (7.5.10), (7.5.12), and (7.5.13) we have

$$\begin{aligned} &|\delta x_{\mathcal{I}_k}| \le \rho_1 |\mu^k_{\mathcal{I}_k}| + \rho_2 |\delta x_{\mathcal{A}_k}|\\ &\le \rho_1\Big(|(\mu^k_{\mathcal{I}_k\cap\mathcal{A}_{k-1}^+})^-| + |(\mu^k_{\mathcal{I}_k\cap\mathcal{A}_{k-1}^-})^+|\Big) + \rho_2\Big(|\delta x_{\mathcal{A}_k^+}| + |\delta x_{\mathcal{A}_k^-}|\Big)\\ &\le \rho_1\Big(|(\mu^k_{\mathcal{I}_k\cap\mathcal{A}_{k-1}^+})^-| + |(\mu^k_{\mathcal{I}_k\cap\mathcal{A}_{k-1}^-})^+|\Big) + \rho_2\Big(|(x^k - \psi)^+_{\mathcal{A}_k^+\cap\mathcal{I}_{k-1}}| + \tfrac{1}{c}|(\mu^k_{\mathcal{A}_k^+\cap\mathcal{A}_{k-1}^-})^+|\\ &+|(x^k - \varphi)^-_{\mathcal{A}_k^-\cap\mathcal{I}_{k-1}}| + \tfrac{1}{c}|(\mu^k_{\mathcal{A}_k^-\cap\mathcal{A}_{k-1}^+})^-|\Big). \end{aligned}$$

This implies

$$|\delta x_{\mathcal{I}_k}| \le 2\max\left(\rho_1, \rho_2, \frac{\rho_2}{c}\right)\mathcal{M}(x^k, \mu^k). \tag{7.5.16}$$

From (7.5.8) further

$$\mu^{k+1}_{\mathcal{A}_k} - (\mu^k_{\mathcal{A}_k} - c\delta x_{\mathcal{A}_k}) = g, \tag{7.5.17}$$

where $g = (cI - A\delta x_{\mathcal{A}_k})\delta x_{\mathcal{A}_k} - A_{\mathcal{A}_k\mathcal{I}_k}A^{-1}_{\mathcal{A}_k}A_{\mathcal{I}_k\mathcal{A}_k}\delta x_{\mathcal{A}_k}$. By (7.5.9), (7.5.12), we have

$$\mu^k_{\mathcal{A}^+_k} - c\,\delta x_{\mathcal{A}^+_k} \ge 0.$$

Similarly, by (7.5.11), (7.5.13)

$$\mu^k_{\mathcal{A}^-_k} - c\,\delta x_{\mathcal{A}^-_k} \le 0.$$

Consequently,

$$\begin{aligned} |(\mu^{k+1}_{\mathcal{A}^+_k})^-| + |(\mu^{k+1}_{\mathcal{A}^-_k})^+| &\le |g_{\mathcal{A}_k}| \le (\rho_3 + \rho_5)|\delta x_{\mathcal{A}_k}| + \rho_4|\mu_{I_k}| \\ &\le 2\max\left(\rho_4, \rho_3 + \rho_5, \frac{\rho_3 + \rho_5}{c}\right)\mathcal{M}(x^k, \mu^k). \end{aligned} \tag{7.5.18}$$

By (7.5.15), (7.5.16), and (7.5.18)

$$\mathcal{M}(x^{k+1}, \mu^{k+1}) \le 2\max\left(\max\left(\rho_1, \rho_2, \frac{\rho_2}{c}\right), \max\left(\rho_4, \rho_3 + \rho_5, \frac{\rho_3 + \rho_5}{c}\right)\right)\mathcal{M}(x^k, \mu^k).$$

It follows that $\mathcal{M}(x^{k+1}, \mu^{k+1}) \le \rho^k\mathcal{M}(x^1, \mu^1)$ and if $\rho < 1$, then $M(x^k, \mu^k) \to 0$ as $k \to \infty$. From the estimates leading to (7.5.16) it follows that x^k is a Cauchy sequence. Moreover μ^k is a Cauchy sequence by (7.5.8). Hence there exist (x^*, μ^*) such that $\lim_{k\to\infty}(x^k, \mu^k) = (x^*, \mu^*)$. By Lebesgue's bounded convergence theorem and since $M(x^k, \mu^k) \to 0$, it follows that $\varphi \le x^* \le \psi$. Clearly $Ax^* + \mu^* = a$ and $(x^* - \psi)(x^* - \varphi)\mu^* = 0$ by (7.5.4). This last equation implies that $\mu^* = 0$ on $\mathcal{I}^* = \{\varphi < x^* < \psi\}$. It remains to show that $\mu^* \ge 0$ on $\mathcal{A}^{*,+} = \{x^* = \psi\}$ and $\mu^* \le 0$ on $\mathcal{A}^{*,-} = \{x^* = \varphi\}$. Let $s \in \mathcal{A}^{*,+}$ be such that $x^k(s)$ and $\mu^k(s)$ converge. Then $\mu^*(s) \ge 0$. If not, then $\mu^*(s) < 0$ and there exists $\bar{k}$ such that $\mu^k(s) + c(x^k(s) - \psi(s)) \le \frac{\mu^*(s)}{2} < 0$ for all $k \ge \bar{k}$. Then $s \in \mathcal{I}^k$ and $\mu^{k+1} = 0$ for $k \ge \bar{k}$, contradicting $\mu^*(s) < 0$. Analogously one shows that $\mu^* \le 0$ on $\mathcal{A}^{*,-}$. □

Conditions (7.5.6) and (7.5.7) are satisfied for additive perturbations of the operator cI, for example. This can be deduced from the following result.

Theorem 7.9. *Assume that $A = cI + K$ with $K \in \mathcal{L}(L^1(\Omega))$ and $\|K\| < c$ and that* (7.5.5), (7.5.6), (7.5.7) *are satisfied. If*

$$\bar{\rho} = 2\max\left(\max\left(\frac{\|K\|}{c}\rho_1, \rho_2\right), \max(\rho_3 + \rho_5, \rho_4), \frac{\rho_3 + \rho_5}{c}\right) < 1,$$

then the conclusions of the previous theorem are valid.

Proof. We follow the proof of Theorem 7.8 and eliminate the overestimate (7.5.14). Let $P = \{x^{k+1}_{\mathcal{I}_k} - \psi > 0\} \cap \mathcal{I}_k$. We find

$$x^{k+1} - \psi \begin{cases} \leq \delta x^k_{P\cap\mathcal{I}_{k-1}} & \text{on} \quad P\cap\mathcal{I}_{k-1}, \\ = \delta x^k_{P\cap\mathcal{A}^+_{k-1}} & \text{on} \quad P\cap\mathcal{A}^+_{k-1}, \\ = \delta x^k_{P\cap\mathcal{A}^-_{k-1}} + (\varphi - \psi)_{P\cap\mathcal{A}^-_{k-1}} & \text{on} \quad P\cap\mathcal{A}^-_{k-1}. \end{cases}$$

This estimate, together with $A^{-1} = \frac{1}{c}I - \frac{1}{c}KA^{-1}$, implies that

$$\int_{\mathcal{I}_k} (x^{k+1} - \psi)^+ \leq \int_P \delta x^k + \int_{P\cap A^-_{k-1}} \varphi - \psi$$

$$\leq \int_P A^{-1}_{\mathcal{I}_k}\mu^k_{\mathcal{I}_k} - \int_P A^{-1}_{\mathcal{I}_k} A_{\mathcal{I}_k\mathcal{A}_k}\delta x_{\mathcal{A}_k} + \int_{P\cap A^-_{k-1}} \varphi - \psi$$

$$= \frac{1}{c}\int_P \mu^k_{\mathcal{I}_k} + \int_{P\cap A^-_{k-1}} \varphi - \psi - \frac{1}{c}\int_P K_{\mathcal{I}_k} A^{-1}_{\mathcal{I}_k}\mu^k_{\mathcal{I}_k} - \int_P A^{-1}_{\mathcal{I}_k} A_{\mathcal{I}_k\mathcal{A}_k}\delta x_{\mathcal{A}_k}$$

$$\leq -\frac{1}{c}\int_P K_{\mathcal{I}_k} A^{-1}_{\mathcal{I}_k}\mu^k_{\mathcal{I}_k} - \int_P A^{-1}_{\mathcal{I}_k} A_{\mathcal{I}_k\mathcal{A}_k}\delta x_{\mathcal{A}_k},$$

and hence

$$\int_{\mathcal{I}_k} (x^{k+1} - \psi)^+ \leq \frac{\|K\|}{c}\rho_1|\mu^k_{\mathcal{I}_k}| + \rho_2|\delta x_{\mathcal{A}_k}|.$$

An analogous estimate can be obtained for $\int_{\mathcal{I}_k}(x^{k+1} - \varphi)^-$ and we find

$$\int_{\mathcal{I}_k} (x^{k+1} - \psi)^+ \int_{\mathcal{I}_k} (x^{k+1} - \varphi)^- \leq \frac{\|K\|}{c}\rho_1|\mu^k_{\mathcal{I}_k}| + \rho_2|\delta x_{\mathcal{A}_k}|.$$

We can now proceed as in the proof of Theorem 7.8. ☐

Example 7.10. We apply Theorem 7.9, $A = I + K \in \mathcal{L}(L^1(\Omega))$. By Neumann series arguments we find $\bar{\rho} = 2\max(\frac{\gamma}{1-\gamma}, \max(\gamma + \frac{\gamma^2}{1-\gamma}, \frac{\gamma}{1-\gamma})) = \frac{\gamma}{1-\gamma}$, where $\gamma = \|K\|$, and $\bar{\rho} < 1$ if $\|K\| < \frac{1}{3}$. If $A = I + K$ is replaced by $A = cI + K$, then $\bar{\rho} < 1$ if $\gamma < \frac{c}{2c+1}$, in case $c \geq 1$, and $\bar{\rho} < 1$ if $\frac{c^2}{c+2}$, in case $c \leq 1$.

Example 7.11. Here we consider the finite-dimensional case $A = I + K \in \mathbb{R}^{n\times n}$, where $\mathbb{R}^n$ is endowed with the ℓ^1-norm. Again Theorem 7.9 is applicable and $\bar{\rho} < 1$ if $\|K\| < \frac{1}{3}$, where $\|\cdot\|$ denotes the matrix-norm subordinate to the ℓ^1-norm of $\mathbb{R}^n$. Recall that this norm is given by the maximum over the column sums of the absolute values of the matrix.

7.6 Nonlinear control problems with bilateral constraints

In this section we consider a special case of bilaterally constrained problems that were already investigated in the previous section, where the operator A is of the form T^*T. This will allow us to obtain improved sufficient conditions for global convergence of the primal-dual active set algorithm. The primary motivation for such problems are optimal control problems, and we therefore slightly change the notation to that which is more common in optimal control.

Let U and Y be Hilbert spaces with $U = L^2(\Sigma)$, where Σ is a bounded measurable set in $\mathbb{R}^d$, and let $T : U \to Y$ be a, possibly nonlinear, continuously differentiable, injective, mapping with Fréchet derivative denoted by T'. Further let $\varphi, \psi \in U$ with $\varphi < \psi$ a.e. in Σ. For $\alpha > 0$ and $z \in Y$ consider

$$\min_{\varphi \le u \le \psi} J(u) = \frac{1}{2} |T(u) - z|_Y^2 + \frac{\alpha}{2} |u|_U^2. \tag{7.6.1}$$

The necessary optimality condition for (7.6.1) is given by

$$\begin{cases} \alpha u + T'(u)^*(T(u) - z) + \mu = 0, \\ \mu = \max(0, \mu + \alpha(u - \psi)) + \min(0, \mu + \alpha(u - \varphi)), \end{cases} \tag{7.6.2}$$

where $(u, \mu) \in U \times U$, and max as well as min are interpreted as pointwise a.e. operations. If the upper or lower constraint are not present, we can set $\psi = \infty$ or $\varphi = -\infty$.

Example 7.12. Let $\Omega \subset \mathbb{R}^n$ be a bounded domain in $\mathbb{R}^n$ with Lipschitz continuous boundary Γ, let $\widetilde{\Omega} \subset \Omega$, $\widetilde{\Gamma} \subset \Gamma$ be measurable subsets, and consider the optimal control problem

$$\min_{\varphi \le u \le \psi} \frac{1}{2} |y - z|_Y^2 + \frac{\alpha}{2} |u|_U^2 \text{ subject to}$$

$$(\nabla y, \nabla v)_\Omega + (y, v)_\Omega = (u, v)_{\widetilde{\Gamma}} \quad \text{for all} \quad v \in H^1(\Omega), \tag{7.6.3}$$

where $z \in L^2(\tilde{\Omega})$. We set $Y = L^2(\widetilde{\Omega})$ and $U = L^2(\widetilde{\Gamma})$. Define $L\,u = y$ with $L\colon L^2(\widetilde{\Gamma}) \to L^2(\Omega)$ as the solution operator to the inhomogeneous Neumann boundary value problem (7.6.3) and set $T = R_{\widetilde{\Omega}} L : L^2(\widetilde{\Gamma}) \to L^2(\widetilde{\Omega})$, where $R_{\widetilde{\Omega}} : L^2(\Omega) \to L^2(\widetilde{\Omega})$ denotes the canonical restriction operator. Then $T^*\colon L^2(\widetilde{\Omega}) \to L^2(\widetilde{\Gamma})$ is given by

$$T^* = R_{\widetilde{\Gamma}} L^* \, E_{\widetilde{\Omega}}$$

with $R_{\widetilde{\Gamma}}$ the restriction operator to $\widetilde{\Gamma}$ and $E_{\widetilde{\Omega}}$ the extension-by-zero operator from $\widetilde{\Omega}$ to Ω. Further the adjoint $L^*\colon L^2(\Omega) \to L^2(\Gamma)$ of L is given by $L^*\tilde{w} = \tau_\Gamma p$, where τ_Γ is the Dirichlet trace operator from $H^1(\Omega)$ to $L^2(\Gamma)$ and p is the solution to

$$(\nabla p, \nabla w)_\Omega + (p, w)_\Omega = (\tilde{w}, w)_\Omega \quad \text{for all} \quad w \in H^1(\Omega). \tag{7.6.4}$$

The fact that L and L^* are adjoint to each other follows by setting $v = p$ in (7.6.3) and $w = y$ in (7.6.4).

We next specify the primal-dual active set algorithm for (7.6.1). The iteration index is denoted by k and an initial choice (u^0, μ^0) is assumed to be available.

Primal-Dual Active Set Algorithm.

(i) Given (u^k, μ^k), determine
$\mathcal{A}_k^+ = \{\mu^k + \alpha(u^k - \psi) > 0\}$,
$\mathcal{I}_k = \{\mu^k + \alpha(u^k - \psi) \le 0 \le \mu^k + \alpha(u^k - \varphi)\}$,
$\mathcal{A}_k^- = \{\mu^k + \alpha(u^k - \varphi) < 0\}$.

(ii) Determine (u^{k+1}, μ^{k+1}) from

$$u^{k+1} = \psi \text{ on } \mathcal{A}_k^+, \quad u^{k+1} = \varphi \text{ on } \mathcal{A}_k^-, \quad \mu^{k+1} = 0 \text{ on } \mathcal{I}_k,$$

and

$$\alpha\, u^{k+1} + T'(u^{k+1})^*(T(u^{k+1}) - z) + \mu^{k+1} = 0. \tag{7.6.5}$$

Note that the equations for (u^{k+1}, μ^{k+1}) in step (ii) of the algorithm constitute the necessary optimality condition for the auxiliary problem

$$\begin{cases} \min \frac{1}{2}\,|T(u) - z|_Y^2 + \frac{\alpha}{2}\,|u|_U^2 \quad \text{over } u \in U \\ \text{subject to} \quad u = \psi \text{ on } A_k^+,\ u = \varphi \text{ on } A_k^-. \end{cases} \tag{7.6.6}$$

The analysis in this section relies on the fact that (7.6.2) can be equivalently expressed as

$$\begin{cases} y = T(u), \\ \lambda = -T'(u)^*(y - z), \\ \alpha\, u - \lambda + \mu = 0, \\ \mu = \max(0, \mu + (u - \psi)) + \min(0, \mu + (u - \varphi)), \end{cases} \tag{7.6.7}$$

where $\lambda = -T'(u)^*(T(u) - z)$ is referred to as the adjoint state. Analogously, for $u^{k+1} \in U$, setting $y^{k+1} = T(u^{k+1})$, $\lambda^{k+1} = -T'(u^{k+1})^*(T(u^{k+1}) - z)$, (7.6.5) can equivalently be expressed as

$$\begin{cases} y^{k+1} = T(u^{k+1}), \text{ where} \\ u^{k+1} = \begin{cases} \psi & \text{on } \mathcal{A}_k^+, \\ \frac{1}{\alpha}\lambda^{k+1} & \text{on } \mathcal{I}_k, \\ \varphi & \text{on } \mathcal{A}_k^-, \end{cases} \\ \lambda^{k+1} = -T'(u^{k+1})^*(y^{k+1} - z), \\ \alpha\, u^{k+1} - \lambda^{k+1} + \mu^{k+1} = 0. \end{cases} \tag{7.6.8}$$

In what follows we give conditions which guarantee convergence of the primal-dual active set strategy for linear and certain nonlinear operators T from arbitrary initial data. The convergence proof is based on an appropriately defined functional which decays when evaluated along the iterates of the algorithm. An a priori estimate for the adjoint variable λ in (7.6.8) will play an essential role.

To specify the condition alluded to in the above let us consider two consecutive iterates of the algorithm. For every $k = 1, 2, \ldots,$ the sets $\mathcal{A}_k^+$, $\mathcal{A}_k^-$, and $\mathcal{I}_k$ give a mutually disjoint decomposition of Σ. According to (i) and (ii) in the form (7.6.8) we find

$$u^{k+1} - u^k = \begin{cases} R^k_{\mathcal{A}^+} & \text{on } \mathcal{A}_k^+, \\ \frac{1}{\alpha}(\lambda^{k+1} - \lambda^k) + R^k_{\mathcal{I}} & \text{on } \mathcal{I}_k, \\ R^k_{\mathcal{A}^-} & \text{on } \mathcal{A}_k^-, \end{cases} \tag{7.6.9}$$

where the residual R^k is given by

$$R^k_{\mathcal{A}^+} = \begin{cases} 0 & \text{on } \mathcal{A}_{k-1}^+ \cap \mathcal{A}_k^+, \\ \psi - \frac{1}{\alpha}\lambda^k = \psi - u^k < 0 & \text{on } \mathcal{I}_{k-1} \cap \mathcal{A}_k^+, \\ \psi - \varphi < \frac{1}{\alpha}\mu_k & \text{on } \mathcal{A}_{k-1}^- \cap \mathcal{A}_k^+, \end{cases} \tag{7.6.10}$$

$$R^k_{\mathcal{I}} = \begin{cases} \frac{1}{\alpha}\mu^k = \frac{1}{\alpha}\lambda^k - \psi \le 0 & \text{on } \mathcal{A}_{k-1}^+ \cap \mathcal{I}_k, \\ 0 & \text{on } \mathcal{I}_{k-1} \cap \mathcal{I}_k, \\ \frac{1}{\alpha}\mu^k = \frac{1}{\alpha}\lambda^k - \varphi \ge 0 & \text{on } \mathcal{A}_{k-1}^- \cap \mathcal{I}_k, \end{cases} \tag{7.6.11}$$

$$R^k_{\mathcal{A}^-} = \begin{cases} \varphi - \psi > \frac{1}{\alpha}\mu_k & \text{on } \mathcal{A}_{k-1}^+ \cap \mathcal{A}_k^-, \\ \varphi - \frac{1}{\alpha}\lambda^k = \varphi - u^k > 0 & \text{on } \mathcal{I}_{k-1} \cap \mathcal{A}_k^-, \\ 0 & \text{on } \mathcal{A}_{k-1}^- \cap \mathcal{A}_k^-. \end{cases} \tag{7.6.12}$$

Here R^k denotes the function defined on Ω whose restrictions to $\mathcal{A}_k^+$, $\mathcal{I}_k$, $\mathcal{A}_k^-$ coincide with $R^k_{\mathcal{A}^+}$, $R^k_{\mathcal{I}}$, and $R^k_{\mathcal{A}^-}$.

We shall utilize the following a priori estimate:

$$\begin{cases} \text{There exists } \rho < \alpha \text{ such that} \\ |\lambda^{k+1} - \lambda^k|_U < \rho\, |R^k|_U \text{ for every } k = 1, 2, \ldots. \end{cases} \tag{7.6.13}$$

Sufficient conditions for (7.6.13) will be given at the end of this section. The convergence proof will be based on the following merit functional $M\colon U \times U \to R$ given by

$$M(u, \mu) = \alpha^2 \int_\Sigma (|(u - \psi)^+|^2 + |(\varphi - u)^+|^2)\, dx + \int_{\mathcal{A}^+(u)} |\mu^-|^2\, dx + \int_{\mathcal{A}^-(u)} |\mu^+|^2\, dx,$$

where $\mathcal{A}^+(u) = \{x : u \ge \psi\}$ and $\mathcal{A}^-(u) = \{x : u \le \varphi\}$. Note that the iterates $(u^k, \mu^k) \in U \times U$ satisfy

$$\mu^k(u^k - \psi)(\varphi - u^k)(x) = 0 \text{ for a.e. } x \in \Sigma, \tag{7.6.14}$$

and hence at most one of the integrands of $M(u^k, \mu^k)$ can be strictly positive at $x \in \Sigma$.

Theorem 7.13. *Assume that* (7.6.13) *holds for the iterates of the primal-dual active set strategy. Then* $M(u^{k+1}, \mu^{k+1}) \le \alpha^{-2}\rho^2 M(u^k, \mu^k)$ *for every* $k = 1, \ldots.$ *Moreover there exist* $(u^*, \mu^*) \in U \times U$*, such that* $\lim_{k\to\infty}(u^k, \mu^k) = (u^*, \mu^*)$ *and* (u^*, μ^*) *satisfies* (7.6.2).

Proof. From (7.6.8) we have

$$\mu^{k+1} = \lambda^{k+1} - \alpha\psi \quad \text{on } \mathcal{A}_k^+,$$

$$u^{k+1} = \frac{1}{\alpha}\lambda^{k+1} \quad \text{on } \mathcal{I}_k,$$

$$\mu^{k+1} = \lambda^{k+1} - \alpha\varphi \quad \text{on } \mathcal{A}_k^-.$$

Using step (ii) of the algorithm in the form of (7.6.8) implies that

$$\mu^{k+1} = \lambda^{k+1} - \lambda^k + \lambda^k - \alpha\psi = \lambda^{k+1} - \lambda^k + \begin{cases} \mu^k > 0 & \text{on } \mathcal{A}_{k-1}^+ \cap \mathcal{A}_k^+, \\ \alpha(u^k - \psi) > 0 & \text{on } \mathcal{I}_{k-1} \cap \mathcal{A}_k^+, \\ \alpha u^k + \mu^k - \alpha\psi \geq 0 & \text{on } \mathcal{A}_{k-1}^- \cap \mathcal{A}_k^+, \end{cases}$$

and therefore

$$|\mu^{k+1,-}(x)| \leq |\lambda^{k+1}(x) - \lambda^k(x)| \text{ for } x \in \mathcal{A}_k^+. \tag{7.6.15}$$

Analogously one derives

$$|\mu^{k+1,+}(x)| \leq |\lambda^{k+1}(x) - \lambda^k(x)| \text{ for } x \in \mathcal{A}_k^-. \tag{7.6.16}$$

Moreover

$$\begin{aligned} u^{k+1} - \psi &= \frac{1}{\alpha}(\lambda^{k+1} - \lambda^k + \lambda^k) - \psi \\ &= \frac{1}{\alpha}(\lambda^{k+1} - \lambda^k) + \begin{cases} \frac{1}{\alpha}\mu^k \leq 0 & \text{on } \mathcal{A}_{k-1}^+ \cap \mathcal{I}_k, \\ u^k - \psi \leq 0 & \text{on } \mathcal{I}_{k-1} \cap \mathcal{I}_k, \\ \frac{1}{\alpha}\mu^k + u - \psi \leq 0 & \text{on } \mathcal{A}_{k-1}^- \cap \mathcal{I}_k, \end{cases} \end{aligned}$$

which implies that

$$|(u^{k+1} - \psi)^+(x)| \leq \frac{1}{\alpha}|\lambda^{k+1}(x) - \lambda^k(x)| \text{ for } x \in \mathcal{I}_k. \tag{7.6.17}$$

Analogously one derives that

$$|(\varphi - u^{k+1})^+(x)| \leq \frac{1}{\alpha}|\lambda^{k+1}(x) - \lambda^k(x)| \text{ for } x \in \mathcal{I}_k. \tag{7.6.18}$$

Due to (ii) of the algorithm we have that

$$(u^{k+1} - \psi)^+ = (\varphi - u^{k+1})^+ = 0 \text{ on } \mathcal{A}_k^+ \cup \mathcal{A}_k^-,$$

which, together with (7.6.17)–(7.6.18), implies that

$$|(u^{k+1} - \psi)^+(x)| + |(\varphi - u^{k+1})^+(x)| \leq \frac{1}{\alpha}|\lambda^{k+1}(x) - \lambda^k(x)| \text{ for } x \in \Sigma. \tag{7.6.19}$$

From (7.6.14)–(7.6.16) and since $\varphi < \psi$ a.e. on Σ we find

$$|\mu^{k+1,-}(x)| \leq |\lambda^{k+1}(x) - \lambda^k(x)| \text{ for } x \in \mathcal{A}^+(u^{k+1}) \tag{7.6.20}$$

and

$$|\mu^{k+1,+}(x)| \leq |\lambda^{k+1}(x) - \lambda^k(x)| \text{ for } x \in \mathcal{A}^-(u^{k+1}). \tag{7.6.21}$$

Combining (7.6.19)–(7.6.21) implies that

$$M(u^{k+1}, \mu^{k+1}) \leq \int_\Sigma |\lambda^{k+1}(x) - \lambda^k(x)|^2 dx. \tag{7.6.22}$$

Since (7.6.13) is supposed to hold we have

$$M(u^{k+1}, \mu^{k+1}) \leq \rho^2 |R^k|_U^2.$$

Moreover, from the definition of R^k we deduce that

$$|R^k|_U^2 \leq \alpha^{-2} M(u^k, \mu^k), \tag{7.6.23}$$

and consequently

$$M(u^{k+1}, \mu^{k+1}) \leq \alpha^{-2}\rho^2 M(u^k, \mu^k) \text{ for } k = 1, 2, \dots. \tag{7.6.24}$$

From (7.6.13), (7.6.23), (7.6.24) it follows that $|\lambda^{k+1} - \lambda^k|_U \leq (\frac{\rho}{\alpha})^k \rho |R^0|_U$. Thus there exists $\lambda^* \in U$ such that $\lim_{k\to\infty} \lambda^k = \lambda^*$.

Note that for $k \geq 1$

$$\mathcal{A}_k^+ = \{x : \lambda^k(x) > \alpha\, \psi(x)\}, \qquad \mathcal{I}_k = \{x : \alpha\, \varphi(x) \leq \lambda^k(x) \leq \alpha\, \psi(x)\},$$
$$\mathcal{A}_k^- = \{x : \lambda^k(x) < \alpha\, \varphi(x)\},$$

and hence

$$\mu^{k+1} = \max(0, \lambda^k - \alpha\, \psi) + \min(0, \lambda^k - \alpha\, \varphi) + (\lambda^{k+1} - \lambda^k)\chi_{A_k^+ \cup A_k^-}.$$

Since $\lim_{k\to\infty}(\lambda^{k+1} - \lambda^k) = 0$ and $\lim_{k\to\infty} \lambda^k$ exists, it follows that there exists $\mu^* \in U$ such that $\lim_{k\to\infty} \mu^k = \mu^*$, and

$$\mu^* = \max(0, \lambda^* - \alpha\, \psi) + \min(0, \lambda^* - \alpha\, \varphi). \tag{7.6.25}$$

From the last equation in (7.6.8) it follows that there exists u^* such that $\lim_{k\to\infty} u^k = u^*$ and $\alpha\, u^* - \lambda^* + \mu^* = 0$. Combined with (7.6.25) the triple (u^*, μ^*) satisfies the complementarity condition given by the second equation in (7.6.2). Passing to the limit with respect to k in (7.6.5) we obtain that the first equation in (7.6.2) is satisfied by (u^*, μ^*). □

We turn to the discussion of (7.6.13) and consider the linear case first.

Proposition 7.14. *If T is linear and $\|T\|^2_{\mathcal{L}(U,Y)} < \alpha$, then* (7.6.13) *holds.*

Proof. From (7.6.8) and (7.6.9) we have, with $\delta u = u^{k+1} - u^k$, $\delta y = y^{k+1} - y^k$, and $\delta\lambda = \lambda^{k+1} - \lambda^k$,

$$\begin{cases} \delta u = R^k + \frac{1}{\alpha}\, \delta\lambda\, \chi_{\mathcal{I}_k}, \\ T\, \delta u = \delta y, \\ T^*\, \delta y + \delta\lambda = 0. \end{cases} \tag{7.6.26}$$

Taking the inner product in L^2 with $\delta\lambda$ in the first equation we arrive at

$$(\delta u, \delta\lambda) \le (R^k, \delta\lambda),$$

and taking the inner product with δu in the third equation,

$$|\delta y|^2 + (\delta\lambda, \delta u) = 0.$$

Combining these two relations implies that

$$|\delta y|^2 \le (R^k, \delta\lambda).$$

Utilizing the third equation in (7.6.26) in this last inequality and the fact that the norms of T and T^* coincide, we have $|\delta\lambda| \le \|T^*\|^2 |R^k|$, from which the desired estimate follows. $\square$

We now turn to a particular case when (7.6.13) holds for a nonlinear operator T. Let $\Sigma = \Omega$ be a bounded domain in $\mathbb{R}^n$, $n = 2$ or 3, with smooth boundary $\partial\Omega$. Further let $\phi\colon \mathbb{R} \to \mathbb{R}$ be a monotone mapping with locally Lipschitzian derivative, satisfying $\phi(0) = 0$, and such that the substitution operator determined by ϕ maps $H^1(\Omega)$ into $L^2(\Omega)$. We choose $U = Y = L^2(\Omega)$ and define $T(u) = y$ as the solution operator to

$$\begin{cases} -\Delta y + \phi(y) = u & \text{in } \Omega, \\ y = 0 & \text{on } \partial\Omega, \end{cases} \tag{7.6.27}$$

where Δ denotes the Laplacian. The adjoint variable λ is the solution to

$$\begin{cases} \Delta\lambda + \phi'(y)\lambda = -(y - z) & \text{in } \Omega, \\ \lambda = 0 & \text{on } \partial\Omega. \end{cases} \tag{7.6.28}$$

Let (u^0, μ^0) be an arbitrary initialization and let $\widetilde{U} = \{u^k : k = 1, 2, \ldots\}$ denote the set of iterates generated by the primal-dual active set algorithm. Since these iterates are solutions to the auxiliary problems (7.6.6), it follows that for every $\bar{\alpha} > 0$ the set $\widetilde{U}$ is bounded in $L^2(\Omega)$ uniformly with respect to $\alpha \ge \bar{\alpha}$.

By monotone operator theory and regularity theory of elliptic partial differential equations it follows that the set of primal states $\{y^k = y(u^k) : k = 1, 2, \ldots\}$ and adjoint states $\{\lambda^k = \lambda(y(u^k)) : k = 1, 2, \ldots\}$ are bounded subsets of $L^\infty(\Omega)$; see [Tr]. Let C denote this bound and let L_C denote the Lipschitz constant of ϕ on the ball $B_C(0)$ with center 0 and radius C in $\mathbb{R}$. Denote by $H_0^1(\Omega) = \{u \in H^1(\Omega) : u = 0 \text{ on } \partial\Omega\}$ the Hilbert space endowed with norm $|\nabla u|_{L^2}$ and let κ stand for the embedding constant from $H_0^1(\Omega)$ into $L^2(\Omega)$.

Proposition 7.15. *Assume that* $0 < \frac{(1+C\,L_C)\,\kappa^4}{\alpha-(1+C\,L_C)\,\kappa^4} < 1$, *where* $\alpha \ge \bar{\alpha}$. *Then* (7.6.13) *holds for the mapping T determined by the solution operator to* (7.6.27).

Proof. From (7.6.8) we have

$$-\Delta(y^{k+1} - y^k) + \phi(y^{k+1}) - \phi(y^k) = \frac{1}{\alpha}(\lambda^{k+1} - \lambda^k)\chi_{I^{k+1}} + R^k, \tag{7.6.29}$$

$$-\Delta(\lambda^{k+1} - \lambda^k) + \phi'(y^{k+1})(\lambda^{k+1} - \lambda^k) + (\phi'(y^{k+1}) - \phi'(y^k))\lambda^k + y^{k+1} - y^k = 0, \tag{7.6.30}$$

both Laplacians with homogeneous Dirichlet boundary conditions. Taking the inner product of (7.6.29) with $y^{k+1} - y^k$ we have, using monotonicity of ϕ,

$$|y^{k+1} - y^k|_1 \le \frac{\kappa^2}{\alpha} |(\lambda^{k+1} - \lambda^k)|_1 + |R^k|_{-1}, \tag{7.6.31}$$

where $|\cdot|_1$ and $|\cdot|_{-1}$ denote the norms in $H_0^1(\Omega)$ and $H^{-1}(\Omega)$, respectively. Note that $\phi'(y^{k+1}) \ge 0$. Hence from (7.6.30) we find

$$|\lambda^{k+1} - \lambda^k|_1^2 \le C\,L_C |y^{k+1} - y^k|_{L^2} |\lambda^{k+1} - \lambda^k|_{L^2} + |(y^{k+1} - y^k, \lambda^{k+1} - \lambda^k)|$$

$$\le (1 + C\,L_C)\,\kappa^2 |y^{k+1} - y^k|_1 |\lambda^{k+1} - \lambda^k|_1.$$

Thus,

$$|\lambda^{k+1} - \lambda^k|_1 \;\le\; (1 + C\,L_C)\,\kappa^2\,|y^{k+1} - y^k|_1$$

and hence from (7.6.31)

$$|y^{k+1} - y^k|_1 \;\le\; \frac{\alpha}{\alpha - (1 + C\,L_C)\,\kappa^4} |R^k|_{-1}.$$

It thus follows that

$$|\lambda^{k+1} - \lambda^k|_{L^2} \;\le\; \frac{\alpha\,(1 + C\,L_C)\,\kappa^4}{\alpha - (1 + C\,L_C)\,\kappa^4}\;|R^k|_{L^2}.$$

This implies (7.6.13) with

$$\rho \;=\; \frac{(1 + C\,L_C)\,\kappa^4}{\alpha - (1 + C\,L_C)\,\kappa^4}. \qquad \square$$

Chapter 8

Semismooth Newton Methods I

8.1 Introduction

In this chapter we study semismooth Newton methods for solving nonlinear nonsmooth equations. These investigations are motivated by complementarity problems, variational inequalities, and optimal control problems with control or state constraints, for example. The operator equation for which we desire to find a solution is typically Lipschitz continuous but not C^1 regular. We shall also establish the relationship between the semismooth Newton method and the primal-dual active set method that was discussed in Chapter 7.

Since semismooth Newton methods are not widely known even for finite-dimensional problems, we consider the finite-dimensional case before we turn to problems in infinite dimensions. In fact, these two cases are distinctly different. In finite dimensions we have Rademacher's theorem, which states that every locally Lipschitz continuous function is differentiable almost everywhere. This result has no counterpart for functions between infinite-dimensional function spaces.

As an example consider the nonlinear complementarity problem

$$g(x) \leq 0, \quad x \leq \psi, \quad \text{and} \quad (g(x), x - \psi)_{\mathbb{R}^n} = 0,$$

where $g : \mathbb{R}^n \to \mathbb{R}^n$ and $\psi \in \mathbb{R}^n$. It can be expressed equivalently as the problem of finding a root to the following equation:

$$F(x) = g(x) + \max(0, -g(x) + x - \psi) = \max(g(x), x - \psi) = 0, \tag{8.1.1}$$

where the max operation just like the inequalities must be interpreted componentwise. Note that F is a locally Lipschitz continuous function if g is locally Lipschitzian, but it is not C^1 if g is C^1. A function is called locally Lipschitz continuous if it is Lipschitz continuous on every bounded subset of its domain.

Let us introduce some of the key concepts for the finite-dimensional case. For a locally Lipschitz continuous function $F : \mathbb{R}^m \to \mathbb{R}^n$ let D_F denote the set of points at which F is differentiable. For $x \in \mathbb{R}^m$ we define $\partial_B F(x)$ as

$$\partial_B F(x) = \left\{ J : J = \lim_{x_i \to x,\ x_i \in D_F} \nabla F(x_i) \right\}, \tag{8.1.2}$$

and we denote by $\partial F(x)$ the generalized derivative at x introduced by Clarke [Cla], i.e.,

$$\partial F(x) = co\ \partial_B F(x), \tag{8.1.3}$$

where co stands for the convex hull.

A generalized Newton iteration for solving the nonlinear equation $F(x) = 0$ with $F : \mathbb{R}^n \to \mathbb{R}^n$ can be defined by

$$x^{k+1} = x^k - V_k^{-1} F(x^k), \text{ where } V_k \in \partial_B F(x^k). \tag{8.1.4}$$

The reason for using $\partial_B F$ rather than ∂F in (8.1.4) is the following: For the convergence analysis we shall require that all $V \in \partial_B F(x^*)$ are nonsingular, where x^* is the thought solution to $F(x) = 0$. This is more readily satisfied for $\partial_B F$ than for ∂F, as can be seen from $F(x) = |x|$, for example. In this case $0 \in \partial F(0)$ but $0 \notin \partial_B F$.

We also introduce the coordinatewise operation

$$\partial_b F(x) = \otimes_{i=1}^m \partial_B F_i(x),$$

where F_i is the ith coordinate of F. From the definition of $\partial_B F(x)$ it follows that $\partial_B F(x) \subset \partial_b F(x)$. For F given in (8.1.1) we have

$$\partial_B F(x) = \partial_b F(x)$$

if $-g'(x) + I$ is surjective. Moreover, in Section 8.4 it will be shown that if we select

$$V(x)\, d = \begin{cases} d & \text{if } -g(x) + x - \psi > 0, \\ g'(x)d & \text{if } -g(x) + x - \psi \le 0, \end{cases}$$

then the generalized Newton method reduces to the primal-dual active set method.

Local convergence of $\{x^k\}$ to x^*, a solution of $F(x) = 0$, is based on the following concepts. The generalized Jacobians $V^k \in \partial_B F(x_k)$ are selected so that their inverses $V(x_k)^{-1}$ are uniformly bounded and that they satisfy the condition

$$|F(x^* + h) - F(x^*) - V h| = o(|h|), \tag{8.1.5}$$

where $V = V(x^* + h) \in \partial_B F(x^* + h)$, for h in a neighborhood of x^*. Then from (8.1.5) with $h = x^k - x^*$ and $V_k = V(x^k)$ we have

$$|x^{k+1} - x^*| = |V_k^{-1}(F(x^k) - F(x^*) - V_k(x^k - x^*))| = o(|x^k - x^*|). \tag{8.1.6}$$

Thus, there exists a neighborhood $B(x^*, \rho)$ of x^* such that if $x^0 \in B(x^*, \rho)$, then $x^k \in B(x^*, \rho)$ and x^k converges to x^* superlinearly. This discussion will be made rigorous for F mapping between finite-dimensional spaces, as above, as well as for the infinite-dimensional case. For the finite-dimensional case we shall rely on the notion of semismooth functions.

Definition 8.1. *$F : \mathbb{R}^m \to \mathbb{R}^n$ is called semismooth at x if F is locally Lipschitz at x and*

$$\lim_{V \in \partial F(x+th'),\ h' \to h,\ t \to 0^+} V h' \text{ exists for all } h \in \mathbb{R}^m. \tag{8.1.7}$$

Semismoothness was originally introduced by Miflin [Mif] for scalar-valued functions. Convex functions and real-valued C^1 functions are examples for such semismooth functions. Definition 8.1 is due to Qi and Sun [Qi, QiSu]. It will be shown that if F is semismooth at x^*, condition (8.1.5) holds. Due to the fact that the notion of Clarke derivative is not available in infinite dimensions, Definition 8.1 does not allow a direct generalization to the infinite-dimensional case. Rather, the notion of Newton differentiability related to the property expressed in (8.1.5) was developed in [CNQ, HiIK]. Alternatively, in [Ulb] the infinite-dimensional case of mappings into L^p spaces was treated by considering the superposition of mappings $F = \Psi(G)$, with G a C^1 mapping into L^r and Ψ the substitution operator $(\Psi y)(s) = \psi(y(s))$ for $y \in L^r$, where ψ is a semismooth function between finite-dimensional spaces.

As an alternative to (8.1.4) the increment d^k for a generalized Newton method $x^{k+1} = x^k + d^k$ can be defined as the solution to

$$F(x^k) + F'(x^k; d) = 0, \tag{8.1.8}$$

where $F'(x; d)$ denotes the directional derivative of F at x in direction d. Note that it may be a nontrivial task to solve (8.1.8) for d^k. This method was investigated in [Pan1, Pan2]. We shall return to (8.1.8) in the context of globalization of the method specified by (8.1.4).

In Section 8.2 we present the finite-dimensional theory for Newton's method for semismooth functions. Section 8.3 is devoted to the discussion of Newton differentiability and solving nonsmooth equations in Banach spaces. In Section 8.4 we exhibit the relationship between the primal-dual active set method and semismooth Newton methods. Section 8.5 is devoted to a class of nonlinear complementarity problems. In Section 8.6 we discuss applications where, for different reasons, semismooth Newton methods are not directly applicable but rather a regularization is necessary as, for instance, in the case of state-constrained optimal control problems.

8.2 Semismooth functions in finite dimensions

8.2.1 Basic concepts and the semismooth Newton algorithm

In this section we discuss properties of semismooth functions and analyze convergence of the generalized Newton method (8.1.4). We follow quite closely the work by Qi and Sun [QiSu]. First we describe the relationship of semismoothness to directional differentiability. A function $F : \mathbb{R}^m \to \mathbb{R}^n$ is called directionally differentiable at $x \in \mathbb{R}^m$ if

$$\lim_{t \to 0^+} \frac{F(x + t h) - F(x)}{t} = F'(x; h)$$

exists for all $h \in \mathbb{R}^d$. Further F is called Bouligand-differentiable (B-differentiable) [Ro4] at x if it is directionally differentiable at x and

$$\lim_{h \to 0} \frac{F(x + h) - F(x) - F'(x; h)}{|h|} = 0.$$

A locally Lipschitzian function F is B-differentiable at x if and only if it is directionally differentiable at x; see [Sha] and the references therein.

Theorem 8.2. (i) *Suppose that* $F : \mathbb{R}^m \to \mathbb{R}^n$ *is locally Lipschitz continuous and directionally differentiable at* x. *Then* $F'(x;\cdot)$ *is Lipschitz continuous and for every* h, *there exists a* $V \in \partial F(x)$ *such that*

$$F'(x;h) = Vh. \tag{8.2.1}$$

(ii) *If* F *is locally Lipschitz continuous, then the following statements are equivalent.*
(1) F *is semismooth at* x.
(2) F *is directionally differentiable at* x *and for every* $V \in \partial F(x+h)$,

$$Vh - F'(x;h) = o(|h|) \text{ as } h \to 0.$$

(3) $\lim_{x+h\in D_F,\, |h|\to 0} \frac{F'(x+h;h)-F'(x;h)}{|h|} = 0.$

Proof. (i) For $h,\ h' \in \mathbb{R}^m$ we have

$$|F'(x;h) - F'(x;h')| = \left| \lim_{t\to 0^+} \frac{F(x+th) - F(x+th')}{t} \right| \le L\,|h-h'|, \tag{8.2.2}$$

where L is the Lipschitz constant of F in a neighborhood of x. Thus $F'(x;\cdot)$ is Lipschitz continuous at x. Since

$$F(y) - F(x) \in co\ \partial F([x,y])(y-x) \tag{8.2.3}$$

for all $x, y \in \mathbb{R}^m$ (see [Cla, p. 72]), there exist a sequence $\{t_k\}$, with $t_k \to 0^+$, and $V_k \in co\ \partial F([x, x+t_k h])(y-x)$ such that

$$F'(x;h) = \lim_{k\to\infty} V_k h.$$

Since F is locally Lipschitz, the sequence $\{V_k\}$ is bounded, and there exists a subsequence of V_k, denoted by the same symbol, such that $V_k \to V$. Moreover ∂F is closed at x; i.e., $x_i \to x$ and $Z_i \to Z$, with $Z_i \in \partial F(x_i)$, imply that $Z \in \partial F(x)$ [Cla, p. 70], and hence $V \in \partial F(x)$. Thus $F'(x;h) = Vh$, as desired.

(ii) We turn to verify the equivalence of (1)–(3).

(1) → (2): First we show that $F'(x;h)$ exists and that

$$F'(x;h) = \lim_{V\in\partial F(x+th),\ t\to 0^+} Vh. \tag{8.2.4}$$

Since F is locally Lipschitz, $\{\frac{F(x+th)-F(x)}{t} : t \to 0\}$ is bounded. Thus there exists a sequence $t_i \to 0^+$ and $\ell \in \mathbb{R}^n$ such that

$$\lim_{i\to\infty} \frac{F(x+t_i h) - F(x)}{t_i} = \ell.$$

We argue that ℓ equals the limit in (8.2.4). By (8.2.3)

$$\frac{F(x+t_i h) - F(x)}{t_i} \in co\,(\partial F([x, x+t_i h]))\,h = co\,(\partial F([x, x+t_i h]\,h).$$

The Carathéodory theorem implies that for each i there exist $t_i^k \in [0, t_i]$, $\lambda_i^k \in [0, 1]$ with $\sum_{k=0}^n \lambda_i^k = 1$, and $V_i^k \in \partial F(x + t_i^k h)$, where $k = 0, \ldots, n$, such that

$$\frac{F(x + t_i h) - F(x)}{t_i} = \sum_{k=0}^{n} \lambda_i^k V_i^k h.$$

By passing to subsequences such that $\lim \lambda_i^k \to \lambda^k$, for $k = 0, \ldots, n$, we have

$$\ell = \sum_{k=0}^{n} \lim_{i\to\infty} \lambda_i^k \left(\lim_{i\to\infty} V_i^k h \right) = \sum_{k=0}^{n} \lambda^k \lim_{V\in\partial F(x+th),\, t\to 0^+} Vh = \lim_{V\in\partial F(x+th),\, t\to 0^+} Vh.$$

Next we prove that the limit in (8.2.4) is uniform for all h with $|h| = 1$. This implies (2). If the claimed uniform convergence in (8.2.4) does not hold, then there exists $\epsilon > 0$, and sequences $\{h^k\}$ in $\mathbb{R}^m$ with $|h^k| = 1$, $\{t_k\}$ with $t_k \to 0^+$, and $V_k \in \partial F(x + t_k h^k)$ such that

$$|V_k h^k - F'(x; h^k)| \geq 2\epsilon.$$

Passing to a subsequence we may assume that h^k converges to some $h \in \mathbb{R}^m$. By Lipschitz continuity of $F'(x; h)$ with respect to h we have

$$|V_k h^k - F'(x; h)| \geq \epsilon$$

for all k sufficiently large. This contradicts the semismoothness of F at x.

(2) $\to$ (1): Suppose that F is not semismooth at x. Then there exist $\epsilon > 0$, $h \in \mathbb{R}^m$, and sequences $\{h^k\}$ in $\mathbb{R}^m$ and $\{t_k\}$, and $h \in \mathbb{R}^m$ satisfying $h^k \to h$, $t_k \to 0^+$, and $V_k \in \partial F(x + t_k h^k)$ such that

$$|V_k h^k - F'(x; h)| \geq 2\epsilon.$$

Since $F'(x; \cdot)$ is Lipschitz continuous,

$$|V_k h^k - F'(x; h^k)| \geq \epsilon$$

for sufficiently large k. This contradicts assumption (2).

(2) $\to$ (3) follows from (8.2.1).

(3) $\to$ (2): For arbitrary $\epsilon > 0$ there exists $\delta > 0$ such that for all h with $|h| < \delta$ and $x + h \in D_F$

$$|F'(x + h; h) - F'(x; h)| \leq \epsilon\, |h|. \tag{8.2.5}$$

Let $V \in \partial F(x + h)$ and $|h| < \frac{\delta}{2}$ with $h \neq 0$. By (8.1.3)

$$Vh \in co \left\{ \lim_{h'\to h,\; x+h'\in D_F} F'(x + h'; h) \right\}.$$

By the Carathéodory theorem, there exist $\lambda^k \geq 0$ with $\sum_{k=0}^n \lambda^k = 1$ and h^k, $k = 1, \ldots,$ satisfying $x + h^k \in D_F$ and $|h^k - h| \leq \min(\frac{\delta}{2}, \frac{\epsilon|h|}{L}, |h|)$, where L is the Lipschitz constant of F near x, such that

$$\left| Vh - \sum_{k=0}^{n} \lambda^k F'(x + h^k; h) \right| \leq \epsilon |h|. \tag{8.2.6}$$

From (8.2.5) and (8.2.2) we find

$$\begin{aligned}\left|\sum_{k=0}^{n} \lambda^k F'(x+h^k;h) - F'(x;h)\right| \\ &\le \sum_{k=0}^{n} \lambda^k\big(|F'(x+h^k;h) - F'(x+h^k;h^k)| + |F'(x+h^k;h^k) - F'(x;h^k)| \\ &\qquad + |F'(x;h^k) - F'(x;h)|\big) \\ &\le \sum_{k=0}^{n} \lambda^k (2L|h^k - h| + \epsilon\,|h^k|) \le 4\epsilon\,|h|.\end{aligned}$$

It thus follows from (8.2.6) that

$$|Vh - F'(x+h;h)| \le 5\epsilon\,|h|.$$

Since $\epsilon > 0$ is arbitrary, (2) follows. □

Theorem 8.3. *Let $F : \mathbb{R}^n \to \mathbb{R}^n$ be locally Lipschitz continuous, let $x \in \mathbb{R}^n$, and suppose that all $V \in \partial_B F(x)$ are nonsingular. Then there exist a neighborhood N of x and a constant C such that for all $y \in N$ and $V \in \partial_B F(y)$*

$$|V^{-1}| \le C. \tag{8.2.7}$$

Proof. First, we claim that there exist a neighborhood N of x and a constant C such that for all $y \in D_F \cap N$, $\nabla F(y)$ is nonsingular and

$$|\nabla F(y)^{-1}| \le C. \tag{8.2.8}$$

If this claim is not true, then there exists a sequence $y^k \to x$, $y^k \in D_F$, such that either all $\nabla F(y^k)$ are singular or $|(\nabla F(y^k))^{-1}| \to \infty$. Since F is locally Lipschitz the set $\{\nabla F(y^k) : k = 1, \ldots\}$ is bounded. Thus there exists a subsequence of $\nabla F(y^k)$ that converges to some V. Then V must be singular and $V \in \partial_B F(x)$. This contradicts the assumption and there exists a neighborhood N of x such that (8.2.8) holds for all $y \in D_F \cap N$. Moreover (8.2.7) follows from (8.1.2), (8.2.8), and continuity of the norm. □

Lemma 8.4. *Suppose that $F : \mathbb{R}^n \to \mathbb{R}^n$ is semismooth at a solution x^* of $F(x) = 0$ and that all $V \in \partial_B F(x^*)$ are nonsingular. Then there exist a neighborhood N of x^* and $\epsilon : \mathbb{R}^+ \to \mathbb{R}^+$ with $\lim_{t\to 0^+} \epsilon(t) = 0$ monotonically such that*

$$|x - V^{-1}F(x) - x^*| \le \epsilon(|x - x^*|)|x - x^*|,$$

$$|F(x - V^{-1}F(x))| \le \epsilon(|x - x^*|)|F(x)|$$

for all $V \in \partial_B F(x)$ and $x \in N$.

Proof. From Theorem 8.3 there exist a neighborhood N of x^* and a constant C such that $|V^{-1}| \le C$ for all $V \in \partial_B F(x)$ with $x \in N$. Thus, if $x \in N$, then it follows from

Theorem 8.2 (2) and B-differentiability of F at x^*, which is implied by semismoothness of F at x^*, that

$$\begin{aligned}
&|x - V^{-1}F(x) - x^*| \le |V^{-1}||F(x) - F(x^*) - V(x - x^*)| \\
&\le |V^{-1}|\big(|F(x) - F(x^*) - F'(x^*; x - x^*)| + |F'(x^*; x - x^*) - V(x - x^*)|\big) \\
&\le \epsilon(|x - x^*|)|x - x^*|.
\end{aligned}$$

This implies the first claim. Let $\hat{x} = x - V^{-1}F(x)$. Since F is B-differentiable at x^* we have

$$|F(\hat{x})| \le |F'(x^*; \hat{x} - x^*)| + \epsilon(|\hat{x} - x^*|)|\hat{x} - x^*|.$$

From the first part of the theorem we obtain for a possibly redefined function ϵ

$$|F(\hat{x})| \le (L + \epsilon)\, \epsilon\, |x - x^*|,$$

where $\epsilon = \epsilon(|x - x^*|)$ and L is the Lipschitz constant of F at x^*. Since

$$\begin{aligned}
|x - x^*| &\le |\hat{x} - x| + |\hat{x} - x^*| \\
&\le |V^{-1}F(x)| + |\hat{x} - x^*| \le C|F(x)| + \epsilon\, |x - x^*|,
\end{aligned}$$

we find

$$|x - x^*| \le \frac{C}{1 - \epsilon}\, |F(x)|$$

for all x sufficiently close to x^* and hence

$$|F(\hat{x})| \le \frac{C\epsilon(L + \epsilon)}{1 - \epsilon} |F(x)|.$$

This implies the second claim. □

We are now prepared for the local superlinear convergence result that was announced in the introduction of this chapter.

Theorem 8.5 (Superlinear Convergence). *Suppose that $F : \mathbb{R}^n \to \mathbb{R}^n$ is semismooth at a solution x^* of $F(x) = 0$ and that all $V \in \partial_B F(x^*)$ are nonsingular. Then the iterates*

$$x^{k+1} = x^k - V_k^{-1}F(x^k), \quad V_k \in \partial_B F(x^k),$$

for $k = 0, 1, \ldots$ are well defined and converge to x^* superlinearly if x^0 is chosen sufficiently close to x^*. Moreover, $|F(x_k)|$ decreases superlinearly to 0.

Proof. We proceed by induction and suppose that $x^0 \in N$ and $\epsilon(|x^0 - x^*|) \le 1$ with N, ϵ defined in Lemma 8.4. Without loss of generality we may assume that N is a ball in $\mathbb{R}^m$. To verify the induction step suppose that $x^k \in N$. Then

$$|x^{k+1} - x^*| \le \epsilon(|x^k - x^*|)|x^k - x^*| \le |x^0 - x^*|.$$

This implies that $x^{k+1} \in N$ and superlinear convergence of $x^k \to x^*$. Superlinear convergence of $F(x^k)$ to 0 easily follows from the second part of Lemma 8.4. □

8.2.2 Globalization

We now discuss globalization of the semismooth Newton method (8.1.4). For this purpose we define the merit function θ by

$$\theta(x) = |F(x)|^2.$$

Throughout we assume that $F : \mathbb{R}^n \to \mathbb{R}^n$ is locally Lipschitz continuous and B-differentiable and that the following assumptions (8.2.9)–(8.2.11) hold:

$$S = \{x \in \mathbb{R}^n : |F(x)| \le |F(x^0)|\} \text{ is bounded.} \tag{8.2.9}$$

There exist $\bar{\sigma}, b > 0$ and a graph Φ from S to the nonempty subsets of $\mathbb{R}^n$ such that $\theta'(x; d) \le -\bar{\sigma}\theta(x)$ and $|d| \le b\,|F(x)|$ for all $d \in \Phi(x)$ and $x \in S$. (8.2.10)

Moreover Φ has the following closure property: $x_k \to \bar{x}$ and $d_k \to \bar{d}$ with $x_k \in S$ and $d_k \in \Phi(x_k)$ imply $\theta^o(\bar{x}; \bar{d}) \le -\bar{\sigma}\theta(\bar{x})$. (8.2.11)

Here $\theta^o(x; d)$ denotes the Clarke generalized directional derivative of θ at x in direction d (see [Cla]) which is defined by

$$\theta^o(x; d) = \limsup_{y \to x,\, t \to 0^+} \frac{\theta(y + t\,d) - \theta(y)}{t}.$$

It is assumed that $b > C$ is an arbitrarily large parameter. It serves the purpose that the iterates $\{d_k\}$ are uniformly bounded. With reference to the closure property of Φ given in (8.2.11), note that it is not required that $\bar{d} \in \Phi(\bar{x})$. Conditions (8.2.10) and (8.2.11) will be discussed in Section 8.2.3 below.

Following [HaPaRa, Qi] we next investigate a globalization strategy for the generalized Newton iteration (8.1.4) and introduce the following algorithm.

Algorithm G.

Let $\beta,\ \gamma \in (0, 1)$ and $\sigma \in (0, \bar{\sigma})$. Choose $x^0 \in \mathbb{R}^n$ and set $k = 0$. Given x^k with $F(x^k) \neq 0$. Then

(i) If there exists a solution h^k to

$$V_k\, h^k = -F(x^k)$$

with $|h^k| \le b|F(x^k)|$, and if further

$$|F(x^k + h^k)| < \gamma\, |F(x^k)|,$$

set $d^k = h^k$, $x^{k+1} = x^k + d^k$, $\alpha_k = 1$, and $m_k = 0$.

(ii) Otherwise choose $d^k \in \Phi(x^k)$ and let $\alpha_k = \beta^{m_k}$, where m_k is the first positive integer m for which

$$\theta(x^k + \beta^m\, d^k) - \theta(x^k) \le -\sigma\beta^m\, \theta(x^k).$$

Finally set $x^{k+1} = x^k + \alpha_k d^k$.

In (ii) the increment $d^k = h^k$ from (i) can be chosen if $\theta'(x^k; h^k) \le -\sigma\theta(x^k)$.

Theorem 8.6. *Suppose that $F : \mathbb{R}^n \to \mathbb{R}^n$ is locally Lipschitz and B-differentiable.*

(a) *Assume that* (8.2.9)–(8.2.11) *hold. Then the sequence $\{x^k\}$ generated by Algorithm G is bounded, it satisfies $|F(x^{k+1})| < |F(x^k)|$ for all $k \ge 0$, and each accumulation point x^* of $\{x^k\}$ satisfies $F(x^*) = 0$.*

(b) *If moreover for one such accumulation point*

$$|h| \le c\,|F'(x^*; h)| \quad \text{for all } h \in \mathbb{R}^n, \tag{8.2.12}$$

then the sequence x^k converges to x^*.

(c) *If in addition to the above assumptions F is semismooth at x^* and all $V \in \partial_B F(x^*)$ are nonsingular, then x^k converges to x^* superlinearly.*

Proof. (a) First we prove that for each $x \in S$ such that $\theta(x) \ne 0$ and d satisfying $\theta'(x; d) \le -\bar{\sigma}\theta(x)$, there exists a $\bar{\tau} > 0$ such that

$$\theta(x + \tau\, d) - \theta(x) \le -\sigma\,\tau\,\theta(x) \text{ for all } \tau \in [0, \bar{\tau}].$$

If this is not the case, then there exists a sequence $\tau_n \to 0^+$ such that

$$\theta(x + \tau_n\, d) - \theta(x) > -\sigma\,\tau_n\,\theta(x).$$

Dividing both sides by τ_n and letting $n \to \infty$, we have by (8.2.10)

$$-\bar{\sigma}\theta(x) \ge \theta'(x; d) \ge -\sigma\,\theta(x).$$

Since $\sigma < \bar{\sigma}$, this shows $\theta(x) = 0$, which contradicts the assumption $\theta(x) \ne 0$. Hence for each level k at which $d^k \in \Phi(x^k)$ is chosen according to the second alternative in Algorithm G, there exists $m^k < \infty$ and $\alpha_k > 0$ such that $|F(x^{k+1})| < |F(x^k)|$. By construction the iterates therefore satisfy $|F(x^{k+1})| < |F(x^k)|$ for each $k \ge 0$.

Assume first that $\limsup \alpha_k > 0$. If the first alternative in Algorithm G with $\alpha_k = 1$ occurs infinitely many times, then using the fact that $\gamma < 1$ we find that $\lim_{k\to 0}\theta(x^k) = 0$. Otherwise, for all k sufficiently large

$$0 \le \theta(x^{k+1}) \le (1 - \sigma\alpha_k)\theta(x^k) \le \theta(x^k).$$

Thus $\theta(x^k)$ is monotonically decreasing, bounded below by 0, and hence convergent. Therefore $\lim_{k\to\infty}(\theta(x^{k+1}) - \theta(x^k)) = 0$ and consequently $\lim_{k\to\infty}\alpha_k\theta(x^k) = 0$. Thus $\limsup \alpha_k > 0$ implies that $\lim_{k\to\infty}\theta(x^k) = 0$. Hence each accumulation point x^* of $\{x^k\}$ satisfies $F(x^*) = 0$. Note that the existence of an accumulation point follows from (8.2.9).

If on the other hand $\limsup \alpha_k = 0$, then $\lim m_k \to \infty$. By the definition of m_k, for $\tau_k := \beta^{m_k - 1}$, we have $\tau_k \to 0$ and

$$\theta(x^k + \tau_k\, d^k) - \theta(x^k) > -\sigma\,\tau_k\,\theta(x^k). \tag{8.2.13}$$

By (8.2.9), (8.2.10) the sequence $\{(x^k, d^k)\}$ is bounded. Let $\{(x^k, d^k)\}_{k\in K}$ be any convergent subsequence with limit (x^*, d). Note that

$$\frac{\theta(x^k + \tau_k d^k) - \theta(x^k)}{\tau_k} = \frac{\theta(x^k + \tau_k d) - \theta(x^k)}{\tau_k} + \frac{\theta(x^k + \tau_k d^k) - \theta(x^k + \tau_k d)}{\tau_k},$$

where

$$\lim_{k\in K, k\to\infty} \frac{\theta(x^k + \tau_k d^k) - \theta(x^k + \tau_k d)}{\tau_k} \to 0,$$

since θ is locally Lipschitz continuous. Since $d^k \in \Phi(x^k)$ for all $k \in K$ it follows from (8.2.11) that $\theta'(x^*; d) \le -\bar\sigma\,\theta(x^*)$. Then from (8.2.13) and (8.2.11) we find

$$-\sigma\theta(x^*) \le \limsup_{k\in K,\, k\to\infty} \frac{\theta(x^k + \tau_k d^k) - \theta(x^k)}{\tau_k} \le \theta^o(x^*; d) \le -\bar\sigma\,\theta(x^*). \tag{8.2.14}$$

It follows that $(\bar\sigma - \sigma)\,\theta(x^*) \le 0$ and thus $\theta(x^*) = 0$.

(b) Since F is B-differentiable at x^* there exists a $\delta > 0$ such that for $|x - x^*| \le \delta$

$$|F(x) - F(x^*) - F'(x^*; x - x^*)| \le \frac{1}{2c}\,|x - x^*|.$$

Thus,

$$|F'(x^*; x - x^*)| \le |F(x)| + |F(x) - F(x^*) - F'(x^*, x - x^*)|$$

$$\le |F(x)| + \frac{1}{2c}\,|x - x^*|.$$

From (8.2.12)

$$|x - x^*| \le c\,|F'(x^*; x - x^*)| \le c\,|F(x)| + \frac{1}{2}\,|x - x^*|$$

and thus

$$|x - x^*| \le 2c\,|F(x)| \quad \text{if } |x - x^*| \le \delta.$$

Given $\epsilon \in (0, \delta)$ define the set

$$N(x^*, \epsilon) = \left\{ x \in \mathbb{R}^n : |x - x^*| \le \epsilon, \quad |F(x)| \le \frac{\epsilon}{2c + b} \right\}.$$

Since x^* is an accumulation point of $\{x^k\}$, there exists an index $\bar k$ such that $x^{\bar k} \in N(x^*, \epsilon)$. Since $|d^k| \le b|F(x^k)|$ for all k we have

$$|x^{\bar k+1} - x^*| \le |x^{\bar k} - x^* + \alpha_{\bar k}\, d^{\bar k}| \le |x^{\bar k} - x^*| + \alpha_{\bar k}\,|d^{\bar k}|$$

$$\le 2c\,|F(x^{\bar k})| + b\,|F(x^{\bar k})| = (2c + b)\,|F(x^{\bar k})| \le \epsilon.$$

Hence $x^{\bar k+1} \in N(x^*, \epsilon)$. By induction, $x^k \in N(x^*, \epsilon)$ for all $k \ge \bar k$ and thus the sequence x^k converges to x^*.

(c) Since $\lim_{k\to\infty} x^k = x^*$ the iterates x^k of Algorithm G enter into the region of attraction for Theorem 8.5. Moreover, referring to the proof of Lemma 8.4, for any $\gamma \in (0, 1)$ there exists k_γ such that the iterates according to (8.1.4) satisfy $|F(x^{k+1})| \le \gamma |F(x^k)|$ for $k \ge k_\gamma$. Hence these iterates coincide with those of the Algorithm G for $k \ge k_\gamma$, and superlinear convergence follows. $\square$

Remark 8.2.1. (i) We point out that the requirement that the graph Φ satisfies the closure property (8.2.11) is used in the proof of Theorem 8.6 only for the case that $\limsup_{k\to\infty} \alpha_k = 0$.

(ii) For part (a) of Theorem 8.6 the condition $|h^k| \le b|F(x^k)|$ in the first alternative of Algorithm G is not required and $|d| \le b|F(x)|$ for all $x \in S$ used in alternative (ii) can be replaced by requiring that the directions are uniformly bounded. These conditions are used in the proof of Theorem 8.6(b).

(iii) Since $h \to F'(x^*; h)$ is positively homogeneous, since we consider the finite-dimensional case here, one can easily argue that (8.2.12) is equivalent to $F'(x^*; h) \ne 0$ for all $h \ne 0$, which is called BD-regularity in [Qi].

8.2.3 Descent directions

We turn to a discussion of conditions (8.2.10) and (8.2.11) required for the descent directions d.

For the Clarke generalized directional derivative $\theta^o(x, d)$ we have by local Lipschitz continuity of F at x that

$$\theta^o(x, d) = \limsup_{y\to x, t\to 0^+} \frac{2(F(x), (F(y + t\,d) - F(y))}{t}$$

and there exists $F^o : \mathbb{R}^n \times \mathbb{R}^n \to \mathbb{R}^n$ such that

$$\theta^o(x, d) = 2(F(x), F^o(x; d)) \quad \text{for } (x, d) \in \mathbb{R}^n \times \mathbb{R}^n.$$

We introduce the notion of quasi-directional derivative.

Definition 8.7. *Let $F : \mathbb{R}^n \to \mathbb{R}^n$ be directionally differentiable. Then $G : S \times \mathbb{R}^n \to \mathbb{R}^n$ is called a quasi-directional derivative of F on $S \subset \mathbb{R}^n$ if*

(i) $(F(x), F'(x; d)) \le (F(x), G(x; d))$,

(ii) $G(x; td) = tG(x; d)$ *for all* $d \in \mathbb{R}^n$, $x \in S$, *and* $t \ge 0$,

(iii) $(F(\bar{x}), F^o(\bar{x}; \bar{d})) \le \limsup_{x\to\bar{x},\, d\to\bar{d}} (F(x), G(x; d))$ *for all* $x \to \bar{x}$, $d \to \bar{d}$ *with* $x, \bar{x} \in S$.

For the special case of optimization subject to box constraints a quasi-directional derivative will be constructed in Section 8.2.5. In the remainder of this section we consider the relationship between well-known choices for descent directions (see, e.g., [HPR, Pan1]) and the concept of quasi-directional derivative of Definition 8.7, and we assume that F is a locally Lipschitz continuous and directionally differentiable function on S where S refers to the set defined in (8.2.9).

(a) *Bouligand direction.* If there exists $\bar{b}$ such that

$$|h| \le \bar{b}\,|F'(x;h)| \text{ for all } x \in S,\ \ h \in \mathbb{R}^n \tag{8.2.15}$$

and if

$$F(x) + F'(x;d) = 0 \tag{8.2.16}$$

admits a solution d for each $x \in S$, then a first choice for the direction is given by the solution d to (8.2.16), i.e., $\Phi(x) = d$; see, e.g., [Pan1]. By (8.2.15) we have $|d| \le \bar{b}|F(x)|$. Moreover

$$\theta'(x,d) = 2\,(F'(x;d), F(x)) = -2\,\theta(x),$$

and therefore the inequalities in (8.2.10) hold with $b = \bar{b}$ and $\bar{\sigma} = 2$. For this choice, however, Φ does not satisfy (8.2.11), in general, unless additional conditions are imposed on the problem data; see Section 8.2.5 below.

(b) *Generalized Bouligand direction.* As a second choice (see [HPR, Pan1]), we assume that G is a quasi-directional derivative of F on S, that

$$|h| \le \bar{b}\,|G(x;h)| \text{ for all } x \in S,\ \ h \in \mathbb{R}^n \tag{8.2.17}$$

holds, and that

$$F(x) + G(x;d) = 0 \tag{8.2.18}$$

admits a solution d which is used as the descent direction for each $x \in S$. We set $\Phi(x) = d$. Then one argues as for the first choice that the inequalities in (8.2.10) hold with $b = \bar{b}$ and $\bar{\sigma} = 2$. Moreover Φ satisfies (8.2.11), since for any $(x,d) \to (\bar{x},\bar{d})$ in $S \times \mathbb{R}$ with $\Phi(x) = d$ we have

$$\theta^o(\bar{x},\bar{d}) \le 2 \limsup_{x\to\bar{x},\,d\to\bar{d}} (F(x), G(x;d)) = -2 \lim_{x\to\bar{x}} |F(x)|^2 = -2\theta(\bar{x}).$$

We refer the reader to Section 8.2.5 for the construction of G for specific applications.

(c) *Generalized gradient direction.* The following choice was discussed in [HPR]. Here d is chosen as the solution to

$$\min_d J(x,d) = 2(F(x), G(x;d)) + \eta\,|d|^2, \tag{8.2.19}$$

where $\eta > 0$ and $x \in S$. Assume that for some $L > 0$

$$\begin{cases} h \to G(x;h) \text{ is continuous and} \\ \\ |G(x;h)| \le L\,|h| \text{ for all } x \in S,\ \ h \in \mathbb{R}^n. \end{cases} \tag{8.2.20}$$

Then, $d \to J(d)$ is coercive, bounded below, and continuous. Thus there exists an optimal solution d to (8.2.19).

Lemma 8.8. *Assume that $F : \mathbb{R}^n \to \mathbb{R}^n$ is Lipschitz continuous, directionally differentiable and that G is a quasi-directional derivative of S satisfying* (8.2.20).

(a) *If d is an optimal solution to* (8.2.19), *then*

$$(F(x), G(x; d)) = -\eta\, |d|^2.$$

(b) *If $d = 0$ is an optimal solution to* (8.2.19), *then $(F(x), G(x; h)) \geq 0$ for all $h \in \mathbb{R}^n$.*

Proof. (a) Let d be an optimal solution to (8.2.19) and consider

$$\min_{\alpha \geq 0} 2(F(x), G(x; \alpha d)) + \alpha^2 \eta\, |d|^2.$$

Then $\alpha = 1$ is optimal and differentiating with respect to α, we have

$$(F(x), G(x; d)) + \eta\, |d|^2 = 0.$$

(b) If $d = 0$ is an optimal solution, then the optimal value of (8.2.19) is zero. It follows that for all $h \in \mathbb{R}^n$ and $\alpha \geq 0$

$$0 \leq 2\alpha(F(x), G(x; h)) + \alpha^2 \eta |h|^2.$$

Dividing by $\alpha > 0$ and letting $\alpha \to 0^+$ we obtain the claim. $\square$

By Lemma 8.8 the optimal value of the cost J in (8.2.19) is given by $-\eta |d|^2$. If this value is negative, then any solution to (8.2.19) provides a decay for θ. The optimal value of the cost is 0 if and only if $d = 0$ is the optimal solution. In this case Lemma 8.8 implies that x is a stationary point in the sense that $(F(x), G(x; h)) \geq 0$ for all $h \in \mathbb{R}^n$.

Let us now turn to the discussion of condition (8.2.10) for the direction given by the solution d to (8.2.19), i.e., $\Phi(x) = d$. We assume (8.2.17) and that (8.2.18) admits a solution for every $x \in S$. Since $J(0) = 0$, we have $2(F(x), G(x; d)) \leq -\eta\, |d|^2$ and therefore

$$\eta\, |d|^2 \leq -2(F(x), G(x; d)) \leq 2|F(x)|\, |G(x; d)| \leq 2L\, |d|\, |F(x)|.$$

Thus,

$$|d| \leq \frac{2L}{\eta} |F(x)|,$$

and the second condition in (8.2.10) holds. Turning to the first condition let $\hat{d}$ satisfy $F(x) + G(x; \hat{d}) = 0$. Then using Lemma 8.8(a) and (8.2.17) we find, at a solution d to (8.2.19),

$$\begin{aligned} J(x, d) &= (F(x), G(x; d)) = -\eta\, |d|^2 \leq 2\,(G(x; \hat{d}), F(x)) + \eta\, |\hat{d}|^2 \\ &\leq -2\,|F(x)|^2 + \eta \bar{b}^2\, |F(x)|^2 = -(2 - \eta \bar{b}^2)\, \theta(x). \end{aligned} \tag{8.2.21}$$

Since G is a quasi-directional derivative of F on S, we have $\theta'(x; d) \leq 2(F(x), G(x; d)) \leq -2(2 - \eta\bar{b}^2)\, \theta(x)$ and thus the direction d defined by (8.2.19) satisfies the first condition in (8.2.10) with $\bar{\sigma} = 2(2 - \eta\bar{b}^2)$, provided that $\eta < \frac{2}{\bar{b}^2}$.

To argue (8.2.11) for this choice of Φ, let $x_k \to \bar{x}$, $d_k \to \bar{d}$, with $d_k = \Phi(x_k)$, $x_k \in S$, and choose $\hat{d}_k$ such that $F(x_k) + G(x_k; \hat{d}_k) = 0$. Then

$$\begin{aligned}
\frac{1}{2}\theta^o(\bar{x}; \bar{d}) &\leq \limsup_{k\to\infty}(F(x_k), G(x_k; d_k)) \\
&\leq \limsup_{k\to\infty}(2(F(x_k), G(x_k; d_k)) + \eta|d_k|^2) \\
&\leq \limsup_{k\to\infty}(2(F(x_k), G(x_k; \hat{d}_k)) + \eta|\hat{d}_k|^2) \\
&\leq -2|F(\bar{x})|^2 + \eta\,\bar{b}^2 \lim_{k\to\infty} |F(x_k)|^2 = -(2 - \eta\bar{b}^2)\theta(\bar{x}),
\end{aligned}$$

and thus (8.2.11) holds if $\eta < \frac{2}{\bar{b}^2}$.

8.2.4 A Gauss–Newton algorithm

In this section we admit the case that the minimum of θ is not attained with value 0, which typically arises in nonlinear least squares problems, and the case where the equation $V_k d = -F(x^k)$ or $F(x^k) + G(x^k; d) = 0$, which would provide candidates for search directions in (ii) of Algorithm G, does not have a solution. For these cases we consider the following Gauss–Newton method.

Gauss–Newton algorithm.

Let $\beta, \gamma \in (0, 1)$, $\alpha > 0$, and $\sigma \in (0, 2\eta)$. Choose $x^0 \in \mathbb{R}^n$ and set $k = 0$. Given x^k with $F(x^k) \neq 0$. Then

(i) If there exists $V_k \in \partial_B F(x^k)$ such that

$$h^k = -(\alpha\, |F(x^k)|\, I + V_k^T V_k)^{-1} V_k^T F(x^k) \tag{8.2.22}$$

with $|h^k| \leq b\,|F(x^k)|$ satisfies $|F(x^k+h^k)| < \gamma\,|F(x^k)|$, let $d^k = h^k$ and set $x^{k+1} = x^k+d^k$, $\alpha^k = 1$, and $m_k = 0$.

(ii) Otherwise, let d^k be defined by (8.2.19) with $x = x^k$. Stop if $d^k = 0$ or $F(x^k) = 0$. Otherwise, set $\alpha_k = \beta^{m_k}$ where m_k is the first positive integer m for which

$$\theta(x^k + \beta^m\, d^k) - \theta(x^k) \leq -\sigma\beta^m\, |d^k|^2, \tag{8.2.23}$$

and set $x^{k+1} = x^k + \alpha^k\, d^k$.

Note that (8.2.22) gives the solution to

$$\min_h \frac{1}{2}|F(x^k) + V_k h|^2 + \frac{\alpha}{2}|F(x^k)|\, |h|^2. \tag{8.2.24}$$

If the Gauss–Newton algorithm terminates after finitely many steps with index $\bar{k}$, then either $F(x^{\bar{k}}) = 0$ or $d^{\bar{k}} = 0$ in the second alternative of the algorithm. In this case $x^{\bar{k}}$ is a stationary point of θ in the sense that $(F(x^{\bar{k}}), G(x^{\bar{k}}; d)) \geq 0$ for all d by Lemma 8.8.

We next discuss global convergence when the Gauss–Newton algorithm takes infinitely many steps.

Theorem 8.9. *Suppose that* $F : \mathbb{R}^n \to \mathbb{R}^n$ *is locally Lipschitz and B-differentiable,* (8.2.9), (8.2.11), *and* (8.2.20) *hold, and* G *is a quasi-directional derivative of* F *on* S.

(a) *If* $\{x^k\}$ *is an infinite sequence generated by the Gauss–Newton algorithm, then* $\{x^k\}$ *is bounded and* $|F(x^{k+1})| < |F(x^k)|$ *for all* k. *If the first alternative occurs infinitely often, then all accumulation points* x^* *satisfy* $F(x^*) = 0$. *Otherwise,* $\lim_{k\to\infty} d^k = 0$ *and*

$$(F(x^*), G(x; h)) \geq 0 \quad \text{for all } h \in \mathbb{R}^n. \tag{8.2.25}$$

(b) *If for some accumulation point* $F(x^*) = 0$ *and*

$$|h| \leq c\,|F'(x^*; h)| \quad \text{for all } h \in \mathbb{R}^n, \tag{8.2.26}$$

then the sequence x^k *converges to* x^*.

(c) *If in addition to the assumptions in* (a) *and* (b), F *is semismooth at* x^* *and all* $V \in \partial_B F(x^*)$ *are nonsingular, then* x^k *converges to* x^* *superlinearly.*

Proof. (a) The algorithm guarantees that $|F(x^{k+1})| < |F(x^k)|$ for all k. Due to (8.2.9) the sequence $\{x^k\}$ is bounded. In case the first alternative is taken we have from (8.2.24) with $h = 0$

$$\alpha |d^k|^2 \leq |F(x^k)|.$$

In case of the second alternative Lemma 8.8(a) and (8.2.20) imply that

$$\eta |d^k|^2 \leq |F(x^k)||G(x^k; d^k)| \leq L|F(x^k)||d^k|.$$

Consequently the sequence $\{d^k\}$ is bounded.

If the first alternative of the Gauss–Newton algorithm occurs infinitely often, then $\lim_{k\to\infty} |F(x^k)| = 0$ and every accumulation point x^* of $\{x^k\}$ satisfies $F(x^*) = 0$.

Let us turn to the case when eventually only the second alternative occurs. Since $|F(x^k)|$ is monotonically decreasing and bounded below, the sequence $F(x^{k+1}) - F(x^k)$ is convergent and hence by (8.2.23)

$$\lim_{k\to\infty} \alpha_k |d^k|^2 = 0.$$

If $\lim_{k\to\infty} d^k \neq 0$, then there exists an index set K such that $\lim_{k\in K,\, k\to\infty} |d^k| \neq 0$ and consequently $\lim_{k\in K,\, k\to\infty} \alpha_k = 0$. For $\tau_k = \beta^{m_k - 1}$ we have

$$-\sigma |d^k|^2 \leq \frac{1}{\tau^k}(\theta(x^k + \tau^k d^k) - \theta(x^k)). \tag{8.2.27}$$

Let $\hat{K} \subset K$ be such that $\{x^k\}_{k\in\hat{K}}$, $\{d^k\}_{k\in\hat{K}}$ are convergent subsequences with limits x^* and d. Note that

$$\begin{aligned}\frac{1}{\tau^k}(\theta(x^k + \tau^k d^k) - \theta(x^k)) &= \frac{1}{\tau^k}(\theta(x^k + \tau^k d) - \theta(x^k)) \\ &\quad + \frac{1}{\tau^k}(\theta(x^k + \tau^k d^k) - \theta(x^k + \tau^k d))\end{aligned} \tag{8.2.28}$$

with

$$\lim_{k\in\hat{K},k\to\infty} \frac{1}{\tau^k}(\theta(x^k+\tau^k d^k)-\theta(x^k+\tau^k d))=0,$$

since θ is locally Lipschitz continuous. By Lemma 8.8(a) we have $(F(x^k), G(x^k; d^k)) = -2\eta|d^k|$ for all $k \in \hat{K}$. Since G is assumed to be a quasi-directional derivative we have $\theta^o(x^*, d) = (F(x^*), F^o(x^*; d)) \le -2\eta|d|$. Passing to the limit in (8.2.27), utilizing (8.2.28), we find

$$-\sigma|d|^2 \le \limsup_{k\in\hat{K},\,k\to\infty} \frac{1}{\tau^k}(\theta(x^k+\tau^k d^k)-\theta(x^k)) = \theta^o(x^*, d) \le -2\eta|d|^2.$$

Since $\sigma \in (0, 2\eta)$ this implies that $d = 0$, which contradicts our assumption. Consequently $\lim_{k\to\infty} d^k = 0$. From Lemma 8.8(a) we have

$$2\alpha(F(x^k), G(x^k; h)) + \eta\alpha^2|h|^2 = J(x^k, \alpha h) \ge J(k^k; d^k) = -\eta|d^k|^2$$

for every $\alpha > 0$ and $h \in \mathbb{R}^n$. Passing to the limit with respect to k and dividing by α we obtain

$$2\,(F(x^*), G(x^*; h)) + \eta\alpha|h|^2 \ge 0,$$

and (8.2.25) follows by letting $\alpha \to 0^+$.

(b) The proof is identical to the one of Theorem 8.6(b).

(c) From Theorem 8.3 there exists a bounded neighborhood $N \subset \{x : |F(x)| \le |F(x^0)|\}$ of x^* and a constant C such that for all $x \in N$ and $V \in \partial_B F(x)$ we have $|V^{-1}| \le C$. Consequently there exists $M > 0$ such that for all $x \in N$ we have

$$\begin{aligned} &\left|\left(\alpha|F(x)|I + V(x)^T V(x)\right)^{-1} V(x)^T - V(x)^{-1}\right| \\ &\quad \le \left|\alpha|F(x)|\left(\alpha|F(x)|I + V(x)^T V(x)\right)^{-1}\right| \le M|F(x)| \le M|F(x^0)|. \end{aligned} \tag{8.2.29}$$

Let $h = -\left(\alpha|F(x)|I + V(x)^T V(x)\right)^{-1} V(x)^T F(x)$. Then by Lemma 8.4, possibly after shrinking N, we have for all $x \in N$

$$\begin{aligned} |x+h-x^*| &= |x - V(x)^{-1}F(x) - x^* + h + V(x)^{-1}F(x)| \\ &\le \epsilon(|x-x^*|)|x-x^*| + M|F(x)|^2. \end{aligned} \tag{8.2.30}$$

Moreover

$$\begin{aligned} &|F(x+h)| \le |F\big(x - V(x)^{-1}F(x)\big)| \\ &\quad + |F[\big(V(x)^{-1} - (\alpha|F(x)|I + V(x)^T V(x))^{-1} V(x)^T\big)F(x)]| \\ &\quad \le \bar{L}\epsilon(|x-x^*|)|F(x)| + M\bar{L}^2|F(x)|^2 \le \bar{L}\big(\epsilon(|x-x^*|) + M\bar{L}^2|x-x^*|\big)|F(x)|, \end{aligned}$$

where we used (8.2.29) and denoted by $\bar{L}$ the Lipschitz constant of F on the bounded set $\{x : |x| \le M|F(x_0)|^2\} \cup \{x - V(x)^{-1}F(x) : x \in N\} \cup N$.

Since $x^k \to x^*$, the last estimate implies the existence of an index $\bar{k}$ such that $x^k \in N$ and

$$|F(x^k + h^k)| \le \gamma |F(x^k)| \quad \text{for all } k \ge \bar{k},$$

where $h^k = -\left(\alpha |F(x^k)| I + V(x^k)^T V(x^k)\right)^{-1} V(x^k)^T F(x^k)$. Thus the first alternative in the Gauss–Newton algorithm is chosen for all $k \ge \bar{k}$, and superlinear convergence follows from (8.2.30) with $(x, h) = (x^k, h^k)$, where we also use that $|F(x)| \le \bar{L}|x|$ for $x \in N$. □

8.2.5 A nonlinear complementarity problem

Consider the complementarity problem

$$\begin{cases} \phi \le x \le \psi, \quad g(x)_i\ (x-\psi)_i\ (x-\phi)_i = 0 \text{ for all } i, \\ g(x) \le 0 \text{ for } x \ge \psi \quad \text{and} \quad g(x) \ge 0 \text{ for } x \le \phi, \end{cases}$$

where $g \in C^1(\mathbb{R}^n, \mathbb{R}^n)$ and $\phi < \psi$. This corresponds to the bilateral constraint case and can equivalently be expressed as

$$F(x) = g(x) + \max(0, -g(x) + x - \psi) + \min(0, -g(x) + x - \phi) = 0;$$

see Section 4.7, Example 4.53. Clearly $\theta(x) = |F(x)|^2$ is locally Lipschitz continuous and directionally differentiable, and hence B-differentiable.

Define

$$\mathcal{A}^+ = \{-g(x) + x - \psi > 0\}, \quad \mathcal{A}^- = \{-g(x) + x - \phi < 0\},$$

$$\mathcal{I}^1 = \{-g(x) + x - \psi = 0\}, \quad \mathcal{I}^2 = \{x - \psi < g(x) < x - \phi\}, \quad \text{and}$$

$$\mathcal{I}^3 = \{-g(x) + x - \phi = 0\}.$$

We obtain

$$F'(x; d) = \begin{cases} d & \text{on } \mathcal{A}^+ \cup \mathcal{A}^-, \\ g'(x)d & \text{on } \mathcal{I}^2, \\ \max(g'(x)d, d) & \text{on } \mathcal{I}^1, \\ \min(g'(x)d, d) & \text{on } \mathcal{I}^3, \end{cases}$$

and the Bouligand direction (8.2.16) is as the solution to

$$\begin{gathered} d + x - \psi = 0 \text{ on } \mathcal{A}^+, \quad d + x - \phi = 0 \text{ on } \mathcal{A}^-, \quad g'(x)d + g(x) = 0 \text{ on } \mathcal{I}^2, \\ \max(g'(x)d, d) + F(x) = 0 \text{ on } \mathcal{I}^1, \quad \min(g'(x)d, c\, d) + F(x) = 0 \text{ on } \mathcal{I}^3. \end{gathered} \tag{8.2.31}$$

Further if $g'(x) - I$ is surjective and $g \in C^2(\mathbb{R}^n, \mathbb{R}^n)$, then $F^o(x, d)$ satisfies

$$F^o(x; d) = \begin{cases} d & \text{on } \mathcal{A}^+ \cup \mathcal{A}^-, \\ g'(x)d & \text{on } \mathcal{I}^2, \\ \max(g'(x)d, d) & \text{on } \{F(x) > 0\} \cup (\mathcal{I}^1 \cup \mathcal{I}^3), \\ \min(g'(x)d, d) & \text{on } \{F(x) \le 0\} \cup (\mathcal{I}^1 \cup \mathcal{I}^3). \end{cases} \tag{8.2.32}$$

To verify this claim we consider $F(x) = g(x) + \max(0, -g(x) + x - \psi)$. The general case then follows with minor modifications. We find

$$\begin{aligned} \theta^o(x, d) &= \limsup_{y \to x, t \to 0^+} \frac{2}{t}(F(x), F(y + td) - F(y)) \\ &= \limsup_{y \to x, t \to 0^+} \frac{2}{t}(F(x), \max(g'(x)(y + td), y + td - \hat{\psi}) - \max(g'(x)y, y - \hat{\psi})), \end{aligned} \tag{8.2.33}$$

where $\hat{\psi} = \psi + g(x) - g'(x)x$. For $d \in \mathbb{R}^n$ we define $r \in \mathbb{R}^n$ by

$$\begin{aligned} r_i > 0 \text{ on } S_1 &= \{i \in \mathcal{I}^1 : F_i(x) > 0, (Ad)_i > d_i\} \cup \{i \in \mathcal{I}^1 : F_i(x) \le 0, (Ad)_i < d_i\}, \\ r_i < 0 \text{ on } S_2 &= \{i \in \mathcal{I}^1 : F_i(x) > 0, (Ad)_i \le d_i\} \cup \{i \in \mathcal{I}^1 : F_i(x) \le 0, (Ad)_i \ge d_i\}, \end{aligned}$$

and r arbitrary otherwise. Here we set $A = g'(x)$. Let $y \in \mathbb{R}^n$ be such that

$$Ay = y - \hat{\psi} + r,$$

and choose $t = t_y$ such that

$$\begin{aligned} (Ay + tAd)_i &> (y + td - \hat{\psi})_i \text{ for } i \in S_1, \\ (Ay + tAd)_i &< (y + td - \hat{\psi})_i \text{ for } i \in S_2 \end{aligned}$$

for $t \in (0, t_y)$. Passing to the limit in (8.2.33) we arrive at (8.2.32).

For arbitrary $\delta > 0$ we claim that the quasi-directional derivative G, defined by

$$G(x; d) = \begin{cases} d \text{ on } \mathcal{A}_\delta = \{-g(x) + x - \psi > \delta\} \cap \{-g(x) + x - \phi < -\delta\}, \\ g'(x)d \text{ on } \mathcal{I}_\delta = \{x - \psi + \delta \le g(x) \le x - \phi - \delta\}, \\ \text{and otherwise} \\ \max(g'(x)d, d) \quad \text{if } \{F(x) > 0\}, \\ \min(g'(x)d, d) \quad \text{if } \{F(x) \le 0\}, \end{cases} \tag{8.2.34}$$

is a quasi-directional derivative for F. In fact, G is clearly positively homogeneous of degree 1, and

$$(F(x), F'(x, d)) \le (F(x), G(x, d)),$$

which shows (i) of Definition 8.7. Moreover we have

$$(F(x), F^o(x,d)) \le (F(x), G(x,d)).$$

If $|x - \bar{x}|$ sufficiently small, then

$$i \in \mathcal{I}_\delta(x) \text{ implies that } i \in \mathcal{I}^2(\bar{x}),$$

$$i \in \mathcal{A}_\delta(x) \text{ implies that } i \in \mathcal{A}(\bar{x}).$$

Thus, (iii) holds by the definition of G.

To argue that (8.2.18) has a solution, let $\mathcal{A}_\delta \cup \mathcal{I}_\delta \cup (\mathcal{A}_\delta \cap \mathcal{I}_\delta)^c$ be a pairwise disjoint partition of the index $\{i = 1, \dots, n\}$. Set $A = g'(x)$, for $x \in S$, and decompose A according to the partition of the index set

$$A = \begin{pmatrix} A_{11} & A_{12} & A_{13} \\ A_{21} & A_{22} & A_{23} \\ A_{31} & A_{32} & A_{33} \end{pmatrix}.$$

Setting $d = (d_1, d_2, d_3)^T$ and $F = (F_1, F_2, F_3)^T$, the equation

$$G(x; d) = -F$$

is equivalent to

$$d_1 = -F_1,$$

$$d_2 = -A_{22}^{-1} A_{23} d_3 + A_{22}^{-1}(-F_2 + A_{21} F_1),$$

and

$$\begin{cases} M d_3 + \tilde{\mu} = -w - F_3, \\ \tilde{\mu} = \max(0, \tilde{\mu} + d_3 + F_3) \quad \text{on } \{F(x) > 0\}, \\ \tilde{\mu} = \min(0, \tilde{\mu} + d_3 + F_3) \quad \text{on } \{F(x) \le 0\}, \end{cases} \tag{8.2.35}$$

where

$$w = -A_{31} F_1 + A_{32} A_{22}^{-1}(-F_2 + A_{21} F_1) \quad \text{and} \quad M = A_{33} - A_{32} A_{22}^{-1} A_{23}.$$

Assume that A is symmetric positive definite for every $x \in S$. Then every Schur complement of A is positive definite as well, and (8.2.35) admits a unique solution $(d_3, \tilde{\mu})$. It follows that $G(x; d) = -F$ admits a unique solution d for every $x \in S$ and $F \in \mathbb{R}^n$. Consequently (8.2.18) is satisfied. Since $g'(x)$ is positive definite for every $x \in S$ and S is closed and bounded by (8.2.9), and hence compact, it follows that $g'(x)$ and M are uniformly positive definite with respect to $x \in S$. This implies that there exists $\bar{b}$ such that $|d| \le \bar{b}|F|$ for some $\bar{b}$ independent of $x \in S$. Consequently (8.2.17) holds.

We remark that (iii) of Definition 8.7 is not satisfied for the choice $G(x; d) = F'(x, d)$ unless $g(x)_i \ne (x - \psi)_i$ and $g(x)_i \ne (x - \phi)_i$ for all i and $x \in S$.

8.3 Semismooth functions in infinite-dimensional spaces

In infinite-dimensional spaces notions of generalized derivatives for functions which are not C^1 cannot rely on Rademacher's theorem. Here, instead, we shall mainly utilize a concept of generalized derivative that is sufficient to guarantee superlinear convergence of Newton's method. We refer the reader to the discussion in Section 8.1, especially (8.1.5) and (8.1.6). This notion of differentiability is called Newton derivative and will be defined below. We refer the reader to [HiIK, CNQ, Ulb] for further discussions of the topics covered in this section.

Let X, Z be real Banach spaces and let $D \subset X$ be an open set.

Definition 8.10. (1) *$F\colon D \subset X \to Z$ is called Newton differentiable at x if there exist an open neighborhood $N(x) \subset D$ and mappings $G\colon N(x) \to \mathcal{L}(X, Z)$ such that*

$$\lim_{|h|\to 0} \frac{|F(x+h) - F(x) - G(x+h)h|_Z}{|h|_X} = 0. \tag{A}$$

The family $\{G(s) : s \in N(x)\}$ is called an N-derivative of F at x.

(2) *F is called semismooth at x if it is Newton differentiable at x and*

$$\lim_{t\to 0^+} G(x+t\,h)h \quad \text{exists uniformly in } |h| = 1.$$

(3) *F is directionally differentiable at $x \in D$ if*

$$\lim_{t\to 0^+} \frac{F(x+t\,h) - F(x)}{t} =: F'(x; h)$$

exists for all $h \in X$.

(5) *F is B-differentiable at $x \in D$ if F is directionally differentiable at x and*

$$\lim_{|h|\to 0} \frac{F(x+h) - F(x) - F'(x; h)}{|h|_X} = 0.$$

Note that differently from the finite-dimensional case we do not require Lipschitz continuity of F as part of the definition of semismoothness.

Lemma 8.11. *Suppose that $F : D \subset X \to Z$ is Newton differentiable at $x \in D$ with N-derivative G.*

(1) *F is directionally differentiable at x if and only if*

$$\lim_{t\to 0^+} G(x+t\,h)h \quad \text{exists for all } h \in X. \tag{8.3.1}$$

In this case $F'(x; h) = \lim_{t\to 0^+} G(x+th)h$ for all $h \in X$.

(2) *F is B-differentiable at x if and only if*

$$\lim_{t\to 0^+} G(x+t\,h)h \quad \text{exists uniformly in } |h| = 1, \tag{8.3.2}$$

i.e., F is semismooth.

Proof. (1) If (8.3.1) holds for $h \in X$ with $|h| = 1$, then

$$\lim_{t \to 0^+} \frac{F(x+th) - F(x)}{t} = \lim_{t \to 0^+} G(x+th)h.$$

Since $h \in X$ with $|h| = 1$ was arbitrary this implies that F is directionally differentiable at x and

$$F'(x; h) = \lim_{t \to 0^+} G(x+th)h.$$

Similarly the converse holds.

(2) If F is directionally differentiable at x, then F is B-differentiable at x, i.e.,

$$\lim_{|h| \to 0} \frac{F(x+h) - F(x) - F'(x; h)}{|h|_X} = 0 \text{ if and only if}$$

$$\lim_{t \to 0^+} \frac{F(x+tv) - F(x)}{t} - F'(x; v) = 0 \text{ and the limit is uniform in } |v|_X = 1.$$

Here we use positive homogeneity of the directional derivative $F'(x; h)$ with respect to the second variable.

If F is B-differentiable at x, then it is differentiable and from (1) we have

$$\lim_{t \to 0} \frac{F(x+tv) - F(x)}{t} = F'(x; v) = \lim_{t \to 0} G(x+tv)v.$$

The Bouligand property and the equivalence stated above imply that the $\lim_{t \to 0} G(x+tv)v$ exists uniformly in $|v| = 1$. The converse easily follows as well. ☐

Example 8.12. Let $\psi : \mathbb{R} \to \mathbb{R}$ be semismooth at every $x \in \mathbb{R}$ in the sense of Definition 8.1 and globally Lipschitz, i.e., there exists a constant L such that

$$|\psi(s) - \psi(t)| \le L|s-t| \text{ for all } s, t \in \mathbb{R}.$$

We first argue that there exists a measurable selection $V : \mathbb{R} \to \mathbb{R}$ such that $V(t) \in \partial\psi(t)$ for a.e. $t \in \mathbb{R}$. Since $\partial\psi(s)$ is a nonempty closed set in $\mathbb{R}$ for every $s \in \mathbb{R}$ (see [Cla, p. 70]), it suffices to prove that the multivalued function $s \to \partial\psi$ is measurable; i.e., for every compact set $C \subset \mathbb{R}$ the preimage $P_C = \{t \in \mathbb{R} : \partial\psi(t) \cap C \neq \emptyset\}$ is measurable. The measurable selection theorem (see [Cla, p. 111]) then ensures the existence of the desired measurable selection V. To verify measurability of $s \to \partial\psi(s)$ let C be a compact set and let $\{t_k\}$ be a convergent sequence in P_C with limit t^*. Choose $v_k \in \partial\psi(t_k) \cap C$ for $k = 1, 2, \ldots$. By compactness of C there exists a convergent subsequence, denoted by the same symbol, with limit $v^* \in C$. Since $t_k \to t^*$, upper semicontinuity of $\partial\psi$ at t^* (see [Cla, p. 70]) implies the existence of a sequence $\tilde{v}_k \in \partial\psi(t^*)$ such that $\lim_{k\to\infty}(\tilde{v}_k - v_k) = 0$. Consequently $\lim_{k\to\infty} \tilde{v}_k = v^*$ and by closedness of $\partial\psi(t^*)$ we have $v^* \in \partial\psi(t^*) \cap C$. Thus P_C is closed and therefore measurable.

Associated to ψ with the properties specified above, we define for $1 \le p \le q \le \infty$ the substitution operator

$$F : L^q(\Omega) \to L^p(\Omega)$$

by

$$F(x)(s) = \psi(x(s)) \text{ for a.e. } s \in \Omega,$$

where $x \in L^q(\Omega)$, and with Ω a bounded domain in $\mathbb{R}^n$.

We now verify that F is Newton differentiable on $\mathbb{R}$ if $1 \le p < q \le \infty$ and that any measurable selection V of $\partial\psi$ provides an N-derivative. Let $D : \mathbb{R} \times \mathbb{R} \to \mathbb{R}$ be given by

$$D(s, v) = |\psi(s+v) - \psi(s) - V(s+v)v|.$$

By Theorem 8.2 we have

$$\lim_{v \to 0} v^{-1} D(s, v) = 0 \text{ for every } s \in \mathbb{R}. \tag{8.3.3}$$

Moreover global Lipschitz continuity of ψ implies that

$$D(s, v) \le 2L|v| \text{ for all } (s, v) \in \mathbb{R}^2. \tag{8.3.4}$$

Let $x \in L^q(\Omega)$ and let $\{h_k\}$ be a sequence in $L^q(\Omega)$ converging to 0. Then there exists a subsequence $\{h_{k'}\}$ converging to 0 a.e. in Ω. By (8.3.3) this implies that $h_{k'}(s)^{-1} D(x(s), h_{k'}(s)) \to 0$ for a.e. $s \in \Omega$. By (8.3.4) Lebesgue's bounded convergence theorem is applicable and implies that $h_{k'}^{-1} D(x, h_{k'})$ converges to 0 in $L^{\hat{p}}$ for every $1 \le \hat{p} < \infty$. Since this is the case for every a.e. convergent subsequence of h_k and since $\{h_k\}$ was arbitrary, we find that

$$|h^{-1}\, D(x, h)|_{L^{\hat{p}}} \to 0 \text{ for every } 1 \le \hat{p} < \infty \text{ as } |h|_{L^q(\Omega)} \to 0. \tag{8.3.5}$$

By the Hölder inequality we obtain

$$|D(x, h)|_{L^p} \le |h^{-1}\, D(x, h)|_{L^r} |h|_{L^q},$$

where $r = \frac{qp}{q-p}$ if $q < \infty$, and $r = p$ if $q = \infty$. Using (8.3.5) this implies Newton differentiability of F at x.

To verify that F is semismooth at $x \in L^q(\Omega)$, we shall show that

$$\frac{|V(x+h)h - \psi'(x; h)|_{L^p}}{|h|_{L^q}} \to 0 \text{ as } |h|_{L^q} \to 0. \tag{8.3.6}$$

Let $\bar{D} : \mathbb{R} \times \mathbb{R} \to \mathbb{R}$ be given by $\bar{D}(s, v) = |V(s+v)v - \psi'(s; v)|$. Then by Theorem 8.2 (3) and Lipschitz continuity of ψ we have

$$\lim_{v \to 0} v^{-1} \bar{D}(s, v) = 0 \text{ and } \bar{D}(s, v) \le 2L|v| \text{ for all } (s, v) \in \mathbb{R}^2.$$

The proof of (8.3.6) can now be completed in the same manner as that for Newton differentiability, by using Lebesgue's bounded convergence theorem. Semismoothness of $F : L^q(\Omega) \to L^p(\Omega)$ follows from (8.3.5), (8.3.6), and Lemma 8.11 (2). The class of mappings F of this example was treated in [Ulb].

Example 8.13. Let X be a Hilbert space. Then the norm functional $F(x) = |x|$ is Newton differentiable. In fact, let $G(x+h)h = (\frac{x+h}{|x+h|}, h)_X$ and $G(0)h = (\lambda, h)_X$ for some λ with $\lambda \in X$. Then

$$|h|^{-1}|F(x+h) - F(x) - G(x+h)h| = |h|^{-1}\left| \frac{(2(x+h), h)_X - |h|^2}{|x| + |x+h|} - \frac{(x+h, h)_X}{|x+h|} \right| \to 0$$

as $h \to 0$. Hence F is Newton differentiable on X. Moreover F is semismooth.

Example 8.14. Let $F : L^q(\Omega) \to L^p(\Omega)$ denote the pointwise max operation $F(x) = \max(0, x)$ and define G_m by

$$G_m(x)(s) = \begin{cases} 0 & \text{if } x(s) < 0, \\ \delta & \text{if } x(s) = 0, \\ 1 & \text{if } x(s) > 0, \end{cases}$$

where $\delta \in [0, 1]$. It follows from Example 8.12 that F is semismooth from $L^q(\Omega)$ into $L^p(\Omega)$ provided that $1 \le p < q \le \infty$, with G_m as N-derivative. It can also be argued that G_m is an N-derivative for F for any choice of $\delta \in \mathbb{R}$ (see [HiIK]). If $p = q$, then F is directionally differentiable at every $x \in L^q(\Omega)$. In fact, for $h \in L^q(\Omega)$ define

$$F'(x; h)(s) = \begin{cases} 0 & \text{if } x(s) < 0 \text{ or } x(s) = 0,\ h(s) \le 0, \\ h(s) & \text{if } x(s) > 0 \text{ or } x(s) = 0,\ h(s) \ge 0. \end{cases}$$

Then we have

$$\left| \frac{F(x(s) + t\,h(s)) - F(x(s))}{t} - F'(x; h)(s) \right| \le 2|h(s)|$$

and

$$\lim_{t \to 0^+} \left| \frac{F(x(s) + t\,h(s)) - F(x(s))}{t} - F'(x; h)(s) \right| \quad \text{for a.e. } s \in \Omega.$$

By the Lebesgue dominated convergence theorem

$$\lim_{t \to 0^+} \left| \frac{F(x + t\,h) - F(x)}{t} - F'(x; h) \right|_{L^p} = 0,$$

i.e., $F'(x; h)$ is the directional derivative of F at x.

However, F is not Newton differentiable with G_m as N-derivative in general. For this purpose consider $x = -|s|$ on $\Omega = (-1, 1)$ and choose $h_n(s)$ as $\frac{1}{n}$ multiplied by the characteristic function of the interval $(-\frac{1}{n}, \frac{1}{n})$. Then $|h_n|_{L^p}^p = \frac{2}{n^{p+1}}$ and

$$\int_{-1}^{1} |F(x + h_n) - F(x) - G_m(x + h_n)h_n|^p \, ds = \int_{-\frac{1}{n}}^{\frac{1}{n}} |x(s)|^p = \frac{2}{p+1}\left(\frac{1}{n}\right)^{p+1}.$$

Thus,

$$\lim_{n \to \infty} \frac{|F(x + h_n) - F(x) - G_m(x + h_n)h_n|_{L^p}}{|h_n|_{L^p}} = \left(\frac{1}{p+1}\right)^{\frac{1}{p}} \neq 0,$$

and hence condition (A) is not satisfied at x for any $p \in [1, \infty)$. To consider the case $p = \infty$ we choose $\Omega = (0, 1)$ and show that (A) is not satisfied at $x(s) = s$. For this purpose define for $n = 2, \dots$

$$h_n(s) = \begin{cases} -(1 + \frac{1}{n})s & \text{on } (0, \frac{1}{n}], \\ (1 + \frac{1}{n})s - \frac{2}{n}(1 + \frac{1}{n}) & \text{on } (\frac{1}{n}, \frac{2}{n}], \\ 0 & \text{on } (\frac{2}{n}, 1]. \end{cases}$$

Observe that $E_n = \{s : x(s) + h_n(s) < 0\} \supset (0, \frac{1}{n}]$. Therefore

$$\lim_{n\to\infty} \frac{1}{|h_n|_{L^\infty(0,1)}} |\max(0, x + h_n) - \max(0, x) - G_m(x + h_n)h_n|_{L^\infty(0,1)}$$

$$= \lim_{n\to\infty} \frac{n^2}{n+1} |x|_{L^\infty(E_n)} \geq \lim_{n\to\infty} \frac{n}{n+1} = 1,$$

and hence (A) cannot be satisfied.

Lemma 8.15. *Suppose that $H : D \subset X \to Y$ is continuously Fréchet differentiable at $x \in D$ and $\phi : Y \to Z$ is Newton differentiable at $H(x)$ with N-derivative G. Then $F = \phi(H)$ is Newton differentiable at x with N-derivative $G(H(x+h))H'(x+h) \in \mathcal{L}(X, Z)$ for h sufficiently small.*

Proof. Let U be a convex neighborhood of x in D such that $H' \in \mathcal{L}(X, Y)$ is continuous in U and $H(U)$ is contained in $N(H(x))$, with $N(H(x))$ defined according to N-differentiability of ϕ at $H(x)$. Let $h \in X$ be such that $x^* + h \in U$ and note that

$$H(x+h) = H(x) + \int_0^1 H'(x + \theta h)h\, d\theta, \tag{8.3.7}$$

where

$$\left\| \int_0^1 H'(x+\theta h)\, d\theta - H'(x+h) \right\| \to 0 \quad \text{as } |h|_X \to 0, \tag{8.3.8}$$

since $H' \in \mathcal{L}(X, Y)$ is continuous at x. By Newton differentiability of ϕ at $H(x)$ and (8.3.7) it follows that

$$\lim_{|h|_X\to 0} \frac{1}{|h|_X} \left| \phi(H(x+h)) - \phi(H(x)) - G(H(x+h)) \int_0^1 H'(x+\theta h)h\, d\theta \right|_Z = 0.$$

From (8.3.8) we deduce that

$$\lim_{|h|_X\to 0} \frac{1}{|h|_X} |\phi(H(x+h)) - \phi(H(x)) - G(H(x+h))H'(x+h)h|_Z = 0.$$

This implies Newton differentiability of $F = \phi(H)$ at x. ☐

Theorem 8.16. *Suppose that x^* is a solution to $F(x) = 0$ and that F is Newton differentiable at x^* with N-derivative G. If G is nonsingular for all $x \in N(x^*)$ and $\{\|G(x)^{-1}\| : x \in N(x^*)\}$ is bounded, then the Newton iteration*

$$x^{k+1} = x^k - G(x^k)^{-1}F(x^k)$$

converges superlinearly to x^* provided that $|x^0 - x^*|$ is sufficiently small.

Proof. Note that the Newton iterates satisfy

$$|x^{k+1} - x^*| \leq \|G(x^k)^{-1}\|\, |F(x^k) - F(x^*) - G(x^k)(x^k - x^*)| \tag{8.3.9}$$

if $x^k \in N(x^*)$. Let $B(x^*, r)$ denote a ball of radius r centered at x^* contained in $N(x^*)$ and let M be such that $\|G(x)^{-1}\| \leq M$ for all $x \in B(x^*, r)$. We apply (A) with $x = x^*$. Let $\eta \in (0, 1]$ be arbitrary. Then there exists $\rho \in (0, r)$ such that

$$|F(x^* + h) - F(x^*) - G(x^* + h)h| < \frac{\eta}{M}\,|h| \leq \frac{1}{M}\,|h| \qquad (8.3.10)$$

for all $|h| < \rho$. Consequently, if we choose x^0 such that $|x^0 - x^*| < \rho$, then by induction from (8.3.9), (8.3.10) with $h = x^k - x^*$ we have $|x^{k+1} - x^*| < \rho$ and in particular $x^{k+1} \in B(x^*, \rho)$. It follows that the iterates are well defined. Moreover, since $\eta \in (0, 1]$ is chosen arbitrarily, $x^k \to x^*$ converges superlinearly. ☐

The following theorem provides conditions which guarantee in an appropriate sense global convergence of the Newton iteration [QiSu].

Theorem 8.17. *Suppose that F is continuous and directionally differentiable on the closed sphere $S = \overline{B(x^0, r)}$. Assume also the existence of bounded operators $G(\cdot) \in \mathcal{L}(X, Z)$ and constants $\beta, \gamma \in \mathbb{R}^+$ such that*

$$\|G(x)^{-1}\| \leq \beta, \quad |G(x)(y - x) - F'(x; y - x)| \leq \gamma\,|y - x|,$$

$$|F(y) - F(x) - F'(x; y - x)| \leq \delta\,|y - x|$$

for all $x, y \in S$, where $\alpha = \beta(\gamma + \delta) < 1$, and

$$\beta\,|F(x^0)| \leq r(1 - \alpha).$$

Then the iterates defined by

$$x^{k+1} = x^k - G(x^k)^{-1}F(x^k) \text{ for } k = 0, 1, \ldots$$

remain in S and converge to the unique solution x^* of $F(x) = 0$ in S. Moreover, we have the error estimate

$$|x^k - x^*| \leq \frac{\alpha}{1 - \alpha}\,|x^k - x^{k-1}|.$$

Proof. First note that

$$|x^1 - x^0| \leq |G(x^0)^{-1}F(x^0)| \leq \beta\,|F(x^0)| \leq r(1 - \alpha).$$

Thus $x^1 \in S$. Suppose that $x^1, \ldots, x^k \in S$. Then

$$\begin{aligned}
|x^{k+1} - x^k| &\leq |G(x^k)^{-1}F(x^k)| \leq \beta\,|F(x^k)| \\
&\leq \beta\,|F(x^k) - F(x^{k-1}) - F'(x^{k-1}; x^k - x^{k-1})| \\
&\quad + \beta\,|G(x^{k-1})(x^k - x^{k-1}) - F'(x^{k-1}; x^k - x^{k-1})| \\
&\leq \beta(\delta + \gamma)\,|x^k - x^{k-1}| = \alpha\,|x^k - x^{k-1}| \leq \alpha^k\,|x^1 - x_0| \leq r(1 - \alpha)\alpha^k.
\end{aligned} \qquad (8.3.11)$$

Since

$$|x^{k+1} - x^0| \le \sum_{j=0}^{k} |x^{j+1} - x^j| \le \sum_{j=0}^{k} r\alpha^j(1-\alpha) \le r,$$

we have $r^{k+1} \in S$ and by induction $x^k \in S$ for all k. For each $m > n$

$$|x^m - x^n| \le \sum_{j=n}^{m-1} |x^{j+1} - x^j| \le \sum_{j=n}^{m-1} r\alpha^j(1-\alpha) \le r\alpha^n.$$

Hence $\{x^k\}$ is a Cauchy sequence in S and there exists $\lim_{k\to\infty} x^k = x^* \in S$. By locally Lipschitz continuity of F and (8.3.11)

$$|F(x^*)| = \lim\ |F(x^k)| \le \lim\ \alpha\beta^{-1}\, |x^k - x^{k-1}| = 0,$$

i.e., $F(x^*) = 0$. For $y^* \in S$ satisfying $F(y^*) = 0$ we find

$$\begin{aligned} |y^* - x^*| &\le \beta\, |G(x^*)(y^* - x^*)| \\ &\le \beta\, |F(y^*) - F(x^*) - F'(x^*; y^* - x^*)| \\ &\quad + \beta\, |G(x^*)(y^* - x^*) - F'(x^*; y^* - x^*)| \\ &\le \alpha\, |y^* - x^*|. \end{aligned}$$

This implies $y^* = x^*$ and hence x^* is the unique solution to $F(x) = 0$ in S. Finally

$$|x^m - x^k| \le \sum_{j=k}^{m-1} |x^{j+1} - x^j| \le \sum_{j=1}^{m-k} \alpha^j\, |x^k - x^{k-1}| \le \frac{\alpha}{1-\alpha}\, |x^k - x^{k-1}|$$

implies the asserted error estimate by letting $m \to \infty$. □

8.4 The primal-dual active set method as a semismooth Newton method

Let us consider the complementarity problem in the unknowns $(x, \mu) \in L^2(\Omega) \times L^2(\Omega)$

$$\begin{aligned} Ax + \mu &= a, \\ \mu &= \max(0, \mu + c(x - \psi)), \end{aligned} \tag{8.4.1}$$

where $A \in \mathcal{L}(L^2(\Omega))$, with Ω a bounded domain in $\mathbb{R}^n$, $c > 0$, and $a \in L^p(\Omega)$, $\psi \in L^p(\Omega)$ for some $p > 2$. Recall that the second equation in (8.4.1) is equivalent to

$$x \le \psi, \quad \mu \ge 0, \quad (\mu, x - \psi)_{L^2(\Omega)} = 0, \tag{8.4.2}$$

where the inequalities and the max operation are understood in the pointwise a.e. sense. In Example 7.1 of Chapter 7 it was shown that such problems arise in the context of constrained optimal control problems with A of the form

$$A = \alpha I + B^*(-\Delta)^{-2}B, \tag{8.4.3}$$

where $\alpha > 0$, B is a bounded operator, and Δ denotes the Laplacian with homogeneous Dirichlet boundary conditions. This suggests and justifies our assumption that

$$A = \alpha I + C, \quad \text{where } C \in \mathcal{L}(L^2(\Omega), L^p(\Omega)) \quad \text{with } p > 2. \tag{8.4.4}$$

We shall show that the primal-dual active set method discussed in Chapter 7 is equivalent to the semismooth Newton method applied to

$$0 = F(x, \mu) = \begin{cases} Ax - a + \mu, \\ \mu - \max(0, \mu + c\,(x - \psi)). \end{cases} \tag{8.4.5}$$

For this purpose we define $G : L^p(\Omega) \to L^2(\Omega)$ by

$$G(x)(s) = \begin{cases} 0 & \text{if } x(s) \le 0, \\ 1 & \text{if } x(s) > 0, \end{cases} \tag{8.4.6}$$

and we recall that the max function is Newton differentiable from $L^p(\Omega)$ to $L^2(\Omega)$ if $p > 2$ with G as an N-derivative. A Newton step applied to the second equation in (8.4.4) results in

$$c\,(x^{k+1} - x^k) + c\,(x^k - \psi) = 0 \quad \text{in } \mathcal{A}_k = \{s : \mu^k(s) + c\,(x^k(s) - \psi(s)) > 0\},$$

$$(\mu^{k+1} - \mu^k) + \mu^k = 0 \quad \text{in } \mathcal{I}_k = \{s : \mu^k(s) + c\,(x^k(s) - \psi(s)) \le 0\}.$$

Hence a Newton step for (8.4.5) is given by

$$\begin{cases} Ax^{k+1} - a + \mu^{k+1} = 0, \\ x^{k+1} = \psi \text{ in } \mathcal{A}_k, \\ \mu^{k+1} = 0 \text{ in } \mathcal{I}_k. \end{cases} \tag{8.4.7}$$

This coincides with the primal-dual active set strategy of Section 7.1. To analyze its local convergence properties note that under assumption (8.4.4), equation (8.4.5) with $c = \alpha$ is equivalent to the reduced problem

$$Ax - a + \max(0, -Cx + a - \alpha\psi) = 0. \tag{8.4.8}$$

Applying a semismooth Newton step to (8.4.8) results in

$$\begin{aligned} &A(x^{k+1} - x^k) - G(-Cx^k + a - \alpha\psi)\,C(x^{k+1} - x^k) \\ &+ Ax^k - a + \max(0, -Cx^k + a - \alpha\psi) = 0. \end{aligned} \tag{8.4.9}$$

Setting $\mu^{k+1} = a - Ax^{k+1}$ the iterates of (8.4.9) coincide with those of (8.4.7), provided that the initialization for the reduced iteration (8.4.9) is chosen such that $\mu^0 = a - Ax^0$. This follows from the fact that $\mu^k + \alpha(x^k - \psi) = -Cx^k + a - \alpha\psi$.

For any partition $\Omega = \mathcal{A} \cup \mathcal{I}$ into measurable sets $\mathcal{A}$ and $\mathcal{I}$ let $R_{\mathcal{I}} : L^2(\mathcal{I}) \to L^2(\mathcal{I})$ denote the canonical restriction operator and $R_{\mathcal{I}}^* : L^2(\mathcal{I}) \to L^2(\mathcal{I})$ its adjoint. Further set

$$A_{\mathcal{I}} = R_{\mathcal{I}}\, A\, R_{\mathcal{I}}^*.$$

Proposition 8.18. *Assume that* (8.4.1) *with* ψ *and* a *in* $L^p(\Omega)$, $p > 2$, *admits a solution* $(x^*, \mu^*) \in L^2(\Omega) \times L^2(\Omega)$ *and that* (8.4.4) *holds. If moreover*

$$\{A_{\mathcal{I}}^{-1} : \Omega = \mathcal{I} \cup \mathcal{A}\} \text{ is uniformly bounded in } \mathcal{L}(L^2(\Omega)), \tag{8.4.10}$$

then the iterates (x^k, μ^k) *defined by* (8.4.7) *converge superlinearly in* $L^2(\Omega) \times L^2(\Omega)$ *to* (x^*, μ^*), *provided that* $|x^* - x^0|$ *is sufficiently small and* $\mu^0 = a - Ax^0$. *Moreover* $(x^*, \mu^*) \in L^p(\Omega) \times L^p(\Omega)$.

Proof. First note that if (8.4.1) admits a solution (x^*, μ^*) for some $c > 0$, then it admits the same solution with $c = \alpha$ and x^* is also a solution of (8.4.8). By (8.4.4) this implies that $x^* \in L^p(\Omega)$ and the first equation in (8.4.1) implies that $\mu^* \in L^p(\Omega)$. By Lemma 8.15 and Example 8.14 the mapping $\tilde{F}x = \alpha Ax - a + \max(0, -Cx + a - \alpha\psi)$ is Newton differentiable from $L^2(\Omega)$ into itself and $\tilde{G}(x) = A - G(-Cx + a - \alpha\psi)C$ is an N-derivative. By (8.4.10) the family of operators $\{\tilde{G}(x)^{-1} : x \in L^2(\Omega)\}$ is uniformly bounded in $\mathcal{L}(L^2(\Omega))$, and hence by Theorem 8.16 the iterates x^k converge superlinearly to x^*, provided that $|x^0 - x^*|$ is sufficiently small. Superlinear convergence of μ^k to μ^* follows from (8.4.5) and (8.4.7). □

Note that (8.4.10) is satisfied for operators of the form given in (8.4.3).

The characterization of inequalities by means of max and min operations in complementarity systems such as (8.4.2) is one of several possibilities. Another frequently used complementarity function is the Fischer–Burmeister function $\sqrt{(\psi - x)^2 + \mu^2} - (\psi - x) - \mu$. Numerical experiments appear to indicate that max-based complementarity functions can be more efficient numerically than the Fischer–Burmeister function; see, e.g., [Kan].

As shown in Section 7.5 a class of optimization problems with bilateral constraints $\varphi \le x \le \psi$ can be expressed as

$$Ax - a + \mu = 0, \qquad \mu = \max(0, \mu + c\,(x - \psi)) + \min(0, \mu + c\,(x - \varphi)) \tag{8.4.11}$$

or, equivalently, as

$$Ax - a + \max(0, -Cx + a - \alpha\psi) + \min(0, -Cx + a - \alpha\psi) = 0, \tag{8.4.12}$$

where $c = \alpha$ and $\varphi < \psi$ with a, φ, and ψ in $L^p(\Omega)$.

The primal-dual active set strategy applied to (8.4.11) can be expressed as

$$\begin{aligned}
&Ax^{k+1} - a + \mu^{k+1} = 0,\\
&x^{k+1} = \psi \text{ in } \mathcal{A}_k^+ = \{s : \mu^k(s) + c\,(x^k(s) - \psi(s)) > 0\},\\
&\mu^{k+1} = 0 \text{ in } \mathcal{I}_k = \{s : \mu^k(s) + c\,(x^k(s) - \varphi(s)) \le 0 \le \mu^k(s) + c\,(x^k(s) - \psi(s))\},\\
&x^{k+1} = \varphi \text{ in } \mathcal{A}_k^- = \{s : \mu^k(s) + c\,(x^k(s) - \varphi(s)) < 0\}.
\end{aligned} \tag{8.4.13}$$

In the bilateral case, the role of c influences the likelihood of switches from being active-above to active-below in one iteration. If, for example, $s \in \mathcal{A}_k^+$, then $\mu^{k+1}(s) + c\,(x^{k+1}(s) - \varphi(s)) = \mu^{k+1}(s) + c\,(\psi(s) - \varphi(s))$ is more likely to be negative and hence $s \in \mathcal{A}_{k+1}^-$ if c is large.

For $c = \alpha$, iteration (8.4.13) is equivalent to applying the semismooth Newton method to (8.4.12) with G as N-derivative for the max function and

$$\tilde{G}(x)(s) = \begin{cases} 0 & \text{if } x(s) \geq 0, \\ 1 & \text{if } x(s) < 0 \end{cases}$$

as N-derivative for the min function. Under the assumptions of Proposition 8.18 local superlinear convergence of the iteration (8.4.13) can be shown.

8.5 Semismooth Newton methods for a class of nonlinear complementarity problems

In this section we consider nonlinear complementarity problems in the Hilbert space $X = L^2(\Omega)$. Let g be a continuous mapping from $L^2(\Omega)$ into itself with Ω a bounded domain in $\mathbb{R}^n$, and consider

$$g(x) + \mu = 0, \quad \mu \geq 0, \ \ x \leq \psi, \quad \text{and} \quad (\mu, x - \psi) = 0, \tag{8.5.1}$$

where $\psi \in L^2(\Omega)$. As discussed in Chapter 7, (8.5.1) can be expressed equivalently as

$$0 = F(x, \mu) = \begin{cases} g(x) + \mu, \\ \mu - \max(0, \mu + c\,(x - \psi)), \end{cases} \tag{8.5.2}$$

for any $c > 0$, where max denotes the pointwise max operation.

If Φ is a continuously differentiable functional on X, then (8.5.1) with $g = \Phi'$ is the necessary optimality condition for

$$\min_{x \in L^2(\Omega)} \ \Phi(x) \quad \text{subject to } x \leq \psi. \tag{8.5.3}$$

We shall employ semismooth Newton methods for solving $F(x, \mu) = 0$. Let $G(x)$ be as defined in Section 8.4.

Applying a primal-dual active set strategy to the second equation in (8.5.2) results in

$$\begin{aligned} & g(x^{k+1}) + \mu^{k+1} = 0, \\ & x^{k+1} = \psi \text{ in } \mathcal{A}_k = \{s : \mu^k(s) + c\,(x^k(s) - \psi(s)) > 0\}, \\ & \mu^{k+1} = 0 \text{ in } \mathcal{I}_k = \{s : \mu^k(s) + c\,(x^k(s) - \psi(s)) \leq 0\} \end{aligned} \tag{8.5.4}$$

for $k = 0, 1, \ldots$. To investigate the convergence properties of (8.5.4) we note that (8.5.2) is equivalent to

$$g(x) + \max(0, -g(x) + c\,(x - \psi)) = 0, \tag{8.5.5}$$

and the iteration (8.5.4) is equivalent to the reduced iteration

$$g(x^{k+1}) + G\big(-g(x^k) + c\,(x^k - \psi)\big)\big(-g(x^{k+1}) + c\,(x^{k+1} - \psi) \\ -(-g(x^k) + c\,(x^k - \psi)) + \max(0, -g(x^k) + c\,(x^k - \psi))\big) = 0, \tag{8.5.6}$$

provided that the initialization for (8.5.4) is chosen such that $\mu^0 = -g(x^0)$. Note that (8.5.5) can be considered as a *partial semismooth Newton iteration*. Separating the generalized differentiation of the max operation from that of the linearization of the nonlinearity g was investigated numerically in [dReKu, GrVo].

Proposition 8.19. *Assume that* (8.5.2) *admits a solution x^*, that $x \to g(x) - c\,(x - \psi)$ is Lipschitz continuous from $L^2(\Omega)$ to $L^p(\Omega)$ in a neighborhood U of x^* for some $c > 0$ and $p > 2$, and that there exists $\alpha > 0$ such that*

$$\alpha|x - y|^2_{L^2(\mathcal{I})} \le \int_{\mathcal{I}} (g(x) - g(y))(s)(x - y)(s)\,ds \tag{8.5.7}$$

for all partitions $\Omega = \mathcal{A} \cup \mathcal{I}$ and $x, y \in L^2(\Omega)$. Then the iterates (x^k, μ^k) defined by (8.5.4) *converge superlinearly to (x^*, μ^*), provided that $|x^* - x^0|$ is sufficiently small and $\mu^0 = -g(x^0)$.*

Proof. From (8.5.5) and (8.5.6) we have

$$g(x^{k+1}) - g(x^*) + G(z^k)(-g(x^{k+1}) + c(x^{k+1} - \psi) + g(x^*) - c(x^* - \psi))$$

$$= G(z^k)\left(-g(x^k) + c(x^k - \psi) + g(x^*) - c(x^* - \psi)\right)$$

$$+ \max(0, -g(x^*) + c(x^* - \psi)) - \max(0, -g(x^k) + c(x^k - \psi)),$$

where $z^k = -g(x^k) + c(x^k - \psi)$. Taking the inner product in $L^2(\Omega)$ with $x^{k+1} - x^*$, using (8.5.7) to estimate the left-hand side from below, and Example 8.14 and Lipschitz continuity of $x \to g(x) + c(x - \psi)$ from $L^2(\Omega)$ to $L^p(\Omega)$ to bound the right-hand side from above we find

$$\sqrt{\min(c, \alpha)}|x^{k+1} - x^*|_{L^2(\Omega)} \le o(|x^k - x^*|_{L^2(\Omega)}).$$

Finally (8.5.2) and (8.5.4) imply that

$$\mu^{k+1} - \mu^* = g(x^*) - g(x^{k+1}),$$

and hence superlinear convergence of μ^k to μ^* follows from superlinear convergence of x^k to x^*. □

A full *semismooth Newton step* applied to (8.5.5) is given by

$$g'(x^k)(x^{k+1} - x^k) + G\left(-g(x^k) + c\,(x^k - \psi)\right)\left(-g'(x)(x^{k+1} - x^k) + c\,(x^{k+1} - x^k)\right) \\ + g(x^k) + \max(0, -g(x^k) + c\,(x^k - \psi)) = 0. \tag{8.5.8}$$

It can be equivalently expressed as

$$\begin{aligned} &g'(x^k)(x^{k+1}-x^k)+g(x^k)+\mu^{k+1}=0,\\ &x^{k+1}=\psi \text{ in } \mathcal{A}_k=\{s: -g(x^k)(s)+c\,(x^k(s)-\psi(s))>0\},\\ &\mu^{k+1}=0 \text{ in } \mathcal{I}_k=\{s: -g(x^k)(s)+c\,(x^k(s)-\psi(s))\le 0\}. \end{aligned} \tag{8.5.9}$$

Remark 8.5.1. Note that, differently from the linear case considered in Section 8.4, if we apply a full semismooth Newton step to (8.5.2) rather than to the reduced equation (8.5.5), then the resulting algorithm differs in the update of the active/inactive sets. Applying a semismooth Newton step to (8.5.2) results in $\mathcal{A}_k=\{s:\mu^k(s)+c\,(x^k(s)-\psi(s))>0\}=\{s:-g(x^{k-1})(s)-g'(x^{k-1})(x^k-x^{k-1})+c\,(x^k(s)-\psi(s))>0\}$.

To investigate local convergence of (8.5.8) we denote, for any partition $\Omega=\mathcal{A}\cup\mathcal{I}$ into measurable sets $\mathcal{I}$ and Ω, by $R_\mathcal{I}: L^2(\Omega)\to L^2(\mathcal{I})$ the canonical restriction operator and by $R_\mathcal{I}^*: L^2(\mathcal{I})\to L^2(\Omega)$ its adjoint. Further we set

$$g'(x)_\mathcal{I}=R_\mathcal{I}\, g'(x)\, R_\mathcal{I}^*.$$

Proposition 8.20. *Assume that* (8.5.2) *admits a solution x^*, that $x\to g(x)-c\,(x-\psi)$ is a C^1 function from $L^2(\Omega)$ to $L^p(\Omega)$ in a neighborhood U of x^* for some $c>0$ and $p>2$, and that*

$$\{g'(x)_\mathcal{I}^{-1}\in\mathcal{L}(L^2(\mathcal{I})): x\in U,\ \Omega=\mathcal{A}\cup\mathcal{I}\} \text{ is uniformly bounded.}$$

Then the iterates x^k defined by (8.5.8) *converge superlinearly to x^*, provided that $|x^*-x^0|$ is sufficiently small.*

Proof. By Lemma 8.15 and Example 8.14 the mapping $x\to\max(0,-g(x)+c(x-\psi))$ is Newton differentiable from $L^2(\Omega)$ into itself and $G(-g(x)+c(x-\psi))(-g'(x)+cI)$ is an N-derivative. Moreover $g'(x)+G(-g(x)+c(x-\psi))(-g'+cI)$ is invertible in $\mathcal{L}(L^2(\Omega))$ with uniformly bounded inverses for $x\in U$. Setting

$$z=-g(x)+c(x-\psi),\quad \mathcal{A}=\{z>0\},\quad \mathcal{I}=\Omega\setminus\mathcal{A},\quad h_\mathcal{I}=\chi_\mathcal{I}h,\quad h_\mathcal{A}=\chi_\mathcal{A}h,$$

this follows from the fact that for given $f\in L^2(\Omega)$ the solution to the equation

$$g'(x)h+G(z)(-g'(x)h+ch)=f$$

is given by

$$ch_\mathcal{A}=f_\mathcal{A}\quad\text{and}\quad h_\mathcal{I}=g'_\mathcal{I}(x)^{-1}\left(f_\mathcal{I}-\frac{1}{c}\chi_\mathcal{I}\,g'(x)f_\mathcal{A}\right).$$

From Theorem 8.16 we conclude that $x^k\to x^*$ superlinearly, provided that $|x^*-x^0|$ is sufficiently small. $\square$

8.6 Semismooth Newton methods and regularization

In this section we discuss the case where the operator A of Section 8.4 is not a continuous but only a closed operator in $X = L^2(\Omega)$. Let V denote a real Hilbert space that is densely and continuously embedded into X and let V^* denote its dual. For $\psi \in V$ let $\mathcal{C} = \{y \in V : y \le \psi\}$.

We identify X^* with X and thus we have the Gel'fand triple framework: $V \subset X = X^* \subset V^*$. Let $a : V \times V \to \mathbb{R}$ be a bounded coercive bilinear form, i.e., for $M, \ \nu > 0$

$$|a(y, v)| \le M\,|y|_V|v|_V \text{ for all } y, v \in V,$$

$$a(v, v) \ge \nu\,|v|_V^2 \text{ for all } v \in V.$$

For given $f \in V^*$ consider the variational inequality for $y \in \mathcal{C}$:

$$a(y, v - y) - \langle f, v - y\rangle \ge 0 \text{ for all } v \in \mathcal{C}. \tag{8.6.1}$$

The existence of solutions to (8.6.1) is established in Theorem 8.26. The solution to (8.6.1) is unique. In fact if $\tilde{y} \in \mathcal{C}$ is a solution to (8.6.1), we have

$$a(y, \tilde{y} - y) - \langle f, \tilde{y} - y\rangle \ge 0,$$

$$a(\tilde{y}, y - \tilde{y}) - \langle f, y - \tilde{y}\rangle \ge 0.$$

Summing these inequalities, we have $a(y - \tilde{y}, y - \tilde{y}) \le 0$ and thus $\tilde{y} = y$.

If a is symmetric, then (8.6.1) is the necessary and sufficient optimality condition for the constrained minimization problem in V

$$\min \frac{1}{2}\,a(y, y) - \langle f, y\rangle \quad \text{over } y \in C. \tag{8.6.2}$$

Let us define $A \in \mathcal{L}(V, V^*)$ by

$$\langle Ay, v\rangle_{V^* \times V} = a(y, v) \quad \text{for } y, v \in V.$$

We note that (8.6.2) is equivalent to

$$Ay + \mu = f, \quad y \le \psi, \quad \mu \in C^+, \quad \langle \mu, y - \psi\rangle_{V^*,V} = 0, \tag{8.6.3}$$

where the first equality holds in V^* and $C^+ = \{\mu \in V^* : \langle \mu, v\rangle_{V^*,V} \le 0 \text{ for all } v \le 0\}$. If μ (or equivalently y) has extra regularity in the sense that $\mu \in L^2(\Omega)$, then (8.6.3) is equivalent to

$$\begin{cases} Ay + \mu = f, \\ \mu = \max(0, \mu + c(y - \psi)) \end{cases} \tag{8.6.4}$$

for each $c > 0$.

Example 8.21. For the obstacle problem in its simplest form we choose $V = H_0^1(\Omega)$ and $a(v, w) = \int_\Omega \nabla v \nabla w \, dx$. In general, the unique solution (y, μ) is in $L^2(\Omega) \times H^{-1}(\Omega)$. If $\partial\Omega$ and ψ are sufficiently regular, then $(y, \mu) \in (H_0^1(\Omega) \cap H^2(\Omega)) \times L^2(\Omega)$ and (8.6.3) is equivalent to (8.6.4).

Example 8.22. Consider the state-constrained optimal control problem with $\beta > 0$ and $\bar{y} \in L^2(\Omega)$

$$\min_{u \in L^2(\Omega)} \quad \frac{1}{2}|y - \bar{y}|^2_{L^2(\Omega)} + \frac{\beta}{2}|u|^2_{L^2(\Omega)} \quad \text{subject to } Ey = u \text{ and } y \in \mathcal{C}, \tag{8.6.5}$$

where E is a closed linear operator in $L^2(\Omega)$. We assume that E^{-1} exists and set $V = \operatorname{dom}(E)$, where $\operatorname{dom}(E)$ is endowed with the graph norm of E. For $E = -\Delta$ with homogeneous Dirichlet boundary conditions, we have $\operatorname{dom} E = H_0^1(\Omega) \cap H^2(\Omega)$. The necessary and sufficient optimality condition for (8.6.5) is given by

$$\begin{aligned} &Ey = u, \quad y \in \mathcal{C}, \quad \beta u = p, \\ &\langle E^* p + y - \bar{y}, v - y \rangle_{V^* \times V} \geq 0 \quad \text{for all } v \in \mathcal{C}, \end{aligned} \tag{8.6.6}$$

where $E^* : X \to V^*$ denotes the conjugate of E as operator from V to X. This optimality system can be written as (8.6.1) with $f = \bar{y}$ and

$$a(v, w) = \beta\,(Ev, Ew)_X + (v, w) \quad \text{for } v, \ w \in V.$$

It can also be written in the form (8.6.3) with $\mu \in V^*$, but differently from Example 8.21, $\mu \notin L^2(\Omega)$ in general; see [BeK3].

In the case of mixed constraints, i.e., when pointwise constraints on the controls and the state are present, it is natural to treat the control constraints with the active set methods presented in Chapter 7 and to relax the state constraints by the technique explained in this section.

Thus the obstacle and the state-constrained optimal control problem differ in that for the former, under weak regularity conditions on the problem data, the Lagrange multiplier is in $L^2(\Omega)$ and complementarity can be expressed in the form of (8.6.4), whereas for the latter this is not the case. Before we can address the applicability of the primal-dual active set strategy or semismooth Newton methods to such problems we also need to consider the candidates for the iterates. In the case of the obstacle problem these are given formally by

$$\begin{aligned} &a(y^{k+1}, v) + (\mu^{k+1}, v)_{L^2(\Omega)} = (f, v)_{L^2(\Omega)} \text{ for all } v \in H_0^1(\Omega), \\ &y^{k+1} = \psi \text{ in } \mathcal{A}_k = \{x : \mu^k + c\,(y^k(x) - \psi(x)) > 0\}, \\ &\mu^{k+1} = 0 \text{ in } \mathcal{I}_k = \{x : \mu^k + c\,(y^k(x) - \psi(x)) \leq 0\}. \end{aligned} \tag{8.6.7}$$

This is the optimality condition for

$$\min \quad \frac{1}{2}\int_\Omega |\nabla y|^2 \, dx - (f, y)_{L^2(\Omega)} \quad \text{over } y \in H_0^1(\Omega) \text{ with } y = \psi \text{ on } \mathcal{A}_k,$$

for which the Lagrange μ^{k+1} is not in $L^2(\Omega)$, and hence an update of $\mathcal{A}_k$ as in (8.6.7) is not feasible. Alternatively, considering the possibility of applying a semismooth Newton approach we observe that the reduced form of (8.6.4) is given by

$$\mu = \max(0, \mu + c(A^{-1}(f - \mu) - \psi)),$$

so that no smoothing of μ under the max operation occurs.

In order to remedy these difficulties related to the regularity properties of the Lagrange multiplier we consider a one-parameter family of regularized problems based on smoothing of the complementarity condition given by

$$\mu = \alpha \max(0, \mu + c(y - \psi)), \quad \text{with } 0 < \alpha < 1. \tag{8.6.8}$$

This is a relaxation of the second equation in (8.6.4) with α as a continuation parameter. Note that an update for μ based on (8.6.8) results in $\mu \in L^2(\Omega)$. Equation (8.6.8) is equivalent to

$$\mu = \max\left(0, \frac{c\alpha}{1-\alpha}(y - \psi)\right), \tag{8.6.9}$$

with $c\alpha/(1-\alpha)$ ranging in $(0, \infty)$ for $\alpha \in (0, 1)$. We shall use a generalization of (8.6.8) and introduce an additional shift parameter $\bar{\mu} \in L^2(\Omega)$ into (8.6.9). Moreover we replace $c\alpha/(1-\alpha)$ by c and arrive at

$$\mu = \max(0, \bar{\mu} + c\,(y - \psi)), \quad \text{with } c \in (0, \infty). \tag{8.6.10}$$

This is exactly the same as the generalized Yosida–Moreau approximation for inequality constraints discussed in Chapter 4.7 and is related to augmented Lagrangian methods as described in Chapter 4.6. Utilizing this regularization in (8.6.4) results in

$$\begin{cases} Ay + \mu = f, \\ \mu = \max(0, \bar{\mu} + c\,(y - \psi)). \end{cases} \tag{8.6.11}$$

Note that for each $c > 0$

$$y \to \max(0, \bar{\mu} + c\,(y - \psi))$$

is Lipschitz continuous and monotone from X to X. Thus existence of a unique $y_c \in V$ satisfying

$$Ay + \max(0, \bar{\mu} + c\,(y - \psi)) = f$$

follows by monotone operator techniques; see, e.g., the discussion above Theorem 4.43. This implies the existence of a unique solution $(y_c, \mu_c) \in V \times L^2(\Omega)$ to (8.6.11). Convergence as $c \to \infty$ will be analyzed at the end of this section.

The semismooth Newton iteration for (8.6.11) is discussed next.

Semismooth Newton Algorithm with Regularization.

(i) Choose μ, c, y_0, set $k = 0$.

(ii) Set $\mathcal{A}_k = \{x : (\bar{\mu} + c\,(y^k - \psi))(x) > 0\}$, $\mathcal{I}_k = \Omega \setminus \mathcal{A}_k$.

(iii) Solve for $y^{k+1} \in V$:

$$a(y, v) + (\bar{\mu} + c\,(y - \psi),\, \chi_{\mathcal{A}_k} v)_{L^2(\Omega)} = (f, v)_{L^2(\Omega)} \quad \text{for all } \ v \in V.$$

(iv) Set

$$\mu^{k+1} = \begin{cases} 0 & \text{on} \quad \mathcal{I}_k, \\ \bar{\mu} + c\,(y^{k+1} - \psi) & \text{on} \quad \mathcal{A}_k. \end{cases}$$

(v) Stop, or set $k = k + 1$, goto (ii).

The iterates μ^k as assigned in step (iv) are not necessary for the algorithm, but they will be useful in the convergence analysis. The practical relevance of $\bar{\mu}$ is given by the fact that for certain problems a proper choice can guarantee feasibility of the iterates, i.e., $y_k \le \psi$, as will be shown next.

Theorem 8.23. *Assume that $a(\phi, \phi^+) \le 0$, for $\phi \in V$, implies that $\phi = 0$. Then if*

$$\bar{\mu} \ge (f - A\psi)^+$$

in the sense that

$$(\bar{\mu}, \phi) \ge \langle f, \phi \rangle - a(\psi, \phi) \tag{8.6.12}$$

for all $\phi \in C$, the solution to (6.11) *is feasible, i.e., $y_c \le \psi$.*

Proof. Since for $\phi = (y_c - \psi)^+$

$$\big(\max 0,\, \bar{\mu} + c(y_c - \psi),\, \phi\big) \ge (\bar{\mu}, \phi),$$

it follows from (8.6.11) and (8.6.12) that

$$a(y - \psi, \phi) = \langle f, \phi \rangle - a(\psi, \phi) - (\bar{\mu}, \phi) \le 0.$$

This implies that $\phi = 0$ and thus $y_c \le \psi$. $\square$

Theorem 8.24. *If $a(\phi, \phi^+) \le 0$, for $\phi \in V$, implies that $\phi = 0$, then the iterates of the semismooth Newton algorithm with regularization satisfy $y^{k+1} \le y^k$ for all $k \ge 1$.*

Proof. Let $\phi = (y^{k+1} - y^k)^+$ and observe that

$$a(y^{k+1} - y^k, \phi) + a(y^k, \phi) - \langle f, \phi \rangle + (\bar{\mu} + c\,(y^k - \psi), \phi\, \chi_{\mathcal{A}_k}) + c(y^{k+1} - y^k, \phi\, \chi_{\mathcal{A}_k}) = 0.$$

We have

$$\begin{aligned} &a(y^k, \phi) - \langle f, \phi \rangle + (\bar{\mu} + c(y^k - \psi), \phi\, \chi_{\mathcal{A}_k}) \\ &= -(\bar{\mu} + c(y^k - \psi), \phi\, \chi_{\mathcal{A}_{k-1}}) + (\bar{\mu} + c(y^k - \psi), \phi\, \chi_{\mathcal{A}_k}) \\ &= (\bar{\mu} + c(y^k - \psi), \phi\, \chi_{\mathcal{A}_k \cap \mathcal{I}_{k-1}}) - (\bar{\mu} + c(y^k - \psi), \phi\, \chi_{\mathcal{I}_k \cap \mathcal{A}_{k-1}}) \ge 0. \end{aligned}$$

This implies that

$$a(y^{k+1}-y^k,\phi)+c(y^{k+1}-y^k,\phi\,\chi_{\mathcal{A}_k})\le 0;$$

thus, $a(y^{k+1}-y^k,\phi)\le 0$ and hence $\phi=0$. $\square$

Let us stress that in a finite-dimensional or discretized setting with $L^2(\Omega)$ replaced by $\mathbb{R}^n$ the need for regularization is not apparent, since the lack of regularity, as exhibited in Examples 8.21 and 8.22 and in the discussion of the iterative step of the semismooth Newton algorithm, does not exist. If the finite-dimensional problems arise from discretization of continuous ones, then the regularity of the Lagrange multipliers, however, certainly influences the convergence and approximation properties. It will typically lead to mesh-size-dependent behavior of the semismooth Newton algorithm.

We turn to convergence of the semismooth Newton algorithm with regularization. Recall that $A^{-1}\in\mathcal{L}(V^*,V)$. Below we shall also denote the restriction of A^{-1} to $L^2(\Omega)$ by the same symbol.

Theorem 8.25. *Assume that $\bar{\mu}-c\psi\in L^p(\Omega)$ and $A^{-1}\in\mathcal{L}(L^2(\Omega),L^p(\Omega))$ for some $p>2$. If $\mu_0\in L^2(\Omega)$ and $|\mu_0-\mu_c|_{L^2(\Omega)}$ is sufficiently small, then $(y^k,\mu^k)\to(y_c,\mu_c)$ superlinearly in $V\times L^2(\Omega)$.*

Proof. First we show superlinear convergence of μ^k to μ_c by applying Theorem 8.16 to $F: L^2(\Omega)\to L^2(\Omega)$ defined by

$$F(\mu)=\mu-\max(0,\bar{\mu}+c(A^{-1}(f-\mu)-\psi)).$$

By Example 8.14 and Lemma 8.15 the mapping F is Newton differentiable with N-derivative given by

$$\tilde{G}(\mu)=I+c\,G(\bar{\mu}+c(A^{-1}(f-\mu)-\psi))A^{-1},$$

where G was defined in (8.4.6). The proof will be completed by showing that $\tilde{G}(\mu)$ has uniformly bounded inverses in $\mathcal{L}(L^2(\Omega))$ for $\mu\in L^2(\Omega)$. We define

$$\mathcal{A}=\{x:(\bar{\mu}+c(A^{-1}(f-\mu)-\psi))(x)>0\},\quad \mathcal{I}=\Omega\setminus\mathcal{A}.$$

Further, let $R_{\mathcal{A}}: L^2(\Omega)\to L^2(\mathcal{A})$ and $R_{\mathcal{I}}: L^2(\Omega)\to L^2(\mathcal{I})$ denote the restriction operators to $\mathcal{A}$ and $\mathcal{I}$. Their adjoints $R^*_{\mathcal{A}}: L^2(\mathcal{A})\to L^2(\Omega)$ and $R^*_{\mathcal{I}}: L^2(\mathcal{I})\to L^2(\Omega)$ are the extension-by-zero operators from $\mathcal{A}$ and $\mathcal{I}$ to Ω, respectively. The mapping $(R_{\mathcal{A}},R_{\mathcal{I}}): L^2(\Omega)\to L^2(\mathcal{A})\times L^2(\mathcal{I})$ determines an isometric isomorphism and every $\mu\in L^2(\Omega)$ can uniquely be expressed as $(R_{\mathcal{A}}\,\mu, R_{\mathcal{I}}\,\mu)$. The operator $\tilde{G}(\mu)$ can equivalently be expressed as

$$\tilde{G}(\mu)=\begin{pmatrix} I_{\mathcal{A}} & 0\\ 0 & I_{\mathcal{I}}\end{pmatrix}+c\begin{pmatrix} R_{\mathcal{A}}A^{-1}R^*_{\mathcal{A}} & R_{\mathcal{A}}A^{-1}R^*_{\mathcal{I}}\\ 0 & 0\end{pmatrix},$$

where $I_{\mathcal{A}}$ and $I_{\mathcal{I}}$ denote the identity operators on $L^2(\mathcal{A})$ and $L^2(\mathcal{I})$. Let $(g_{\mathcal{A}},g_{\mathcal{I}})\in L^2(\mathcal{A})\times L^2(\mathcal{I})$ be arbitrary and consider the equation

$$\tilde{G}(\mu)((\delta\mu)_{\mathcal{A}},(\delta\mu)_{\mathcal{I}})=(g_{\mathcal{A}},g_{\mathcal{I}}).\tag{8.6.13}$$

Then necessarily $(\delta\mu)_{\mathcal{I}} = g_{\mathcal{I}}$ and (8.6.13) is equivalent to

$$(\delta\mu)_{\mathcal{A}} + c\, R_{\mathcal{A}} A^{-1} R^*_{\mathcal{A}}\, (\delta\mu)_{\mathcal{A}} = g_{\mathcal{A}} - c\, R_{\mathcal{A}} A^{-1} R^*_{\mathcal{I}}\, g_{\mathcal{I}}. \qquad (8.6.14)$$

The Lax–Milgram theorem and nonnegativity of A^{-1} imply the existence of a unique solution $(\delta\mu)_{\mathcal{A}}$ to (8.6.14) and consequently (8.6.13) has a unique solution for every $(g_{\mathcal{A}}, g_{\mathcal{I}})$ and every μ. Moreover these solutions are uniformly bounded with respect to $\mu \in L^2$ since $(\delta\mu)_{\mathcal{I}} = g_{\mathcal{I}}$ and

$$|\delta\mu_{\mathcal{A}}|_{L^2(\mathcal{A})} \le |g_{\mathcal{A}}|_{L^2(\Omega)} + c\|A^{-1}\|_{\mathcal{L}(L^2(\Omega))}\, |g_{\mathcal{I}}|_{L^2(\mathcal{I})}.$$

This proves superlinear convergence $\mu^k \to \mu_c$ in $L^2(\Omega)$. Superlinear convergence of y^k to y_c in V follows from $Ay^k + \mu^k = f$ and the fact that $A : V^* \to V$ is a homeomorphism. ☐

Convergence of the solutions (y_c, μ_c) to (8.6.11) as $c \to \infty$ is addressed next.

Theorem 8.26. *The solution $(y_c, \mu_c) \in V \times X$ of the regularized problem* (8.6.11) *converges to the solution $(y^*, \mu^*) \in V \times V^*$ of* (8.6.3) *in the sense that $y_c \to y^*$ strongly in V and $\mu_c \to \mu^*$ strongly in V^* as $c \to \infty$.*

Proof. Recall that $y_c \in V$ satisfies

$$\begin{cases} a(y_c, v) + (\mu_c, v) = \langle f, v\rangle \quad \text{for all } v \in V, \\ \mu_c = \max(0, \bar{\mu} + c(y_c - \psi)). \end{cases} \qquad (8.6.15)$$

Since $\mu_c \ge 0$, for $y \in \mathcal{C}$

$$(\mu_c, y_c - y) = (\mu_c, y_c - \psi - (y - \psi)) \ge \frac{c}{2}|(y_c - \psi)^+|^2_X - \frac{1}{2c}|\bar{\mu}|^2_X.$$

Letting $v = y_c - y$ in (8.6.15),

$$a(y_c, y_c - y) + \frac{c}{2}|(y_c - \psi)^+|^2_X \le \langle f, y_c - y\rangle + \frac{1}{2c}|\bar{\mu}|^2_X. \qquad (8.6.16)$$

Since a is coercive, this implies that

$$\nu\, |y_c|^2_V + \frac{c}{2}|(y_c - \psi)^+|^2_X \text{ is bounded uniformly in } c.$$

Thus there exist a subsequence y_c, denoted by the same symbol, and $y^* \in V$ such that $y_c \to y^*$ weakly in V. For all $\phi \ge 0$

$$((y_c - \psi)^+, \phi)_X \ge (y_c - \psi, \phi)_X.$$

Since $(y_c - \psi)^+ \to 0$ in X and $y_c \to y^*$ weakly in X as $c \to \infty$, this yields

$$(y^* - \psi, \phi)_X \le 0 \text{ for all } \phi \ge 0$$

which implies $y^* - \psi \leq 0$ and thus $y^* \in \mathcal{C}$. Since $\phi \to \sqrt{a(\phi, \phi)}$ defines an equivalent norm on V and norms are w.l.s.c., letting $c \to 0$ in (8.6.16) we obtain

$$a(y^*, y^* - y) \leq \langle f, y^* - y \rangle \text{ for all } y \in \mathcal{C}$$

which implies that y^* is the solution to (8.6.1). Now, setting $y = y^*$ in (8.6.16)

$$a(y_c - y^*, y_c - y^*) + a(y^*, y_c - y^*) - \langle f, y_c - y^* \rangle \leq \frac{1}{2c}|\bar{\mu}|_X^2.$$

Since $y_c \to y^*$ weakly in V, this implies that $\lim_{c\to\infty} |y_c - y^*|_V = 0$. Hence $(y_c, \mu_c) \to (y^*, \mu^*)$ strongly in $V \times V^*$. $\square$

Chapter 9

Semismooth Newton Methods II: Applications

In the previous chapter semismooth Newton methods in function spaces were investigated. In was demonstrated that in certain cases the semismooth Newton method is equivalent to the primal-dual active set method. The application to nonlinear complementarity problems was discussed and the necessity of introducing regularization in cases where the Lagrange multiplier associated to the inequality condition has low regularity was demonstrated. In this chapter applications of semismooth Newton methods to nondifferentiable variational problems in function spaces will be treated. They concern image restoration problems regularized by bounded variation functionals in Section 9.1 and frictional contact problems in elasticity in Section 9.2.

We shall make use of the Fenchel duality theorem which we recall for further reference; see, e.g., Section 4.3 and [BaPe, EkTe] for details. Let V and Y be Banach spaces with topological duals $V^\star$ and $Y^\star$, respectively. Further, let $\Lambda \in \mathcal{L}(V, Y)$ and let $\mathcal{F} : V \longrightarrow \mathbb{R} \cup \{\infty\}$, $\mathcal{G} : Y \longrightarrow \mathbb{R} \cup \{\infty\}$ be convex, proper, and l.s.c. functionals such that there exists $v_0 \in V$ with $\mathcal{F}(v_0) < \infty$, $\mathcal{G}(\Lambda v_0) < \infty$ and $\mathcal{G}$ is continuous at Λv_0. Then

$$\inf_{v \in V} \{\mathcal{F}(v) + \mathcal{G}(\Lambda v)\} = \sup_{q \in Y^\star} \left\{-\mathcal{F}^\star(-\Lambda^\star q) - \mathcal{G}^\star(q)\right\}, \tag{9.0.1}$$

where $\Lambda^\star \in \mathcal{L}(Y^\star, V^\star)$ is the adjoint of Λ. See, e.g., Section 4.3, Theorem 4.30 and Example 4.33 with $\Phi(v, w) = \mathcal{F}(v) + \mathcal{G}(\Lambda v + w)$. The convex conjugates $\mathcal{F}^\star : V^\star \longrightarrow \mathbb{R} \cup \{\infty\}$ and $\mathcal{G}^\star : Y^\star \longrightarrow \mathbb{R} \cup \{\infty\}$ of $\mathcal{F}$ and $\mathcal{G}$, respectively, are defined by

$$\mathcal{F}^\star(v^\star) = \sup_{v \in V} \left\{\langle v, v^\star\rangle_{V,V^\star} - \mathcal{F}(v)\right\},$$

and analogously for $\mathcal{G}^\star$. The conditions imposed on $\mathcal{F}$ and $\mathcal{G}$ guarantee that the dual problem, i.e., the problem on the right-hand side of (9.0.1), admits a solution. Furthermore, $\bar{v} \in V$ and $\bar{q} \in Y^\star$ are solutions to the two optimization problems in (9.0.1) if and only if the extremality conditions

$$\begin{aligned} -\Lambda^\star \bar{q} &\in \partial\mathcal{F}(\bar{u}), \\ \bar{q} &\in \partial\mathcal{G}(\Lambda \bar{u}) \end{aligned} \tag{9.0.2}$$

hold, where $\partial\mathcal{F}$ denotes the subdifferential of $\mathcal{F}$.

9.1 BV-based image restoration problems

In this section we consider the nondifferentiable optimization problem

$$\begin{cases} \min \frac{1}{2}\int_\Omega |\mathcal{K}u - f|^2 dx + \frac{\alpha}{2}\int_\Omega |u|^2 dx + \beta \int_\Omega |Du| \\ \text{over } u \in \mathrm{BV}(\Omega), \end{cases} \tag{9.1.1}$$

where Ω is a simply connected domain in $\mathbb{R}^2$ with Lipschitz continuous boundary $\partial\Omega$, $f \in L^2(\Omega)$, $\beta > 0, \alpha \geq 0$ are given, and $\mathcal{K} \in \mathcal{L}(L^2(\Omega))$. We assume that $\mathcal{K}^*\mathcal{K}$ is invertible or $\alpha > 0$. Further $\mathrm{BV}(\Omega)$ denotes the space of functions of bounded variation. A function u is in $\mathrm{BV}(\Omega)$ if the BV-seminorm defined by

$$\int_\Omega |Du| = \sup\left\{\int_\Omega u \,\operatorname{div} \vec{v} : \vec{v} \in (C_0^\infty(\Omega))^2,\ |\vec{v}(x)|_{\ell^\infty} \leq 1\right\}$$

is finite. Here $|\cdot|_{\ell^\infty}$ denotes the supremum norm on $\mathbb{R}^2$. It is well known that $\mathrm{BV}(\Omega) \subset L^2(\Omega)$ for $\Omega \subset \mathbb{R}^2$ (see [Giu]) and that $u \mapsto |u|_{L^2} + \int_\Omega |Du|$ defines a norm on $\mathrm{BV}(\Omega)$. If $\mathcal{K}$ = identity and $\alpha = 0$, then (9.1.1) is the well-known image restoration problem with BV-regularization. It consists of recovering the true image u from the noisy image f. BV-regularization is known to be preferable to regularization by $\int_\Omega |\nabla u|^2\, dx$, for example, due to its ability to preserve edges in the original image during the reconstruction process. Since the pioneering work in [ROF], the literature on (9.1.1) has grown tremendously. We give some selected references [AcVo, CKP, ChLi, ChK, DoSa, GeYa, IK12] and refer the reader to the monograph [Vog] for additional ones.

Despite its favorable properties for reconstruction of images, and especially images with blocky structure, problem (9.1.1) poses some severe difficulties. On the analytical level these are related to the fact that (9.1.1) is posed in a nonreflexive Banach space, the dual of which is difficult to characterize [Giu, IK18], and on the numerical level the optimality system related to (9.1.1) consists of a nonlinear partial differential equation, which is not directly amenable to numerical implementations.

Following [HinK1] we shall show that the predual of (9.1.1) is a bilaterally constrained optimization problem in a Hilbert space, for which the primal-dual active set strategy can advantageously be applied and which can be analyzed as a semismooth Newton method. We require some facts from vector-valued function spaces, which we summarize next. Let $\mathbb{L}^2(\Omega) = L^2(\Omega) \times L^2(\Omega)$ be endowed with the Hilbert space inner product structure and norm. If the context suggests to do so, then we shall distinguish between vector fields $\vec{v} \in \mathbb{L}^2(\Omega)$ and scalar functions $v \in L^2(\Omega)$ by using an arrow on top of the letter. Analogously we set $\mathbb{H}_0^1(\Omega) = H_0^1(\Omega) \times H_0^1(\Omega)$. We set $L_0^2(\Omega) = \{v \in L^2(\Omega) : \int_\Omega v dx = 0\}$ and $H_0(\operatorname{div}) = \{\vec{v} \in \mathbb{L}^2(\Omega) : \operatorname{div} \vec{v} \in L^2(\Omega),\ \vec{v} \cdot n = 0 \text{ on } \partial\Omega\}$, where n is the outer normal to $\partial\Omega$. The space $H_0(\operatorname{div})$ is endowed with $|\vec{v}|^2_{H_0(\operatorname{div})} = |\vec{v}|^2_{\mathbb{L}^2(\Omega)} + |\operatorname{div} \vec{v}|^2_{L^2}$ as norm. Further we put $H_0(\operatorname{div} 0) = \{\vec{v} \in H_0(\operatorname{div}) : \operatorname{div} \vec{v} = 0 \text{ a.e. in } \Omega\}$. It is well known that

$$\mathbb{L}^2(\Omega) = \operatorname{grad} H^1(\Omega) \oplus H_0(\operatorname{div} 0); \tag{9.1.2}$$

cf. [DaLi, p. 216], for example. Moreover,

$$H_0(\operatorname{div}) = H_0(\operatorname{div} 0)^\perp \oplus H_0(\operatorname{div} 0), \tag{9.1.3}$$

with

$$H_0(\operatorname{div} 0)^{\perp} = \{\vec{v} \in \operatorname{grad} H^1(\Omega) : \operatorname{div} \vec{v} \in L^2(\Omega),\ \vec{v} \cdot n = 0 \text{ on } \partial\Omega\},$$

and $\operatorname{div} : H_0(\operatorname{div} 0)^{\perp} \subset H_0(\operatorname{div}) \to L_0^2(\Omega)$ is a homeomorphism. In fact, it is injective by construction and for every $f \in L_0^2(\Omega)$ there exists, by the Lax–Milgram lemma, $\varphi \in H^1(\Omega)$ such that

$$\operatorname{div} \nabla\varphi = f \text{ in } \Omega, \qquad \nabla\varphi \cdot n = 0 \text{ on } \partial\Omega,$$

with $\nabla\varphi \in H_0(\operatorname{div} 0)^{\perp}$. Hence, by the closed mapping theorem we have

$$\operatorname{div} \in \mathcal{L}(H_0(\operatorname{div} 0)^{\perp}, L_0^2(\Omega)).$$

Finally, let $\mathrm{P}_{\mathrm{div}}$ and $\mathrm{P}_{\mathrm{div}^{\perp}}$ denote the orthogonal projections in $\mathbb{L}^2(\Omega)$ onto $H_0(\operatorname{div} 0)$ and $\operatorname{grad} H^1(\Omega)$, respectively. Note that the restrictions of $\mathrm{P}_{\mathrm{div}}$ and $\mathrm{P}_{\mathrm{div}^{\perp}}$ to $H_0(\operatorname{div} 0)$ coincide with the orthogonal projections in $H_0(\operatorname{div})$ onto $H_0(\operatorname{div} 0)$ and $H_0(\operatorname{div} 0)^{\perp}$.

Let $\vec{1}$ denote the two-dimensional vector field with 1 in both coordinates, set $B = \alpha I + \mathcal{K}^*\mathcal{K}$, and consider

$$\begin{cases} \min \frac{1}{2}|\operatorname{div} \vec{p} + \mathcal{K}^* f|_B^2 \quad \text{over } \vec{p} \in H_0(\operatorname{div}), \\ \text{such that } -\beta\vec{1} \le \vec{p} \le \beta\vec{1}, \end{cases} \tag{9.1.4}$$

where for $v \in L^2(\Omega)$ we put $|v|_B^2 = (v, B^{-1}v)_{L^2}$. It is straightforward to argue that (9.1.4) admits a solution.

Theorem 9.1. *The Fenchel dual to* (9.1.4) *is given by* (9.1.1) *and the solutions* u^* *of* (9.1.1) *and* $\vec{p}^*$ *of* (9.1.4) *are related by*

$$Bu^* = \operatorname{div} \vec{p}^* + \mathcal{K}^* f, \tag{9.1.5}$$

$$\langle(-\operatorname{div})^* u^*, \vec{p} - \vec{p}^*\rangle_{H_0(\operatorname{div})^*, H_0(\operatorname{div})} \le 0 \quad \text{for all } \vec{p} \in H_0(\operatorname{div}), \tag{9.1.6}$$

with $-\beta\vec{1} \le \vec{p} \le \beta\vec{1}$.

Alternatively (9.1.4) can be considered as the predual of the original problem (9.1.1).

Proof. We apply Fenchel duality as recalled at the beginning of the chapter with $V = H_0(\operatorname{div})$, $Y = Y^* = L^2(\Omega)$, $\Lambda = -\operatorname{div}$, $\mathcal{G} : Y \to \mathbb{R}$ given by $\mathcal{G}(v) = \frac{1}{2}|v - \mathcal{K}^* f|_B^2$, and $\mathcal{F} : V \to \mathbb{R}$ defined by $\mathcal{F}(\vec{p}) = \mathrm{I}_{[-\beta\vec{1}, \beta\vec{1}]}(\vec{p})$, where

$$\mathrm{I}_{[-\beta\vec{1}, \beta\vec{1}]}(\vec{p}) = \begin{cases} 0 & \text{if } -\beta\vec{1} \le \vec{p}(x) \le \beta\vec{1} \text{ for a.e. } x \in \Omega, \\ \infty & \text{otherwise.} \end{cases}$$

The convex conjugate $\mathcal{G}^* : L^2(\Omega) \to \mathbb{R}$ of $\mathcal{G}$ is given by

$$\mathcal{G}^*(v) = \frac{1}{2}|\mathcal{K}v + f|^2 + \frac{\alpha}{2}|v|^2 - \frac{1}{2}|f|^2.$$

Further the conjugate $\mathcal{F}^* : H_0(\operatorname{div})^* \to \mathbb{R}$ of $\mathcal{F}$ is given by

$$\mathcal{F}^*(\vec{q}) = \sup_{\vec{p} \in S_1} \langle \vec{q}, \vec{p} \rangle_{H_0(\operatorname{div})^*, H_0(\operatorname{div})} \quad \text{for } \vec{q} \in H_0(\operatorname{div})^*, \tag{9.1.7}$$

where $S_1 = \{\vec{p} \in H_0(\mathrm{div}) : -\beta\vec{1} \le \vec{p} \le \beta\vec{1}\}$. Let us set

$$S_2 = \{\vec{p} \in C_0^1(\Omega) \times C_0^1(\Omega) : -\beta\vec{1} \le \vec{p} \le \beta\vec{1}\}.$$

The set S_2 is dense in the topology of $H_0(\mathrm{div})$ in S_1. In fact, let $\vec{p}$ be an arbitrary element of S_1. Since $(\mathcal{D}(\Omega))^2$ is dense in $H_0(\mathrm{div})$ (see, e.g., [GiRa, p. 26]), there exists a sequence $\vec{p}_n \in (\mathcal{D}(\Omega))^2$ converging in $H_0(\mathrm{div})$ to $\vec{p}$. Let $\mathcal{P}$ denote the canonical projection in $H_0(\mathrm{div})$ onto the closed convex subset S_1 and note that, since $\vec{p} \in S_1$,

$$\begin{aligned}|\vec{p} - \mathcal{P}\vec{p}_n|_{H_0(\mathrm{div})} &\le |\vec{p} - \vec{p}_n|_{H_0(\mathrm{div})} + |\vec{p}_n - \mathcal{P}\vec{p}_n|_{H_0(\mathrm{div})} \\ &\le 2|\vec{p} - \vec{p}_n|_{H_0(\mathrm{div})} \to 0 \quad \text{for } n \to \infty.\end{aligned}$$

Hence $\lim_{n\to\infty} |\vec{p} - \mathcal{P}\vec{p}_n|_{H_0(\mathrm{div})} = 0$ and S_2 is dense in S_1. Returning to (9.1.7) we have for $v \in L^2(\Omega)$ and $(-\mathrm{div})^* \in \mathcal{L}(L^2(\Omega), V^*)$,

$$\mathcal{F}^*((-\mathrm{div})^* v) = \sup_{\vec{p}\in S_2} (v, -\mathrm{div}\,\vec{p}),$$

which can be $+\infty$. By the definition of the functions of bounded variation it is finite if and only if $v \in \mathrm{BV}(\Omega)$ (see [Giu, p. 3]) and

$$\mathcal{F}^*((-\mathrm{div})^* v) = \beta \int_\Omega |\mathrm{D}v| < \infty \quad \text{for } v \in \mathrm{BV}(\Omega).$$

The dual problem to (9.1.4) is found to be

$$\min \frac{1}{2}|\mathcal{K}u - f|^2 + \frac{\alpha}{2}|u|^2 + \beta \int_\Omega |\mathrm{D}u| \quad \text{over } u \in \mathrm{BV}(\Omega).$$

From (9.0.2) moreover we find

$$\langle (-\mathrm{div})^* u^*, \vec{p} - \vec{p}^{\,*} \rangle_{H_0(\mathrm{div})^*, H_0(\mathrm{div})} \le 0 \quad \text{for all } p \in S_1$$

and

$$Bu^* = \mathrm{div}\,\vec{p}^{\,*} + \mathcal{K}^* f. \qquad \square$$

We obtain the following optimality system for (9.1.4).

Corollary 9.2. *Let $\vec{p}^{\,*} \in H_0(\mathrm{div})$ be a solution to (9.1.4). Then there exists $\vec{\lambda}^* \in H_0(\mathrm{div})^*$ such that*

$$\mathrm{div}^* B^{-1} \mathrm{div}\,\vec{p}^{\,*} + \mathrm{div}^* B^{-1}\mathcal{K}^* f + \vec{\lambda}^* = 0, \tag{9.1.8}$$

$$\langle \vec{\lambda}^*, \vec{p} - \vec{p}^{\,*} \rangle_{H_0(\mathrm{div})^*, H_0(\mathrm{div})} \le 0 \quad \textit{for all } \vec{p} \in H_0(\mathrm{div}), \tag{9.1.9}$$

with $-\beta\vec{1} \le \vec{p} \le \beta\vec{1}$.

Proof. Apply $\mathrm{div}^* B^{-1}$ to (9.1.5) and set $\vec{\lambda}^* = -\mathrm{div}^* u^* \in H_0(\mathrm{div})^*$ to obtain (9.1.8). For this choice of $\vec{\lambda}^*$, (9.1.9) follows from (9.1.6). $\square$

To guarantee uniqueness of the solutions we replace (9.1.4) by

$$\begin{cases} \min \frac{1}{2}|\operatorname{div} \vec{p} + \mathcal{K}^* f|_B^2 + \frac{\gamma}{2}|\vec{p}|^2 \quad \text{over } \vec{p} \in H_0(\operatorname{div}) \\ \text{such that } -\beta\vec{1} \le \vec{p} \le \beta\vec{1} \end{cases} \tag{9.1.10}$$

with $\gamma > 0$. Clearly (9.1.10) admits a unique solution.

Our goal now is to investigate the primal-dual active set strategy for (9.1.10). If we were to simply apply this method to the first order necessary optimality condition for (9.1.10), then the resulting nonlinear operator arising in the complementarity system is not Newton differentiable; cf. Section 8.6. The numerical results which are obtained in this way are competitive from the point of view of image reconstruction [HinK1], but due to lack of Newton differentiability the iteration numbers are mesh-dependent. We therefore approximate (9.1.10) by a sequence of more regular problems where the constraints $-\beta\vec{1} \le \vec{p} \le \beta\vec{1}$ are realized by penalty terms and an $\mathbb{H}_0^1(\Omega)$ smoothing guarantees superlinear convergence of semismooth Newton methods applied to the first order optimality conditions of the approximating problems:

$$\begin{cases} \min \frac{1}{2c}|\nabla\vec{p}|^2 + \frac{1}{2}|\operatorname{div} \vec{p} + \mathcal{K}^* f|_B^2 + \frac{\gamma}{2}|\vec{p}|^2 \\ \qquad + \frac{1}{2c}|\max(0, c(\vec{p} - \beta\vec{1}))|^2 + \frac{1}{2c}|\min(0, c(\vec{p} + \beta\vec{1}))|^2 \text{ over } \vec{p} \in \mathbb{H}_0^1(\Omega). \end{cases} \tag{9.1.11}$$

Let $\vec{p}_c$ denote the unique solution to (9.1.11). It satisfies the optimality condition

$$-\frac{1}{c}\Delta\vec{p}_c - \nabla B^{-1}\operatorname{div} \vec{p}_c - \nabla B^{-1}\mathcal{K}^* f + \gamma\vec{p}_c + \vec{\lambda}_c = 0, \tag{9.1.12a}$$

$$\vec{\lambda}_c = \max(0, c(\vec{p}_c - \beta\vec{1})) + \min(0, c(\vec{p}_c + \beta\vec{1})). \tag{9.1.12b}$$

Next we address convergence as $c \to \infty$.

Theorem 9.3. *The family $\{(\vec{p}_c, \vec{\lambda}_c)\}_{c>0}$ converges weakly in $H_0(\operatorname{div}) \times \mathbb{H}_0^1(\Omega)^*$ to the unique solution $(\vec{p}^*, \vec{\lambda}^*) \in H_0(\operatorname{div}) \times H_0(\operatorname{div})^*$ of the optimality system associated to* (9.1.10) *given by*

$$\operatorname{div}^* B^{-1}\operatorname{div} \vec{p}^* + \operatorname{div}^* B^{-1}\mathcal{K}^* f + \gamma\vec{p}^* + \vec{\lambda}^* = 0, \tag{9.1.13}$$

$$\langle\vec{\lambda}^*, \vec{p} - \vec{p}^*\rangle_{H_0(\operatorname{div})^*, H_0(\operatorname{div})} \le 0 \quad \textit{for all } \vec{p} \in H_0(\operatorname{div}). \tag{9.1.14}$$

Moreover, the convergence of $\vec{p}_c$ to $\vec{p}^*$ is strong in $H_0(\operatorname{div})$.

Proof. The proof is related to that of Theorem 8.26. The variational form of (9.1.13) is given by

$$(\operatorname{div} \vec{p}^*, \operatorname{div} \vec{v})_B + (\mathcal{K}^* f, \operatorname{div} \vec{v})_B + \gamma(\vec{p}^*, \vec{v}) + \langle\vec{\lambda}^*, \vec{v}\rangle_{H_0(\operatorname{div})^*, H_0(\operatorname{div})} = 0 \tag{9.1.15}$$

for all $\vec{v} \in H_0(\operatorname{div})$. To verify uniqueness, let us suppose that $(\vec{p}_i, \vec{\lambda}_i) \in H_0(\operatorname{div}) \times H_0(\operatorname{div})^*$, $i = 1, 2$, are two solution pairs to (9.1.13), (9.1.14). For $\delta\vec{p} = \vec{p}_2 - \vec{p}_1$, $\delta\vec{\lambda} = \vec{\lambda}_2 - \vec{\lambda}_1$ we have

$$(B^{-1}\operatorname{div} \delta\vec{p}, \operatorname{div} \vec{v}) + \gamma(\delta\vec{p}, \vec{v}) + \langle\delta\vec{\lambda}, \vec{v}\rangle_{H_0(\operatorname{div})^*, H_0(\operatorname{div})} = 0 \tag{9.1.16}$$

for all $\vec{v} \in H_0(\text{div})$, and

$$\langle \delta\vec{\lambda}, \delta\vec{p}\rangle_{H_0(\text{div})^*, H_0(\text{div})} \geq 0.$$

With $\vec{v} = \delta\vec{p}$ in (9.1.16) we obtain

$$|B^{-1} \operatorname{div} \delta\vec{p}|^2 + \gamma |\delta\vec{p}|^2 \leq 0$$

and hence $\vec{p}_1 = \vec{p}_2$. From (9.1.15) we deduce that $\vec{\lambda}_1 = \vec{\lambda}_2$. Thus uniqueness is established and we can henceforth rely on subsequential arguments.

In the following computation we consider the coordinates $\vec{\lambda}_c^i$, $i = 1, 2$, of $\vec{\lambda}_c$. We have for a.e. $x \in \Omega$

$$\begin{aligned} \vec{\lambda}_c^i \vec{p}_c^{\,i} &= \left(\max(0, c(\vec{p}_c^{\,i} - \beta)) + \min(0, c(\vec{p}_c^{\,i} + \beta))\right) \vec{p}_c^{\,i} \\ &= \begin{cases} c(\vec{p}_c^{\,i} - \beta)\vec{p}_c^{\,i} & \text{if } \vec{p}_c^{\,i} \geq \beta, \\ 0 & \text{if } |\vec{p}_c^{\,i}| = \beta, \\ c(\vec{p}_c^{\,i} + \beta)\vec{p}_c^{\,i} & \text{if } \vec{p}_c^{\,i} \leq \beta. \end{cases} \end{aligned}$$

It follows that

$$(\vec{\lambda}_c^i, \vec{p}_c^{\,i})_{\mathbb{L}^2(\Omega)} \geq \frac{1}{c} |\vec{\lambda}_c^i|^2_{\mathbb{L}^2(\Omega)} \quad \text{for } i = 1, 2,$$

and consequently

$$(\vec{\lambda}_c, \vec{p}_c)_{\mathbb{L}^2(\Omega)} \geq \frac{1}{c} |\vec{\lambda}_c|^2_{\mathbb{L}^2(\Omega)} \quad \text{for every } c > 0. \tag{9.1.17}$$

From (9.1.12) and (9.1.17) we deduce that

$$\frac{1}{c} |\nabla \vec{p}_c|^2 + |\operatorname{div} \vec{p}_c|_B^2 + \gamma |\vec{p}_c|^2 \leq |\operatorname{div} \vec{p}_c|_B |\mathcal{K}^* f|_B$$

and hence

$$\frac{1}{c} |\nabla \vec{p}_c|^2 + \frac{1}{2} |\operatorname{div} \vec{p}_c|_B^2 + \gamma |\vec{p}_c|^2 \leq \frac{1}{2} |\mathcal{K}^* f|_B. \tag{9.1.18}$$

We further estimate

$$\begin{aligned} |\vec{\lambda}_c|_{\mathbb{H}_0^1(\Omega)^*} &= \sup_{|\vec{v}|_{\mathbb{H}_0^1(\Omega)} = 1} \langle \vec{\lambda}_c, \vec{v}\rangle_{\mathbb{H}_0^1(\Omega)^*, \mathbb{H}_0^1(\Omega)} \\ &\leq \sup_{|\vec{v}|_{\mathbb{H}_0^1(\Omega)} = 1} \left\{ \frac{1}{c} |\nabla \vec{p}_c| |\nabla \vec{v}| + |\operatorname{div} \vec{p}_c|_B |\operatorname{div} \vec{v}|_B + |\mathcal{K}^* f|_B |\operatorname{div} \vec{v}|_B + \gamma |\vec{p}_c| \, |\vec{v}| \right\}. \end{aligned}$$

From (9.1.18) we deduce the existence of a constant K, independent of $c \geq 1$, such that

$$|\vec{\lambda}_c|_{\mathbb{H}_0^1(\Omega)^*} \leq K. \tag{9.1.19}$$

Combining (9.1.18) and (9.1.19) we can assert the existence of $(\vec{p}^*, \vec{\lambda}^*) \in H_0(\text{div}) \times \mathbb{H}_0^1(\Omega)^*$ such that for a subsequence denoted by the same symbol

$$(\vec{p}_c, \vec{\lambda}_c) \rightharpoonup (\vec{p}^*, \vec{\lambda}^*) \text{ weakly in } H_0(\text{div}) \times \mathbb{H}_0^1(\Omega)^*. \tag{9.1.20}$$

We recall the variational form of (9.1.12), i.e.,

$$\frac{1}{c}(\nabla \vec{p}_c, \nabla \vec{v}) + (\operatorname{div} \vec{p}_c, \operatorname{div} \vec{v})_B + (\mathcal{K}^* f, \operatorname{div} \vec{v})_B + \gamma(\vec{p}_c, \vec{v}) + (\vec{\lambda}_c, \vec{v}) = 0$$

for all $\vec{v} \in \mathbb{H}_0^1(\Omega)$. Passing to the limit $c \to \infty$, using (9.1.18) and (9.1.20) we have

$$\begin{aligned}(\operatorname{div} \vec{p}^*, \operatorname{div} \vec{v})_B + &(\mathcal{K}^* f, \operatorname{div} \vec{v})_B + \gamma(\vec{p}^*, \vec{v}) \\ &+ \langle \vec{\lambda}^*, \vec{v}\rangle_{\mathbb{H}_0^1(\Omega)^*, \mathbb{H}_0^1(\Omega)} = 0 \quad \text{for all } \vec{v} \in \mathbb{H}_0^1(\Omega).\end{aligned} \tag{9.1.21}$$

Since $\mathbb{H}_0^1(\Omega)$ is dense in $H_0(\operatorname{div})$ and $\vec{p}^* \in H_0(\operatorname{div})$ we have that (9.1.21) holds for all $\vec{v} \in H_0(\operatorname{div})$. Consequently $\vec{\lambda}^*$ can be identified with an element in $H_0(\operatorname{div})^*$ and $\langle \cdot, \cdot \rangle_{\mathbb{H}_0^1(\Omega)^*, \mathbb{H}_0^1(\Omega)}$ in (9.1.21) can be replaced by $\langle \cdot, \cdot \rangle_{H_0(\operatorname{div})^*, H_0(\operatorname{div})}$. We next verify that $\vec{p}^*$ is feasible. For this purpose note that

$$(\vec{\lambda}_c, \vec{p} - \vec{p}_c) = \big(\max(0, c(\vec{p}_c - \beta\vec{1})) + \min(0, c(\vec{p}_c + \beta\vec{1}), \vec{p} - \vec{p}_c)\big) \le 0 \tag{9.1.22}$$

for all $-\beta\vec{1} \le \vec{p} \le \beta\vec{1}$. From (9.1.11) we have

$$\frac{1}{c}|\nabla \vec{p}_c|^2 + |\operatorname{div} \vec{p}_c + \mathcal{K}^* f|_B^2 + \gamma|\vec{p}_c|^2 + \frac{1}{c}|\vec{\lambda}_c|^2 \le |\mathcal{K}^* f|_B^2. \tag{9.1.23}$$

Consequently, $\frac{1}{c}|\vec{\lambda}_c|^2 \le |\mathcal{K}^* f|_B^2$ for all $c > 0$. Note that

$$\frac{1}{c}|\vec{\lambda}_c|^2_{\mathbb{L}^2(\Omega)} = c|\max(0, \vec{p}_c - \beta\vec{1})|^2_{\mathbb{L}^2(\Omega)} + c|\min(0, \vec{p}_c + \beta\vec{1})|^2_{\mathbb{L}^2(\Omega)}$$

and thus

$$|\max(0, (\vec{p}_c - \beta\vec{1}))|^2_{\mathbb{L}^2(\Omega)} \xrightarrow{c\to\infty} 0 \text{ and } |\min(0, (\vec{p}_c + \beta\vec{1}))|^2_{\mathbb{L}^2(\Omega)} \xrightarrow{c\to\infty} 0. \tag{9.1.24}$$

Recall that $\vec{p}_c \rightharpoonup \vec{p}^*$ weakly in $\mathbb{L}^2(\Omega)$. Weak lower semicontinuity of the convex functional $\vec{p} \mapsto |\max(0, \vec{p} - \beta\vec{1})|_{\mathbb{L}^2(\Omega)}$ and (9.1.24) imply that

$$\int_\Omega |\max(0, \vec{p}^* - \beta\vec{1})|^2 dx \le \liminf_{c\to\infty} \int_\Omega |\max(0, \vec{p}_c - \beta\vec{1})|^2 dx = 0.$$

Consequently, $\vec{p}^* \le \beta\vec{1}$ and analogously one verifies that $-\beta\vec{1} \le \vec{p}^*$. In particular $\vec{p}^*$ is feasible and from (9.1.22) we conclude that

$$\langle \vec{\lambda}_c, \vec{p}^* - \vec{p}_c \rangle_{H_0(\operatorname{div})^*, H_0(\operatorname{div})} \le 0 \quad \text{for all } c > 0. \tag{9.1.25}$$

By optimality of $\vec{p}_c$ for (9.1.11) we have

$$\limsup_{c\to\infty} \left(\frac{1}{2}|\operatorname{div} \vec{p}_c + \mathcal{K}^* f|_B^2 + \frac{\gamma}{2}|\vec{p}_c|^2\right) \le \frac{1}{2}|\operatorname{div} \vec{p} + \mathcal{K}^* f|_B^2 + \frac{\gamma}{2}|\vec{p}|^2 \tag{9.1.26}$$

for all $\vec{p} \in S_2 = \{\vec{p} \in (C_0^1(\Omega))^2 : -\beta\vec{1} \le \vec{p} \le \beta\vec{1}\}$. Density of S_2 in $S_1 = \{\vec{p} \in H_0(\text{div}) : -\beta\vec{1} \le \vec{p} \le \beta\vec{1}\}$ in the norm of $H_0(\text{div})$ implies that (9.1.26) holds for all $\vec{p} \in S_1$ and consequently

$$\limsup_{c\to\infty}\left(\frac{1}{2}|\operatorname{div}\vec{p}_c + \mathcal{K}^* f|_B^2 + \frac{\gamma}{2}|\vec{p}_c|^2\right) \le \frac{1}{2}|\operatorname{div}\vec{p}^* + \mathcal{K}^* f|_B^2 + \frac{\gamma}{2}|\vec{p}^*|^2$$
$$\le \liminf_{c\to\infty}\left(\frac{1}{2}|\operatorname{div}\vec{p}_c + \mathcal{K}^* f|_B^2 + \frac{\gamma}{2}|\vec{p}_c|^2\right),$$

where for the last inequality weak lower semicontinuity of norms is used. The above inequalities, together with weak convergence of $\vec{p}_c$ to $\vec{p}^*$ in $H_0(\text{div})$, imply strong convergence of $\vec{p}_c$ to $\vec{p}^*$ in $H_0(\text{div})$. Finally we aim at passing to the limit in (9.1.25). This is impeded by the fact that we only established $\vec{\lambda}_c \rightharpoonup \vec{\lambda}^*$ in $\mathbb{H}_0^1(\Omega)^*$. Note from (9.1.12) that $\{-\frac{1}{c}\Delta\vec{p}_c + \vec{\lambda}_c\}_{c\ge 1}$ is bounded in $H_0(\text{div})$. Hence there exists $\vec{\mu}^* \in H_0(\text{div})^*$ such that

$$-\frac{1}{c}\Delta\vec{p}_c + \vec{\lambda}_c \rightharpoonup \vec{\mu}^* \quad \text{weakly in } H_0(\text{div})^*,$$

and consequently also in $\mathbb{H}_0^1(\Omega)^*$. Moreover, $\{\frac{1}{\sqrt{c}}|\nabla\vec{p}_c|\}_{c\ge 1}$ is bounded and hence

$$-\frac{1}{c}\Delta\vec{p}_c \rightharpoonup 0 \quad \text{weakly in } \mathbb{H}_0^1(\Omega)^*$$

as $c \to \infty$. Since $\vec{\lambda}_c \rightharpoonup \vec{\lambda}^*$ weakly in $\mathbb{H}_0^1(\Omega)^*$ it follows that

$$\langle\vec{\lambda}^* - \vec{\mu}^*, \vec{v}\rangle_{\mathbb{H}_0^1(\Omega)^*, \mathbb{H}_0^1(\Omega)} = 0 \quad \text{for all } \vec{v} \in \mathbb{H}_0^1(\Omega).$$

Since both $\vec{\lambda}^*$ and $\vec{\mu}^*$ are elements of $H_0(\text{div})^*$ and since $\mathbb{H}_0^1(\Omega)$ is dense in $H_0(\text{div})$, it follows that $\vec{\lambda}^* = \vec{\mu}^*$ in $H_0(\text{div})^*$. For $\vec{p} \in S_2$ we have

$$\begin{aligned}
\langle\vec{\lambda}^*, \vec{p} - \vec{p}^*\rangle_{H_0(\text{div})^*, H_0(\text{div})} &= \langle\mu^*, \vec{p} - \vec{p}^*\rangle_{H_0(\text{div})^*, H_0(\text{div})}\\
&= \lim_{c\to\infty}\left\langle -\frac{1}{c}\Delta\vec{p}_c + \vec{\lambda}_c, \vec{p} - \vec{p}_c\right\rangle_{H_0(\text{div})^*, H_0(\text{div})}\\
&= \lim_{c\to\infty}\left(\frac{1}{c}(\nabla\vec{p}_c, \nabla(\vec{p} - \vec{p}_c)) + (\vec{\lambda}_c, \vec{p} - \vec{p}_c)\right)\\
&\le \lim_{c\to\infty}\left(\frac{1}{c}(\nabla\vec{p}_c, \nabla\vec{p}) + (\vec{\lambda}_c, \vec{p} - \vec{p}_c)\right) \le 0
\end{aligned}$$

by (9.1.22) and (9.1.23). Since S_2 is dense in S_1 we find

$$\langle\vec{\lambda}^*, \vec{p} - \vec{p}^*\rangle_{H_0(\text{div})^*, H_0(\text{div})} \le 0 \quad \text{for all } \vec{p} \in S_1. \qquad \square$$

Remark 9.1.1. For $\gamma = 0$ problems (9.1.10) and (9.1.11) admit a solution, which, however, is not unique. From the proof of Theorem 9.3 it follows that $\{(\vec{p}_c, \vec{\lambda}_c)\}_{c>0}$ contains a weak accumulation point in $H_0(\text{div}) \times \mathbb{H}_0^1(\Omega)^*$ and every weak accumulation point is a solution of (9.1.10).

Semismooth Newton treatment of (9.1.11).

We turn to the algorithmic treatment of the infinite-dimensional problem (9.1.11) for which we propose the following algorithm.

Algorithm.

(1) Choose $\vec{p}_0 \in \mathbb{H}_0^1(\Omega)$ and set $k = 0$.

(2) Set, for $i = 1, 2$,

$$\mathcal{A}_{k+1}^{+,i} = \{x : (\vec{p}_k^{\,i} + \beta\vec{1})(x) > 0\},$$
$$\mathcal{A}_{k+1}^{-,i} = \{x : (\vec{p}_k^{\,i} + \beta\vec{1})(x) < 0\},$$
$$\mathcal{I}_{k+1}^{i} = \Omega \setminus (\mathcal{A}_{k+1}^{+,i} \cup \mathcal{A}_{k+1}^{-,i}).$$

(3) Solve for $\vec{p} \in \mathbb{H}_0^1(\Omega)$ and set $\vec{p}_{k+1} = \vec{p}$ where

$$\begin{aligned} &\frac{1}{c}(\nabla\vec{p}, \nabla\vec{v}) + (\operatorname{div}\vec{p}, \operatorname{div}\vec{v})_B + (\mathcal{K}^* f, \operatorname{div}\vec{v})_B + \gamma(\vec{p}, \vec{v}) \\ &\quad + (c(\vec{p} - \beta\vec{1})\chi_{\mathcal{A}_{k+1}^+}, \vec{v}) + (c(\vec{p} + \beta\vec{1})\chi_{\mathcal{A}_{k+1}^-}, \vec{v}) = 0 \end{aligned} \tag{9.1.27}$$

for all $\vec{v} \in \mathbb{H}_0^1(\Omega)$.

(4) Set

$$\vec{\lambda}_{k+1}^{i} = \begin{cases} 0 & \text{on } \mathcal{I}_{k+1}^{i}, \\ c(\vec{p}_{k+1}^{\,i} - \beta\vec{1}) & \text{on } \mathcal{A}_{k+1}^{+,i}, \\ c(\vec{p}_{k+1}^{\,i} + \beta\vec{1}) & \text{on } \mathcal{A}_{k+1}^{-,i} \end{cases}$$

for $i = 1, 2$.

(5) Stop, or set $k = k + 1$, and goto (2).

Above $\chi_{\mathcal{A}_{k+1}^+}$ stands for

$$\chi_{\mathcal{A}_{k+1}^+}^{i} = \begin{cases} 1 & \text{if } x \in \mathcal{A}_{k+1}^{+,i}, \\ 0 & \text{if } x \notin \mathcal{A}_{k+1}^{+,i}, \end{cases}$$

and analogously for $\mathcal{A}_{k+1}^{-}$. The superscript i, $i = 1, 2$, refers to the respective component. We note that (9.1.27) admits a solution $\vec{p}_{k+1} \in \mathbb{H}_0^1(\Omega)$. Step (4) is included for the sake of the analysis of the algorithm. Let $C : \mathbb{H}_0^1(\Omega) \to H^{-1}(\Omega) \times H^{-1}(\Omega)$ stand for the operator

$$C = -\frac{1}{c}\Delta - \nabla B^{-1}\operatorname{div} + \gamma\,\mathrm{id}.$$

It is a homeomorphism for every $c > 0$ and allows us to express (9.1.12) as

$$C\vec{p} - \nabla B^{-1}\mathcal{K}^* f + c\max(0, \vec{p} - \beta\vec{1}) + c\min(0, \vec{p} + \beta\vec{1}) = 0, \tag{9.1.28}$$

where we drop the index in the notation for $\vec{p}_c$. For $\varphi \in L^2(\Omega)$ we define

$$\mathrm{Dmax}(0,\varphi)(x) = \begin{cases} 1 & \text{if } \varphi(x) > 0, \\ 0 & \text{if } \varphi(x) \le 0 \end{cases} \tag{9.1.29}$$

and

$$\mathrm{Dmin}(0,\varphi)(x) = \begin{cases} 1 & \text{if } \varphi(x) < 0, \\ 0 & \text{if } \varphi(x) \ge 0. \end{cases} \tag{9.1.30}$$

Using (9.1.29), (9.1.30) as Newton derivatives for the max and the min operations in (9.1.28) the semismooth Newton step can be expressed as

$$C\vec{p}_{k+1} + c(\vec{p}_{k+1} - \beta\vec{1})\chi_{\mathcal{A}^+_{k+1}} + c(\vec{p}_{k+1} + \beta\vec{1})\chi_{\mathcal{A}^-_{k+1}} - \nabla B^{-1}\mathcal{K}^* f = 0, \tag{9.1.31}$$

and $\vec{\lambda}_{k+1}$ from step (4) of the algorithm is given by

$$\vec{\lambda}_{k+1} = c(\vec{p}_{k+1} - \beta\vec{1})\chi_{\mathcal{A}^+_{k+1}} + c(\vec{p}_{k+1} + \beta\vec{1})\chi_{\mathcal{A}^-_{k+1}}. \tag{9.1.32}$$

The iteration of the algorithm can also be expressed with respect to the variable $\vec{\lambda}$ rather than $\vec{p}$. For this purpose we define

$$F(\vec{\lambda}) = \vec{\lambda} - c\max(0, C^{-1}(\nabla\hat{f} - \vec{\lambda}) - \beta\vec{1}) - c\min(0, C^{-1}(\nabla\hat{f} - \vec{\lambda}) + \beta\vec{1}), \tag{9.1.33}$$

where we put $\hat{f} = B^{-1}\mathcal{K}^* f$. Setting $\vec{p}_k = C^{-1}(\nabla\hat{f} - \vec{\lambda}_k)$, the semismooth Newton step applied to $F(\vec{\lambda}) = 0$ at $\vec{\lambda} = \vec{\lambda}_k$ results in

$$\vec{\lambda}_{k+1} = c(C^{-1}(\nabla\hat{f} - \vec{\lambda}_{k+1}) - \beta\vec{1})\chi_{\mathcal{A}^+_{k+1}} + c(C^{-1}(\nabla\hat{f} - \vec{\lambda}_{k+1}) + \beta\vec{1})\chi_{\mathcal{A}^-_{k+1}}$$

which coincides with (9.1.32). Therefore the semismooth Newton iterations according to the algorithm and that for $F(\vec{\lambda}) = 0$ coincide, provided that the initializations are related by $C\vec{p}_0 - \nabla\hat{f} + \vec{\lambda}_0 = 0$. The mapping F is Newton differentiable, i.e., for every $\vec{\lambda} \in \mathbb{L}^2(\Omega)$

$$|F(\vec{\lambda} + \vec{h}) - F(\vec{\lambda}) - DF(\vec{\lambda} + \vec{h})h|_{\mathbb{L}^2(\Omega)} = o(|\vec{h}|_{\mathbb{L}^2(\Omega)}) \tag{9.1.34}$$

for $|\vec{h}|_{\mathbb{L}^2(\Omega)} \to 0$; see Example 8.14. Here D denotes the Newton derivative of F defined by means of (9.1.29) and (9.1.30). For (9.1.34) to hold, the smoothing property of C^{-1} in the sense of an embedding from $\mathbb{L}^2(\Omega)$ into $\mathbb{L}^p(\Omega)$ for some $p > 2$ is essential. The following result now follows from Theorem 8.16.

Theorem 9.4. *If $|\vec{\lambda}_c - \vec{\lambda}_0|_{\mathbb{L}^2(\Omega)}$ is sufficiently small, then the iterates $\{(\vec{p}_k, \vec{\lambda}_k)\}_{k=1}^\infty$ of the algorithm converge superlinearly in $\mathbb{H}^1_0(\Omega) \times \mathbb{L}^2(\Omega)$ to the solution $(\vec{p}_c, \vec{\lambda}_c)$ of (9.1.11).*

In Theorem 9.3 convergence as $c \to \infty$ is established and Theorem 9.4 asserts convergence of the iterates $(\vec{p}_k, \vec{\lambda}_k)$ for each fixed c. The combination of these two limit processes was addressed in [HinK2], where a path concept with respect to the variable c is developed and the iterations with respect to k are controlled to remain in a neighborhood around the path.

Let us compare the algorithm of this section to the general framework of augmented Lagrangians presented in Section 4.6 for nonsmooth problems. We again introduce in (9.1.10) a diffusive regularization and realize the inequality constraints by a generalized Yosida–Moreau approximation. This suggests considering the Lagrangian $L(\vec{p},\vec{\mu}) : \mathbb{H}_0^1(\Omega) \times \mathbb{L}^2(\Omega) \to \mathbb{R}$ defined by

$$L_c(\vec{p},\vec{\lambda}) = \frac{1}{2\bar{c}}|\nabla\vec{p}|^2 + \frac{1}{2}|\operatorname{div}\vec{p} + \mathcal{K}^* f|_B^2 + \frac{\gamma}{2}|\vec{p}|^2 + \phi_c(\vec{p},\vec{\lambda}), \tag{9.1.35}$$

where ϕ_c is the generalized Yosida–Moreau approximation of the indicator function ϕ of the set $\{\vec{p} \in \mathbb{L}^2(\Omega) : -\beta\vec{1} \le \vec{p} \le \beta\vec{1}\}$, and $c > 0$, $\bar{c} > 0$. Here we choose c differently from $\bar{c}$ since in the limit of the augmented Lagrangian iteration the constraint $-\beta\vec{1} \le \vec{p} \le \beta\vec{1}$ is satisfied for any fixed $c > 0$. We have

$$\phi_c(\vec{p},\vec{\lambda}) = \inf_{\vec{q}\in\mathbb{L}^2(\Omega)} \phi(\vec{p}-\vec{q}) + (\vec{\mu},\vec{q})_{\mathbb{L}^2(\Omega)} + \frac{c}{2}|\vec{q}|^2_{\mathbb{L}^2(\Omega)}$$

for $c > 0$ and $\vec{\mu} \in \mathbb{L}^2(\Omega)$, which can equivalently be expressed as

$$\begin{aligned}\phi_c(\vec{p},\vec{\lambda}) = &\frac{1}{2c}\left|\max\left(0,\vec{\lambda} + c(\vec{p}-\beta\vec{1})\right)\right|^2_{\mathbb{L}^2(\Omega)} \\ &+ \frac{1}{2c}\left|\min\left(0,\vec{\lambda} + c(\vec{p}+\beta\vec{1})\right)\right|^2_{\mathbb{L}^2(\Omega)} - \frac{1}{2c}|\vec{\lambda}|^2_{\mathbb{L}^2(\Omega)}.\end{aligned}$$

The auxiliary problems in step 2 of the augmented Lagrangian method of Section 4.6 with L_c given in (9.1.35) coincide with (9.1.11) except for the shift by $\vec{\lambda}_k$ in the max/min operations. Each of these auxiliary problems can efficiently be solved by the semismooth Newton algorithm presented in this section if $\vec{p}_k \mp \beta\vec{1}$ is replaced by $\vec{\lambda}_k + c(\vec{p}_k \mp \beta\vec{1})$. Conversely, one can think of introducing augmented Lagrangian steps to the algorithm of this section, with the goal of avoiding possible ill-conditioning as $c \to \infty$. For numerical experience with the algorithmic concepts of this section we refer the reader to [HinK1].

9.2 Friction and contact problems in elasticity

9.2.1 Generalities

This chapter is concerned with the application of semismooth Newton methods to contact problems with Tresca and Coulomb friction in two spatial dimensions. In contact problems, also known as Signorini problems, one has to detect the contact zone between an elastic body and a rigid foundation. At the contact boundary, frictional forces are often too large to be neglected. Thus, besides the nonpenetration condition the frictional behavior in the contact zone has to be taken into account. Modeling friction involves a convex, nondifferentiable functional which, by means of Fenchel duality theory, can be related to a bilateral obstacle problem for which the primal-dual active set concept can advantageously be applied.

We commence with a formulation of the Signorini contact problem with Coulomb friction in two dimensions. A generalization to the three-dimensional case is given in [HuStWo]. We consider an elastic body that occupies, in its initial configuration, the open and bounded domain $\Omega \subset \mathbb{R}^2$ with $C^{1,1}$-boundary $\Gamma = \partial\Omega$. Let this boundary be divided into three disjoint parts, namely, the Dirichlet part Γ_d, further the part Γ_n with prescribed surface load $\boldsymbol{h} \in \mathbf{L}^2(\Gamma_n) := \left(L^2(\Gamma_n)\right)^2$, and the part Γ_c, where contact and friction with a rigid foundation may occur. For simplicity we assume that $\bar{\Gamma}_c \cap \bar{\Gamma}_d = \emptyset$ to avoid working with the space $H_{00}^{\frac{1}{2}}(\Gamma_c)$. We are interested in the deformation $\boldsymbol{y} = (y_1, y_2)^\top$ of the elastic body which is also subject to a given body force $\boldsymbol{f} \in \mathbf{L}^2(\Omega) := \left(L^2(\Omega)\right)^2$. The gap between the elastic body and the rigid foundation is $d := \tau_N \boldsymbol{d} \geq 0$, where $\boldsymbol{d} \in \mathbf{H}^1(\Omega) := \left(H^1(\Omega)\right)^2$ and $\tau_N \boldsymbol{y}$ denotes the normal component of the trace along Γ_c. As usual in linear elasticity, the linearized strain tensor is

$$\boldsymbol{\varepsilon}(\boldsymbol{y}) = \frac{1}{2}\left(\nabla \boldsymbol{y} + (\nabla \boldsymbol{y})^\top\right).$$

Using Hooke's law for the stress-strain relation, the linearized stress tensor

$$\boldsymbol{\sigma}(\boldsymbol{y}) := \mathbb{C}\boldsymbol{\varepsilon}(\boldsymbol{y}) := \lambda \mathrm{tr}(\boldsymbol{\varepsilon}(\boldsymbol{y}))\mathrm{Id} + 2\mu\boldsymbol{\varepsilon}(\boldsymbol{y})$$

is obtained, where λ and μ are the Lamé parameters. These parameters are given by

$$\lambda = (E\nu)/\big((1+\nu)(1-2\nu)\big) \quad \text{and} \quad \mu = E/\big(2(1+\nu)\big)$$

with Young's modulus $E > 0$ and the Poisson ration $\nu \in (0, 0.5)$. Above, $\mathbb{C}$ denotes the fourth order isotropic material tensor for linear elasticity.

The Signorini problem with Coulomb friction is then given as follows:

$$
\begin{aligned}
-\mathrm{Div}\,\boldsymbol{\sigma}(\boldsymbol{y}) &= \boldsymbol{f} && \text{in } \Omega, && (9.2.1a)\\
\boldsymbol{\tau}\boldsymbol{y} &= 0 && \text{on } \Gamma_d, && (9.2.1b)\\
\boldsymbol{\sigma}(\boldsymbol{y})\boldsymbol{n} &= \boldsymbol{h} && \text{on } \Gamma_n, && (9.2.1c)\\
\tau_N \boldsymbol{y} - d \leq 0,\ \sigma_N(\boldsymbol{y}) \leq 0,\ (\tau_N \boldsymbol{y} - d)\sigma_N(\boldsymbol{y}) &= 0 && \text{on } \Gamma_c, && (9.2.1d)\\
|\sigma_T(\boldsymbol{y})| &< \mathfrak{F}|\sigma_N(\boldsymbol{y})| && \text{on } \{x \in \Gamma_c : \tau_T \boldsymbol{y} = 0\}, && (9.2.1e)\\
\sigma_T(\boldsymbol{y}) &= -\mathfrak{F}\frac{|\sigma_N(\boldsymbol{y})|}{|\tau_T \boldsymbol{y}|}\tau_T \boldsymbol{y} && \text{on } \{x \in \Gamma_c : \tau_T \boldsymbol{y} \neq 0\}, && (9.2.1f)
\end{aligned}
$$

where the sets $\{x \in \Gamma_c : \tau_T \boldsymbol{y} = 0\}$ and $\{x \in \Gamma_c : \tau_T \boldsymbol{y} \neq 0\}$ are referred to as the sticky and the sliding regions, respectively. Above, Div denotes the rowwise divergence operator and $\boldsymbol{\tau} : \mathbf{H}^1(\Omega) \to \mathbf{H}^{\frac{1}{2}}(\Gamma) := \left(H^{\frac{1}{2}}(\Gamma)\right)^2$ the zero order trace mapping. The corresponding scalar-valued normal and tangential component mappings are denoted by $\tau_N, \tau_T : \mathbf{H}^1(\Omega) \to H^{\frac{1}{2}}(\Gamma_c)$; i.e., for $\boldsymbol{y} \in \mathbf{H}^1(\Omega)$ we have the splitting $\boldsymbol{\tau}\boldsymbol{y} = (\tau_N \boldsymbol{y})\boldsymbol{n} + (\tau_T \boldsymbol{y})\boldsymbol{t}$ with $\boldsymbol{n}$ and $\boldsymbol{t}$ denoting the unit normal and tangential vector along Γ_c, respectively. Similarly, using (9.2.1a) we can, following [KO], decompose the stress along the boundary, namely, $\boldsymbol{\sigma}(\boldsymbol{y})\boldsymbol{n} = \sigma_N(\boldsymbol{y})\boldsymbol{n} + \sigma_T(\boldsymbol{y})\boldsymbol{t}$ with mappings $\sigma_N, \sigma_T : \mathbf{Y} \to H^{-\frac{1}{2}}(\Gamma_c)$. Moreover, $\mathfrak{F} : \Gamma_c \to \mathbb{R}$ denotes the friction coefficient which is supposed to be uniformly Lipschitz continuous.

There are major mathematical difficulties inherent in the problem (9.2.1). For instance, in general $\sigma_N(\boldsymbol{y})$ in (9.2.1e), (9.2.1f) is not pointwise a.e. defined. Replacing the Coulomb friction in the above model by Tresca friction means replacing $|\sigma_N(\boldsymbol{y})|$ by a given friction g. The resulting system can then be analyzed and the existence of a unique solution can be proved. For a review we refer the reader to [Rao], for example.

9.2.2 Contact problem with Tresca friction

We now give a variational statement of the Signorini problem with given friction. The set of admissible deformations is defined as

$$\mathbf{Y} := \{\boldsymbol{v} \in \mathbf{H}^1(\Omega) : \boldsymbol{\tau v} = 0 \text{ a.e. on } \Gamma_d\},$$

where $\mathbf{H}^1(\Omega) := (H^1(\Omega))^2$. To incorporate the nonpenetration condition between the elastic body and the rigid foundation we introduce the cone

$$\mathbf{K} := \{\boldsymbol{v} \in \mathbf{Y} : \tau_N \boldsymbol{v} \le 0 \text{ a.e. on } \Gamma_c\}.$$

We define the symmetric bilinear form $a(\cdot\,,\cdot)$ on $\mathbf{Y} \times \mathbf{Y}$ and the linear form $L(\cdot)$ on $\mathbf{Y}$ by

$$a(\boldsymbol{y}, \boldsymbol{z}) := \int_\Omega (\boldsymbol{\sigma y}) : (\boldsymbol{\varepsilon z})\, dx, \qquad L(\boldsymbol{y}) = \int_\Omega \boldsymbol{f y}\, dx + \int_{\Gamma_n} \boldsymbol{h \tau y}\, dx,$$

where : denotes the sum of the componentwise products.

For the friction g we assume that $g \in H^{-\frac{1}{2}}(\Gamma_c)$ and $g \ge 0$, i.e., $\langle g, h\rangle_{\Gamma_c} \ge 0$ for all $h \in H^{\frac{1}{2}}(\Gamma_c)$ with $h \ge 0$. Since the friction coefficient $\mathfrak{F}$ is assumed to be uniformly Lipschitz continuous, it is a factor on $H^{\frac{1}{2}}(\Gamma_c)$, i.e., the mapping

$$\lambda \in H^{\frac{1}{2}}(\Gamma_c) \mapsto \mathfrak{F}\lambda \in H^{\frac{1}{2}}(\Gamma_c)$$

is well defined and bounded; see [Gri, p. 21]. By duality it follows that $\mathfrak{F}$ is a factor on $H^{-\frac{1}{2}}(\Gamma_c)$ as well. Consequently, the nondifferentiable functional

$$j(\boldsymbol{y}) := \int_{\Gamma_c} \mathfrak{F}g|\tau_T \boldsymbol{y}|\, dx$$

is well defined on $\mathbf{Y}$. After these preparations we can state the contact problem with given friction as

$$\min_{\boldsymbol{y} \in \boldsymbol{d}+\mathbf{K}} J(\boldsymbol{y}) := \frac{1}{2}a(\boldsymbol{y}, \boldsymbol{y}) - L(\boldsymbol{y}) + j(\boldsymbol{y}) \tag{$\mathcal{P}$}$$

or, equivalently, as elliptic variational inequality [Glo]:

$$\begin{cases} \text{Find } \boldsymbol{y} \in \boldsymbol{d} + \mathbf{K} \text{ such that} \\ a(\boldsymbol{y}, \boldsymbol{z} - \boldsymbol{y}) + j(\boldsymbol{z}) - j(\boldsymbol{y}) \ge L(\boldsymbol{z} - \boldsymbol{y}) \ \text{ for all } \boldsymbol{z} \in \boldsymbol{d} + \mathbf{K}. \end{cases} \tag{9.2.2}$$

Due to the Korn inequality, the functional $J(\cdot)$ is uniformly convex; further it is l.s.c. This implies that $(\mathcal{P})$ and equivalently (9.2.2) admit a unique solution $\boldsymbol{y}^* \in \boldsymbol{d} + \mathbf{K}$.

To derive the dual problem corresponding to $(\mathcal{P})$, we apply the Fenchel calculus to the mappings $\mathcal{F} : \mathbf{Y} \to \mathbb{R}$ and $\mathcal{G} : V \times H^{\frac{1}{2}}(\Gamma_c) \to \mathbb{R}$ given by

$$\mathcal{F}(\boldsymbol{y}) := \begin{cases} -L(\boldsymbol{y}) & \text{if } \boldsymbol{y} \in \boldsymbol{d} + \mathbf{K}, \\ \infty & \text{else;} \end{cases} \qquad \mathcal{G}(\boldsymbol{q}, \nu) := \frac{1}{2}\int_\Omega \boldsymbol{q} : \mathbb{C}\boldsymbol{q}\, dx + \int_{\Gamma_c} \mathfrak{F}g|\nu|\, dx,$$

where

$$\mathbf{V} = \{\boldsymbol{p} \in (L^2(\Omega))^{2\times 2} : p_{12} = p_{21}\}.$$

Furthermore, $\Lambda \in \mathcal{L}(\mathbf{Y}, \mathbf{V} \times H^{\frac{1}{2}}(\Gamma_c))$ is given by

$$\Lambda \boldsymbol{y} := (\Lambda_1 \boldsymbol{y}, \Lambda_2 \boldsymbol{y}) = (\boldsymbol{\varepsilon}\boldsymbol{y}, \tau_T \boldsymbol{y}),$$

which allows us to express $(\mathcal{P})$ as

$$\min_{y\in\mathbf{Y}} \left\{\mathcal{F}(y) + \mathcal{G}(\Lambda y)\right\}.$$

Endowing $\mathbf{V} \times H^{\frac{1}{2}}(\Gamma_c)$ with the usual product norm, $\mathcal{F}$ and $\mathcal{G}$ satisfy the conditions for the Fenchel duality theorem, which was recalled in the preamble of this chapter. For the convex conjugate one derives that $\mathcal{F}^\star(-\Lambda^\star(\boldsymbol{p}, \mu))$ equals $+\infty$ unless

$$-\operatorname{Div} \boldsymbol{p} = \boldsymbol{f}, \;\; \boldsymbol{p}\cdot\boldsymbol{n} = \boldsymbol{h} \text{ in } \mathbf{L}^2(\Gamma_n), \quad \text{and} \quad p_T + \mu = 0 \text{ in } H^{-\frac{1}{2}}(\Gamma_c), \tag{9.2.3}$$

where $p_T = (\boldsymbol{n}^\top \boldsymbol{p})\cdot \boldsymbol{t} \in H^{-\frac{1}{2}}(\Gamma_c)$. Further one obtains that

$$\mathcal{F}^\star(-\Lambda^\star(\boldsymbol{p},\mu)) = \begin{cases} -\langle p_N, d\rangle_{\Gamma_c} & \text{if (9.2.3) and } p_N \le 0 \text{ in } H^{-\frac{1}{2}}(\Gamma_c) \text{ hold,} \\ \infty & \text{else.} \end{cases}$$

Evaluating the convex conjugate for $\mathcal{G}$ yields that

$$\mathcal{G}^\star(\boldsymbol{p},\mu) = \begin{cases} \frac{1}{2}\int_\Omega \mathbb{C}^{-1}\boldsymbol{p} : \boldsymbol{p}\,dx & \text{if } \langle \mathfrak{F}g, |\nu|\rangle_{\Gamma_c} - \langle \nu, \mu\rangle_{\Gamma_c} \ge 0 \text{ for all } \nu \in H^{\frac{1}{2}}(\Gamma_c), \\ \infty & \text{else.} \end{cases}$$

Thus, following (9.0.1) we derive the dual problem corresponding to $(\mathcal{P})$:

$$\sup_{\substack{(\boldsymbol{p},\mu)\in\mathbf{V}\times H^{-\frac{1}{2}}(\Gamma_c) \\ \text{s.t. (9.2.3)}, \, p_N \le 0 \text{ in } H^{-\frac{1}{2}}(\Gamma_c), \\ \text{and } \langle \mathfrak{F}g, |\nu|\rangle_{\Gamma_c} - \langle \nu,\mu\rangle_{\Gamma_c} \ge 0 \\ \text{for all } \nu\in H^{\frac{1}{2}}(\Gamma_c).}} -\frac{1}{2}\int_\Omega \mathbb{C}^{-1}\boldsymbol{p} : \boldsymbol{p}\,dx + \langle p_N, d\rangle_{\Gamma_c}. \tag{$\mathcal{P}^\star$}$$

This problem is an inequality-constrained maximization problem of a quadratic functional, while the primal problem $(\mathcal{P})$ involves the minimization of a nondifferentiable functional. Evaluating the extremality conditions (9.0.2) for the above problems one obtains the following lemma.

Lemma 9.5. *The solution $\boldsymbol{y}^* \in \boldsymbol{d} + \mathbf{K}$ of $(\mathcal{P})$ and the solution $(\boldsymbol{p}^*, \mu^*)$ of $(\mathcal{P}^\star)$ are related by $\boldsymbol{\sigma}\boldsymbol{y}^* = \boldsymbol{p}^*$ and by the existence of $\lambda^* \in H^{-\frac{1}{2}}(\Gamma_c)$ such that*

$$a(\boldsymbol{y}^*, \boldsymbol{z}) - L(\boldsymbol{z}) + \langle \mu^*, \tau_T \boldsymbol{z}\rangle_{\Gamma_c} + \langle \lambda^*, \tau_N \boldsymbol{z}\rangle_{\Gamma_c} = 0 \textit{ for all } \boldsymbol{z} \in \mathbf{Y}, \tag{9.2.4a}$$

$$\langle \lambda^*, \tau_N \boldsymbol{z}\rangle_{\Gamma_c} \le 0 \textit{ for all } \boldsymbol{z} \in \mathbf{K}, \tag{9.2.4b}$$

$$\langle \lambda^*, \tau_N \boldsymbol{y}^* - d\rangle_{\Gamma_c} = 0, \tag{9.2.4c}$$

$$\langle \mathfrak{F}g, |\nu|\rangle_{\Gamma_c} - \langle \mu^*, \nu\rangle_{\Gamma_c} \ge 0 \textit{ for all } \nu \in H^{\frac{1}{2}}(\Gamma_c), \tag{9.2.4d}$$

$$\langle \mathfrak{F}g, |\tau_T \boldsymbol{y}^*|\rangle_{\Gamma_c} - \langle \mu^*, \tau_T \boldsymbol{y}^*\rangle_{\Gamma_c} = 0. \tag{9.2.4e}$$

Proof. The extremality condition $-\Lambda^\star(\boldsymbol{p}^*, \mu^*) \in \partial\mathcal{F}(\boldsymbol{y}^*)$ results in $\boldsymbol{y}^* \in \boldsymbol{d} + \mathbf{K}$ and

$$(\boldsymbol{p}^*, \boldsymbol{\varepsilon}(\boldsymbol{z} - \boldsymbol{y}^*)) - L(\boldsymbol{z} - \boldsymbol{y}^*) + \langle \mu^*, \tau_T(\boldsymbol{z} - \boldsymbol{y}^*)\rangle_{\Gamma_c} \geq 0 \text{ for all } \boldsymbol{z} \in \boldsymbol{d} + \mathbf{K}. \tag{9.2.5}$$

The condition $(\boldsymbol{p}^*, \mu^*) \in \partial\mathcal{G}(\Lambda \boldsymbol{y}^*)$ yields that $\boldsymbol{p}^* = \boldsymbol{\sigma}\,\boldsymbol{y}^*$ and (9.2.4d) and (9.2.4e). Introducing the multiplier λ^* for the variational inequality (9.2.5) leads to (9.2.4a), (9.2.4b), and (9.2.4c). □

According to (9.2.3) the multiplier $\mu^* \in H^{-\frac{1}{2}}(\Gamma_c)$ corresponding to the nondifferentiability of the primal functional $J(\cdot)$ has the mechanical interpretation $\mu^* = -\sigma_T \boldsymbol{y}^*$. Using Green's theorem in (9.2.4a) one finds

$$\lambda^* = -\sigma_N \boldsymbol{y}^*, \tag{9.2.6}$$

i.e., λ^* is the negative stress in normal direction.

We now briefly comment on the case that the given friction g is more regular, namely, $g \in L^2(\Gamma_c)$. In this case we can define $\mathcal{G}$ on the larger set $\mathbf{V} \times L^2(\Gamma_c)$. One can verify that the assumptions for the Fenchel duality theorem hold, and thus obtain higher regularity for the dual variable μ corresponding to the nondifferentiability of the cost functional in $(\mathcal{P}^\star)$, in particular $\mu \in L^2(\Gamma_c)$. This implies that the dual problem can be written as follows:

$$\sup_{\substack{(\boldsymbol{p},\mu)\in\mathbf{V}\times L^2(\Gamma_c)\\ \text{s.t. (9.2.3)}, p_N\leq 0 \text{ in } H^{-\frac{1}{2}}(\Gamma_c),\\ \text{and } |\mu|\leq\mathfrak{F}g \text{ a.e. on } \Gamma_c.}} -\frac{1}{2}\int_\Omega \mathbb{C}^{-1}\boldsymbol{p} : \boldsymbol{p}\,dx + \langle p_N, d\rangle_{\Gamma_c}. \tag{9.2.7}$$

Utilizing the relation $\boldsymbol{p} = \boldsymbol{\sigma}\,\boldsymbol{y}$ and (9.2.6), one can transform (9.2.7) into

$$\begin{cases} - \min\limits_{\substack{(\lambda,\mu)\in H^{-\frac{1}{2}}(\Gamma_c)\in\times L^2(\Gamma_c)\\ \lambda\geq 0 \text{ in } H^{-\frac{1}{2}}(\Gamma_c)\\ |\mu|\leq\mathfrak{F}g \text{ a.e. on } \Gamma_c}} \dfrac{1}{2}a(\boldsymbol{y}_{\lambda,\mu}, \boldsymbol{y}_{\lambda,\mu}) + \langle \lambda, d\rangle_{\Gamma_c}, \\ \text{where } \boldsymbol{y}_{\lambda,\mu} \text{ satisfies} \\ a(\boldsymbol{y}_{\lambda,\mu}, \boldsymbol{z}) - L(\boldsymbol{z}) + \langle \lambda, \tau_N \boldsymbol{z}\rangle_{\Gamma_c} + (\mu, \tau_T \boldsymbol{z})_{\Gamma_c} = 0 \text{ for all } \boldsymbol{z} \in \mathbf{Y}. \end{cases} \tag{9.2.8}$$

Problem (9.2.8) is an equivalent form for the dual problem (9.2.7), now written in the variables λ and μ. The primal variable $\boldsymbol{y}_{\lambda,\mu}$ appears only as an auxiliary variable determined from λ and μ. Since $g \in L^2(\Gamma_c)$, also the extremality conditions corresponding to $(\mathcal{P})$ and (9.2.7) can be given more explicitly. First, (9.2.4d) is equivalent to

$$|\mu^*| \leq \mathfrak{F}g \text{ a.e. on } \Gamma_c, \tag{9.2.4d$'$}$$

and a brief computation shows that (9.2.4e) is equivalent to

$$\begin{cases} \tau_T \boldsymbol{y}^* = 0 \text{ or} \\ \tau_T \boldsymbol{y}^* \neq 0 \text{ and } \mu^* = \mathfrak{F}g\frac{\tau_T \boldsymbol{y}^*}{|\tau_T \boldsymbol{y}^*|}. \end{cases} \tag{9.2.4e$'$}$$

Moreover (9.2.4d′) and (9.2.4e′) can equivalently be expressed as

$$\sigma\,\tau_T \boldsymbol{y}^* - \max(0, \sigma\,\tau_T \boldsymbol{y}^* + \mu^* - \mathfrak{F}g) - \min(0, \sigma\,\tau_T \boldsymbol{y}^* + \mu^* + \mathfrak{F}g) = 0 \tag{9.2.9}$$

with arbitrary $\sigma > 0$.

We now introduce and analyze a regularized version of the contact problem with given friction that allows the application of the semismooth Newton method. In what follows we assume that $g \in L^2(\Gamma_c)$. We start our consideration with a regularized version of the dual problem (9.2.7) written in the form (9.2.8). Let $\gamma_1, \gamma_2 > 0$, $\hat{\lambda} \in L^2(\Gamma_c)$, $\hat{\lambda} \geq 0$, and $\hat{\mu} \in L^2(\Gamma_c)$, and define the functional $J^\star_{\gamma_1,\gamma_2} : L^2(\Gamma_c) \times L^2(\Gamma_c) \longrightarrow \mathbb{R}$ by

$$\begin{aligned} J^\star_{\gamma_1,\gamma_2}(\lambda, \mu) := \frac{1}{2}a(y_{\lambda,\mu}, y_{\lambda,\mu}) + (\lambda, d)_{\Gamma_c} + \frac{1}{2\gamma_1}\|\lambda - \hat{\lambda}\|^2_{\Gamma_c} \\ + \frac{1}{2\gamma_2}\|\mu - \hat{\mu}\|^2_{\Gamma_c} - \frac{1}{2\gamma_1}\|\hat{\lambda}\|^2_{\Gamma_c} - \frac{1}{2\gamma_2}\|\hat{\mu}\|^2_{\Gamma_c}, \end{aligned}$$

where $y_{\lambda,\mu} \in \mathbf{Y}$ satisfies

$$a(y_{\lambda,\mu}, z) - L(z) + (\lambda, \tau_N z)_{\Gamma_c} + (\mu, \tau_T z)_{\Gamma_c} = 0 \text{ for all } z \in \mathbf{Y}. \tag{9.2.10}$$

The regularized dual problem with given friction is defined as

$$\max_{\substack{(\lambda,\mu)\in L^2(\Gamma_c)\times L^2(\Gamma_c)\\ \lambda\geq 0,\ |\mu|\leq \mathfrak{F}g \text{ a.e. on } \Gamma_c}} -J^\star_{\gamma_1,\gamma_2}(\lambda, \mu). \tag{$\mathcal{P}^\star_{\gamma_1,\gamma_2}$}$$

Obviously, the last two terms in the definition of $J^\star_{\gamma_1,\gamma_2}$ are constants and can thus be neglected in the optimization problem ($\mathcal{P}^\star_{\gamma_1,\gamma_2}$). However, they are introduced with regard to the primal problem corresponding to ($\mathcal{P}^\star_{\gamma_1,\gamma_2}$), which we turn to next. We define the functional $J_{\gamma_1,\gamma_2} : \mathbf{Y} \to \mathbb{R}$ by

$$\begin{aligned} J_{\gamma_1,\gamma_2}(y) := \frac{1}{2}a(y, y) - L(y) + \frac{1}{2\gamma_1}\|\max(0, \hat{\lambda} + \gamma_1(\tau_N y - d))\|^2_{\Gamma_c} \\ + \frac{1}{\gamma_2}\int_{\Gamma_c} \mathfrak{F}gh(\tau_T y(x), \hat{\mu}(x))\, dx, \end{aligned}$$

where $h(\cdot\,,\cdot) : \mathbb{R} \times \mathbb{R} \longrightarrow \mathbb{R}$ is a local smoothing of the absolute value function, given by

$$h(x, \alpha) := \begin{cases} |\gamma_2 x + \alpha| - \frac{1}{2}\mathfrak{F}g & \text{if } |\gamma_2 x + \alpha| \geq \mathfrak{F}g, \\ \frac{1}{2\mathfrak{F}g}|\gamma_2 x + \alpha|^2 & \text{if } |\gamma_2 x + \alpha| < \mathfrak{F}g. \end{cases} \tag{9.2.11}$$

Then the primal problem corresponding to ($\mathcal{P}^\star_{\gamma_1,\gamma_2}$) is

$$\min_{y\in\mathbf{Y}} J_{\gamma_1,\gamma_2}(y). \tag{$\mathcal{P}_{\gamma_1,\gamma_2}$}$$

This can be verified similarly as for the original problem using Fenchel duality theory; see [Sta1] for details. Clearly, both ($\mathcal{P}_{\gamma_1,\gamma_2}$) and ($\mathcal{P}^\star_{\gamma_1,\gamma_2}$) admit unique solutions y_{γ_1,γ_2} and $(\lambda_{\gamma_1,\gamma_2}, \mu_{\gamma_1,\gamma_2})$, respectively. Note that the regularization turns the primal problem into the unconstrained minimization of a continuously differentiable functional, while the corresponding dual problem is still a constrained minimization of a quadratic functional. To shorten notation we henceforth mark all variables of the regularized problems only by the index γ instead of γ_1, γ_2. It can be shown that the extremality conditions relating $(\mathcal{P}_{\gamma_1,\gamma_2})$ and $(\mathcal{P}^\star_{\gamma_1,\gamma_2})$ are

$$a(\boldsymbol{y}_\gamma, \boldsymbol{z}) - L(\boldsymbol{z}) + (\mu_\gamma, \tau_T \boldsymbol{z})_{\Gamma_c} + (\lambda_\gamma, \tau_N \boldsymbol{z})_{\Gamma_c} = 0 \text{ for all } \boldsymbol{z} \in \mathbf{Y}, \tag{9.2.12a}$$

$$\lambda_\gamma - \max(0, \hat{\lambda} + \gamma_1(\tau_N \boldsymbol{y}_\gamma - d)) = 0 \quad \text{on } \Gamma_c, \tag{9.2.12b}$$

$$\begin{cases} \gamma_2(\xi_\gamma - \tau_T \boldsymbol{y}_\gamma) + \mu_\gamma - \hat{\mu} = 0, \\ \xi_\gamma - \max(0, \xi_\gamma + \sigma(\mu_\gamma - \mathfrak{F}g)) - \min(0, \xi_\gamma + \sigma(\mu_\gamma + \mathfrak{F}g)) = 0 \end{cases} \tag{9.2.12c}$$

for any $\sigma > 0$. Here, ξ_γ is the Lagrange multiplier associated to the constraint $|\mu| \le \mathfrak{F}g$ in $(\mathcal{P}^\star_{\gamma_1,\gamma_2})$. By setting $\sigma = \gamma_2^{-1}$, ξ_γ can be eliminated from (9.2.12c), which results in

$$\gamma_2 \tau_T \boldsymbol{y}_\gamma + \hat{\mu} - \mu_\gamma - \max(0, \gamma_2 \tau_T \boldsymbol{y}_\gamma + \hat{\mu} - \mathfrak{F}g) - \min(0, \gamma_2 \tau_T \boldsymbol{y}_\gamma + \hat{\mu} + \mathfrak{F}g) = 0. \tag{9.2.13}$$

While (9.2.13) and (9.2.12c) are equivalent, they will motivate slightly different active set algorithms due to the parameter σ in (9.2.12c).

Next we investigate the convergence of the primal variable $\boldsymbol{y}_\gamma$ as well as the dual variables $(\lambda_\gamma, \mu_\gamma)$ as the regularization parameters γ_1, γ_2 tend to infinity. For this purpose we denote by $\boldsymbol{y}^*$ the solution of $(\mathcal{P})$ and by (λ^*, μ^*) the solution to $(\mathcal{P}^\star)$.

Theorem 9.6. *For all $\hat{\lambda} \in L^2(\Gamma_c)$, $\hat{\lambda} \ge 0$, $\hat{\mu} \in L^2(\Gamma_c)$, and $g \in L^2(\Gamma_c)$, the primal variable $\boldsymbol{y}_\gamma$ converges to $\boldsymbol{y}^*$ strongly in $\mathbf{Y}$ and the dual variables $(\lambda_\gamma, \mu_\gamma)$ converge to (λ^*, μ^*) weakly in $H^{-\frac{1}{2}}(\Gamma_c) \times L^2(\Gamma_c)$ as $\gamma_1 \to \infty$ and $\gamma_2 \to \infty$.*

Proof. The proof of this theorem is related to that of Theorem 8.26 in Chapter 8 and can be found in [KuSt]. ☐

An active set algorithm for solving (9.2.12) is presented next. The interpretation as generalized Newton method is discussed later. In the following we drop the index γ.

Algorithm SSN.

1. Initialize $(\lambda^0, \xi^0, \mu^0, \boldsymbol{y}^0) \in L^2(\Gamma_c) \times L^2(\Gamma_c) \times L^2(\Gamma_c) \times \mathbf{Y}$, $\sigma > 0$ and set $k := 0$.

2. Determine the active and inactive sets

$$\begin{aligned}
\mathcal{A}_c^{k+1} &= \{x \in \Gamma_c : \hat{\lambda} + \gamma_1(\tau_N \boldsymbol{y}^k - d) > 0\}, \\
\mathcal{I}_c^{k+1} &= \Gamma_c \setminus \mathcal{A}_c^{k+1}, \\
\mathcal{A}_{f,-}^{k+1} &= \{x \in \Gamma_c : \xi^k + \sigma(\mu^k + \mathfrak{F}g) < 0\}, \\
\mathcal{A}_{f,+}^{k+1} &= \{x \in \Gamma_c : \xi^k + \sigma(\mu^k - \mathfrak{F}g) > 0\}, \\
\mathcal{I}_f^{k+1} &= \Gamma_c \setminus (\mathcal{A}_{f,-}^{k+1} \cup \mathcal{A}_{f,+}^{k+1}).
\end{aligned}$$

3. If $k \ge 1$, $\mathcal{A}_c^{k+1} = \mathcal{A}_c^k$, $\mathcal{A}_{f,-}^{k+1} = \mathcal{A}_{f,-}^k$, and $\mathcal{A}_{f,+}^{k+1} = \mathcal{A}_{f,+}^k$ stop. Else

4. Solve

$$a(y^{k+1}, z) - L(z) + (\mu^{k+1}, \tau_T z)_{\Gamma_c} + (\lambda^{k+1}, \tau_N z)_{\Gamma_c} = 0 \text{ for all } z \in \mathbf{Y},$$
$$\lambda^{k+1} = 0 \text{ on } \mathcal{I}_c^{k+1}, \quad \lambda^{k+1} = \hat{\lambda} + \gamma_1(\tau_N y^{k+1} - d) \text{ on } \mathcal{A}_c^{k+1},$$
$$\mu^{k+1} - \hat{\mu} - \gamma_2 \tau_T y^{k+1} = 0 \text{ on } \mathcal{I}_f^{k+1},$$
$$\mu^{k+1} = -\mathfrak{F}g \text{ on } \mathcal{A}_{f,-}^{k+1}, \quad \mu^{k+1} = \mathfrak{F}g \text{ on } \mathcal{A}_{f,+}^{k+1}.$$

5. Set

$$\xi^{k+1} := \begin{cases} \tau_T y^{k+1} + \gamma_2^{-1}(\hat{\mu} + \mathfrak{F}g) & \text{on } \mathcal{A}_{f,-}^{k+1}, \\ \tau_T y^{k+1} + \gamma_2^{-1}(\hat{\mu} - \mathfrak{F}g) & \text{on } \mathcal{A}_{f,+}^{k+1}, \\ 0 & \text{on } \mathcal{I}_f^{k+1}, \end{cases}$$

$k := k+1$ and goto step 2.

Note that there exists a unique solution to the system in step 4, since it represents the necessary and sufficient optimality conditions for the equality-constrained auxiliary problem

$$\min_{\substack{\lambda = 0 \text{ on } \mathcal{I}_c^{k+1}, \\ \mu = -\mathfrak{F}g \text{ on } \mathcal{A}_{f,-}^{k+1},\ \mu = \mathfrak{F}g \text{ on } \mathcal{A}_{f,+}^{k+1}}} J^\star_{\gamma_1,\gamma_2}(\lambda, \mu),$$

with $J^\star_{\gamma_1,\gamma_2}$ as defined in (9.2.10) that clearly has a unique solution. If the algorithm stops at step 3 then y^k is the solution to the primal problem $(\mathcal{P}_{\gamma_1,\gamma_2})$ and (λ^k, μ^k) solves the dual problem $(\mathcal{P}^\star_{\gamma_1,\gamma_2})$.

Provided that we choose $\sigma = \gamma_2^{-1}$, the above algorithm can be interpreted as a semismooth Newton method in infinite-dimensional spaces. To show this assertion, we consider a reduced system instead of (9.2.12). Thereby, as in the dual problem $(\mathcal{P}^\star_{\gamma_1,\gamma_2})$, the primal variable y only acts as an auxiliary variable that is calculated from the dual variables (λ, μ). We introduce the mapping $F : L^2(\Gamma_c) \times L^2(\Gamma_c) \longrightarrow L^2(\Gamma_c) \times L^2(\Gamma_c)$ by

$$F(\lambda, \mu) = \begin{pmatrix} \lambda - \max(0, \hat{\lambda} + \gamma_1(\tau_N y_{\lambda,\mu} - d)) \\ \gamma_2 \tau_T y_{\lambda,\mu} + \hat{\mu} - \mu - \max(0, \gamma_2 \tau_T y_{\lambda,\mu} + \hat{\mu} - \mathfrak{F}g) \cdots \\ \cdots - \min(0, \gamma_2 \tau_T y_{\lambda,\mu} + \hat{\mu} + \mathfrak{F}g) \end{pmatrix}, \tag{9.2.14}$$

where for given λ and μ we denote by $y_{\lambda,\mu}$ the solution to

$$a(y, z) - L(z) + (\mu, \tau_T z)_{\Gamma_c} + (\lambda, \tau_N z)_{\Gamma_c} = 0 \text{ for all } z \in \mathbf{Y}. \tag{9.2.15}$$

For $(\lambda, \mu) \in L^2(\Gamma_c) \times L^2(\Gamma_c)$ we have that $\tau_N y_{\lambda,\mu}, \tau_T y_{\lambda,\mu} \in H^{\frac{1}{2}}(\Gamma_c)$. Since $H^{\frac{1}{2}}(\Gamma_c)$ embeds continuously into $L^p(\Gamma_c)$ for every $p < \infty$, we have the following composition of the first component of F, for each $2 < p < \infty$:

$$\begin{cases} L^2(\Gamma_c) \times L^2(\Gamma_c) & \rightarrow & L^p(\Gamma_c) & \overset{\Theta}{\rightarrow} & L^2(\Gamma_c), \\ (\lambda, \mu) & \mapsto & \tau_N y_{\lambda,\mu} & \mapsto & \max(0, \hat{\lambda} + \gamma_1(\tau_N y_{\lambda,\mu} - d)). \end{cases} \tag{9.2.16}$$

Since the mapping Θ involves a norm gap under the max functional, it is Newton differentiable (see Example 8.14), and thus the first component of F is Newton differentiable. A similar observation holds for the second component as well, and thus the whole mapping F is Newton differentiable. Hence, we can apply the semismooth Newton method to the equation $F(\lambda, \mu) = 0$. Calculating the explicit form of the Newton step leads to Algorithm SSN with $\sigma = \gamma_2^{-1}$.

Theorem 9.7. *Suppose that there exists a constant $g_0 > 0$ with $\mathfrak{F}g \geq g_0$, and further suppose that $\sigma \geq \gamma_2^{-1}$ and that $\|\lambda^0 - \lambda_\gamma\|_{\Gamma_c}$, $\|\mu^0 - \mu_\gamma\|_{\Gamma_c}$ are sufficiently small. Then the iterates $(\lambda^k, \xi^k, \mu^k, \mathbf{y}^k)$ of Algorithm* SSN *converge superlinearly to $(\lambda_\gamma, \xi_\gamma, \mu_\gamma, \mathbf{y}_\gamma)$ in $L^2(\Gamma_c) \times L^2(\Gamma_c) \times L^2(\Gamma_c) \times \mathbf{Y}$.*

Proof. The proof consists of two steps. First we prove the assertion for $\sigma = \gamma_2^{-1}$ and then we utilize this result for the general case $\sigma \geq \gamma_2^{-1}$.

Step 1. For $\sigma = \gamma_2^{-1}$ Algorithm SSN is a semismooth Newton method for the equation $F(\lambda, \mu) = 0$ (F as defined in (9.2.14)). We already argued Newton differentiability of F. To apply Theorem 8.16, it remains to show that the generalized derivatives have uniformly bounded inverses, which can be achieved similarly as in the proof of Theorem 8.25 in Chapter 8. Clearly, the superlinear convergence of (λ^k, μ^k) carries over to the variables ξ^k and $\mathbf{y}^k$.

Step 2. For $\sigma > \gamma_2^{-1}$ we cannot use the above argument directly. Nevertheless, one can prove superlinear convergence of the iterates by showing that in a neighborhood of the solution the iterates of Algorithm SSN with $\sigma > \gamma_2^{-1}$ coincide with those of Algorithm SSN with $\sigma = \gamma_2^{-1}$. The argument for this fact exploits the smoothing properties of the Neumann-to-Dirichlet mapping for the elasticity equation. First, we again consider the case $\sigma = \gamma_2^{-1}$. Clearly, for all $k \geq 1$ we have $\lambda^k - \lambda^{k-1} \in L^2(\Gamma_c)$ and $\mu^k - \mu^{k-1} \in L^2(\Gamma_c)$. The corresponding difference $\mathbf{y}^k - \mathbf{y}^{k-1}$ of the primal variables satisfies

$$a(\mathbf{y}^k - \mathbf{y}^{k-1}, z) + (\mu^k - \mu^{k-1}, \tau_T z)_{\Gamma_c} + (\lambda^k - \lambda^{k-1}, \tau_N z)_{\Gamma_c} = 0 \quad \text{for all } z \in \mathbf{Y}.$$

From regularity results for elliptic variational equalities it follows that there exists a constant $C > 0$ such that

$$\|\tau_T \mathbf{y}^k - \tau_T \mathbf{y}^{k-1}\|_{C^0(\Gamma_c)} \leq C\big(\|\lambda^k - \lambda^{k-1}\|_{\Gamma_c} + \|\mu^k - \mu^{k-1}\|_{\Gamma_c}\big). \tag{9.2.17}$$

We now show that (9.2.17) implies

$$\mathcal{A}_{f,-}^k \cap \mathcal{A}_{f,+}^{k+1} = \mathcal{A}_{f,+}^k \cap \mathcal{A}_{f,-}^{k+1} = \emptyset \tag{9.2.18}$$

provided that $\|\lambda^0 - \lambda_\gamma\|_{\Gamma_c}$ and $\|\mu^0 - \mu_\gamma\|_{\Gamma_c}$ are sufficiently small. If $\mathcal{B} := \mathcal{A}_{f,-}^k \cap \mathcal{A}_{f,+}^{k+1} \neq \emptyset$, then it follows that $\tau_T \mathbf{y}^{k-1} + \gamma_2^{-1}(\hat{\mu} + \mathfrak{F}g) < 0$ and $\tau_T \mathbf{y}^k + \gamma_2^{-1}(\hat{\mu} - \mathfrak{F}g) > 0$ on $\mathcal{B}$, which implies that $\tau_T \mathbf{y}^k - \tau_T \mathbf{y}^{k-1} > 2\gamma_2^{-1}\mathfrak{F}g \geq 2\gamma_2^{-1}\mathfrak{F}g_0 > 0$ on $\mathcal{B}$. This contradicts (9.2.17) provided that $\|\lambda^0 - \lambda_\gamma\|_{\Gamma_c}$ and $\|\mu^0 - \mu_\gamma\|_{\Gamma_c}$ are sufficiently small. Analogously, one can show that $\mathcal{A}_{f,+}^k \cap \mathcal{A}_{f,-}^{k+1} = \emptyset$.

We now choose an arbitrary $\sigma \geq \gamma_2^{-1}$ and assume that (9.2.18) holds for Algorithm SSN if $\sigma = \gamma_2^{-1}$. Then we can argue that in a neighborhood of the solution the iterates of Algorithm SSN are independent of $\sigma \geq \gamma_2^{-1}$. To verify this assertion we separately consider the sets $\mathcal{I}_f^k$, $\mathcal{A}_{f,-}^k$, and $\mathcal{A}_{f,+}^k$. On $\mathcal{I}_f^k$ we have that $\xi^k = 0$ and thus σ has no influence when determining the new active and inactive sets. On the set $\mathcal{A}_{f,-}^k$ we have $\mu^k = -\mathfrak{F}g$. Here, we consider two types of sets. First, sets where $\xi^k < 0$ belong to $\mathcal{A}_{f,-}^{k+1}$ for the next iteration independently of σ. And, second, if $\xi^k \geq 0$, we use

$$\xi^k + \sigma(\mu^k - \mathfrak{F}g) = \xi^k - 2\sigma\mathfrak{F}g.$$

Sets where $\xi^k - 2\sigma\mathfrak{F}g \leq 0$ are transferred to $\mathcal{I}_f^{k+1}$, and those where $0 < \xi^k - 2\sigma\mathfrak{F}g \leq \xi^k - 2\gamma_2^{-1}\mathfrak{F}g$ belong to $\mathcal{A}_{f,+}^{k+1}$ for the next iteration. However, the case that $x \in \mathcal{A}_{f,-}^k \cap \mathcal{A}_{f,+}^{k+1}$ cannot occur for $\sigma \geq \gamma_2^{-1}$, since it is already ruled out by (9.2.18) for $\sigma = \gamma_2^{-1}$. On the set $\mathcal{A}_{f,+}^k$ one argues analogously.

This shows that in a neighborhood of the solution the iterates are the same for all $\sigma \geq \gamma_2^{-1}$, and thus the superlinear convergence result from Step 1 carries over to the general case $\sigma \geq \gamma_2^{-1}$, which ends the proof. □

Aside from the assumption that $\|\lambda^0 - \lambda_\gamma\|_{\Gamma_c}$ and $\|\mu^0 - \mu_\gamma\|_{\Gamma_c}$ are sufficiently small, σ controls the probability that points are moved from the lower active set to the upper, or vice versa, in one iteration. Smaller values for σ make it more likely that points belong to $\mathcal{A}_{f,-}^k \cap \mathcal{A}_{f,+}^{k+1}$ or $\mathcal{A}_{f,+}^k \cap \mathcal{A}_{f,-}^{k+1}$. In the numerical realization of Algorithm SSN it turns out that choosing small values for σ may not be optimal, since this may lead to the situation that points which are active with respect to the upper bound become active with respect to the lower bound in the next iteration, and vice versa. This in turn may lead to cycling of the iterates. Such undesired behavior can be overcome by choosing larger values for σ. If the active set strategy is based on (9.2.13), one cannot take advantage of a parameter which helps avoid points from changing from $\mathcal{A}_{f,+}$ to $\mathcal{A}_{f,-}$, or vice versa, in one iteration.

Remark 9.2.1. So far we have not remarked on the choice of $\bar{\lambda}$ and $\bar{\mu}$. One possibility is to choose them according to first order augmented Lagrangian updates. In practice this will result in carrying out some steps of Algorithm SSN and then updating $(\bar{\lambda}, \bar{\mu})$ as the current values of (λ^k, μ^k).

9.2.3 Contact problem with Coulomb friction

We give a short description of how the contact problem with Coulomb friction can be treated with the methods of the previous subsection. As pointed out earlier, one of the main difficulties of such problems is related to the lack of regularity of $\sigma_N(y)$ in (9.2.1). The following formulation utilizes the contact problem with given friction $g \in H^{-\frac{1}{2}}(\Gamma_c)$ and a fixed point idea. We define the cone of nonnegative functionals over $H^{\frac{1}{2}}(\Gamma_c)$ as

$$H_+^{-\frac{1}{2}}(\Gamma_c) := \left\{\xi \in H^{-\frac{1}{2}}(\Gamma_c) : \langle \xi, \eta \rangle_{\Gamma_c} \geq 0 \text{ for all } \eta \in H^{\frac{1}{2}}(\Gamma_c), \eta \geq 0\right\}$$

and consider the mapping $\Psi : H_+^{-\frac{1}{2}}(\Gamma_c) \to H_+^{-\frac{1}{2}}(\Gamma_c)$ defined by $\Psi(g) := \lambda_g$, where λ_g is the unique multiplier for the contact condition in (9.2.4) for the problem with given friction g. Property (9.2.4b) implies that Ψ is well defined. With (9.2.6) in mind, $y \in \mathbf{Y}$ is called a weak solution of the Signorini problem with Coulomb friction if its negative normal boundary stress $-\sigma_N(y)$ is a fixed point of the mapping Ψ. In general, such a fixed point for the mapping Ψ does not exist unless $\mathfrak{F}$ is sufficiently small; see, e.g., [EcJa, Has, HHNL].

The regularization for the Signorini contact problem with Coulomb friction that we consider here corresponds to the regularization in $(\mathcal{P}_{\gamma_1,\gamma_2})$ and reflects the fact that the Lagrangian for the contact condition relates to the negative stress in the normal direction. It is given by

$$
\begin{aligned}
&a(y, z-y) + \big(\max(0, \hat\lambda + \gamma_1(\tau_N y - d)), \tau_N(z-y)\big)_{\Gamma_c} - L(z-y)\\
&+\frac{1}{\gamma_2}\int_{\Gamma_c} \mathfrak{F}\max(0, \hat\lambda + \gamma_1(\tau_N y - d))\left\{h(\tau_\tau z, \hat\mu) - h(\tau_\tau y, \hat\mu)\right\} dx \ge 0
\end{aligned}
\tag{9.2.19}
$$

for all $z \in \mathbf{Y}$, with $h(\cdot\,,\cdot)$ as defined in (9.2.11). Existence for (9.2.19) is obtained by means of a fixed point argument for the regularized Tresca friction problem. For this purpose we set $L^2_+(\Gamma_c) := \{\xi \in L^2(\Gamma_c) : \xi \ge 0 \text{ a.e.}\}$ and define the mapping $\Psi_\gamma : L^2_+(\Gamma_c) \to L^2_+(\Gamma_c)$ by $\Psi_\gamma(g) := \lambda_\gamma = \max(0, \hat\lambda + \gamma_1(\tau_N y_\gamma - d))$ with y_γ the unique solution of the regularized contact problem with friction $g \in L^2_+(\Gamma_c)$. In a first step Lipschitz continuity of the mapping $\Phi_\gamma : L^2_+(\Gamma_c) \to \mathbf{Y}$ which assigns to a given friction $g \in L^2_+(\Gamma_c)$ the corresponding solution y_γ of $(\mathcal{P}_{\gamma_1,\gamma_2})$ is investigated.

Lemma 9.8. *For every $\gamma_1, \gamma_2 > 0$ and $\hat\lambda \in L^2(\Gamma_c)$, $\hat\mu \in L^2(\Gamma_c)$ the mapping Φ_γ defined above is Lipschitz continuous with constant*

$$
\mathfrak{L} = \frac{\|\mathfrak{F}\|_{L^\infty(\Gamma_c)}\, c_1}{\kappa},
\tag{9.2.20}
$$

where $\|\mathfrak{F}\|_\infty$ denotes the essential supremum of $\mathfrak{F}$, κ the coercivity constant of $a(\cdot\,,\cdot)$, and c_1 the continuity constant of the trace mapping from $\mathbf{Y}$ to $L^2(\Gamma_c)$. In particular, the Lipschitz constant $\mathfrak{L}$ does not depend on the regularization parameters γ_1, γ_2.

For the proof we refer the reader to [Sta1]. We next address properties of the mapping Ψ_γ.

Lemma 9.9. *For every $\gamma_1, \gamma_2 > 0$ and $\hat\lambda \in L^2(\Gamma_c)$, $\hat\mu \in L^2(\Gamma_c)$ the mapping Ψ_γ : $L^2_+(\Gamma_c) \to L^2_+(\Gamma_c)$ is compact and Lipschitz continuous with constant*

$$
\mathfrak{L} = \frac{c\gamma}{\kappa}\|\mathfrak{F}\|_\infty,
$$

where c is a constant resulting from trace theorems.

Proof. We consider the following composition of mappings.

$$
\begin{array}{ccccccc}
L^2_+(\Gamma_c) & \xrightarrow{\Phi_\gamma} & \mathbf{Y} & \xrightarrow{\Theta} & L^2(\Gamma_c) & \xrightarrow{\Upsilon} & L^2_+(\Gamma_c),\\
g & \mapsto & y & \mapsto & \tau_N y & \mapsto & \max(0, \hat\lambda + \gamma_1(\tau_N y - d)).
\end{array}
\tag{9.2.21}
$$

From Lemma 9.8 it is known that Φ_γ is Lipschitz continuous. The mapping Θ consists of the linear trace mapping from $\mathbf{Y}$ into $H^{\frac{1}{2}}(\Gamma_c)$ and the compact embedding of this space into $L^2(\Gamma_c)$. Therefore, it is compact and linear, in particular Lipschitz continuous with a constant $c_2 > 0$. Since

$$\|\max(0, \hat{\lambda} + \gamma_1(\xi - d)) - \max(0, \hat{\lambda} + \gamma_1(\tilde{\xi} - d))\|_{\Gamma_c} \leq \gamma_1 \|\xi - \tilde{\xi}\|_{\Gamma_c} \tag{9.2.22}$$

for all $\xi, \tilde{\xi} \in L^2(\Gamma_c)$, the mapping Υ is Lipschitz continuous with constant γ_1. This implies that Ψ_γ is Lipschitz continuous with constant

$$\mathfrak{L} = \frac{c_1 c_2 \gamma_1}{\kappa} \|\mathfrak{F}\|_\infty, \tag{9.2.23}$$

where c_1, c_2 are constants from trace theorems. Concerning compactness the composition of Θ and Φ_γ is compact. From (9.2.22) it then follows that Ψ_γ is compact. This ends the proof. ☐

We can now show that the regularized contact problem with Coulomb friction has a solution.

Theorem 9.10. *The mapping Ψ_γ admits at least one fixed point, i.e., the regularized Coulomb friction problem* (9.2.19) *admits a solution. If $\|\mathfrak{F}\|_\infty$ is such that $\mathfrak{L}$ as defined in* (9.2.23) *is smaller than* 1, *the solution is unique.*

Proof. We apply the Leray–Schauder fixed point theorem to the mapping $\Psi_\gamma : L^2(\Gamma_c) \to L^2(\Gamma_c)$. Using Lemma 9.9, it suffices to show that λ is bounded in $L^2(\Gamma_c)$ independently of g. This is clear taking into account the dual problem $(\mathcal{P}^\star_{\gamma_1,\gamma_2})$. Indeed,

$$\min_{\lambda \geq 0,\ |\mu| \leq \mathfrak{F}g \text{ a.e. on } \Gamma_c} J^\star_{\gamma_1,\gamma_2}(\lambda, \mu) \leq \min_{\lambda \geq 0} J^\star_{\gamma_1,\gamma_2}(\lambda, 0) < \infty.$$

Hence, the Leray–Schauder theorem guarantees the existence of a solution to the regularized Coulomb friction problem. Uniqueness of the solution holds if $\mathfrak{F}$ is such that $\mathfrak{L}$ is smaller than 1, since in this case Ψ_γ is a contraction. ☐

In the following algorithm the fixed point approach is combined with an augmented Lagrangian concept to solve (9.2.19).

Algorithm ALM-FP.

1. Initialize $(\hat{\lambda}^0, \hat{\mu}^0) \in L^2(\Gamma_c) \times L^2(\Gamma_c)$ and $g^0 \in L^2(\Gamma_c)$, $m := 0$.
2. Choose $\gamma_1^m, \gamma_2^m > 0$ and determine the solution (λ^m, μ^m) to problem $(\mathcal{P}^\star_{\gamma_1,\gamma_2})$ with given friction g^m and $\hat{\lambda} := \hat{\lambda}^m$, $\hat{\mu} := \hat{\mu}^m$.
3. Update $g^{m+1} := \lambda^m$, $\hat{\lambda}^{m+1} := \lambda^m$, $\hat{\mu}^{m+1} := \mu^m$, and $m := m + 1$. Unless an appropriate stopping criterion is met, goto step 2.

The auxiliary problems $(\mathcal{P}^\star_{\gamma_1,\gamma_2})$ in step (2) can be solved by Algorithm SSN. Numerical experiments with this algorithm are given in [KuSt]. For the following brief discussion of the above algorithm, we assume that the mapping Ψ admits a fixed point λ^* in $L^2(\Gamma_c)$. Then, with the variables $\boldsymbol{y}^* \in \mathbf{Y}$ and $\mu^* \in L^2(\Gamma_c)$ corresponding to λ^*, we have for $\gamma_1, \gamma_2 > 0$ that

$$a(\boldsymbol{y}^*, z) - L(z) + (\lambda^*, \tau_N z)_{\Gamma_c} + (\mu^*, \tau_T z)_{\Gamma_c} = 0 \text{ for all } z \in \mathbf{Y},$$
$$\lambda^* - \max(0, \lambda^* + \gamma_1(\tau_N \boldsymbol{y}^* - d)) = 0 \text{ on } \Gamma_c,$$
$$\gamma_2 \tau_T \boldsymbol{y}^* - \max(0, \gamma_2 \tau_T \boldsymbol{y}^* + \mu^* - \mathfrak{F}\lambda^*) - \min(0, \gamma_2 \tau_T \boldsymbol{y}^* + \mu^* + \mathfrak{F}\lambda^*) = 0 \text{ on } \Gamma_c.$$

Provided that Algorithm ALM-FP has a limit point, it also satisfies this system of equations; i.e., the limit satisfies the original, nonregularized contact problem with Coulomb friction.

We end the section with a numerical example taken from [KuSt].

Example 9.11. We solve the contact problem with Tresca as well as with Coulomb friction using Algorithms **SSN** and **ALM-FP**, respectively. We choose $\Omega = [0, 3] \times [0, 1]$, the gap function $d = \max\left(0.0015, 0.003(x_1 - 1.5)^2 + 0.001\right)$ and $E = 10000$, $\nu = 0.45$, and $f = 0$. The boundary of possible contact and friction is $\Gamma_c := [0, 3] \times \{0\}$, and we assume traction-free boundary conditions on $\Gamma_n := [0, 3] \times \{1\} \cup \{0\} \times [0, 0.2] \cup \{3\} \times [0, 0.2]$. On $\Gamma_d := \{0\} \times [0.2, 1] \cup \{3\} \times [0.2, 1]$ we prescribe the deformation as follows:

$$\boldsymbol{\tau y} = \begin{cases} \begin{pmatrix} 0.003(1 - x_2) \\ -0.004 \end{pmatrix} & \text{on } \{0\} \times [0.2, 1], \\ \begin{pmatrix} -0.003(1 - x_2) \\ -0.004 \end{pmatrix} & \text{on } \{3\} \times [0.2, 1]. \end{cases}$$

Further $\gamma_1 = \gamma_2 = 10^8$, $\sigma = 1$, $\bar{\lambda} = \bar{\mu} = 0$, and Algorithm **SSN** is initialized by $\lambda^0 = \mu^0 = 0$. Algorithm **ALM-FP** is initialized with the solution of the pure contact problem and $g^0 = 0$, $\hat{\lambda}^0 = \hat{\mu}^0 = 0$. The MATLAB code is based on [ACFK], which that uses linear and bilinear finite elements for the discretization of the elasticity equations without friction and contact.

The semismooth Newton method detects the solution for given friction $g \equiv 1$ and $\mathfrak{F} = 1$ after 7 iterations. The corresponding deformed mesh and the elastic shear energy density are shown in Figure 9.1. As expected, in a neighborhood of the points (0, 0.2) and

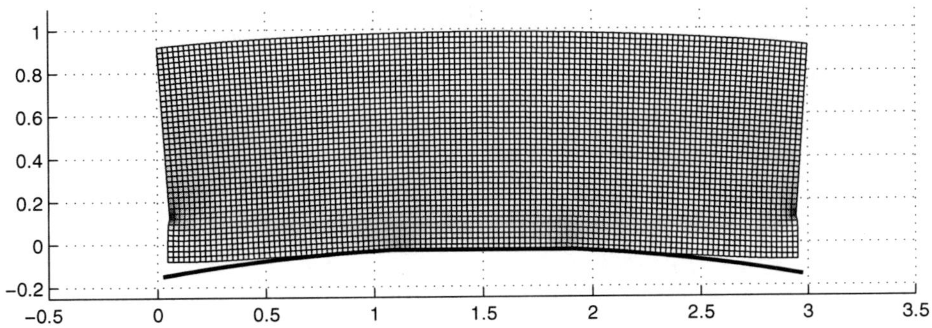

Figure 9.1. *Deformed mesh for $g \equiv 1$; gray tones visualize the elastic shear energy density.*

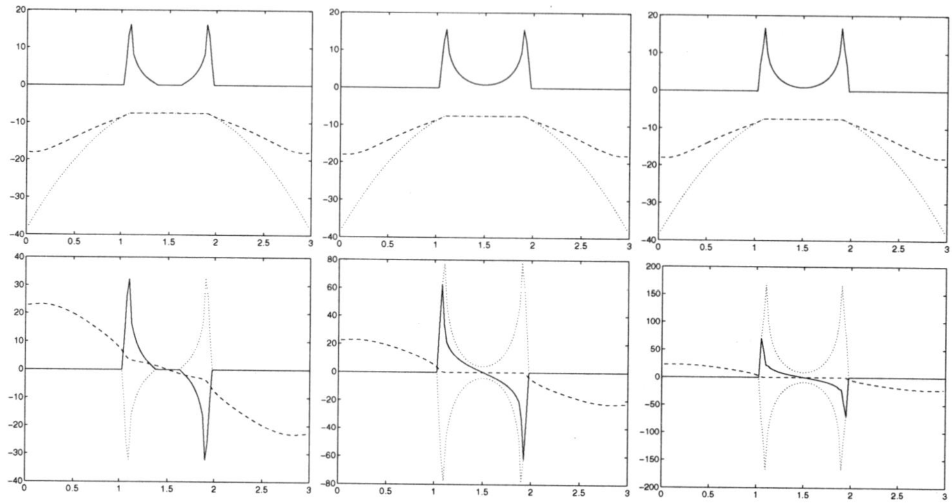

Figure 9.2. *Upper row, left: multiplier λ^* (solid), rigid foundation (multiplied by $5 \cdot 10^3$, dotted), and normal displacement $\tau_N y^*$ (multiplied by $5 \cdot 10^3$, dashed). Lower row, left figure: dual variable μ^* (solid) with bounds $\pm\mathfrak{F}\lambda^*$ (dotted) and tangential displacement y^* (multiplied by $5 \cdot 10^3$, dashed) for $\mathfrak{F} = 2$. Middle column: same as left, but with $\mathfrak{F} = 5$. Right column: same as first column, but with $\mathfrak{F} = 10$.*

$(3, 0.2)$, i.e., the points where the boundary conditions change from Neumann to Dirichlet, we observe a stress concentration due to a local singularity of the solution. Further, we also observe a (small) stress concentration close to the points where the rigid foundation has the kinks.

We turn to the Coulomb friction problem and investigate the performance of Algorithm **ALM-FP**. In Figure 9.2 the normal and the tangential displacement with corresponding multipliers for $\mathfrak{F} = 2, 5, 10$ are depicted. One observes that the friction coefficient significantly influences the deformation. For instance, in the case $\mathfrak{F} = 2$ the elastic body is in contact with the foundation in the interval $[1.4, 1.6]$, but it is not for $\mathfrak{F} = 5$ and $\mathfrak{F} = 10$. These large values of $\mathfrak{F}$ may, however, be physically of little relevance. Algorithm **ALM-FP** requires overall between 20 and 25 linear solves to stop with $\frac{|g^m - g^{m-1}|_{\Gamma_c}}{|g^m|_{\Gamma_c}} \le 10^{-7}$. For further information on the numerical performance we refer the reader to [Sta1].

Chapter 10

Parabolic Variational Inequalities

In this section we discuss the Lagrange multiplier approach to parabolic variational inequalities in the Hilbert space $H = L^2(\Omega)$ which are of the type

$$\left\langle \frac{d}{dt} y^*(t) + Ay^*(t) - f(t), y - y^*(t) \right\rangle \geq 0, \quad y^*(t) - \psi \in C, \qquad (10.0.1)$$
$$y^*(0) = y_0$$

for all $y - \psi \in C$, where the closed convex set C of H is defined by

$$C = \{y \in H : y \geq 0\},$$

A is a closed operator on H, Ω denotes an open domain in $\mathbb{R}^n$, and $y \geq 0$ is interpreted in the pointwise a.e. sense.

We consider the Black–Scholes model for American options, which is a variational inequality of the form

$$-\frac{d}{dt} v(t, S) - \left(\frac{\sigma^2}{2} S^2 v_{SS} + r S v_S - r v + Bv \right) \geq 0 \ \perp v(t, S) \geq \psi(S), \qquad (10.0.2)$$
$$v(T, S) = \psi(S)$$

for a.e. $(t, S) \in (0, T) \times (0, \infty)$, where $\perp$ denotes complementarity, i.e., $a \geq 0 \perp b \geq \psi$ if $a \geq 0$, $b \geq \psi$, and $a(b - \psi) = 0$. In (10.0.2) the reward function $\psi(S) = (K - S)^+$ for the put option and $\psi(S) = (S - K)^+$ for the call option. Here $S \geq 0$ denotes the price, v the value of the share, $r > 0$ is the interest rate, $\sigma > 0$ is the volatility of the market, and K is the strike price. Further T is the maturity date. The integral operator B is defined by

$$Bv(S) = -\lambda \int_0^\infty \big((z - 1)Sv_S + (v(t, S) - v(t, zS))\big) d\nu(z).$$

Note that (10.0.2) is a backward equation with respect to the time variable. Setting $y(t, S) = v(T - t, S)$ we arrive at (10.0.1), and (10.0.2) has the following interpretation

[Kou] in mathematical finance. The price process S_t is governed by the Ito's stochastic differential equation

$$dS_t/S_{t\_} = r\,dt + \sigma\,dB_t + (J_t - 1)\,d\pi_t,$$

where B_t denotes a standard Brownian motion, π_t is a counting Poisson process, $J_t - 1$ is the magnitude of the jump ν, and $\lambda \geq 0$ is the rate. The value function v is represented by

$$v(t, S) = \sup_\tau E^{t,x}[e^{-r(\tau-t)}\psi(S_\tau)] \quad \text{over all stopping times } \tau \leq T. \tag{10.0.3}$$

It will be shown that for the put case, $(0, \infty)$ can be replaced by $\Omega = (\bar{S}, \infty)$ with certain $\bar{S} > 0$. Thus, (10.0.2) can be formulated as (10.0.1) by defining a bounded bilinear form a on $V \times V$ by

$$a(v, \phi) = \int_{\bar{S}}^\infty \left(\frac{\sigma^2}{2} S^2 v_S + (r - \sigma^2) S\, v\right) \phi_S + (2r - \sigma^2)\, v\phi - Bv\,\phi)\, dS \tag{10.0.4}$$

for $v, \phi \in V$, where V is defined by

$$V = \left\{\phi \in H : \phi \text{ is absolutely continuous on } (\bar{S}, \infty), \right.$$

$$\left. \int_{\bar{S}}^\infty S^2|\phi_S|^2\, dS < \infty \text{ and } \phi(S) \to 0 \text{ as } S \to \infty \text{ and } \psi(\bar{S}) = 0\right\}$$

equipped with

$$|\phi|_V^2 = \int_{\bar{S}}^\infty (S^2|\phi_S|^2 + |\phi|^2)\, dS.$$

Now

$$a(v, \phi) \leq \frac{\sigma^2}{2}|v|_V|\phi|_V + |r - \sigma^2||v|_H|\phi|_V + |2r - \sigma^2||v|_H|\phi|_H$$

and

$$a(v, v) \geq \frac{\sigma^2}{2}|v|_V^2 + \left(2r - \frac{3}{2}\sigma^2\right)|v|_H^2 - |r - \sigma^2||v|_V|v|_H$$

$$\geq \frac{\sigma^2}{4}|v|_V^2 + \left(2r - \frac{3}{2}\sigma^2 - \frac{(r - \sigma^2)^2}{\sigma^2}\right)|v|_H^2.$$

Note that if $Av \in L^2(\Omega)$, then

$$(Av, \phi) = a(v, \phi) \quad \text{for all } \phi \in V.$$

Then, the solution $v(T - t, S)$ satisfies (10.0.1) with $f(S) = \lambda \int_0^{\frac{\bar{S}}{S}} \psi(zS)\, d\nu(z)$. Let $v^+ = \sup(0, v)$. Since $v^+ \geq v$ a.e. in Ω,

$$(-Bv, v^+) \geq (-Bv^+, v^+).$$

Thus,

$$Av = -\left(\frac{\sigma^2}{2} S^2 v_{SS} + rS\, v_S - r\, v + Bv\right)$$

satisfies

$$\langle Av, v^+\rangle \geq \langle Av^+, v^+\rangle.$$

Motivated by this example we make the following assumptions. Let X be a Hilbert space that is continuously embedded into H, and let V be a separable closed linear subspace of X endowed with the induced norm and dense in H. Assume that

$$\psi \in X, \quad f \in L^2(0, T; V^*),$$

and that

$$\phi^+ = \sup(0, \phi) \in V \quad \text{for all } \phi \in V.$$

The following assumptions will be used for the operator A.

(1) $A \in \mathcal{L}(X, V^*)$, i.e., there exists $\bar{M}$ such that

$$|\langle Ay, \phi\rangle_{V^*\times V}| \leq \bar{M}|y|\,|\phi| \quad \text{for all } y \in X \text{ and } \phi \in V,$$

and A is closed in H with

$$\mathrm{dom}(A) = \{y \in X : Ay \in H\} \subset X,$$

where $\mathrm{dom}(A)$ is a Hilbert space equipped with the graph norm.

(2) There exist $\omega > 0$ and $\rho \in \mathbb{R}$ such that for all $\phi \in V$

$$\langle A\phi, \phi\rangle \geq \omega|\phi|_V^2 - \rho|\phi|_H^2.$$

(3) For all $\phi \in V$,

$$\langle A\phi, \phi^+\rangle \geq \langle A\phi^+, \phi^+\rangle.$$

(4) There exists $\bar{\lambda} \in H$ satisfying $\bar{\lambda} \leq 0$ a.e. such that

$$\langle \bar{\lambda} + A\psi - f(t), \phi\rangle \leq 0$$

for a.e. t and all $\phi \in V$ satisfying $\phi \geq 0$ a.e.

(5) There exists an $\bar{\psi} \in \mathrm{dom}(A)$ such that $\bar{\psi} - \psi \in V \cap C = \mathcal{C}$.

(6) Let a_s be the symmetric form on $V \times V$ defined by

$$a_s(y, \phi) = \frac{1}{2}(\langle Ay, \phi\rangle + \langle A\phi, y\rangle)$$

for $y, \phi \in V$ and assume that the skew-symmetric form satisfies

$$\frac{1}{2}|\langle Ay, \phi\rangle - \langle A\phi, y\rangle| \leq M|y|_V|\phi|_H$$

for a constant M independent of $y, \phi \in V$.

(7) $(\phi - \gamma)^+ \in V$ for any $\gamma \in \mathbb{R}^+$ and $\phi \in V$, and $\langle A1, (\phi - \gamma)^+\rangle \geq 0$.

Assumptions (1)–(5) apply to (10.0.2) and to second order elliptic differential operators. Assumption (6) applies to the biharmonic operator Δ^2 and to self-adjoint operators. For the biharmonic operator and systems of equations as the elasticity system, for instance, the monotone property (3) does not hold.

In this chapter we discuss (10.0.1) without assuming that V is embedded compactly into H. In the latter case, one can use the Aubin lemma, which states that $W(0,T) = L^2(0,T;V) \cap H^1(0,T;V^*)$ is compactly embedded into $L^2(0,T;H)$. This ensures that the weak limit of certain approximating sequences defines the solution; see, e.g., [GLT, IK23]. Instead, our analysis uses the monotone trick for variational inequalities. From, e.g., [Tan, p. 151], we recall that $W(0,T)$ embeds continuously into $C([0,T];H)$.

We commence with the definitions of strong and weak solutions to (10.0.1).

Definition 10.1 (Strong Solution). *Given $y_0 - \psi \in \mathcal{C}$ and $f \in L^2(0,T;H)$, an X-valued function $y^*(t)$, with $y^* - \psi \in L^2(0,T;V) \cap H^1(0,T;V^*)$ is called a strong solution of* (10.0.1) *if $y^*(0) = y_0$, $y^* \in H^1(\delta,T;H)$ for every $\delta > 0$, $y^*(t,x) \geq \psi(x)$ a.e. in $(0,T) \times \Omega$, and for a.e. $t \in (0,T)$,*

$$\left\langle \frac{d}{dt} y^*(t) + Ay^*(t) - f(t), y - y^*(t) \right\rangle \geq 0$$

for a.e. $t \in (0,T)$ and for all $y - \psi \in \mathcal{C}$.

Defining $\lambda^* = -\frac{d}{dt} y^* - Ay^* + f(t) \in L^2(0,T;V^*)$, we have in the distributional sense that y^* satisfies

$$\begin{cases} \dfrac{d}{dt} y^*(t) + Ay^*(t) + \lambda^*(t) = f(t), \quad y^*(0) = y_0, \\[2ex] \langle \lambda^*(t), y - \psi \rangle \leq 0 \text{ for all } y - \psi \in \mathcal{C} \quad \text{and} \quad \langle \lambda^*, y^* - \psi \rangle = 0. \end{cases} \tag{10.0.5}$$

Moreover, in case that y^* is a strong solution we have $y^* \in L^2(\delta,T;\text{dom}(A))$ for every $\delta > 0$. Further $\lambda^* \in L^2(\delta,T;H)$ and (10.0.1) can equivalently be written as a variational inequality in the form

$$\begin{cases} \dfrac{d}{dt} y^*(t) + Ay^*(t) + \lambda^*(t) = f(t), \quad y^*(0) = y_0, \\[2ex] \lambda^*(t) \leq 0, \quad y^*(t) \geq \psi, \quad (y^*(t) - \psi, \lambda^*(t))_H = 0 \quad \text{for a.e. } t > 0. \end{cases} \tag{10.0.6}$$

Definition 10.2 (Weak Solution). *Assume that $y_0 \in H$ and $f \in L^2(0,T,V^*)$. Then a function $y^* - \psi \in L^2(0,T;V)$ satisfying $y^*(t,x) \geq \psi(x)$ a.e. in $(0,T) \times \Omega$ is called a weak solution to* (10.0.1) *if*

$$\begin{aligned} &\int_0^T \left[\left\langle \frac{d}{dt} y(t), y(t) - y^*(t) \right\rangle + \langle Ay^*(t), y(t) - y^*(t) \rangle - \langle f(t), y(t) - y^*(t) \rangle \right] dt \\ &\qquad + \frac{1}{2} |y(0) - y_0|_H^2 \geq 0 \end{aligned} \tag{10.0.7}$$

is satisfied for all $y - \psi \in \mathcal{K}$, *where*

$$\mathcal{K} = \{y \in W(0,T) : y(t,x) \geq 0 \text{ a.e. in } (0,T) \times \Omega\} \tag{10.0.8}$$

and

$$W(0,T) = L^2(0,T;V) \cap H^1(0,T;V^*).$$

Since for y^* and y in $W(0,T)$

$$\int_0^T \left\langle \frac{d}{dt} y(t) - y^*(t), y(t) - y^*(t) \right\rangle dt = \frac{1}{2}(|y(T) - y^*(T)|_H^2 - |y(0) - y_0|_H^2),$$

it follows that a strong solution to (10.0.1) is also a weak solution.

Let us briefly outline this chapter. Section 10.1 is devoted to proving existence and uniqueness of strong solutions. Extra regularity of solutions is obtained in Section 10.2. Continuous dependence of the solution with respect to parameters in A is investigated in Section 10.3. Section 10.4 focuses on weak solutions obtained as the limit of approximating difference schemes. In Section 10.5 monotonic behavior of solutions with respect to initial conditions and the forcing function is proved. We refer to [Sey] and the literature cited there for an introduction to numerical aspects in mathematical finance.

10.1 Strong solutions

In this section we establish the existence of the strong solution to (10.0.1) under assumptions (1)–(5) and (1)–(2), (5)–(6), respectively.

For $\bar{\lambda} \in H$ satisfying $\bar{\lambda} \leq 0$ we consider the regularized equations of the form

$$\begin{cases} \frac{d}{dt} y_c + A y_c + \min(0, \bar{\lambda} + c\,(y_c - \psi)) = f, \quad c > 0, \\ y_c(0) = y_0. \end{cases} \tag{10.1.1}$$

Proposition 10.3. *If assumptions* (1)–(2) *hold and* $y_0 \in H$, $f \in L^2(0,T;V^*)$, *and* $\bar{\lambda} \in H$, *then* (10.1.1) *has a unique solution* y_c *satisfying* $y_c - \psi \in L^2(0,T;V) \cap H^1(0,T;V^*)$.

Proof. Existence and uniqueness of the solution to (10.1.1) follow with monotone techniques; see [ItKa, Lio3], for instance. Define $\mathcal{A} : V \to V^*$ by

$$\mathcal{A}\phi = A\phi + \min(-\bar{\lambda}, c\phi).$$

Then (10.1.1) can equivalently be expressed as

$$\frac{d}{dt} v + \mathcal{A} v = f - \bar{\lambda} - A\psi \in L^2(0,T;V^*), \tag{10.1.2}$$

with $v = y_c - \psi$ and $v(0) = y_0 - \psi \in H$. We note that $\mathcal{A}$ is hemicontinuous, i.e., $s \to \langle \mathcal{A}(\phi_1 + s\phi_2), \phi_3 \rangle$ is continuous from $\mathbb{R} \to \mathbb{R}$ for all $\phi_i \in V$, $i = 1, \ldots, 3$, and

$$|\mathcal{A}\phi|_{V^*} \leq |A\phi|_{V^*} + c|\phi|_H \text{ for all } \phi \in V,$$

$$\langle \mathcal{A}\phi_1 - \mathcal{A}\phi_2, \phi_1 - \phi_2 \rangle \geq \omega|\phi_1 - \phi_2|_V^2 - \rho|\phi_1 - \phi_2|_H^2 \text{ for all } \phi_1, \phi_2 \in V,$$

$$\langle \mathcal{A}\phi, \phi \rangle \geq \omega|\phi|_V^2 - \rho|\phi|_H^2 \text{ for all } \phi \in V.$$

It follows that (10.1.2) admits a unique solution $v \in L^2(0,T;V) \cap H^1(0,T;V^*)$ and this gives the desired solution $y_c = v + \psi$ of (10.1.1); cf. [ItKa, Theorem 8.7], [Lio3, Theorem II.1.2]. $\square$

Theorem 10.4. (1) *If in addition to the assumptions in Proposition* 10.3 *assumptions* (3)–(4) *hold,* $y_0 - \psi \in C$, *then* $y_c(t) - \psi \in C$ *and* $y_c(t) \geq y_{\hat{c}}(t)$ *for* $\hat{c} \geq c$. *Moreover,* $y_c - \psi \to y^* - \psi$ *strongly in* $L^2(0,T;V)$ *and weakly in* $H^1(0,T;V^*)$ *as* $c \to \infty$, *where* y^* *is the unique solution of* (10.0.1) *in the sense that* $y^* - \psi \in \mathcal{K}$, (10.0.6) *is satisfied with* $\lambda^* \in L^2(0,T;H)$, *and the estimate*

$$\frac{1}{2} e^{-2\rho t} |y_c(t) - y^*(t)|_H^2 + \int_0^t e^{-2\rho s} \omega \, |y_c(s) - y^*(s)|_V^2 \, ds \leq \frac{1}{c} \int_0^t e^{-2\rho s} |\bar{\lambda}|^2 \, ds \to 0$$

for $t \in [0,T]$ *holds. If in addition assumption* (7) *is satisfied and* $\bar{\lambda} \in L^\infty(\Omega)$, *then*

$$|y_c(t) - y^*(t)|_{L^\infty} \leq \frac{1}{c} |\bar{\lambda}|_{L^\infty}.$$

(2) *If assumptions* (1)–(4) *hold,* $y_0 - \psi \in C$, *and* $f \in L^2(0,T;H)$, *then* y^* *is the unique strong solution to* (10.0.1).

Proof. (1) From (10.1.1) it follows that y_c satisfies

$$\begin{cases} \frac{d}{dt} y_c + A(y_c - \psi) + \lambda_c = f - A\psi, \\ y_c(0) = y_0, \end{cases} \tag{10.1.3}$$

where

$$\lambda_c = \min(0, \bar{\lambda} + c\,(y_c - \psi)).$$

If $y_0 - \psi \in C$, then $y_c(t) - \psi \in C$. In fact, let $\phi = \min(0, y_c - \psi) = -(y_c - \psi)^- \in -C \cap V$. Since $\bar{\lambda} \leq 0$ it follows that

$$\left\langle \frac{d}{dt} y_c + A(y_c - \psi), \phi \right\rangle + \langle A\psi - f + \bar{\lambda} + c\,\phi, \phi \rangle = 0,$$

where by assumptions (3) and (2)

$$\langle A(y_c - \psi), \phi \rangle \geq \langle A\phi, \phi \rangle \geq \omega |\phi|^2 - \rho |\phi|_H^2$$

and by (4)

$$\langle A\psi - f(t) + \bar{\lambda}, \phi \rangle \geq 0.$$

Thus,

$$\frac{1}{2} \frac{d}{dt} |\phi|_H^2 \leq \rho \, |\phi|_H^2$$

and consequently

$$e^{-2\rho t}|\phi|_H^2 \le |\phi(0)|_H^2 = 0. \tag{10.1.4}$$

Since

$$0 \ge \lambda_c = \min(0, \bar{\lambda} + c\,(y_c - \psi)) \ge \bar{\lambda},$$

we have

$$|\lambda_c(t)|_H \le |\bar{\lambda}|_H$$

for all $t \in [0, T]$. From (10.1.3) we deduce that $\{y_c\}$ is bounded in $L^2(0, T; V)$. By assumption (1) and again by (10.1.3) it follows that $\{Ay_c\}$ and $\{\frac{d}{dt} y_c\}$ are bounded in $L^2(0, T; V^*)$. Thus, there exist $\lambda^* \in L^2(0, T; H)$ satisfying $\lambda^* \le 0$ a.e. and y^* satisfying $y^* - \psi \in \mathcal{K}$, such that for a subsequence denoted again by c,

$$\begin{aligned} &\lambda_c \to \lambda^* \text{ weakly in } L^2(0, T; H),\\ &Ay_c \to Ay^* \text{ and } \frac{d}{dt} y_c \to \frac{d}{dt} y^* \text{ weakly in } L^2(0, T; V^*) \end{aligned} \tag{10.1.5}$$

as $c \to \infty$. Taking the limit in (10.1.3) implies that

$$\frac{d}{dt} y^* + Ay^* - f = -\lambda^*, \qquad y^*(0) = y_0, \tag{10.1.6}$$

with equality in the differential equation holding in the sense of $L^2(0, T; V^*)$.

For $\phi = -(y_c - y_{\hat{c}})^-$ with $c \le \hat{c}$ we deduce from (10.1.3) that

$$\left\langle \frac{d}{dt}(y_c - y_{\hat{c}}) + A(y_c - y_{\hat{c}}), \phi \right\rangle + (\lambda_c - \lambda_{\hat{c}}, \phi) = 0,$$

where

$$\begin{aligned} (\lambda_c - \lambda_{\hat{c}}, \phi) = \big(&\min(0, \bar{\lambda} + c(y_{\hat{c}} - \psi)) - \min(0, \bar{\lambda} + \hat{c}(y_{\hat{c}} - \psi))\\ &+ \min(0, \bar{\lambda} + c(y_c - \psi)) - \min(0, \bar{\lambda} + c(y_{\hat{c}} - \psi)), \phi\big) \ge 0, \end{aligned}$$

since $y_{\hat{c}} \ge \psi$. Hence, using the same arguments as those leading to (10.1.4), we have $|\phi(t)|_H = 0$ and thus

$$y_c \ge y_{\hat{c}} \quad \text{for} \quad c \le \hat{c}.$$

By Lebesgue dominated convergence theorem and the theorem of Beppo Levi, $y_c \to y^*$ strongly to $L^2(0, T; H)$ and pointwise a.e. in $(0, T) \times \Omega$. Since

$$0 \ge \int_0^T (\lambda_c, y_c - \psi)_H \, dt \ge -\frac{1}{c} \int_0^T |\bar{\lambda}|_H^2 \, dt \to 0$$

as $c \to \infty$, we have

$$\int_0^T (\lambda^*, y^* - \psi) \, dt = 0.$$

That is, (y^*, λ^*) satisfies (10.0.6), where the first equation is satisfied in the sense of $L^2(0, T; V^*)$. Suppose that $y \in \mathcal{K}$ satisfies (10.0.1). Then it follows that

$$\frac{1}{2}\frac{d}{dt}|y^*(t) - y(t)|_H^2 + \langle A(y^* - y(t)), y^*(t) - y(t)\rangle \le 0$$

and thus $e^{-2\rho t}|y^*(t) - y(t)|_H^2 \le |y_0 - y(0)|_H^2$. This implies that y^* is the unique solution to (10.0.1) in $\mathcal{K}$ and that the whole family $\{(y_c, \lambda_c)\}$ converges in the sense specified in (10.1.5). From (10.0.1) and (10.1.1)

$$\left\langle \frac{d}{dt}y^*(t) + Ay^*(t) - f(t), y_c(t) - y^*(t)\right\rangle \ge 0,$$

$$\left\langle \frac{d}{dt}y_c(t) + Ay_c(t) - f(t), y^*(t) - y_c(t)\right\rangle \ge (\lambda_c, y_c - \psi)_H.$$

Since $y_c \ge \psi$, we have $(\lambda_c, y_c - \psi) \ge -\frac{1}{c}|\bar\lambda|_H^2$. Summing the above inequalities and multiplying by $e^{-2\rho t}$ give

$$\frac{1}{2}\frac{d}{dt}(e^{-2\rho t}|y_c(t) - y^*(t)|_H^2) + e^{-2\rho t}\langle A(y_c(t) - y^*(t)), y_c(t) - y^*(t)\rangle$$
$$+\rho\,|y_c(t) - y^*(t)|_H^2 \le \frac{1}{c}\,e^{-2\rho t}|\bar\lambda|^2,$$

which implies the first estimate and in particular that $y_y \to y^*$ strongly in $L^2(0, T; V)$.

Suppose next that in addition $\bar\lambda \in L^\infty(\Omega)$. Let $k \in \mathbb{R}^+$ and $\phi = (y_c - y^* - k)^+$. By assumption $\phi \in V$. From (10.0.6) and (10.1.1)

$$\left\langle \frac{d}{dt}y^* + Ay^* - f, \phi\right\rangle \ge 0$$

and

$$\left\langle \frac{d}{dt}y_c + Ay_c + \lambda_c - f, \phi\right\rangle = 0.$$

If $k \ge \frac{1}{c}|\bar\lambda|_{L^\infty}$, then

$$(\lambda_c, \phi) = (\min(0, \bar\lambda + c\,(y_c - \psi)), (y_c - y^* - k)^+) = 0,$$

where we use that $y_c \ge y^* \ge \psi$. Hence, we obtain

$$\left\langle \frac{d}{dt}(y_c - y^* - k) + A(y_c - y^* - k) + A\,k, \phi\right\rangle \le 0.$$

By assumption (3) and since $\langle A1, \phi\rangle \ge 0$,

$$\frac{1}{2}\frac{d}{dt}|\phi|^2 + \langle A\phi, \phi\rangle \le 0,$$

which implies the second estimate.

(2) Now suppose that $f \in L^2(0, T; H)$ and that assumptions (1)–(4) hold. Consider (10.1.3) in the form

$$\begin{cases} \frac{d}{dt} y_c + A y_c = f - \lambda_c, \\ y(0) = y_0. \end{cases} \tag{10.1.7}$$

We decompose $y_c = y_{c,i} + y_h$, where $y_{c,i}$ and y_h are the solutions to (10.1.7) with initial condition and forcing functions set to zero, respectively. Note that $\{\lambda_c\}$ is bounded in $L^2(0, T; H)$ uniformly with respect to c. Hence by the following lemma $\{Ay_{c,i}\}$ and $\{\frac{d}{dt} y_{c,i}\}$ are bounded in $L^2(0, T; H)$ uniformly for $c > 0$. Moreover, $Ay_h \in L^2(\delta, T; H)$ and $\frac{d}{dt} y_h \in L^2(\delta, T; H)$ for every $\delta > 0$. Thus y_c is bounded in $H^1(\delta, T; H) \cap L^2(\delta, T; \mathrm{dom}(A))$ and converges weakly in $H^1(\delta, T; H) \cap L^2(\delta, T; \mathrm{dom}(A))$ to y^* for $c \to \infty$. □

Lemma 10.5. *Under the assumptions of the previous theorem $-A$ generates an analytic semigroup on H. If $\frac{d}{dt}x + Ax = g \in L^2(0, T; H)$, with $x_0 = 0$, then $\frac{d}{dt}x(t)$ and $Ax(t) \in L^2(0, T; H)$, and*

$$|Ax|_{L^2(0,T;H)} \le \bar{k}\, |g|_{L^2(0,T;H)},$$

with $\bar{k}$ independent of $g \in L^2(0, T; H)$.

Proof. Let $B = A - \rho I$. Further for $u \in \mathrm{dom}(A)$ and $\lambda \in \mathbb{C}$ with $\mathrm{Re}\lambda \ge 0$ set $\bar{g} = \lambda u + Bu$. Then, since

$$\mathrm{Re}\, \lambda\, (u, u)_H + \langle Bu, u\rangle \le |\bar{g}|_H |u|_H,$$

assumption (2) implies that

$$\omega |u|_V^2 \le |f|_H |u|_H. \tag{10.1.8}$$

From assumption (1) and (10.1.8)

$$|\lambda|\, |u|_H^2 \le |\bar{g}|_H |u|_H + \bar{M} |u|_V^2 \le \left(1 + \frac{\bar{M}}{\omega}\right) |\bar{g}|_H |u|_H,$$

and thus

$$|u|_H = |(\lambda I + B)^{-1} \bar{g}|_H \le \left(1 + \frac{\bar{M}}{\omega}\right) \frac{1}{|\lambda|} |\bar{g}|_H. \tag{10.1.9}$$

It thus follows from [ItKa, Paz, Tan] that $-B$ and hence $-A$ generate analytic semigroups on H related by $e^{-Bt} = e^{(\rho - A)t}$.

For $g \in L^2(0, \infty; H)$ with $e^{\rho \cdot} g \in L^2(0, \infty; H)$ consider

$$\frac{d}{dt} x + Ax = g \text{ with } x(0) = 0.$$

This is related to

$$\frac{d}{dt} z + Bz = g_\rho := g e^{\rho \cdot} \text{ with } x(0) = 0 \tag{10.1.10}$$

by $z(t) = e^{\rho t}x$. Taking the Laplace transform of (10.1.10), we obtain

$$\lambda \hat{z} + B\hat{z} = \hat{g}_\rho, \qquad \text{where } \hat{z} = \int^{\infty} z_0 e^{-\lambda s} z(s)\, ds,$$

and thus by (10.1.9)

$$|B\hat{z}|_H \le |B(\lambda I + B)^{-1}\hat{g}_\rho|_H \le \left(2 + \frac{\bar{M}}{\omega}\right) |\hat{g}_\rho|_H.$$

From the Fourier–Plancherel theorem we have

$$\int_0^{\infty} |Az(t)|_H^2\, dt \le \left(2 + \frac{\bar{M}}{\omega}\right) \int_0^{\infty} |g_\rho|^2\, dt.$$

This implies that $|Ax|_{L^2(0,T;H)} \le e^{\rho T}(2 + \frac{\bar{M}}{\omega})|g|_{L^2(0,T;H)}$ choosing $g = 0$ for $t \ge T$. $\square$

To allow $\delta = 0$ for the strong solutions in the previous theorem we let $\bar{y}$ denote the solution to

$$\begin{cases} \frac{d}{dt}\bar{y} + A\bar{y} = 0, \\ \bar{y}(0) = y_0, \end{cases}$$

and we consider

$$\begin{cases} \frac{d}{dt}(y_c - \bar{y}) + A(y_c - \bar{y}) = f - \lambda_c, \\ y(0) - \bar{y}(0) = 0. \end{cases}$$

Arguing as in (2) of Theorem 10.4 we obtain the following corollary.

Corollary 10.6. *Under the assumptions of Theorem* 10.4 *we have* $y^* - \bar{y} \in H^1(0, T; H) \cap L^2(0, T; \text{dom}(A))$.

Next we turn to verify existence under a different set of assumptions which, in particular, does not involve the monotone assumption (3). For $\bar{\lambda} = 0$ in (10.1.1), let $\hat{y}_c$ denote the corresponding solution, i.e.,

$$\frac{d}{dt}\hat{y}_c + A\hat{y}_c + c\, \min(0, c\,(y_c - \psi)) = f, \quad c > 0, \tag{10.1.11}$$

which exists by Theorem 10.4 (1).

Theorem 10.7. *If assumptions* (1)–(2) *and* (5)–(6) *hold,* $y_0 - \psi \in C \cap V$, *and* $f \in L^2(0, T; H)$, *then* (10.0.1) *has a unique, strong solution* $y^*(t)$ *in* $H^1(0, T; H) \cap L^2(0, T; \text{dom}(A))$, *and* $\hat{y}_c \to y^*$ *strongly in* $L^2(0, T; V) \cap C(0, T; H)$ *as* $c \to \infty$. *Moreover* $t \to y^*(t) \in V$ *is right continuous. If in addition assumptions* (3)–(4) *hold, then* $\hat{y}_c \le \hat{y}_{\hat{c}}$ *for* $c \le \hat{c}$ *and* $\hat{y}_c(t) \to y^*(t)$ *strongly in* H *for each* $t \in [0, T]$ *and pointwise a.e. in* Ω.

Proof. For $\bar{\lambda} = 0$ we have

$$\frac{1}{2}\frac{d}{dt}|\hat{y}_c - \psi|_H^2 + \langle A(\hat{y}_c - \psi), \hat{y}_c - \psi\rangle + \langle A\psi - f, \hat{y}_c - \psi\rangle + c\,|(\hat{y}_c - \psi)^-|^2 = 0.$$

From assumptions (1)–(2) we have

$$\begin{aligned}|\hat{y}_c(t) - \psi|_H^2 &+ \int_0^t (\omega\,|\hat{y}_c - \psi|_V^2 + c\,|(\hat{y}_c - \psi)^-|_H^2)\,ds \\ &\le |y_0 - \psi|_H^2 + \int_0^t \left(2\rho\,|\hat{y}_c - \psi|_H^2 + \frac{1}{\omega}|A\psi - f|_{V^*}^2\right) ds\end{aligned}$$

and thus

$$\begin{aligned}|\hat{y}_c(t) - \psi|_H^2 &+ \int_0^t (\omega\,|\hat{y}_c - \psi|_V^2 + c\,|(\hat{y}_c - \psi)^-|_H^2)\,ds \\ &\le e^{2\rho t}\left(|y_0 - \psi|_H^2 + \frac{1}{\omega}\int_0^t |A\psi - f|_{V^*}^2\,ds\right).\end{aligned} \tag{10.1.12}$$

With assumptions (5) and (6) holding, we have that $\hat{y}_c \in H^1(0, T; H)$,

$$\left(\frac{d}{dt}\hat{y}_c + A\hat{y}_c + c\min(0, \hat{y}_c - \psi) - f(t), \frac{d}{dt}\hat{y}_c\right) = 0$$

for a.e. $t \in (0, T)$. Then

$$\left|\frac{d}{dt}\hat{y}_c\right|_H^2 + \frac{d}{dt}(a_s(\hat{y}_c - \bar{\psi}, \hat{y}_c - \bar{\psi}) + c|(\hat{y}_c - \psi)^-|^2) \le 2(M^2|\hat{y}_c - \bar{\psi}|_V^2 + |A\bar{\psi} - f|_H^2)$$

and hence

$$\begin{aligned}a_s(\hat{y}_c(t) - \bar{\psi}, \hat{y}_c(t) - \bar{\psi}) &+ c\,|(\hat{y}_c(t) - \psi)^-|_H^2 + \int_0^t \left|\frac{d}{dt}\hat{y}_c\right|_H^2 ds \\ &\le a_s(y_0 - \bar{\psi}, y_0 - \bar{\psi}) + \int_0^t 2(M^2|\hat{y}_c(s) - \bar{\psi}|_V^2 + |A\bar{\psi} - f(s)|_H^2)\,ds,\end{aligned} \tag{10.1.13}$$

where we used the fact that $y_0 - \psi \in C$. It thus follows from (10.1.12)–(10.1.13) that

$$\begin{aligned}|\hat{y}_c(t) - \bar{\psi}|_V^2 &+ c\,|(\hat{y}_c(t) - \psi)^-|_H^2 + \int_0^t \left|\frac{d}{dt}\hat{y}_c\right|_H^2 ds \\ &\le K\left(|y_0 - \bar{\psi}|_V + |\psi - \bar{\psi}|_V + \int_0^t |A\bar{\psi} - f(s)|_H^2\,ds\right)\end{aligned} \tag{10.1.14}$$

for a constant K independent of $c > 0$ and $t \in [0, T]$.

Hence there exists a y^* such that $y^* - \bar{\psi} \in H^1(0, T; H) \cap B(0, T; V)$ and on a subsequence

$$\frac{d}{dt}\hat{y}_c \to \frac{d}{dt}y^*, \quad A\hat{y}_c \to Ay^*$$

weakly in $L^2(0, T; H)$ and $L^2(0, T; V^*)$, respectively. In particular this implies that $y_c \to y^*$ weakly in $L^2(0, T; H)$. Above $B(0, T; V)$ denotes the space of all everywhere bounded measurable functions from $[0, T]$ to V. By assumption (5) we have that $y^* - \psi \in B(0, T; V)$ as well.

Since

$$\int_0^T (\hat{y}_c - \psi, \phi)_H \, dt \geq -\int_0^T ((\hat{y}_c - \psi)^-, \phi)_H \, dt$$

for all $\phi \in L^2(0, T; H)$ with $\phi(t) \in C$ for a.e. t, and since $\lim_{c \to 0} \int_0^T |(\hat{y}_c(t) - \psi)^-|_H^2 \, dt = 0$, by (10.1.14) we have

$$\int_0^T (y^*(t) - \psi, \phi)_H \, dt \geq 0 \text{ for all } \phi \in L^2(0, T; H) \text{ with } \phi(t) \in C. \tag{10.1.15}$$

This implies that $y^*(t) - \psi \in C$ for a.e. $t \in [0, T]$. For $y - \psi \in \mathcal{C}$

$$-(c(\hat{y}_c(t) - \psi)^-, y - \hat{y}_c(t)) = -\big(c(\hat{y}_c(t) - \psi)^-, y - \psi - (\hat{y}_c(t) - \psi)\big) \leq 0$$

for a.e. $t \in (0, T)$. It therefore follows from (10.1.11) that for $y \in \mathcal{K}$

$$\left\langle \frac{d}{ds}(\hat{y}_c(t) - y(t) + y(t)) + A(\hat{y}_c(t) - y(t)) + Ay(t) - f(t), y(t) - \hat{y}_c(t) \right\rangle \geq 0. \tag{10.1.16}$$

Hence

$$\int_0^t e^{-2\rho s} \left\langle \frac{d}{ds}(\hat{y}_c(s) - y(s)), \hat{y}_c(s) - y(s) \right\rangle ds$$
$$+ \int_0^t e^{-2\rho s} \left\langle \frac{d}{ds} y(s) + A(\hat{y}_c(s) - y(s)), \hat{y}_c(s) - y(s) \right\rangle$$
$$\leq \int_0^t e^{-2\rho s} \langle Ay(s) - f(s), y(s) - \hat{y}_c(s) \rangle \, ds.$$

For $z \in W(0, T)$ we have

$$\int_0^t e^{-2\rho s} \left\langle \frac{d}{ds} z(s), z(s) \right\rangle ds = \frac{1}{2} (e^{-2\rho t} |z(t)|_H^2 - |z(0)|_H^2) + \int_0^t \rho \, e^{-2\rho s} |z(s)|_H^2 \, ds. \tag{10.1.17}$$

This, together with (10.1.16), implies for $y \in \mathcal{K}$

$$\frac{1}{2} (e^{-2\rho t} |\hat{y}_c(t) - y(t)|_H^2 - |y_0 - y(0)|_H^2)$$
$$+ \int_0^t e^{-2\rho s} \left(\left\langle \frac{d}{ds} y(s) + A(\hat{y}_c(s) - y(s)), \hat{y}_c(s) - y(s) \right\rangle + \rho \, |\hat{y}_c(s) - y(s)|_H^2 \right) ds$$
$$\leq \int_0^t e^{-2\rho s} \langle Ay(s) - f(s), y(s) - \hat{y}_c(s) \rangle \, ds$$
$$\to \int_0^t e^{-2\rho s} \langle Ay - f(s), y(s) - y^*(s) \rangle \, ds$$

as $c \to \infty$. Since norms are w.l.s.c., we obtain

$$\frac{1}{2}\left(e^{-2\rho t}|y^*(t) - y(t)|_H^2 - |y_0 - y(0)|_H^2\right)$$

$$+\int_0^t e^{-2\rho s}\left(\left\langle \frac{d}{ds}y(s) + A(y^*(s) - y(s)), y^*(s) - y(s)\right\rangle + \rho\,|y^*(s) - y(s)|_H^2\right) ds$$

$$\le \int_0^t e^{-2\rho s}\langle Ay(s) - f(s), y(s) - y^*(s)\rangle\, ds,$$

or equivalently, using (10.1.17)

$$\int_0^t e^{-2\rho s}\left(\left\langle \frac{d}{ds}y^*(s) + Ay^*(s) - f(s), y(s) - y^*(s)\right\rangle\right) ds \ge 0 \tag{10.1.18}$$

for all $y \in \mathcal{K}$ and $t \in [0, T]$. If $y(t) - \psi \in H^1(0, T; H) \cap B(0, T; V)$ also satisfies (10.1.18), then it follows from (10.1.18) that

$$\int_0^t e^{-2\rho s}\left\langle \frac{d}{ds}(y(s) - y^*(s)) + A(y(s) - y^*(s)), y(s) - y^*(s)\right\rangle ds \le 0.$$

Using (10.1.17) this implies

$$\frac{1}{2}e^{-2\rho t}|y(t)-y^*(t)|_H^2+\int_0^t e^{-2\rho s}(\langle A(y(s)-y^*(s)), y(s)-y^*(s)\rangle+\rho\,|y^*(s)-y(s)|_H^2)\, ds \le 0$$

and thus $y(t) = y^*(t)$. Hence the solution to (10.1.18) is unique. Integrating (10.1.16) on (τ, t) with $0 \le \tau < t \le T$ we obtain with the arguments that lead to (10.1.18)

$$\int_\tau^t e^{-2\rho s}\left\langle \frac{d}{ds}y^*(s) + Ay^*(s) - f(s), y(s) - y^*(s)\right\rangle ds \ge 0, \tag{10.1.19}$$

and thus y^* satisfies (10.0.1).

To argue that $\hat{y}_c - \psi \to y^* - \psi$ strongly in $L^2(0, T; V) \cap C(0, T; H)$, note that $\hat{\lambda}_c = c\min(0, \hat{y}_c - \psi)$ converges weakly in $L^2(0, T; V^*)$ to λ^*. From (10.1.11) and (10.0.5) we have

$$\frac{1}{2}\frac{d}{dt}|\hat{y}_c - y^*|^2 + \langle A(\hat{y}_c - y^*), \hat{y}_c - y^*\rangle = \langle \lambda^* - \hat{\lambda}_c, \hat{y}_c - y^*\rangle$$
$$\le \langle \lambda^* - \hat{\lambda}_c, \hat{y}_c - \psi + \psi - y^*\rangle \le \langle \lambda^*, \hat{y}_c - \psi\rangle + \langle \hat{\lambda}_c, y^* - \psi\rangle =: \eta_c,$$

where $|\eta_c|_{L^1(0,T;\mathbb{R})} \to 0$ for $c \to \infty$. By assumption (2)

$$\frac{1}{2}\frac{d}{dt}|\hat{y}_c - y^*|_H^2 + \omega|\hat{y}_c - y^*|_V^2 - \rho|\hat{y}_c - y^*|_H^2 \le \eta_c,$$

and hence

$$\frac{d}{dt}[e^{-2\rho t}|\hat{y}_c - y^*|_H^2] + \omega\, e^{-2\rho t}\,|\hat{y}_c - y^*|_V^2 \le 2e^{-2\rho t}\eta_c,$$

which implies that

$$e^{-2\rho t}|\hat{y}_c(t) - y^*(t)|_H^2 + \omega \int_0^t e^{-2\rho s}|\hat{y}_c(s) - y^*(s)|_V^2\, ds \le \int_0^t e^{-2\rho s}\eta_c(s)\, ds,$$

and the desired convergence of y_c to y^* in $L^2(0, T; V) \cap C(0, T; H)$ follows.

It remains to argue right continuity of $t \to y^*(t) \in V$. From (10.1.19) it follows that

$$\int_\tau^t \left\langle e^{-2\rho s}\left(\frac{d}{ds}y^*(s) + Ay^*(s) - f(s)\right),\ \frac{y^*(s-h) - y^*(s)}{-h}\right\rangle ds \le 0, \qquad (10.1.20)$$

where $h > 0$. Using $a(b-a) = \frac{b^2-a^2}{2} - \frac{1}{2}(a-b)^2$ we find

$$\begin{aligned}
&\liminf_{h\to 0} \frac{1}{-h}\int_\tau^t e^{-2\rho s}a_s(y^*(s) - \bar{\psi}, y^*(s-h) - y^*(s))\\
&\ge 2\rho\int_\tau^t e^{-2\rho s}a_s(y^*(s) - \bar{\psi}, y^*(s) - \bar{\psi})\, ds\\
&+\frac{1}{2}e^{-2\rho t}a_s(y^*(t) - \bar{\psi}, y^*(t) - \bar{\psi}) - \frac{1}{2}e^{-2\rho\tau}a_s(y^*(\tau) - \bar{\psi}, y^*(\tau) - \bar{\psi}).
\end{aligned}$$

This estimate, together with assumption (6) and the fact that $\{\frac{y^*(\cdot-h)-y^*}{-h}\}_{h>0}$ is weakly bounded in $L^2(0, T; H)$, allows us to pass to the limit in (10.1.20) to obtain

$$\begin{aligned}
&e^{-2\rho t}a_s(y^*(t) - \bar{\psi}, y^*(t) - \bar{\psi}) - e^{-2\rho\tau}a_s(y^*(\tau) - \bar{\psi}, y^*(\tau) - \bar{\psi})\\
&+\textstyle\int_\tau^t e^{-2\rho s}\left|\frac{d}{dt}y^*(s)\right|_H^2\, ds\\
&\le M\int_\tau^t |y^*(s) - \bar{\psi}|_V \left|\frac{d}{dt}y^*(s)\right| ds + \int_\tau^t e^{-2\rho s}|A\bar{\psi} - f(s)|_H \left|\frac{d}{dt}y^*(s)\right| ds.
\end{aligned}$$

Consequently, we have

$$\begin{aligned}
e^{-2\rho t}a_s(y^*(t) - \bar{\psi}, y^*(t) - \bar{\psi}) &\le e^{-2\rho\tau}a_s(y^*(\tau) - \bar{\psi}, y^*(\tau) - \bar{\psi})\\
&\quad + \int_\tau^t e^{-2\rho s}(M^2|y^*(s) - \bar{\psi}|_V^2 + |A\bar{\psi} - f(s)|_H^2)\, ds
\end{aligned}$$

for $0 \le \tau \le t \le T$. This implies that

$$a_s(y^*(\tau)-\bar{\psi}, y^*(\tau)-\bar{\psi})+|y^*(\tau)-\bar{\psi}|_H^2 \ge \limsup_{t\to\tau} \left(a_s(y^*(t)-\bar{\psi}, y^*(t)-\bar{\psi})+|y^*(t)-\bar{\psi}|_H^2\right).$$

Since $a_s(\phi, \phi)+|\phi|_H^2$ defines an equivalent norm on the Hilbert space V, $|y^*(t)-y^*(\tau)|_V \to 0$ as $t \downarrow \tau$. Hence y^* is right continuous.

Now, in addition assumptions (3)–(4) are supposed to hold. Let

$$\hat{\lambda}_c(t) = c\min(0, \hat{y}_c(t) - \psi).$$

Then for $c \le \hat{c}$ and $\phi = (\hat{y}_c - \hat{y}_{\hat{c}})^+$

$$(\hat{\lambda}_c - \hat{\lambda}_{\hat{c}}, \phi) = \left((c - \hat{c})\min(0, \hat{y}_c - \psi) + \hat{c}(\min(0, \hat{y}_c - \psi) - \min(0, \hat{y}_{\hat{c}} - \psi)), \phi\right) \ge 0.$$

Hence, using the arguments leading to (10.1.4), we have $\hat{y}_c \le \hat{y}_{\hat{c}}$ for $c \le \hat{c}$. Then $\hat{y}_c(t) \to y^*(t)$ strongly in H and pointwise a.e. in Ω. Moreover,

$$\hat{\lambda}_c(t) = c\min(0, \hat{y}_c - \psi) \to \lambda^* \text{ weakly in } L^2(0,T;V^*). \qquad \square$$

Summarizing Theorems 10.4 and 10.7 we have the following corollary.

Corollary 10.8. *If assumptions* (1)–(6) *hold,* $y_0 - \psi \in C \cap V$, *and* $f \in L^2(0,T;H)$, *then for every* $t \in [0,T]$

$$\hat{y}_c(t) \le y^*(t) \le y_c(t)$$

and

$$\hat{y}_c(t) \uparrow y^*(t) \quad \text{and} \quad y_c(t) \downarrow y^*(t)$$

pointwise a.e., monotonically in Ω *as* $c \to \infty$. *Moreover,* $y^* - \psi \in H^1(0,T;H) \cap L^2(0,T;\mathrm{dom}(A)) \cap L^2(0,T;V)$.

10.2 Regularity

In this section we discuss additional regularity of the solution y^* to (10.0.1) under the assumptions of Theorem 10.4 (3). For $h > 0$ we have, suppressing the superscripts "*",

$$\frac{d}{dt}\left(\frac{(y(t+h)-y(t)}{h}\right) + A\left(\frac{y(t+h)-y(t)}{h}\right) + \frac{\lambda(t+h)-\lambda(t)}{h} = \frac{f(t+h)-f(t)}{h}.$$

From (10.0.5)

$$(\lambda(t+h)-\lambda(t), y(t+h)-y(t)) = -(\lambda(t+h), y(t)-\psi) - (\lambda(t), y(t+h)-\psi) \ge 0$$

and thus

$$\begin{aligned}
&\left\langle \frac{d}{dt}\left(\frac{(y(t+h)-y(t)}{h}\right), \frac{(y(t+h)-y(t)}{h}\right\rangle \\
&+\left\langle A\left(\frac{y(t+h)-y(t)}{h}\right), \frac{y(t+h)-y(t)}{h}\right\rangle \\
&\le \left\langle \frac{f(t+h)-f(t)}{h}, \frac{y(t+h)-y(t)}{h}\right\rangle.
\end{aligned}$$

Multiplying this by $t > 0$ we obtain

$$\begin{aligned}
\frac{d}{dt}\left(\frac{t}{2}\left|\frac{y(t+h)-y(t)}{h}\right|_H^2\right) + \frac{t\omega}{2}\left|\frac{y(t+h)-y(t)}{h}\right|_V^2 \le \frac{1}{2}\left|\frac{(y(t+h)-y(t)}{h}\right|^2 \\
+\rho t\left|\frac{y(t+h)-y(t)}{h}\right|_H^2 + \frac{t}{2\omega}\left|\frac{(f(t+h)-f(t)}{h}\right|_{V^*}^2.
\end{aligned}$$

Integrating in time,

$$t\left|\frac{y(t+h)-y(t)}{h}\right|_H^2+\omega\int_0^t s\left|\frac{y(s+h)-y(s)}{h}\right|_V^2 ds$$

$$\le e^{2\rho t}\int_0^t\left(\left|\frac{y(s+h)-y(s)}{h}\right|_H^2+\frac{s}{\omega}\left|\frac{f(s+h)-f(s)}{h}\right|_{V^*}^2\right)ds,$$

and letting $h\to 0^+$, we obtain

$$t\left|\frac{d}{dt}y\right|_H^2+\omega\int_0^t s\left|\frac{d}{dt}y\right|_V^2 ds\le e^{2\rho t}\int_0^t\left(\left|\frac{d}{dt}y\right|_H^2+\frac{s}{\omega}\left|\frac{d}{dt}f\right|_H^2\right)ds. \tag{10.2.1}$$

Hence we obtain the following theorem.

Theorem 10.9. *Suppose that assumptions* (1)–(5) *hold and that* $y_0-\psi\in C\cap V$, $f\in L^2(0,T;H)\cap H^1(0,T,V^*)$. *Then the strong solution satisfies* (10.2.1) *and thus* $y(t)\in$ dom (A) *for all* $t>0$.

The conclusion of Theorem 10.9 remains correct under the assumptions of the first part of Theorem 10.7, i.e., under assumptions (1)–(2) and (5)–(6), $y_0-\psi\in C\cap V$, and $f\in L^2(0,T;H)\cap H^1(0,T,V^*)$.

10.3 Continuity of $q\to y(q)\in L^\infty(\Omega)$

In this section we analyze the continuous dependence of the strong solution to (10.0.1) with respect to parameters in the operator A. Let U denote the normed linear space of parameters and let $\tilde{U}\subset U$ be a bounded subset such that for each $q\in\tilde{U}$ the operator $A(q)$ satisfies the assumptions (1)–(5) specified at the beginning of this chapter with $\rho=0$. We assume further that dom$(A(q))=D$ is independent of $q\in\tilde{U}$ and that

$$q\in\tilde{U}\to A(q)\in\mathcal{L}(X,V^*)$$

is Lipschitz continuous with Lipschitz constant κ. Let $y(q)$ be the solution to (10.0.1) corresponding to $A=A(q)$, $q\in\tilde{U}$. Then for $q_1,q_2\in\tilde{U}$, we have

$$\Big\langle\frac{d}{dt}(y(q_1)-y(q_2))+A(q_1)(y(q_1)-y(q_2))$$
$$+(A(q_1)-A(q_2))y(q_2),\ y(q_1)-y(q_2)\Big\rangle\le 0,$$

and therefore

$$|y(q_1)(T)-y(q_2)(T)|_H^2+\omega\int_0^T|y(q_1)-y(q_2)|_V^2\,dt$$
$$\le\frac{1}{\omega}\int_0^T|(A(q_2)-A(q_1))y(q_2)|_{V^*}^2\,dt=e(q_1,q_2)^2,$$

where

$$e(q_1, q_2)^2 \le \frac{\kappa}{\omega} |q_1 - q_2|_U^2 |y(q_2)|_{L^2(0,T;V)}^2.$$

Since from Theorem 10.9, $y(q)(T) \in D$ for $q \in \tilde{U}$, it follows by interpolation that

$$|y(q_1)(T) - y(q_2)(T)|_{W^\alpha} \le \bar{C} e(q_1, q_2)^{1-\alpha} \tag{10.3.1}$$

where $W^\alpha = [H, D]_\alpha$ is the interpolation space between D and H (see, e.g., [Fat, Chapter 8]), and $\bar{C}$ is an embedding constant. If $L^\infty(\Omega) \subset W^\alpha$, with $\alpha \in (\alpha_0, 1)$ for some α_0, then Hölder's continuity of $q \in \tilde{U} \to y(q)(T) \in L^\infty(\Omega)$ follows.

Next we prove Lipschitz continuity of $q \in \tilde{U} \to y(q) \in L^\infty(0, T; L^\infty(\Omega))$. Some prerequisites are established first. We assume that $A(q)$ generates an analytic semigroup $S(t) = S(t; q)$ on H for every $q \in \tilde{U}$ [Paz]. Then for each $q \in \tilde{U}$ there exists M such that

$$\|A^\alpha S(t)\| \le \frac{M}{t^\alpha} \text{ for all } t > 0, \tag{10.3.2}$$

where A^α denote the fractional powers of A, with $\alpha \in (0, 1)$. We assume that M is independent of $q \in \tilde{U}$. We shall further assume that

$$\operatorname{dom}(A^{\frac{1}{2}}) = V \tag{10.3.3}$$

for all $q \in \tilde{U}$, which is the case for a large class of second order elliptic differential operators; see, e.g., [Fat, Chapter 8]. We assume that

$$\|A(q)\|_{L(V,V)^*} \le \bar{\omega} \text{ for all } q \in \tilde{U}. \tag{10.3.4}$$

Let $r > 2$ be such that

$$V \subset L^r(\Omega),$$

and let $\bar{M}$ denote the embedding constant so that

$$|\zeta|_{L^r(\Omega)} \le \bar{M} |\zeta|_V \text{ for all } \zeta \in V. \tag{10.3.5}$$

For

$$p = \frac{2r}{r-2} \in (2, \infty),$$

we shall utilize the assumption

$$\begin{gathered} |A^{-\frac{1}{2}}(q_1)(A(q_1) - A(q_2))y|_{L^p(\Omega)} \le \bar{\kappa} |q_1 - q_2|_U |A(q_2)^\alpha y|_H \\ \text{for some } \alpha \in (0, 1) \text{ and all } q_1, q_2 \in \tilde{U},\ y \in D. \end{gathered} \tag{10.3.6}$$

This assumption is applicable, for example, if the parameter enters as a constant into the leading differential term of $A(q)$ or if it enters into the lower order terms.

Theorem 10.10. *Let $A(q)$ generate an analytic semigroup for every $q \in \tilde{U}$, and let assumptions* (1)–(5) *and* (7) *hold. If further* (10.3.3)–(10.3.6) *are satisfied and $f \in L^\infty(0, T; H)$, $y_0 \in \mathcal{C} \cap D$, then $q \to y(q)$ is Lipschitz continuous from $\tilde{U} \to L^\infty(0, T; L^\infty(\Omega))$.*

Proof. (1) Let $q \in \tilde{U}$ and $A = A(q)$. Let $f^1, f^2 \in L^\infty(0, T; H)$ with

$$A^{-\frac{1}{2}} f^i \in L^\infty(0, T; L^p(\Omega)), \; i = 1, 2,$$

and let y^1, y^2 denote the corresponding strong solutions to (10.0.1). For $k > 0$ let

$$\phi_k = \max(0, y^1 - y^2 - k)$$

and

$$\Omega_k = \{x \in \Omega : \phi_k > 0\}.$$

By assumption $\phi_k \in V$ and $\phi_k \geq 0$. Note that

$$(\min(0, \bar{\lambda} + c\,(z^1 - \psi)) - \min(0, \bar{\lambda} + c\,(z^2 - \psi)), (z^1 - z^2 - k)^+) \geq 0$$

for all $z_1, z_2 \in H$ and $k > 0$. Thus, it follows from (10.0.5) and Theorem 10.4 that for $\phi_k = (y^1 - y^2 - k)^+ \in V$

$$\left\langle \frac{d}{dt}(y^1 - y^2), \phi_k \right\rangle + \langle A(y^1 - y^2), \phi_k \rangle \leq (f^1 - f^2, \phi_k) \tag{10.3.7}$$

for a.e. $t \in (0, T)$. By assumption we have

$$\langle A\zeta, (\zeta - k)^+ \rangle \geq \omega |(\zeta - k)^+|_V^2$$

for $\zeta \in V$ and hence it follows from (10.3.7) that

$$\omega \int_0^T |\phi_k|_V^2 \, dt \leq \int_0^T |(f, \phi_k)| \, dt \leq \|A(q)\|_{\mathcal{L}(V, V^*)} \int_0^T |A^{-\frac{1}{2}} f|_{L^2(\Omega_k)} |\phi_k|_V \, dt,$$

where $\phi_k(0) = 0$, $f = f^1 - f^2$, and thus for any $\beta > 1$

$$\begin{aligned} \omega \left(\int_0^T |\phi_k|_V^2 \, dt \right)^{\frac{1}{2}} &\leq \bar{\omega} \left(\int_0^T |A^{-\frac{1}{2}} f|_{L^2(\Omega_k)}^2 \, dt \right)^{\frac{1}{2}} \\ &\leq \bar{\omega}\, C^{1-\beta} \left(\int_0^T |A^{-\frac{1}{2}} f|_{L^2(\Omega_k)}^2 \, dt \right)^{\frac{\beta}{2}} \end{aligned} \tag{10.3.8}$$

with

$$C = \left(\int_0^T |A^{-\frac{1}{2}} f|_{L^2}^2 \, dt \right)^{\frac{1}{2}}.$$

For $p > q = 2$ we have

$$\int_{\Omega_k} |A^{-\frac{1}{2}} f|^q \, dx \le \left(\int_{\Omega_k} |A^{-\frac{1}{2}} f|^p \right)^{q/p} |\Omega_k|^{(p-q)/p}. \tag{10.3.9}$$

We denote by h and k arbitrary real numbers satisfying $0 < k < h < \infty$ and we find for $r > 2$

$$|\phi_k|^r_{L^r} \ge \int_{\Omega_k} |\phi - k|^r \, dx > \int_{\Omega_h} |\phi - k|^r \, ds \ge |\Omega_h||h - k|^r. \tag{10.3.10}$$

It thus follows from (10.3.5) and (10.3.8)–(10.3.10) that for $\beta > 1$

$$\left(\int_0^T |\Omega_h|^{\frac{2}{r}} \, dt \right)^{\frac{1}{2}} \le \frac{\bar{M}}{|h-k|} \left(\int_0^T |\phi_k|^2_V \right)^{\frac{1}{2}} \le \frac{\bar{\omega}\bar{M}C^{1-\beta}}{\omega|h-k|} \left(\int_0^T |A^{-\frac{1}{2}} f|^2_{L^2(\Omega_k)} \, dt \right)^{\frac{\beta}{2}}$$

$$\le \frac{\bar{\omega}\bar{M}C^{1-\beta}}{\omega|h-k|} \left(\int_0^T |A^{-\frac{1}{2}} f|^2_{L^p} |\Omega_k|^{\frac{p-2}{p}} \, dt \right)^{\frac{\beta}{2}}.$$

For $\frac{1}{P} + \frac{1}{Q} = 1$ this implies that

$$\left(\int_0^T |\Omega_h|^{\frac{2}{r}} \, dt \right)^{\frac{1}{2}} \le \frac{\bar{\omega}\bar{M}C^{1-\beta}}{\omega|h-k|} \left(\int_0^T |A^{-\frac{1}{2}} f|^{2P}_{L^p} \, dt \right)^{\frac{\beta}{2P}} \left(\int_0^T |\Omega_k|^{\frac{Q(p-2)}{p}} \, dt \right)^{\frac{\beta}{2Q}}.$$

For $P = \infty$ and $Q = 1$ this implies, using that $p = \frac{2r}{r-2}$,

$$\left(\int_0^T |\Omega_h|^{\frac{2}{r}} \, dt \right)^{\frac{1}{2}} \le \frac{K}{|h-k|} \left(\int_0^T |\Omega_k|^{\frac{2}{r}} \, dt \right)^{\frac{\beta}{2}}, \tag{10.3.11}$$

where $K = \frac{\bar{\omega}\bar{M}C^{1-\beta}}{\omega} |A^{-\frac{1}{2}} f|^\beta_{L^\infty(0,T;L^p(\Omega))}$.

Now, we use the following fact [Tr]: Let $\varphi : (k_1, h_1) \to \mathbb{R}$ be a nonnegative, nonincreasing function and suppose that there are positive constants K, s, and $\beta > 1$ such that

$$\varphi(h) \le K(h-k)^{-s} \varphi(k)^\beta \quad \text{for } k_1 < k < h < h_1.$$

Then, if $\hat{k} = K^{\frac{1}{s}} 2^{\frac{\beta}{\beta-1}} \varphi(k_1)^{\frac{\beta-1}{r}}$ satisfies $k_1 + \hat{k} < h_1$, it follows that $\varphi(k_1 + \hat{k}) = 0$. Here we set

$$\varphi(k) = \left(\int_0^T |\Omega_k|^{\frac{r}{2}} \, dt \right)^{\frac{1}{2}}$$

on $(0, \infty)$, $s = 1$, $\beta > 1$, and

$$k_1 = \sup_{t \in (0,T)} |A^{-\frac{1}{2}} (f^1(t) - f^2(t))|_{L^p(\Omega)}.$$

It then follows from (10.3.8)–(10.3.10), as in the computation below (10.3.10), that

$$\varphi(k_1) \le \frac{\bar{M}\bar{\omega}C}{\omega k_1}.$$

From the definition of $\hat{k}$ we have in case $C \ge k_1$

$$\hat{k} \le 2^{\frac{\beta}{\beta-1}} \left(\frac{\bar{\omega}\bar{M}}{\omega}\right)^{\beta} C^{1-\beta} k_1^{\beta} C^{\beta-1} k_1^{1-\beta} = 2^{\frac{\beta}{\beta-1}} \left(\frac{\bar{\omega}\bar{M}}{\omega}\right)^{\beta} k_1.$$

The same estimate can also be obtained in the case that $C \le k_1$, and consequently $k_1 + \hat{k} \le \ell k_1$, where $\ell = 1 + 2^{\frac{\beta}{\beta-1}} (\frac{\bar{\omega}\bar{M}}{\omega})^{\beta}$. Hence we obtain $y^1 - y^2 \le \ell k_1$ a.e. in $(0,T) \times \Omega$. Analogously a uniform lower bound for $y^1 - y^2$ is obtained by using $\phi_k = \min(0, y^1 - y^2 - k) \le 0$ and thus

$$|y^1 - y^2|_{L^\infty(0,T;L^\infty(\Omega))} \le \ell \sup_{t\in(0,T)} |A^{-\frac{1}{2}}(f^1(t) - f^2(t))|_{L^p(\Omega)}. \tag{10.3.12}$$

(2) We use the estimate of step (1) to obtain Lipschitz continuous dependence of the solution on the parameter q. Let $q_1, q_2 \in \tilde{U}$ with corresponding solutions $y(q_1)$ and $y(q_2)$. Since

$$\frac{d}{dt} y(q_2) + A(q_1)y(q_2) + (A(q_2) - A(q_1))y(q_2) + \lambda(q_2) = f(t),$$

$y(q_2)$ is the solution to (10.0.1) with $A = A(q_1)$ and $\tilde{f}(t) = f - (A(q_2) - A(q_1))y(q_2) \in L^2(0,T;H)$. Hence we can apply the estimate of (1) with $A = A(q_1)$ and $f^1 - f^2 = (A(q_2) - A(q_1))y(q_2)$ and obtain

$$|y^1 - y^2|_{L^\infty(0,T;L^\infty(\Omega))} \le \ell \sup_{t\in(0,T)} |A(q_1)^{-\frac{1}{2}}(A(q_1) - A(q_2))y(t;q_2)|_{L^p(\Omega)}.$$

Utilizing (10.3.6) this implies that

$$|y^1 - y^2|_{L^\infty(0,T;L^\infty(\Omega))} \le \ell \bar{\kappa}\, |q_1 - q_2|_U \sup_{t\in(0,T)} |A^{\alpha}(q_2)y(t;q_2)|_H. \tag{10.3.13}$$

To estimate $A^{\alpha}(q_2)y(t;q_2)$ recall from Theorem 10.4 that $\bar{\lambda} \le \lambda(t;q) \le 0$ and thus $\{f - \lambda(q) : q \in \tilde{U}\}$ is uniformly bounded in $L^\infty(0,T;H)$. From (10.0.6) we have that

$$A(q_2)^{\alpha} y(t;q_2) = A(q_2)^{\alpha} S(t;q_2) y_0 + \int_0^t A(q_2)^{\alpha} S(t-s;q_2)(f(s) - \lambda(s;q_2))\, ds$$
$$\in L^\infty(0,T;H).$$

From (10.3.2) and since $y_0 \in D$ it follows that $\{A(q_2)^{\alpha} y(q_2) : q_2 \in \tilde{U}\}$ is bounded in $L^\infty(0,T;H)$ as desired. $\square$

10.4 Difference schemes and weak solutions

In this section we establish existence of weak solutions to (10.0.1) based on finite difference schemes (10.4.1). This difference approximation is also used to establish uniqueness of the weak solution and for proving monotonicity properties of the solution in the following section. For the sake of the simplicity of presentation we assume that $\rho = 0$. Consequently $\langle A\phi, \phi\rangle = a_s(\phi, \phi)$ defines an equivalent norm on V. For $h > 0$ consider the discretized (in time) variational inequality: Find $y^k - \psi \in \mathcal{C},\ k = 1, \dots, N$, satisfying

$$\left\langle \frac{y^k - y^{k-1}}{h} + Ay^k - f^k, y - y^k \right\rangle \geq 0, \quad y^0 = y_0 \tag{10.4.1}$$

for all $y - \psi \in \mathcal{C}$, where

$$f^k = \frac{1}{h} \int_{(k-1)h}^{kh} f(t)\, dt$$

and $Nh = T$. Throughout this section we assume that $y_0 \in H$ and $f \in L^2(0, T; V^*)$.

Theorem 10.11. *Assume that $y_0 \in H$ and $f \in L^2(0, T, V^*)$ and that assumptions* (1)–(2) *hold. Then there exists a unique solution $\{y^k\}_{k=1}^N$ to* (10.4.1).

Proof. To establish existence of solutions to (10.4.1), we proceed by induction with respect to k and assume that existence has been proven up to $k - 1$. To verify the induction step consider the regularized problems

$$\frac{y_c^k - y^{k-1}}{h} + Ay_c^k + c\, \min(0, y_c^k - \psi) - f^k = 0. \tag{10.4.2}$$

Since

$$y \in H \to \min c\,(0, y - \psi)$$

is Lipschitz continuous and monotone, the operator $B : V \to V^*$ defined by

$$B(y) = \frac{y}{h} + Ay + c\, \min(0, y - \psi)$$

is coercive, monotone, and continuous for all $h > 0$. Hence by the theory of maximal monotone operators (10.4.2) admits a unique solution; cf., e.g., [ItKa, Chapter I.5], [Ba, Chapter II.1]. For each $c > 0$ and $k = 1, \dots, N$ we find

$$\frac{1}{2h}(|y_c^k - \psi|_H^2 - |y^{k-1} - \psi|_H^2 + |y_c^k - y^{k-1}|_H^2)$$
$$+\langle A(y_c^k - \psi) + A\psi - f^k, y_c^k - \psi\rangle + c\,|(y_c^k - \psi)^-|_H^2 = 0.$$

Thus the families $|y_c^k - \psi|_V^2$ and $c\,|(y_c^k - \psi)^-|_H^2$ are bounded in $c > 0$ and there exists a subsequence of $\{y_c^k - \psi\}$ that converges to some $y^k - \psi$ weakly in V as $c \to \infty$. As argued in the proof of Theorem 10.7 (cf. (10.1.15)), $y^k - \psi \in C$ and hence $y^k - \psi \in \mathcal{C}$. Note that

$$(-(y_c^k - \psi)^-, y - y_c^k) = (-(y_c^k - \psi)^-, y - \psi - (y_c^k - \psi)) \le 0 \quad \text{for all } y - \psi \in C \quad (10.4.3)$$

and

$$\begin{aligned} &\liminf_{c\to\infty} \langle Ay_c^k, y_c^k - y\rangle \\ &= \liminf_{c\to\infty} \left(\langle A(y_c^k - \psi), y_c^k - \psi\rangle + \langle A(y_c^k - \psi), \psi - y\rangle + \langle A\psi, y_c^k - \psi\rangle\right) \\ &\ge \langle A(y^k - \psi), y^k - \psi\rangle + \langle A(y^k - \psi), \psi - y\rangle + \langle A\psi, y^k - \psi\rangle \\ &= \langle Ay^k, y^k - y\rangle. \end{aligned} \quad (10.4.4)$$

Passing to the limit in (10.4.2) utilizing (10.4.3) and (10.4.4) we obtain

$$\begin{aligned} &\left\langle \frac{y^k - y^{k-1}}{h} + Ay^k - f^k, y^k - y\right\rangle \\ &\le \liminf_{c\to\infty} \left(\left\langle \frac{y_c^k - y^{k-1}}{h}, y_c^k - y\right\rangle + \langle Ay_c^k, y_c^k - y\rangle - \langle f^k, y_c^k - y\rangle\right) \le 0, \end{aligned}$$

and hence y^k satisfies (10.4.1).

To verify uniqueness, let $\tilde{y}^k$ be another solution to (10.4.1). Then, from (10.4.1),

$$\left\langle \frac{(y^k - \tilde{y}^{k-1}) - (y^{k-1} - \tilde{y}^{k-1})}{h} + A(y^k - \tilde{y}^k), y^k - \tilde{y}^k\right\rangle \le 0$$

and thus

$$\frac{1}{2h}|y^k - \tilde{y}^k|_H^2 + \langle A(y^k - \tilde{y}^k), y^k - \tilde{y}^k\rangle \le \frac{1}{2h}|y^{k-1} - \tilde{y}^{k-1}|_H^2.$$

Since $y^0 = \tilde{y}^0 = y_0$, this implies that $y^k = \tilde{y}^k$ for all $k \ge 1$. □

Next we discuss existence and uniqueness of weak solutions to (10.0.1) by passing to the limit in the piecewise defined functions

$$y_h^{(1)} = y^k + \frac{t - kh}{h}(y^{k+1} - y^k), \quad y_h^{(2)} = y^{k+1} \text{ on } (kh, (k+1)h] \quad (10.4.5)$$

for $k = 0, \ldots, N-1$.

Theorem 10.12. *Suppose that the assumptions of Theorem* 10.11 *hold. Then there exists a unique weak solution y^* of* (10.0.1). *Moreover $t \to y^*(t) \in H$ is right continuous, $y^* \in B(0, T; H)$, and $y_h^{(2)} - \psi \to y^* - \psi$ strongly in $L^2(0; T; V)$.*

Proof. Setting $y = \psi$ in (10.4.1), we obtain

$$|y^k - \psi|_H^2 - |y^{k-1} - \psi|_H^2 + |y^k - y^{k-1}|_H^2 + h\omega\,|y^k - \psi|_V^2 \le \frac{h}{\omega}|A\psi - f^k|_{V^*}^2.$$

Thus,

$$|y^m - \psi|_H^2 + \sum_{k=\ell+1}^{m} (|y^k - y^{k-1}|_H^2 + \omega h\,|y^k - \psi|_V^2) \le |y^\ell - \psi|_H^2 + \frac{1}{\omega}\sum_{k=\ell+1}^{m} |A\psi - f^k|_{V^*}^2\, h \tag{10.4.6}$$

for all $0 \le \ell < m \le N$. Since

$$\int_0^T |y_h^{(1)} - y_h^{(2)}|_H^2\, h \le \frac{h}{3}\sum_{k=1}^{N} |y^k - y^{k-1}|_H^2\, h \to 0 \text{ as } h \to 0^+.$$

From the above estimates it follows that there exist subsequences of $y_h^{(1)}$, $y_h^{(2)}$ (denoted by the same symbols) and $y^*(t) \in L^2(0, T; V)$ such that

$$y_h^{(1)}(t),\ \ y_h^{(2)}(t) \to y^*(t) \text{ weakly in } L^2(0, T; V) \quad \text{as } h \to 0^+. \tag{10.4.7}$$

Note that

$$\frac{d}{dt}y_h^{(1)} = \frac{y^{k+1} - y^k}{h} \text{ on } (k\,h, (k+1)\,h].$$

Thus, we have from (10.4.1) for every $y \in \mathcal{K}$

$$\left\langle \frac{d}{dt}y + Ay_h^{(2)} - f_h, y - y_h^{(2)} \right\rangle + \left\langle \frac{d}{dt}y_h^{(1)} - \frac{d}{dt}y, y - y_h^{(2)} \right\rangle \ge 0 \tag{10.4.8}$$

a.e. in $(0, T)$. Here

$$\left\langle \frac{d}{dt}(y_h^{(1)} - y), y - y_h^{(2)} \right\rangle = \left\langle \frac{d}{dt}y_h^{(1)} - \frac{d}{dt}y, y - y_h^{(1)} \right\rangle + \left\langle \frac{d}{dt}y_h^{(1)} - \frac{d}{dt}y, y_h^{(1)} - y_h^{(2)} \right\rangle \tag{10.4.9}$$

with

$$\int_0^T \left\langle \frac{d}{dt}y_h^{(1)} - \frac{d}{dt}y, y - y_h^{(1)} \right\rangle dt \le \frac{1}{2}|y(0) - y_0|_H^2 \tag{10.4.10}$$

and

$$\int_0^T \left(\frac{d}{dt}y_h^{(1)}, y_h^{(1)} - y_h^{(2)} \right) dt = -\frac{1}{2}\sum_{k=1}^{N} |y^k - y^{k-1}|_H^2. \tag{10.4.11}$$

Since

$$\int_0^T \langle Ay^*, y^* - y\rangle\, dt \le \liminf_{h \to 0^+} \int_0^T \langle Ay_h^{(2)}, y_h^{(2)} - y\rangle\, dt,$$

which can be argued as in (10.4.4), it follows from (10.4.7)–(10.4.11) that every weak cluster point y^* of $y_h^{(2)}$ satisfies

$$\int_0^T \left\langle \frac{d}{dt}y + Ay^* - f, y - y^* \right\rangle dt + \frac{1}{2}|y(0) - y_0|_H^2 \geq 0 \tag{10.4.12}$$

for all $y \in \mathcal{K}$ and a.e. $t \in (0, T)$. Hence $y^* \in L^2(0, T; V)$ is a weak solution of (10.0.1) and $y^* \in B(0, T; H)$.

Moreover, from (10.4.6)

$$|y^*(t) - \psi|_H^2 \leq |y^*(\tau) - \psi|_H^2 + \frac{1}{\omega}\int_\tau^t |A\psi - f(s)|_{V^*}^2\, ds$$

for all $0 \leq \tau \leq t \leq T$. Thus,

$$\limsup_{t \downarrow \tau} |y^*(t) - \psi|_H^2 \leq |y^*(\tau) - \psi|_H^2.$$

which implies that $t \to y^*(t) \in H$ is right continuous.

Let y^* be a weak solution. Setting $y = y_h^{(1)} \in \mathcal{K}$ in (10.4.12) and $y = y^*(t)$ in (10.4.8) we have

$$\int_0^T \left\langle \frac{d}{dt}y_h^{(1)} + Ay^* - f, y_h^{(1)} - y^* \right\rangle dt \geq 0, \tag{10.4.13}$$

$$\int_0^T \left\langle \frac{d}{dt}y_h^{(1)} + Ay_h^{(2)} - f_h, y^* - y_h^{(2)} \right\rangle dt \geq 0, \tag{10.4.14}$$

where we used that

$$\int_0^T \left\langle \frac{d}{dt}(y_h^{(1)} - y^*), y^* - y_h^{(2)} \right\rangle dt \leq 0$$

from (10.4.9)–(10.4.11).

Summing up (10.4.13) and (10.4.14) and using (10.4.11) implies that

$$\int_0^T \left(\langle Ay^*, y_h^{(1)} - y_h^{(2)} \rangle - \langle f, y_h^{(1)} - y^* \rangle - \langle f_h, y^* - y_h^{(2)} \rangle \right) dt$$

$$\geq \frac{1}{2}\sum_{k=1}^N |y^k - y^{k-1}|_H^2 + \int_0^T \langle A(y^* - y_h^{(2)}), y^* - y_h^{(2)} \rangle\, dt.$$

Letting $h \to 0^+$ we obtain $0 \geq \langle A(y^*(t) - \hat{y}(t)), y^*(t) - \hat{y}(t) \rangle$ a.e. on $(0, T)$ for every weak cluster point $\hat{y}$ of $y_h^{(2)}$ in $L^2(0, T; V)$. This implies that the weak solution is unique and that

$$\int_0^T \langle A(y^* - y_h^{(2)}), y^* - y_h^{(2)} \rangle\, dt \to 0$$

as $h \to 0^+$. $\square$

Corollary 10.13. *Let $y = y(y_0, y)$ denote the weak solution to* (10.0.1), *given $y_0 \in H$ and $f \in L^2(0, T; V^*)$. Then for all $t \in [0, T]$*

$$|y(y_0, f)(t) - y(\tilde{y}_0, \tilde{f})(t)|_H + \omega \int_0^T |y(y_0, f) - y(\tilde{y}_0, \tilde{f})|_V^2\, ds$$

$$\le |y_0 - \tilde{y}_0|_H^2 + \frac{1}{\omega} \int_0^t |f - \tilde{f}|_{V^*}^2\, ds.$$

Proof. Let y^k and $\tilde{y}^k$ be the solution to (10.4.1) corresponding to (y_0, f) and $(\tilde{y}_0, \tilde{f})$. It then follows from (10.4.1) that

$$\left\langle \frac{(y^k - \tilde{y}^k) - (y^{k-1} - \tilde{y}^{k-1})}{h} + A(y^k - \tilde{y}^k) - (f^k - \tilde{f}^k), y^k - \tilde{y}^k \right\rangle \le 0.$$

Thus,

$$|y^k - \tilde{y}^k|_H^2 + \omega h\, |y^k - \tilde{y}^k|_V^2 \le |y^{k-1} - \tilde{y}^{k-1}|_H^2 + \frac{h}{\omega} |f^k - \tilde{f}^k|_{V^*}^2.$$

Summing this in k, we have

$$|y^m - \tilde{y}^m|_H^2 + \omega \sum_{k=1}^m h\, |y^k - \tilde{y}^k|_V^2 \le |y_0 - \tilde{y}_0|_H^2 + \frac{1}{\omega} \sum_{k=1}^m h\, |f^k - \tilde{f}^k|_{V^*}^2,$$

which implies the desired estimate by letting $h \to 0^+$. ☐

Corollary 10.14. *Let $\bar{\lambda} \in H$ satisfy $\bar{\lambda} \le 0$ and let $y_c \in W(0, T)$ be the solution to*

$$\frac{d}{dt} y_c(t) + A y_c(t) + \min(0, \bar{\lambda} + c\,(y_c - \psi)) = f. \tag{10.4.15}$$

Then $y_c \to y^*$ weakly in $L^2(0, T; V)$ and $y_c(T) \to y^*(T)$ weakly in H as $c \to \infty$, where y^* is the unique weak solution to (10.0.1). *In addition, if $y^* \in W(0, T)$, then*

$$|y_c - y^*|_{L^2(0,T;V)} + |y_c - y^*|_{L^\infty(0,T;H)} \to 0$$

as $c \to \infty$.

Proof. Note that

$$(\min(0, \bar{\lambda} + c\,(y_c - \psi)), y_c - \psi) \ge \frac{c}{2}\, |(y_c - \psi)^-|_H^2 - \frac{1}{2c} |\bar{\lambda}|_H^2.$$

Thus, we have

$$\frac{1}{2} \frac{d}{dt} |y_c - \psi|_H^2 + \langle A(y - \psi), y_c - \psi\rangle + c\, |(y_c - \psi)^-|_H^2 \le \langle f - A\psi, y_c - \psi\rangle + \frac{1}{c} |\bar{\lambda}|^2$$

and

$$|y_c(t) - \psi|_H^2 + \omega \int_0^t (|y_c - \psi|_V^2 + c\, |(y_c - \psi)_H^2)\, ds \le \frac{1}{\omega} \int_0^t \left(|f - A\psi|_{V^*}^2 + \frac{1}{c} |\bar{\lambda}|^2 \right) ds.$$

Hence $\int_0^T |y_c - \psi|_H^2\,dt \to 0$ as $c \to 0$ and $\{y_c - \psi\}_{c\geq 1}$ is bounded is $L^2(0, T; V)$. Using the same arguments as in the proof of Theorem 10.7, there exist y^* and a subsequence of $\{y_c - \psi\}_{c\geq 1}$ that converges weakly to $y^* - \psi \in L^2(0, T; V)$, and $y^* - \psi \geq 0$ a.e. in $(0, T) \times \Omega$. For $y(t) \in \mathcal{K}$

$$\int_0^T \left[\left\langle \frac{d}{dt}y(t) - \frac{d}{dt}(y(t) - y_c), y(t) - y_c(t)\right\rangle + \langle Ay_c(t) - f(t), y(t) - y_c(t)\rangle\right.$$

$$\left. +(\min(0, \bar{\lambda} + c\,(y_c - \psi)), y(t) - \psi - (y_c - \psi))\right] dt = 0,$$

where

$$\int_0^T \left\langle -\frac{d}{dt}(y(t) - y_c), y(t) - y_c(t)\right\rangle dt = \frac{1}{2}(|y(0) - y_0|_H^2 - |y(T) - y_c(T)|^2), \quad (10.4.16)$$

$$(\min(0, \bar{\lambda} + c\,(y_c - \psi)), y(t) - \psi - (y_c - \psi)) \leq \frac{1}{2c}|\bar{\lambda}|_H^2. \quad (10.4.17)$$

Hence, we have

$$\int_0^T [\left\langle \frac{d}{dt}y(t), y(t) - y_c(t)\right\rangle dt + \langle Ay_c(t) - f(t), y(t) - y_c(t)\rangle + \frac{1}{2}|y(0) - y_0|_H^2$$

$$\geq \frac{1}{2}|y(T) - y_c(T)|_H^2 - \frac{1}{2c}\int_0^T |\bar{\lambda}|_H^2\,ds.$$

Letting $c \to \infty$, y^* satisfies (10.0.7) and thus y^* is the weak solution of (10.0.1). Suppose that $y^* \in W(0, T)$. Then by (10.4.15),

$$\left\langle \frac{d}{dt}(y_c - y^*) + A(y_c - y^*) + \frac{d}{dt}y^* + Ay^* - f, y^* - y_c\right\rangle$$

$$+(\min(0, \bar{\lambda} + c\,(y_c - \psi)), y^*(t) - \psi - (y_c - \psi)) = 0.$$

From (10.4.16)–(10.4.17), and since $y_c \to y^*$ weakly in $L^2(0, T; V)$,

$$\frac{1}{2}\,|y_c(t) - y^*(t)|_H^2 + \int_0^t \langle A(y_c - y^*), y_c - y^*\rangle\,ds \to 0,$$

and this convergence is uniform with respect to $t \in [0, T]$. □

10.5 Monotone property

In this section we establish monotone properties of the weak solution to (10.0.1). As in the previous section assumption (2) is used with $\rho = 0$.

Corollary 10.15. *If assumptions* (1)–(3) *hold, then*

$$y(y_0, f) \geq y(\tilde{y}_0, \tilde{f})$$

provided that $y_0 \geq \tilde{y}_0$ *and* $f \geq \tilde{f}$, *with* $y_0, \tilde{y}_0 \in H$ *and* $f, \tilde{f} \in L^2(0,T;V^*)$. *As a consequence, if* $y_0 = \psi$ *and* $f(t) \geq f(s)$ *for a.e.* $t > s$, *then the weak solution satisfies* $y(t) \geq y(s)$ *for* $t \geq s$.

Proof. Assume that $y^{k-1} \geq \tilde{y}^{k-1}$. For $\phi = -(y_c^k - \tilde{y}_c^k)^-$ it follows from (10.4.2) that

$$\left(\frac{y_c^k - \tilde{y}_c^k - (y^{k-1} - \tilde{y}^{k-1})}{h}, \phi\right) + \langle A(y_c^k - \tilde{y}_c^k) - (f^k - \tilde{f}^k), \phi\rangle - c\left((y_c^k - \psi)^- \right.$$
$$\left. - (\tilde{y}_c^k - \psi)^-, \phi\right) = 0.$$

Since

$$\left\langle -\frac{y^{k-1} - \tilde{y}^{k-1}}{h} - (f^k - \tilde{f}^k), \phi\right\rangle - c\left((y_c^k - \psi)^- - (\tilde{y}_c^k - \psi)^-, \phi\right) \geq 0,$$

we have from assumption (3) that

$$\frac{|\phi|_H^2}{h} + \langle A\phi, \phi\rangle \leq 0$$

and thus $y_c^k - \tilde{y}_c^k \geq 0$ for sufficiently small $h > 0$. From the proof of Theorem 10.11 it follows that we can take the limit with respect to c and obtain $y^k - \tilde{y}^k \geq 0$. By induction this holds for all $k \geq 0$. The first claim of the theorem now follows from (10.4.2) and Theorem 10.12. The second one follows from

$$y(t;\psi, f(\cdot)) = y(s; y(t-s;\psi, f), f(\cdot + t - s)) \geq y(s;\psi, f(\cdot)). \qquad \square$$

Corollary 10.16. *Let assumptions* (1)–(3) *hold and suppose that the stationary variational inequality*

$$\langle Ay - f, \phi - y\rangle \geq 0 \quad \textit{for all } \phi - \psi \in \mathcal{C} \tag{10.5.1}$$

with $f \in V^*$ *has a solution* $y - \psi \in \mathcal{C}$. *Then if* $y(0) = \psi$ *and* $f(t) = f$, *we have* $y(t) \uparrow \hat{y}$, *where* $\hat{y}$ *is the minimum solution to* (10.5.1).

Proof. Suppose $\bar{y}$ is a solution to (10.5.1). Since $\bar{y}(t) := \bar{y}$, $t \geq 0$, is also the unique solution to (10.0.1) with $y_0 = \bar{y} \geq \psi$, it follows from Corollary 10.16 that $y(t) \leq \bar{y}$ for all $t \in [0,T]$. On the other hand, it follows from Theorem 10.4 (2) that

$$\left\langle y(\tau+1) - y(\tau) + \int_\tau^{\tau+1} (Ay(s) - f)\, ds,\ \phi - y(\tau)\right\rangle \geq 0 \quad \text{for all } \phi - \psi \in \mathcal{C} \tag{10.5.2}$$

since $y(s) \geq y(\tau)$, $s \geq \tau$. By the Lebesgue monotone convergence theorem $\hat{y} = \lim_{\tau\to\infty} y(\tau) \in \mathcal{C}$ exists. Letting $\tau \to \infty$ in (10.5.2), we obtain that $\hat{y}$ satisfies (10.5.1) and thus $\hat{y} \leq \bar{y}$. $\square$

Corollary 10.17 (Perturbation). *Let assumptions* (1)–(3) *hold, and let* $\psi^1,\ \psi^2 \in H$ *and* $f \in L^2(0,T;V^*)$. *Denote by* y_c^1 *and* y_c^2 *the solutions to* (10.1.1) *with* y_0 *equal to* ψ^1 *and* ψ^2, *respectively, and let* y^1 *and* y^2 *be the corresponding weak solutions to* (10.0.1). *Assume that* $(\phi-\gamma)^+ \in V$ *for any* $\gamma \in \mathbb{R}^+$, *that* $\phi \in V$, *and that* $\langle A1, (\phi-\gamma)^+\rangle \geq 0$. *Then for* $\alpha = \max(0, \sup_{x\in\Omega}(\psi^1-\psi^2))$ *and* $\beta = \min(0, \inf_{x\in\Omega}(\psi^1-\psi^2))$ *we have*

$$\beta \leq y_c^1 - y_c^2 \leq \alpha,$$

$$\beta \leq y^1 - y^2 \leq \alpha.$$

Proof. Note that

$$c(y_c^1 - y_c^2 - \alpha) = \bar\lambda + c(y_c^1 - \psi^1) - (\bar\lambda + c(y_c^2 - \psi^2)) - c(\psi^1 - \psi^2 - \alpha).$$

Thus, an elementary calculation shows that

$$(\min(0, \bar\lambda + c(y_c^1 - \psi^1)) - \min(0, \bar\lambda + c(y_c^2 - \psi^2)), (y_c^1 - y_c^2 - \alpha)^+) \geq 0.$$

As in the proof of Theorem 10.4 it follows that $\phi = (y_c - y_c^2 - \alpha)^+$ satisfies

$$\frac{1}{2}\frac{d}{dt}|\phi|_H^2 \leq \rho|\phi|_H^2.$$

Since $\phi(0) = 0$, this implies that $\phi(t) = 0,\ t \geq 0$, and thus $y_c^1 - y_c^2 \leq \alpha$ a.e. on $(0,T)\times\Omega$. Similarly, letting $\phi = (y_c^1 - y_c^2 - \beta)^-$ we obtain $y_c^1 - y_c^2 \geq \beta$ a.e. on $(0,T)\times\Omega$. Since from Corollary 10.14, $y_c^1 \to y^1$ and $y_c^2 \to y^2$ weakly in $L^2(0,T;H)$ for $c\to\infty$, we obtain the desired estimates. ☐

Chapter 11

Shape Optimization

11.1 Problem statement and generalities

We apply the framework that we developed in Section 5.5 for weakly singular problems to calculate the shape derivative of constrained optimization problems of the form

$$\begin{cases} \min J(y, \Omega, \Gamma) \\ \text{subject to } e(y, \Omega) = 0 \end{cases} \tag{11.1.1}$$

over admissible domains Ω and manifolds Γ. Here e denotes a partial differential equation depending on the state variable y and the domain Ω, and J stands for a cost functional depending, besides y and Ω, on Γ, which constitutes the variable part of the boundary of Ω.

We consider as an example the problem of determining an unknown interior domain $D \subset U$ in an elliptic equation

$$\begin{cases} \Delta y = 0, \quad \text{in } \Omega = U \setminus D, \\ y = 0 \text{ on } \partial D \end{cases}$$

from the Cauchy data

$$y = f, \quad \frac{\partial y}{\partial n} = g \text{ on the outer boundary } \partial U.$$

This problem can be formulated as the shape optimization problem

$$\begin{cases} \min \int_{\partial U} |y - f|^2 \, ds \\ \text{subject to} \\ \Delta y = 0 \quad \text{in } \Omega \quad \text{and} \quad y = 0 \text{ on } \partial D, \quad \frac{\partial}{\partial n} y = g \text{ on } \partial U, \end{cases}$$

and it is a special case by the theory to be presented in Section 11.2.

Shape sensitivity calculus has been widely analyzed in the past and is covered in the well-known lecture notes [MuSi] and in the monographs [DeZo, HaMä, HaNe, SoZo, Zo], for example. The most commonly used approach relies on differentiating the reduced functional $\hat{J}(\Omega) = J(y(\Omega), \Omega, \Gamma(\Omega))$ using the chain rule. As a consequence shape differentiability of y with respect to variations of the domain are essential in this method. In an alternative approach the partial differential equation is realized in a Lagrangian formulation; see [DeZo], for example.

The method for computing the shape derivative that we describe here is quite different and elementary. In short, it can be described as follows. First we embed $e(y, \Omega) = 0$ to an equation in a fixed domain Ω_0 by a coordinate transformation which is called the method of mapping. Then we combine the Lagrange multiplier method to realize the constraint $e(y, \Omega) = 0$, using the shape derivative of functionals to calculate the shape derivative of $\hat{J}(\Omega)$. In this process, differentiability of the state with respect to the geometric quantities is not used. In fact, we require only Hölder continuity with exponent greater than or equal to $\frac{1}{2}$ of y with respect to the geometric data. We refer the reader to [IKPe2] for an example in which the reduced cost functional is shape differentiable whereas the state variable of the constraining partial differential equation is not.

For comparison we briefly discuss an example using the "chain rule" approach. But first we require some notions from shape calculus, which we introduce only formally here. Let Ω be a reference domain in $\mathbb{R}^d$, and let $h : U \to \mathbb{R}^d$, with $\bar{\Omega} \subset U$, denote a mapping describing perturbations of Ω by means of

$$\Omega_t = F_t(\Omega),$$

where $F_t : \Omega \to \mathbb{R}^d$ is a perturbation of the identity, given by $F_t = id + th$, for example. Let $\{z_t : \Omega_t \to \mathbb{R} | t \leq \tau\}$, for some $\tau > 0$, denote a family of mappings, and consider the associated family $z^t = z_t \circ F_t : \Omega \to \mathbb{R}$ which transports z_t back to Ω. Then the shape derivative of $\{z_t\}$ at Ω in direction h is defined as

$$z'(x) = \lim_{t \to 0} \frac{1}{t}(z_t(x) - z_0(x)) \text{ for } x \in \Omega,$$

and the material derivative is given by

$$\dot{z}(x) = \lim_{t \to 0} \frac{1}{t}(z^t(x) - z_0(x)) \text{ for } x \in \Omega.$$

Under appropriate regularity conditions we have

$$z' = \dot{z} - \nabla z \cdot F_0.$$

For a functional $\Omega \to J(\Omega)$ the shape derivative at Ω with respect to the perturbation h is defined as

$$J'(\Omega)h = \lim_{t \to 0} \frac{1}{t}\big(J(\Omega_t) - J(\Omega)\big).$$

Consider now the cost functional

$$\min J(y, \Omega, \Gamma) = \frac{1}{2} \int_\Gamma y^2 d\Gamma \tag{11.1.2}$$

subject to the constraint $e(y, \Omega) = 0$ which is given by the mixed boundary value problem

$$-\Delta y = f \qquad \text{in } \Omega, \tag{11.1.3}$$
$$y = 0 \qquad \text{on } \Gamma_0, \tag{11.1.4}$$
$$\frac{\partial y}{\partial n} = g \qquad \text{on } \Gamma. \tag{11.1.5}$$

Here the boundary of the domain Ω is the disjoint union of a fixed part Γ_0 and the unknown part Γ, and f and g are given functions. A formal differentiation leads to the shape derivative of the reduced cost functional

$$\hat{J}'(\Omega)h = \int_\Gamma y y'_\Omega \, d\Gamma + \frac{1}{2}\int_\Gamma \left(\frac{\partial y^2}{\partial n} + \kappa y^2\right) h \cdot n \, d\Gamma, \tag{11.1.6}$$

where y'_Ω denotes the shape derivative of the solution y of (11.1.3) at Ω with respect to a deformation field h, and κ stands for the curvature of Γ. For a thorough discussion of the details we refer the reader to [DeZo, SoZo]. Differentiating formally the constraint $e(y, \Omega) = 0$ with respect to the domain, one obtains that y'_Ω satisfies

$$\begin{aligned} -\Delta y'_\Omega &= 0 && \text{in } \Omega,\\ y'_\Omega &= 0 && \text{on } \Gamma_0,\\ \frac{\partial y'_\Omega}{\partial n} &= \operatorname{div}_\Gamma(h \cdot n \, \nabla_\Gamma y) + \left(f + \frac{\partial g}{\partial n} + \kappa g\right) h \cdot n && \text{on } \Gamma, \end{aligned} \tag{11.1.7}$$

where $\operatorname{div}_\Gamma$, ∇_Γ stand for the tangential divergence and tangential gradient, respectively, on the boundary Γ. Introducing a suitably defined adjoint variable and using (11.1.7), the first term on the right-hand side of (11.1.6) can be manipulated in such a way that $\hat{J}'(\Omega)h$ can be represented in the form required by the Zolesio–Hadamard structure theorem (see [DeZo])

$$\hat{J}'(\Omega)h = \int_\Gamma G h \cdot n \, d\Gamma.$$

We emphasize that the kernel G does not involve the shape derivative y'_Ω anymore. Although y'_Ω is only an intermediate quantity, a rigorous analysis requires justifying the formal steps in the preceding discussion. In addition one has to verify that the solution of (11.1.7) actually is the shape derivative of y in the sense of the definition in, e.g., [SoZo]. Furthermore, since the trace of y'_Ω on Γ_0 is used in (11.1.7) one needs $y'_\Omega \in H^1(\Omega)$. However, $y \in H^2(\Omega)$ is not sufficient to allow for an interpretation of the Neumann condition in (11.1.7) in $H^{-1/2}(\Gamma)$. Hence $y'_\Omega \in H^1(\Omega)$ requires more regularity of the solution y of (11.1.3) than $H^2(\Omega)$. In the approach of this chapter we utilize only $y \in H^2(\Omega)$ for the characterization of the shape derivative of $\hat{J}(\Omega)$. We return to this example in Section 11.3.

In Section 11.2 we present the proposed general framework to compute the shape derivative for (11.1.1). Section 11.3 contains applications to shape optimization constrained by linear elliptic systems, inverse interface problems, the Bernoulli problem, and shape optimization for the Navier–Stokes equations.

11.2 Shape derivative

Consider the shape optimization problem

$$\min J(y,\Omega,\Gamma) \equiv \int_{\Omega} j_1(y)\,dx + \int_{\Gamma} j_2(y)\,ds + \int_{\partial\Omega\setminus\Gamma} j_3(y)ds \tag{11.2.1}$$

subject to the constraint

$$e(y,\Omega) = 0, \tag{11.2.2}$$

which represents a partial differential equation posed on the domain $\Omega \subset \mathbb{R}^d$ with boundary $\partial\Omega$. We focus on sensitivity analysis of the reduced cost functional in (11.2.1)–(11.2.2) with respect to Ω.

To describe the admissible class of geometries, let $U \subset \mathbb{R}^d$ be a fixed bounded domain with $C^{1,1}$-boundary ∂U, or a convex domain with Lipschitzian boundary, and let D be a domain with $C^{1,1}$-boundary $\Gamma := \partial D$ satisfying $\bar{D} \subset U$. For the reference domain Ω we admit any of the three cases

(i) $\Omega = D$,

(ii) $\Omega = U$,

(iii) $\Omega = U \setminus \bar{D}$.

Note that

$$\partial\Omega = (\partial\Omega \cap \Gamma)\,\dot{\cup}\,(\partial\Omega \setminus \Gamma) \subset U \,\dot{\cup}\, \partial U. \tag{11.2.3}$$

Thus the boundary $\partial\Omega$ for the cases (i)–(iii) is given by

(i)′ $\partial\Omega = \Gamma \cup \emptyset = \Gamma$,

(ii)′ $\partial\Omega = \emptyset \cup \partial U = \partial U$,

(iii)′ $\partial\Omega = \Gamma \cup \partial U$.

To introduce the admissible class of perturbations let $h \in C^{1,1}(\bar{U})$ with $h = 0$ on $\partial U = 0$ and define, for $t \in \mathbb{R}$, the mappings $F_t : U \to \mathbb{R}^d$ by the perturbation of identity

$$F_t = \mathrm{id} + t\,h. \tag{11.2.4}$$

Then there exists $\tau > 0$ such that $F_t(U) = U$ and F_t is a diffeomorphism for $|t| < \tau$. Defining the perturbed domains

$$\Omega_t = F_t(\Omega)$$

and the perturbed manifolds as

$$\Gamma_t = F_t(\Gamma),$$

it follows that Γ_t is of class $C^{1,1}$ and $\bar{\Omega}_t \subset U$ for $|t| < \tau$. Note that since $h|\partial U = 0$ the boundary of U remains fixed as t varies, and hence by (11.2.3)

$$(\partial\Omega)_t \setminus \Gamma_t = \partial\Omega \setminus \Gamma \quad \text{for} \quad |t| < \tau.$$

Alternatively to (11.2.4) the perturbations could be described as the flow determined by the initial value problem

$$\frac{d}{dt}\chi(t) = h(\chi(t)), \quad \chi(0;x) = x,$$

with $F_t(x) = \chi(t;x)$, i.e., by the velocity method.

Let $\hat{J}(\Omega_t)$ be the functional defined by $\hat{J}(\Omega_t) = J(y_t, \Omega_t, \Gamma_t)$, where y_t satisfies the constraint

$$e(y_t, \Omega_t) = 0. \tag{11.2.5}$$

The shape derivative of $\hat{J}$ at Ω in the direction of the deformation field h is defined as

$$\hat{J}'(\Omega)h = \lim_{t\to 0} \frac{1}{t}\big(\hat{J}(\Omega_t) - \hat{J}(\Omega)\big).$$

The functional $\hat{J}$ is called shape differentiable at Ω if $\hat{J}'(\Omega)h$ exists for all $h \in C^{1,1}(U, \mathbb{R}^d)$ and defines a continuous linear functional on $C^{1,1}(\bar{U}, \mathbb{R}^d)$. Using the method of mappings one transforms the perturbed state constraint (11.2.5) to the fixed domain Ω. For this purpose define

$$y^t = y_t \circ F_t.$$

Then $y^t : \Omega \to \mathbb{R}^l$ satisfies an equation on the reference domain Ω, which we express as

$$\tilde{e}(y^t, t) = 0, \quad |t| < \tau. \tag{11.2.6}$$

We suppress the dependence of $\tilde{e}$ on h, because h will denote a fixed vector field throughout. Because $F_0 = \mathrm{id}$ one obtains $y^0 = y$ and

$$\tilde{e}(y^0, 0) = e(y, \Omega). \tag{11.2.7}$$

We axiomatize the above description and impose the following assumptions on $\tilde{e}$, respectively, e.

(H1) There is a Hilbert space X and a C^1-function $\tilde{e} : X \times (-\tau, \tau) \to X^*$ such that $e(y_t, \Omega_t) = 0$ is equivalent to

$$\tilde{e}(y^t, t) = 0 \text{ in } X^*,$$

with $\tilde{e}(y, 0) = e(y, \Omega)$ for all $y \in X$.

(H2) There exists $0 < \tau_0 \le \tau$ such that for $|t| < \tau_0$ there exists a unique solution $y^t \in X$ to $\tilde{e}(y^t, t) = 0$ and

$$\lim_{t\to 0} \frac{|y^t - y^0|_X}{t^{1/2}} = 0.$$

(H3) $e_y(y,\Omega) \in \mathcal{L}(X, X^*)$ satisfies

$$\langle e(v,\Omega) - e(y,\Omega) - e_y(y,\Omega)(v-y), \psi\rangle_{X^*\times X} = \mathcal{O}(|v-y|_X^2)$$

for every $\psi \in X$, where $y, v \in X$.

(H4) $\tilde{e}$ and e satisfy

$$\lim_{t\to 0}\frac{1}{t}\langle \tilde{e}(y^t, t) - \tilde{e}(y,t) - e(y^t,\Omega) + e(y,\Omega), \psi\rangle_{X^*\times X} = 0$$

for every $\psi \in X$, where y^t and y are the solutions of (11.2.6) and (11.2.2), respectively.

In applications (H4) typically results in an assumption on the regularity of the coefficients in the partial differential equation and on the vector field h. We assume throughout that $X \hookrightarrow L^2(\Omega, \mathbb{R}^l)$ and, in the case that j_2, j_3 are nontrivial, that the elements of X admit traces in $L^2(\Gamma, \mathbb{R}^l)$, respectively, $L^2(\partial\Omega \setminus \Gamma, \mathbb{R}^l)$. Typically X will be a subspace of $H^1(\Omega, \mathbb{R}^l)$ for some $l \in \mathbb{N}$.

With regards to the cost functional J we require

(H5) $j_i \in C^{1,1}(\mathbb{R}^l, \mathbb{R})$, $i = 1, 2, 3$.

As a consequence of (H1)–(H2) we infer that (11.2.5) has a unique solution y_t which is given by

$$y_t = y^t \circ F_t^{-1}.$$

Condition (H5) implies that $j_1(y) \in L^2(\Omega)$, $j_1'(y) \in L^2(\Omega)^l$, $j_2(y) \in L^2(\Gamma)$, $j_2'(y) \in L^2(\Gamma)^l$, and $j_3(y) \in L^2(\partial U)$, $j_3'(y) \in L^2(\partial U)^l$ for $y \in X$. Hence the cost functional $J(y, \Omega, \Gamma)$ is well defined for every $y \in X$.

Lemma 11.1. *There is a constant $c > 0$, such that*

$$|j_i(v) - j_i(y) - j_i'(y)(v-y)|_{L^1} \le c|v-y|_X^2$$

holds for all $y, v \in X$, $i = 1, 2, 3$.

Proof. For j_1 the claim follows from

$$\begin{aligned}
&\int_\Omega |j_1(v) - j_1(y) - j_1'(y)(v-y)|\,dx \\
&\le \int_\Omega \int_0^1 |j_1'(y(x) + s(v(x)-y(x))) - j_1'(y(x))|\,ds\,|v(x)-y(x)|\,dx \\
&\le \frac{L}{2}|v-y|_{L^2}^2 \le c|v-y|_X^2,
\end{aligned}$$

where $L > 0$ is the Lipschitz constant for j_1'. The same argument is valid also for j_2 and j_3. $\square$

Subsequently we use the following notation:

$$I_t = \det DF_t, \quad A_t = (DF_t)^{-T}, \quad w_t = I_t |A_t n|,$$

where DF_t is the Jacobian of F_t and n denotes the outer normal unit vector to Ω. We require additional regularity properties of the transformation F_t which we specify next. Let $\mathfrak{I} = [-\tau_0, \tau_0]$ with $\tau_0 \le \tau$ sufficiently small.

$$\begin{aligned}
&F_0 = \mathrm{id}, && t \to F_t \in C(\mathfrak{I}, C^2(\bar{U}, \mathbb{R}^d)),\\
&t \to F_t \in C^1(\mathfrak{I}, C^1(\bar{U}, \mathbb{R}^d)), && t \to F_t^{-1} \in C(\mathfrak{I}, C^1(\bar{U}, \mathbb{R}^d)),\\
&t \to I_t \in C^1(\mathfrak{I}, C(\bar{U})), && t \to A_t \in C(\mathfrak{I}, C(\bar{U}, \mathbb{R}^{d\times d})),\\
&t \to w_t \in C(\mathfrak{I}, C(\Gamma)), && \\
&\frac{d}{dt} F_t|_{t=0} = h, && \frac{d}{dt} F_t^{-1}|_{t=0} = -h,\\
&\frac{d}{dt} DF_t|_{t=0} = Dh, && \frac{d}{dt} DF_t^{-1}|_{t=0} = \frac{d}{dt}(A_t)^T|_{t=0} = -Dh,\\
&\frac{d}{dt} I_t|_{t=0} = \operatorname{div} h, && \frac{d}{dt} w_t|_{t=0} = \operatorname{div}_\Gamma h.
\end{aligned} \tag{11.2.8}$$

The surface divergence $\operatorname{div}_\Gamma$ is defined by

$$\operatorname{div}_\Gamma h = \operatorname{div} h|_\Gamma - (Dh\, n) \cdot n.$$

The properties (11.2.8) are easily verified if F_t is specified by perturbation of the identity as in (11.2.4). As a consequence of (11.2.8) there exists $\alpha > 0$ such that

$$I_t(x) \ge \alpha, \qquad x \in \bar{U}. \tag{11.2.9}$$

We furthermore recall the following transformation theorem where we already utilize (11.2.9).

Lemma 11.2. (1) *Let $\varphi_t \in L^1(\Omega_t)$; then $\varphi_t \circ F_t \in L^1(\Omega)$ and*

$$\int_{\Omega_t} \varphi_t \, dx_t = \int_\Omega \varphi_t \circ F_t \, \det DF_t \, dx.$$

(2) *Let $h_t \in L^1(\Gamma_t)$; then $h_t \circ F_t \in L^1(\Gamma)$ and*

$$\int_{\Gamma_t} h_t \, d\Gamma_t = \int_\Gamma h_t \circ F_t \, \det DF_t \, |(DF_t)^{-T} n| \, d\Gamma.$$

We now formulate the representation of the shape derivative of $\hat{J}$ at Ω in direction h.

Theorem 11.3. *Assume that* (H1)–(H5) *hold, that F satisfies* (11.2.8), *and that the adjoint equation*

$$\langle e_y(y,\Omega)\psi, p\rangle_{X^*\times X} - (j_1'(y), \psi)_\Omega - (j_2'(y), \psi)_\Gamma - (j_3'(y), \psi)_{\Omega\setminus\Gamma} = 0, \qquad \psi \in X, \tag{11.2.10}$$

admits a unique solution $p \in X$, *where* y *is the solution to* (11.2.2). *Then the shape derivative of* $\hat{J}$ *at* Ω *in the direction* h *exists and is given by*

$$\hat{J}'(\Omega)h = -\frac{d}{dt}\langle \tilde{e}(y,t), p\rangle_{X^*\times X}|_{t=0} + \int_\Omega j_1(y)\operatorname{div} h\,dx + \int_\Gamma j_2(y)\operatorname{div}_\Gamma h\,ds. \quad (11.2.11)$$

Proof. Referring to (H2) let y^t, $y \in X$ satisfy

$$\tilde{e}(y^t,t) = e(y,\Omega) = 0 \quad (11.2.12)$$

for $|t| < \tau_0$. Then $y_t = y^t \circ F_t$ is the solution of (11.2.5). Utilizing Lemma 11.2 one therefore obtains

$$\begin{aligned}
\frac{1}{t}(\hat{J}(\Omega_t) - \hat{J}(\Omega)) &= \frac{1}{t}\int_\Omega \big(I_t j_1(y^t) - j_1(y)\big)\,dx + \frac{1}{t}\int_\Gamma \big(w_t j_2(y^t) - j_2(y)\big)\,ds \\
&\quad + \frac{1}{t}\int_{\partial U}(j_3(y^t) - j_3(y))ds \\
&= \frac{1}{t}\int_\Omega \big(I_t(j_1(y^t) - j_1(y) - j_1'(y)(y^t - y)) + (I_t - 1)j_1'(y)(y^t - y) \\
&\quad + j_1'(y)(y^t - y) + (I_t - 1)j_1(y)\big)\,dx \\
&\quad + \frac{1}{t}\int_\Gamma \big(w_t(j_2(y^t) - j_2(y) - j_2'(y)(y^t - y)) + (w_t - 1)j_2'(y)(y^t - 1) \\
&\quad + j_2'(y)(y^t - y) + (w_t - 1)j_2(y)\big)\,ds \\
&\quad + \frac{1}{t}\int_{\partial U}(j_3(y^t) - j_3(y) - j_3'(y)(y^t - y))ds + \frac{1}{t}\int_{\partial U} j_3'(y)(y^t - y)ds.
\end{aligned} \quad (11.2.13)$$

Lemma 11.1 and (11.2.8) result in the estimates

$$\begin{aligned}
\left|\int_\Omega I_t(j_1(y^t) - j_1(y) - j_1'(y)(y^t - y))\,dx\right| &\le c|y^t - y|_X^2, \\
\left|\int_\Gamma w_t(j_2(y^t) - j_2(y) - j_2'(y)(y^t - y))\,ds\right| &\le c|y^t - y|_X^2, \\
\left|\int_{\partial U}(j_3(y^t) - j_3(y) - j_3'(y)(y^t - y))ds\right| &\le c|y^t - y|_X^2,
\end{aligned} \quad (11.2.14)$$

where $c > 0$ does not depend on t. Employing the adjoint state p one obtains

$$\begin{aligned}
(j_1'(y), y^t - y)_\Omega + (j_2'(y), y^t - y)_\Gamma + (j_3'(y), y^t - y)_{\partial U} &= \langle e_y(y,\Omega)(y^t - y), p\rangle_{X^*\times X} \\
&= -\langle e(y^t,\Omega) - e(y,\Omega) - e_y(y,\Omega)(y^t - y), p\rangle_{X^*\times X} \\
&\quad - \langle \tilde{e}(y^t,t) - \tilde{e}(y,t) - e(y^t,\Omega) + e(y,\Omega), p\rangle_{X^*\times X} \\
&\quad - \langle \tilde{e}(y,t) - \tilde{e}(y,0), p\rangle_{X^*\times X},
\end{aligned} \quad (11.2.15)$$

where we used (11.2.12). We estimate the ten additive terms on the right-hand side of (11.2.13). Terms one, five, and nine converge to zero by (11.2.14) and (H2). Terms two and six converge to 0 by (11.2.8) and (H2). For terms four and eight one uses (11.2.8). The claim (11.2.11) now follows by passing to the limit in terms three, seven, and ten using (11.2.15), (H3), (H2), (H4), and (H1). □

To check (H2) in specific applications the following result will be useful. It relies on

(H6) $\begin{cases} \text{the linearized equation} \\ \langle E_y(y,\Omega)\delta y, \psi\rangle_{X^*\times X} = \langle f, \psi\rangle_{X^*\times X}, \qquad \psi \in X, \\ \text{admits a unique solution } \delta y \in X \text{ for every } f \in X^*. \end{cases}$

Note that this condition is more stringent than the assumption of solvability of the adjoint equation in Theorem 11.3, which requires solvability only for a specific right-hand side.

Proposition 11.4. *Assume that* (11.2.2) *admits a unique solution y and that* (H6) *is satisfied. Then* (H2) *holds.*

Proof. Let $y \in X$ be the unique solution of (11.2.2). In view of

$$\tilde{e}_y(y,0) = e_y(y,\Omega)$$

Assumption (H6) implies that $\tilde{e}_y(y,0)$ is bijective. The claim follows from the implicit function theorem. □

Computing the derivative $\frac{d}{dt}\langle \tilde{e}(y,t), p\rangle_{X^*\times X}|_{t=0}$ in (11.2.11), and subsequently arguing that $\hat{J}$ is in fact a shape derivative at Ω, can be facilitated by transforming the equation $\langle \tilde{e}(y,t), p\rangle = 0$ back to $\langle e(y \circ F_t^{-1}, \Omega_t), p \circ F_t^{-1}\rangle = 0$ together with the following well-known differentiation rules of the functionals.

Lemma 11.5 (see [DeZo]). (1) *Let* $f \in C(\mathfrak{I}, W^{1,1}(U))$ *and assume that* $f_t(0)$ *exists in* $L^1(U)$. *Then*

$$\frac{d}{dt}\int_{\Omega_t} f(t,x)\,dx|_{t=0} = \int_\Omega f_t(0,x)\,dx + \int_\Gamma f(0,x)\,h\cdot n\,ds.$$

(2) *Let* $f \in C(\mathfrak{I}, W^{2,1}(U))$ *and assume that* $f_t(0)$ *exists in* $W^{1,1}(U)$. *Then*

$$\frac{d}{dt}\int_{\Gamma_t} f(t,x)\,ds|_{t=0} = \int_\Gamma f_t(0,x)\,ds + \int_\Gamma \left(\frac{\partial}{\partial n} f(0,s) + \kappa f(0,s)\right) h\cdot n\,ds,$$

where κ *stands for the additive curvature of* Γ.

The first part of the lemma is valid also for domains Ω with Lipschitz continuous boundary.

In the examples below $f(t)$ will be typically given by expressions of the form

$$\mu v \circ F_t^{-1}, \ \mu \partial_i (v \circ F_t^{-1}) \partial_j (w \circ F_t^{-1}), \ v \circ F_t^{-1} w \circ F_t^{-1} \partial_i (z \circ F_t^{-1}),$$

where $\mu \in H^1(U)$ and v, z, and $w \in H^2(U)$ are extensions of elements in $H^2(\Omega)$. The assumptions of Lemma 11.5 can be verified using the following result.

Lemma 11.6 (see [SoZo]). (1) *If $y \in L^p(U)$, then $t \to y \circ F_t^{-1} \in C(\mathfrak{I}, L^p(U))$, $1 \le p < \infty$.*

(2) *If $y \in H^2(U)$, then $t \to y \circ F_t^{-1} \in C(\mathfrak{I}, H^2(U))$.*

(3) *If $y \in H^2(U)$, then $\frac{d}{dt}(y \circ F_t^{-1})|_{t=0}$ exists in $H^1(U)$ and is given by*

$$\frac{d}{dt}(y \circ F_t^{-1})|_{t=0} = -(Dy)\, h.$$

As a consequence we note that $\frac{d}{dt}\partial_i\big((y \circ F_t^{-1})\big)|_{t=0}$ exists in $L^2(U)$ and is given by

$$\frac{d}{dt}\partial_i\big((y \circ F_t^{-1})\big)|_{t=0} = -\partial_i (Dy\, h), \qquad i = 1, \dots, d.$$

In the next section ∇y stands for $(Dy)^T$, where y is either a scalar- or vector-valued function. To enhance readability we use two symbols for the inner product in $\mathbb{R}^d$, (x, y), respectively, $x \cdot y$. The latter will be utilized only in the case of nested inner products.

11.3 Examples

Throughout this section it is assumed that (H5) is satisfied and that the regularity assumptions of Section 11.2 for D, Ω, and U hold. If J does not depend on Γ, we write $J(y, \Omega)$ in place of $J(y, \Omega, \Gamma)$.

11.3.1 Elliptic Dirichlet boundary value problem

As a first example we consider the volume functional

$$J(y, \Omega) = \int_\Omega j_1(y)\, dx$$

subject to the constraint

$$(\mu \nabla y, \nabla \psi)_\Omega - (f, \psi)_\Omega = 0, \tag{11.3.1}$$

where $X = H_0^1(\Omega)$, $f \in H^1(U)$, and $\mu \in C^1(\bar{U}, \mathbb{R}^{d \times d})$ such that $\mu(x)$ is symmetric and uniformly positive definite. Here $\Omega = D$ and $\Gamma = \partial\Omega$. Thus $e(y, \Omega) : X \to X^*$ is given by

$$\langle e(y, \Omega), \psi \rangle_{X^* \times X} = (\mu \nabla y, \nabla \psi)_\Omega - (f, \psi)_\Omega.$$

The equation on the perturbed domain is determined by

$$\begin{aligned}\langle e(y_t,\Omega_t),\psi_t\rangle_{X_t^*\times X_t} &= \int_{\Omega_t}(\mu\nabla y_t,\nabla\psi_t)\,dx_t - \int_{\Omega_t} f\psi\,dx_t\\ &= \int_\Omega (\mu^t A_t\nabla y^t, A_t\nabla\psi^t)I_t\,dx - \int_\Omega f^t\psi^t\,dx \equiv \langle\tilde e(y^t,t),\psi^t\rangle_{X^*,X}\end{aligned} \tag{11.3.2}$$

for any $\psi_t\in X_t$, with $y^t = y_t\circ F_t$, $\mu^t=\mu\circ F_t$, $f^t = f\circ F_t$, and $X_t = H_0^1(\Omega_t)$. Here we used that $\nabla y_t = (A_t\nabla y^t)\circ F_t^{-1}$ and Lemma 11.5. (H1) is a consequence of (11.2.8), (11.3.2), and the smoothness of μ and f. Since (11.3.1) admits a unique solution and (H6) holds, Proposition 11.4 implies (H2). Since $\tilde e$ is linear in y, assumption (H3) follows. For the verification of (H4) observe that

$$\begin{aligned}&\langle\tilde e(y^t,t)-\tilde e(y,t)-e(y^t,\Omega)+e(y,\Omega),\psi\rangle_{X^*\times X}\\ &\quad= ((\mu^t I_t A_t-\mu)\nabla(y^t-y), A_t\nabla\psi)_\Omega + (\mu\nabla(y^t-y),(A_t-I)\nabla\psi).\end{aligned}$$

Hence (H4) follows from differentiability of μ, (11.2.8), and (H2). In view of Theorem 11.3 we have to compute $\frac{d}{dt}\langle\tilde e(y,t),p\rangle_{X^*\times X}|_{t=0}$ for which we use the representation on Ω_t in (11.3.2). Recall that the solution y of (11.3.1) as well as the adjoint state p, defined by

$$(\mu\nabla p,\nabla\psi)_\Omega = (j_1'(y),\psi)_\Omega,\qquad \psi\in H_0^1(\Omega), \tag{11.3.3}$$

belong to $H^2(\Omega)\cap H_0^1(\Omega)$. Since $\Omega\in C^{1,1}$ (actually Lipschitz continuity of the boundary would suffice), y as well as p can be extended to functions in $H^2(U)$, which we again denote by the same symbol. Therefore by Lemmas 11.5 and 11.6

$$\begin{aligned}&\frac{d}{dt}\langle\tilde e(y,t),p\rangle_{X^*\times X}|_{t=0}\\ &= \frac{d}{dt}\left(\int_{\Omega_t}(\mu\nabla(y\circ F_t^{-1}),\nabla(p\circ F_t^{-1}))\,dx_t - \int_{\Omega_t} fp\circ F_t^{-1}dx_t\right)|_{t=0}\\ &= \int_\Gamma(\mu\nabla y,\nabla p)\,(h,n)\,ds + \int_\Omega\big((\mu\nabla(-\nabla y\cdot h),\nabla p)\\ &\quad + (\mu\nabla y,\nabla(-\nabla p\cdot h)) + f(\nabla p,h)\big)\,dx.\end{aligned} \tag{11.3.4}$$

Note that $\nabla y\cdot h$ as well as $\nabla p\cdot h$ do not belong to $H_0^1(\Omega)$ but they are elements of $H^1(\Omega)$. Therefore Green's theorem implies

$$\begin{aligned}&\int_\Omega\big((\mu\nabla(-\nabla y\cdot h),\nabla p) + (\mu\nabla y,\nabla(-\nabla p\cdot h)) + f(\nabla p,h)\big)\,dx\\ &\quad= \int_\Omega \operatorname{div}(\mu\nabla p)\,(\nabla y,h)\,dx - \int_\Gamma(\mu\nabla p,n)(\nabla y,h)\,ds\\ &\quad\quad+ \int_\Omega(\operatorname{div}(\mu\nabla y)+f)\,(\nabla p,h)\,dx - \int_\Gamma(\mu\nabla y,n)\,(\nabla p,h)\,ds\\ &\quad= -\int_\Omega j_1'(y)\,(\nabla y,h)\,dx - 2\int_\Gamma(\mu n,n)\frac{\partial y}{\partial n}\frac{\partial p}{\partial n}\,(h,n)\,ds.\end{aligned} \tag{11.3.5}$$

Above we used the strong form of (11.3.1) and (11.3.3) in $L^2(\Omega)$ as well as the identities

$$(\mu\nabla y, n) = (\mu n, n)\frac{\partial y}{\partial n}, \qquad (\nabla y, h) = \frac{\partial y}{\partial n}(h, n)$$

(together with the ones with y and p interchanged) which follow from $y, p \in H_0^1(\Omega)$. Applying Theorem 11.3 results in

$$\begin{aligned}\hat{J}'(\Omega)h &= -\frac{d}{dt}\langle\tilde{e}(y,t), p\rangle_{X^*,X}|_{t=0} + \int_\Omega j_1(y)\,\mathrm{div}\,h\,dx\\ &= \int_\Gamma (\mu n, n)\frac{\partial y}{\partial n}\frac{\partial p}{\partial n}\,(h,n)\,ds + \int_\Omega (j_1'(y)(\nabla y, h) + j_1(y)\,\mathrm{div}\,h)\,dx\\ &= \int_\Gamma (\mu n, n)\frac{\partial y}{\partial n}\frac{\partial p}{\partial n}\,(h,n)\,ds + \int_\Omega \mathrm{div}(j_1(y)h)\,dx,\end{aligned}$$

and the Stokes theorem yields the final result,

$$J'(\Omega)h = \int_\Gamma \left((\mu n, n)\frac{\partial y}{\partial n}\frac{\partial p}{\partial n} + j_1(y)\right)(h,n)\,ds.$$

Remark 11.3.1. If we were to be content with a representation of the shape variation in terms of volume integrals we could take the expression for $\frac{d}{dt}\tilde{e}(y,t)|_{t=0}$ given in (11.3.4) and bypass the use of Green's theorem in (11.3.5). The regularity requirement on the domain then results from $y \in H^2(\Omega)$, $p \in H^2(\Omega)$. In [Ber] the shape derivative in terms of the volume integral is referred to as the weak shape derivative, whereas the final form in terms of the boundary integrals is called the strong shape derivative.

11.3.2 Inverse interface problem

We consider an inverse interface problem which is motivated by electrical impedance tomography. Let $U = \Omega = (-1,1)\times(-1,1)$ and $\partial U = \partial\Omega$. Further let the domain $D = \Omega^-$, with $\bar{\Omega}^- \subset U$, represent the inhomogeneity of the conducting medium and set $\Omega^+ = U \setminus \bar{\Omega}^-$. We assume that Ω^- is a simply connected domain of class C^2 with boundary Γ which represents the interface between Ω^- and Ω^+. The inverse problem consists of identifying the unknown interface Γ from measurements z which are taken on the boundary ∂U. This can be formulated as

$$\min J(y,\Omega) \equiv \int_{\partial U} (y-z)^2\,ds \tag{11.3.6}$$

subject to the constraints

$$\begin{aligned}-\mathrm{div}(\mu\nabla y) &= 0 && \text{in } \Omega^-\cup\Omega^+,\\ [y] = 0, \quad \left[\mu\frac{\partial y}{\partial n^-}\right] &= 0 && \text{on } \Gamma,\\ \frac{\partial y}{\partial n} &= g && \text{on } \partial U,\end{aligned} \tag{11.3.7}$$

where $g \in H^{1/2}(\partial U)$, $z \in L^2(\partial U)$, with $\int_{\partial U} g = \int_{\partial U} z = 0$, $[v] = v^+ - v^-$ on Γ, and $n^{+/-}$ standing for the unit outer normals to $\Omega^{+/-}$. The conductivity μ is given by

$$\mu(x) = \begin{cases} \mu^-, & x \in \Omega^-, \\ \mu^+, & x \in \Omega^+, \end{cases}$$

for some positive constants μ^- and μ^+. In the context of the general framework of Section 11.2 we have $j_1 = j_2 = 0$ and $j_3 = (y - z)^2$. Clearly (11.3.7) admits a unique solution $y \in H^1(U)$ with $\int_{\partial U} y = 0$. Its restrictions to Ω^+ and Ω^- will be denoted by y^+ and y^-, respectively. It turns out that the regularity of $y^\pm$ is better than the one of y.

Proposition 11.7. *Let Ω and $\Omega^\pm$ be as described above. Then the solution $y \in H^1(U)$ of* (11.3.7) *satisfies*

$$y^\pm \in H^2(\Omega^\pm).$$

Proof. Let Γ_H be the smooth boundary of a domain Ω_H with

$$\overline{\Omega^-} \subset \Omega_H \subset \overline{\Omega_H} \subset U.$$

Then $y|\Gamma_H \in H^{3/2}(\Gamma_H)$. The problem

$$\begin{cases} -\text{div}(\mu^+ \nabla y_H) = 0 \quad \text{in} \quad U \backslash \overline{\Omega_H}, \\ \frac{\partial y_H}{\partial n} = g \quad \text{on} \quad \partial U, \quad y_H = y|\Gamma_H \quad \text{on} \quad \Gamma_H \end{cases}$$

has a unique solution $y_H \in H^2(U \backslash \overline{\Omega_H})$ with $y^H = y^+|_{\Omega_H}$. Therefore,

$$b := y|\partial U = y_H|\partial U \in H^{3/2}(\partial U).$$

Then the solution y to (11.3.7) coincides with the solution to

$$\begin{cases} -\text{div}(\mu \nabla y) = 0 & \text{in } \Omega^- \cup \Omega^+, \\ [y] = 0, \ [\mu \frac{\partial y}{\partial n^-}] = 0 & \text{on } \Gamma, \\ y = b & \text{on } \partial U. \end{cases}$$

We now argue that $y^\pm \in H^2(\Omega^\pm)$. Let $y_b \in H^2(U)$ denote the solution to

$$\begin{cases} -\Delta y_b = 0 & \text{in } U, \\ y_b = b & \text{on } \partial U. \end{cases}$$

Define $w \in H_0^1(U)$ as the unique solution to the interface problem

$$\begin{cases} -\text{div}(\mu \nabla w) = 0 & \text{in } \Omega, \\ [w] = 0, \quad [\mu \frac{\partial w}{\partial n^-}] = -[\mu \frac{\partial y_b}{\partial n^-}] & \text{on } \Gamma, \\ w = 0 & \text{on } \partial U. \end{cases} \tag{11.3.8}$$

Then $y_b \in H^2(\Omega)$ implies $[\mu \frac{\partial y_b}{\partial n^-}] \in H^{1/2}(\Gamma)$. By [ChZo], (11.3.8) has a unique solution $w \in H_0^1(U)$ with the additional regularity $w^\pm \in H^2(\Omega^\pm)$. Consequently $y = w + y_b$ satisfies $y|\partial\Omega = g$ and $y^\pm \in H^2(\Omega^\pm)$, as desired. In an analogous way $p^\pm \in H^2(\Omega^\pm)$. □

To consider the inverse problem (11.3.6), (11.3.7) within the general framework of Section 11.2 we set $X = \{v \in H^1(U)\colon \int_{\partial U} v = 0\}$ and define

$$\langle e(y,\Omega), \psi\rangle_{X^*\times X} = (\mu\nabla y, \nabla\psi)_U - (g,\psi)_{\partial U},$$

respectively,

$$\begin{aligned}\langle \tilde e(y,t), \psi\rangle_{X^*\times X} &= (\mu^t A_t\nabla y, A_t\nabla\psi I_t)_U - (g,\psi)_{\partial U}\\ &= (\mu^+\nabla(y\circ F_t^{-1}), \nabla(\psi\circ F_t^{-1}))_{\Omega_t^+} + (\mu^-\nabla(y\circ F_t^{-1}), \nabla(\psi\circ F_t^{-1}))_{\Omega_t^-} - (g,\psi)_{\partial U}.\end{aligned}$$

Note that the boundary term is not affected by the transformation F_t since the deformation field h vanishes on ∂U. The adjoint state is given by

$$\begin{aligned} -\operatorname{div}(\mu\nabla p) &= 0 && \text{in } \Omega^-\cup\Omega^+,\\ [p] = 0,\quad \left[\mu\frac{\partial p}{\partial n^-}\right] &= 0 && \text{on } \Gamma,\\ \frac{\partial p}{\partial n} &= 2(y-z) && \text{on } \partial U,\end{aligned} \tag{11.3.9}$$

respectively,

$$(\mu\nabla p, \nabla\psi)_U = 2(y-z,\psi)_{\partial U} \quad \text{for } \psi\in X. \tag{11.3.10}$$

Assumption (H4) requires us to consider

$$\begin{aligned}&\frac{1}{t}|\langle\tilde e(y^t-y,t) - e(y^t-y,\Omega),\psi\rangle|\\ &\le \frac{1}{t}\int_{\Omega^+} |(\mu^+ I_t A_t\nabla(y^t-y), A_t\nabla\psi) - (\mu^+\nabla(y^t-y),\nabla\psi)|\,dx\\ &+\frac{1}{t}\int_{\Omega^-} |(\mu^- I_t A_t\nabla(y^t-y), A_t\nabla\psi) - (\mu^+\nabla(y^t-y),\nabla\psi)|\,dx\\ &\le \mu^+\int_{\Omega^+}\left|\left(\frac{1}{t}(I_tA_t - I)\nabla(y^t-y), A_t\nabla\psi\right)\right|\,dx\\ &+\mu^+\int_{\Omega^+}\left|\left(\nabla(y^t-y), \frac{1}{t}(A_t-I)\nabla\psi\right)\right|\,dx\\ &+\mu^-\int_{\Omega^-}\left|\left(\frac{1}{t}(I_tA_t - I)\nabla(y^t-y), A_t\nabla\psi\right)\right|\,dx\\ &+\mu^-\int_{\Omega^-}\left|\left(\nabla(y^t-y), \frac{1}{t}(A_t-I)\nabla\psi\right)\right|\,dx.\end{aligned}$$

The right-hand side of this inequality converges to 0 as $t\to 0^+$ by (11.2.8). The remaining assumptions can be verified as in Section 11.3.1 for the Dirichlet problem and thus Theorem

11.3 is applicable. By Proposition 11.7 the restrictions $y^\pm = y|_{\Omega^\pm}$, $p^\pm = p|_{\Omega^\pm}$ satisfy $y^\pm$, $p^\pm \in H^2(\Omega^\pm)$. Using Lemma 11.5 we find that

$$
\begin{aligned}
&\frac{d}{dt}\langle \tilde{e}(y,t), p\rangle_{X^*\times X}|_{t=0} \\
&\quad = \int_{\partial\Omega^+} (\mu^+\nabla y^+, \nabla p^+)(h, n^+)\, ds - \int_{\Omega^+} (\mu^+\nabla(\nabla y^+ \cdot h), \nabla p^+)\, dx \\
&\quad - \int_{\Omega^+} (\mu^+\nabla y^+, \nabla(\nabla p^+ \cdot h))\, dx + \int_{\partial\Omega^-} (\mu^-\nabla y^-, \nabla p^-)(h, n^-)\, ds \\
&\quad - \int_{\Omega^-} (\mu^-\nabla(\nabla y^- \cdot h), \nabla p^-)\, dx - \int_{\Omega^-} (\mu^-\nabla y^-, \nabla(\nabla p^- \cdot h))\, dx \\
&\quad = \int_\Gamma [\mu\nabla y, \nabla p](h, n^+)\, ds - \int_{\Omega^+} \mu^+(\nabla(\nabla y^+ \cdot h), \nabla p^+) + (\nabla y^+, \nabla(\nabla p^+ \cdot h))\, dx \\
&\quad - \int_{\Omega^-} \mu^-(\nabla(\nabla y^- \cdot h), \nabla p^-) + (\nabla y^-, \nabla(\nabla p^- \cdot h))\, dx.
\end{aligned}
$$

Applying Green's formula as in Example 11.3.1 (observe that $(\nabla y, h), (\nabla p, h) \notin H^1(U)$) together with (11.3.9) results in

$$
\begin{aligned}
&- \int_{\Omega^+} (\mu^+\nabla(\nabla y^+ \cdot h), \nabla p^+))\, dx - \int_{\Omega^-} (\mu^-\nabla(\nabla y^- \cdot h), \nabla p^-))\, dx \\
&\qquad = \int_{\Omega^+} \operatorname{div}(\mu^+\nabla p^+)(\nabla y^+, h)\, dx + \int_{\Omega^-} \operatorname{div}(\mu^-\nabla p^-)(\nabla y^-, h)\, dx \\
&\qquad - \int_{\partial\Omega^+} (\mu^+\nabla p^+, n^+)(\nabla y^+, h)\, ds - \int_{\partial\Omega^-} (\mu^-\nabla p^-, n^-)(\nabla y^-, h)\, ds \\
&\qquad = - \int_\Gamma \left[\mu \frac{\partial p}{\partial n^+}(\nabla y, h)\right] ds.
\end{aligned}
$$

In the last step we utilize $h = 0$ on ∂U. Similarly we obtain

$$
\begin{aligned}
&- \int_{\Omega^+} (\mu^+\nabla y^+, \nabla(\nabla p^+ \cdot h)))\, dx - \int_{\Omega^-} (\mu^-\nabla y^-, \nabla(\nabla p^- \cdot h)))\, dx \\
&\qquad = - \int_\Gamma \left[\mu \frac{\partial y}{\partial n^+}(\nabla p, h)\right] ds.
\end{aligned}
$$

Collecting terms results in

$$
\hat{J}'(\Omega)h = - \int_\Gamma [\mu(\nabla y, \nabla p)]\,(h, n^+)\, ds + \int_\Gamma \left(\left[\mu \frac{\partial p}{\partial n^+}(\nabla y, h)\right] + \left[\mu \frac{\partial y}{\partial n^+}(\nabla p, h)\right]\right) ds.
$$

The identity

$$
[ab] = [a]b^+ + a^-[b] = a^+[b] + [a]b^-
$$

implies

$$
[ab] = 0 \qquad \text{if } [a] = [b] = 0.
$$

Hence the transition conditions

$$\left[\mu\frac{\partial y}{\partial n^+}\right]=\left[\frac{\partial y}{\partial \tau}\right]=0,$$
$$\left[\mu\frac{\partial p}{\partial n^+}\right]=\left[\frac{\partial p}{\partial \tau}\right]=0, \tag{11.3.11}$$

where $\frac{\partial}{\partial\tau}$ stands for the tangential derivative imply

$$\begin{aligned}\left[\mu\frac{\partial p}{\partial n^+}(\nabla y,h)\right]&=\left[\mu\frac{\partial p}{\partial n^+}\frac{\partial y}{\partial n^+}(h,n^+)+\mu\frac{\partial p}{\partial n^+}\frac{\partial y}{\partial \tau}(h,\tau)\right]\\&=\left[\mu\frac{\partial p}{\partial n^+}\frac{\partial y}{\partial n^+}(h,n^+)\right]+\left[\mu\frac{\partial p}{\partial n^+}\frac{\partial y}{\partial \tau}(h,\tau)\right]\\&=\left[\mu\frac{\partial p}{\partial n^+}\frac{\partial y}{\partial n^+}\right](h,n^+),\end{aligned}$$

and analogously

$$\left[\mu\frac{\partial y}{\partial n^+}(\nabla p,h)\right]=\left[\mu\frac{\partial p}{\partial n^+}\frac{\partial y}{\partial n^+}\right](h,n^+),$$

which entails

$$\begin{aligned}\hat{J}'(\Omega)h&=-\int_\Gamma[\mu(\nabla y,\nabla p)]\,(h,n^+)\,ds+2\int_\Gamma\left[\mu\frac{\partial p}{\partial n^+}\frac{\partial y}{\partial n^+}\right](h,n^+)\,ds\\&=-\int_\Gamma\left[\mu\frac{\partial y}{\partial\tau}\frac{\partial p}{\partial\tau}\right](h,n^+)\,ds+\int_\Gamma\left[\mu\frac{\partial y}{\partial n^+}\frac{\partial p}{\partial n^+}\right](h,n^+)\,ds\\&=-\int_\Gamma[\mu]\frac{\partial y}{\partial\tau}\frac{\partial p}{\partial\tau}(h,n^+)\,ds+\int_\Gamma\left[\mu\frac{\partial y}{\partial n^+}\frac{\partial p}{\partial n^+}\right](h,n^+)\,ds.\end{aligned}$$

In view of (11.3.11) this can be rearranged as

$$\begin{aligned}&-[\mu]\frac{\partial y}{\partial\tau}\frac{\partial p}{\partial\tau}+\left[\mu\frac{\partial y}{\partial n^+}\frac{\partial p}{\partial n^+}\right]\\&=-\mu^+\frac{\partial y}{\partial\tau}\frac{\partial p}{\partial\tau}+\mu^-\frac{\partial y}{\partial\tau}\frac{\partial p}{\partial\tau}+\mu^+\frac{\partial y^+}{\partial n^+}\frac{\partial p^+}{\partial n^+}-\mu^-\frac{\partial y^-}{\partial n^+}\frac{\partial p^-}{\partial n^+}\\&=-\mu^+\left(\frac{\partial y}{\partial\tau}\frac{\partial p}{\partial\tau}+\frac{1}{2}\left(\frac{\partial y^+}{\partial n^+}\frac{\partial p^-}{\partial n^+}+\frac{\partial y^-}{\partial n^+}\frac{\partial p^+}{\partial n^+}\right)\right)\\&\quad+\mu^-\left(\frac{\partial y}{\partial\tau}\frac{\partial p}{\partial\tau}+\frac{1}{2}\left(\frac{\partial y^-}{\partial n^+}\frac{\partial p^+}{\partial n^+}+\frac{\partial y^+}{\partial n^+}\frac{\partial p^-}{\partial n^+}\right)\right)\\&=-\frac{1}{2}[\mu]\big((\nabla y^+,\nabla p^-)+(\nabla y^-,\nabla p^+)\big),\end{aligned}$$

which gives the representation

$$\begin{aligned}\hat{J}'(\Omega)h&=-\frac{1}{2}\int_\Gamma[\mu]\big((\nabla y^+,\nabla p^-)+(\nabla y^-,\nabla p^+)\big)\,(h,n^+)\,ds\\&=-\int_\Gamma[\mu](\nabla y^+,\nabla p^-)\,(h,n^+)\,ds.\end{aligned}$$

11.3.3 Elliptic systems

Here we consider a domain $\Omega = U \setminus D$, where $\bar{D} \subset U$ and the boundaries ∂U and $\Gamma = \partial D$ are assumed to be C^2 regular.

We consider the optimization problem

$$\min J(y, \Omega, \Gamma) \equiv \int_\Omega j_1(y)\, dx + \int_\Gamma j_2(y)\, ds,$$

where y is the solution of the elliptic system

$$\langle e(y, \Omega), \psi \rangle_{X^* \times X} = \int_\Omega \big(a(x, \nabla y, \nabla \psi) - (f, \psi)\big)\, dx - \int_\Gamma (g, \psi)\, ds = 0 \qquad (11.3.12)$$

in $X = \{v \in H^1(\Omega)^l : v|_{\partial U} = 0\}$. Above ∇y stands for $(Dy)^T$. We require that $f \in H^1(U)^l$ and that g is the trace of a given function $G \in H^2(U)^l$. Furthermore we assume that $a : \bar{U} \times \mathbb{R}^{d \times d} \times \mathbb{R}^{d \times d}$ satisfies

(1) $a(\cdot, \xi, \eta)$ is continuously differentiable for every $\xi, \eta \in \mathbb{R}^{d \times d}$,

(2) $a(x, \cdot, \cdot)$ defines a bilinear form on $\mathbb{R}^{d \times d} \times \mathbb{R}^{d \times d}$ which is uniformly bounded in $x \in \bar{U}$,

(3) $a(x, \cdot, \cdot)$ is uniformly coercive for all $x \in \bar{U}$.

In the case of linear elasticity a is given by

$$a(x, \nabla y, \nabla \psi) = \lambda \operatorname{tr} e(y) \operatorname{tr} e(\psi) + 2\mu\, e(y) : e(\psi),$$

where $e(y) = \frac{1}{2}(\nabla y + (\nabla y)^T)$, and λ, μ are the positive Lamé coefficients. In this case a is symmetric, and (11.3.12) admits a unique solution in $X \cap H^2(\Omega)^l$ for every $f \in L^2(\Omega)^l$ and $g \in H^{\frac{1}{2}}(\partial U)^l$; see, e.g., [Ci].

The method of mapping suggests defining

$$\begin{aligned}\langle \tilde{e}(y, t), \psi \rangle_{X^* \times X} &= \int_\Omega \big(a(F_t(x), A_t \nabla y, A_t \nabla \psi) - (f^t, \psi)\big) I_t\, dx - \int_\Gamma (g^t, \psi) w_t\, ds \\ &= \int_{\Omega_t} \big(a(x, \nabla(y \circ F_t^{-1}), \nabla(\psi \circ F_t^{-1})) - (f, \psi \circ F_t^{-1})\big)\, dx - \int_{\Gamma_t} (g, \psi \circ F_t^{-1})\, ds.\end{aligned} \qquad (11.3.13)$$

The adjoint state is determined by the equation

$$\langle e_y(y, \Omega)\psi, p \rangle_{X^* \times X} = \int_\Omega \big(a(x, \nabla \psi, \nabla p) - j_1'(y)\psi\big)\, dx - \int_\Gamma j_2'(y)\psi\, ds = 0, \quad (11.3.14)$$

$\psi \in X$. Under the regularity assumptions on a, (11.3.12) admits a unique solution in $X \cap H^2(\Omega)^l$ and the adjoint equation admits a solution for any right-hand side in X^* so that Proposition 11.4 is applicable. All these properties are satisfied for the linear elasticity case. Assumptions (H1)–(H4) can then be argued as in Section 11.3.1.

Employing Lemma 11.5 we obtain

$$\begin{aligned}\frac{d}{dt}\langle \tilde{e}(y,t), p\rangle_{X^* \times X}|_{t=0} &= -\int_\Omega \big(a(x, \nabla(\nabla y^T h), \nabla p) + a(x, \nabla y, \nabla(\nabla p^T h))\big)\, dx \\ &+ \int_\Gamma a(x, \nabla y, \nabla p)\,(h,n)\, ds + \int_\Omega (f, \nabla p^T h)\, dx - \int_\Gamma (f,p)\,(h,n)\, ds \\ &+ \int_\Gamma (g, \nabla p^T h)\, ds - \int_\Gamma \left(\frac{\partial}{\partial n}(g,p) + \kappa(g,p)\right)(h,n)\, ds.\end{aligned}$$

Since $\nabla y^T h \in X$ and $\nabla p^T h \in X$, this expression can be simplified using (11.3.12) and (11.3.14):

$$\begin{aligned}\frac{d}{dt}\langle \tilde{e}(y,t), p\rangle_{X^* \times X}|_{t=0} &= -\int_\Omega j_1'(y)\nabla y^T h\, dx - \int_\Gamma j_2'(y)\nabla y^T h\, ds \\ &+ \int_\Gamma \big(a(x, \nabla y, \nabla p) - (f,p)\big)(h,n)\, ds - \int_\Gamma \left(\frac{\partial}{\partial n}(g,p) + \kappa(g,p)\right)(h,n)\, ds,\end{aligned}$$

which implies

$$\begin{aligned}\hat{J}'(\Omega)h &= \int_\Omega j_1'(y)\nabla y^T h\, dx + \int_\Omega j_1(y)\,\mathrm{div}\, h\, dx \\ &+ \int_\Gamma j_2'(y)\nabla y^T h\, ds + \int_\Gamma j_2(y)\,\mathrm{div}_\Gamma\, h\, ds \\ &+ \int_\Gamma \big(-a(x, \nabla y, \nabla p) + (f,p) + \frac{\partial}{\partial n}(g,p) + \kappa(g,p)\big)(h,n)\, ds.\end{aligned}$$

For the third and fourth terms the tangential Green's formula (see, e.g., [DeZo]) yields

$$\int_\Gamma j_2'(y)\nabla y^T h\, ds + \int_\Gamma j_2(y)\,\mathrm{div}_\Gamma\, h\, ds = \int_\Gamma \left(\frac{\partial}{\partial n} j_2(y) + \kappa j_2(y)\right)(h,n)\, ds.$$

The first and second terms can be combined using the Stokes theorem. Summarizing we finally obtain

$$\begin{aligned}\hat{J}'(\Omega)h &= \int_\Gamma \big(-a(x, \nabla y, \nabla p) + (f,p) + j_1(y) \\ &+ \frac{\partial}{\partial n}\big(j_2(y) + (g,p)\big) + \kappa\big(j_2(y) + (g,p)\big)\big)(h,n)\, ds.\end{aligned} \tag{11.3.15}$$

This example also comprises the shape optimization problem of Bernoulli type:

$$\min J(y, \Omega, \Gamma) \equiv \min_\Gamma \int_\Gamma y^2\, ds,$$

where y is the solution of the mixed boundary value problem

$$\begin{aligned} -\Delta y &= f && \text{in } \Omega, \\ y &= 0 && \text{on } \partial U, \\ \frac{\partial y}{\partial n} &= g && \text{on } \Gamma, \end{aligned}$$

which was analyzed with a similar approach in [IKP]. Here the boundary $\partial\Omega$ of the domain $\Omega \subset \mathbb{R}^2$ is the disjoint union of a fixed part ∂U and an unknown part Γ both with nonempty relative interior. Let the state space X be given by

$$X = \{\varphi \in H^1(\Omega) \colon \varphi = 0 \text{ on } \partial U\}.$$

Then the Eulerian derivative of J is given by (11.3.15), which reduces to

$$\hat{J}'(\Omega)h = \int_\Gamma \left(-(\nabla y, \nabla p) + fp + \frac{\partial}{\partial n}(y^2 + gp) + \kappa(y^2 + gp) \right) (h, n)\, ds.$$

This result coincides with the representation obtained in [IKP]. The present derivation, however, is considerably simpler due to a better arrangement of terms in the proof of Theorem 11.3. It is straightforward to adapt the framework to shape optimization problems associated with the exterior Bernoulli problem.

11.3.4 Navier–Stokes system

Consider the stationary Navier–Stokes equations

$$\begin{aligned} -\nu\, \Delta y + (y \cdot \nabla) y + \nabla p &= f && \text{in } \Omega, \\ \operatorname{div} y &= 0 && \text{in } \Omega, \\ y &= 0 && \text{on } \partial\Omega \end{aligned} \tag{11.3.16}$$

on a bounded domain $\Omega \subset \mathbb{R}^d$, $d = 2, 3$, with $\nu > 0$ and $f \in H^1(U)$. In the context of the general framework we set $\Omega = D$ with C^2-boundary $\Gamma = \partial\Omega$. The variational formulation of (11.3.16) is given by

Find $(y, p) \in X \equiv H_0^1(\Omega)^d \times L^2(\Omega)/\mathbb{R}$ such that

$$\begin{aligned} \big\langle e\big((y, p), \Omega\big), (\psi, \chi)\big\rangle_{X^* \times X} &\equiv \nu(\nabla y, \nabla\psi)_\Omega + ((y \cdot \nabla) y, \psi)_\Omega \\ &- (p, \operatorname{div} \psi)_\Omega - (f, \psi)_\Omega + (\operatorname{div} y, \chi)_\Omega = 0 \end{aligned} \tag{11.3.17}$$

holds for all $(\psi, \chi) \in X$. Let the cost functional J be given by

$$J(y, \Omega) = \int_\Omega j_1(y)\, dx.$$

Considering (11.3.17) on a perturbed domain Ω_t mapping the equation back to the reference domain Ω yields the form of $\tilde{e}(y, t)$. Concerning the transformation of the divergence we note that for $\psi_t \in H_0^1(\Omega_t)^d$ and $\psi^t = \psi_t \circ F_t \in H_0^1(\Omega)^d$, one obtains

$$\operatorname{div} \psi_t = (D\psi_i^t A_t^T e_i) \circ F_t^{-1} = ((A_t)_i \nabla \psi_{t,i}) \circ F_t^{-1},$$

where e_i stands for the ith canonical basis vector in $\mathbb{R}^d$ and $(A_t)_i$ denotes the ith row of $A_t = (DF_t)^{-T}$. We follow the convention to sum over indices which occur at least twice in a term. Thus one obtains

$$\begin{aligned}\big\langle \tilde{e}\big((y^t, p^t), t\big), (\psi, \chi)\big\rangle_{X^* \times X} &= \nu(I_t A_t \nabla y^t, A_t \nabla \psi)_\Omega + \big((y^t \cdot A_t \nabla) y^t, I_t \psi\big)_\Omega \\ &- \big(p^t, I_t (A_t)_k \nabla \psi_k\big)_\Omega - (f^t I_t, \psi)_\Omega + \big(I_t (A_t)_k \nabla y^t_k, \chi\big)_\Omega = 0\end{aligned}$$

for all $(\psi, \chi) \in X$.

The adjoint state $(\lambda, q) \in X$ is given by the solution to

$$\big\langle e'\big((y, p), \Omega\big)(\psi, \chi), (\lambda, q)\big\rangle_{X^* \times X} = (j_1'(y), \psi)_\Omega,$$

which amounts to

$$\begin{aligned}&\nu(\nabla \psi, \nabla \lambda)_\Omega + ((\psi \cdot \nabla) y + (y \cdot \nabla)\psi, \lambda)_\Omega \\ &\qquad - (\chi, \operatorname{div} \lambda)_\Omega + (\operatorname{div} \psi, q)_\Omega = (j_1'(y), \psi)_\Omega\end{aligned} \tag{11.3.18}$$

for all $(\psi, \chi) \in X$. Integrating by parts one obtains

$$\begin{aligned}((y \cdot \nabla)\psi, \lambda)_\Omega &= -\int_\Omega \psi \cdot \lambda \operatorname{div} y \, dx - \int_\Omega \psi \cdot ((y \cdot \nabla)\lambda) \, dx \\ &+ \int_\Gamma (\psi \cdot \lambda)(y \cdot n) \, ds = -(\psi, (y \cdot \nabla)\lambda)_\Omega\end{aligned}$$

because $y \in H_0^1(\Omega)^d$ and $\operatorname{div} y = 0$. Therefore

$$((\psi \cdot \nabla) y + (y \cdot \nabla)\psi, \lambda)_\Omega = (\psi, (\nabla y)\lambda - (y \cdot \nabla)\lambda)_\Omega \tag{11.3.19}$$

holds for all $\psi \in H^1(\Omega)^d$. As a consequence the adjoint equation can be interpreted as

$$\begin{aligned}-\nu \Delta \lambda + (\nabla y)\lambda - (y \cdot \nabla)\lambda - \nabla q &= j_1'(y), \\ \operatorname{div} \lambda &= 0,\end{aligned} \tag{11.3.20}$$

where the first equation holds in $L^2(\Omega)^d$ and the second one in $L^2(\Omega)$.

For the evaluation of $\frac{d}{dt}\langle \tilde{e}\big((y, p), t\big), (\lambda, q)\rangle_{X^* \times X}|_{t=0}$, (y, p), $(\lambda, q) \in X$ being the solution of (11.3.17), respectively, (11.3.18), we transform this expression back to Ω_t, which gives

$$\begin{aligned}\big\langle \tilde{e}\big((y, p), t\big), (\lambda, q)\big\rangle_{X^* \times X} &= \nu(\nabla(y \circ F_t^{-1}), \nabla(\lambda \circ F_t^{-1}))_{\Omega_t} \\ &+ \big((y \circ F_t^{-1} \cdot \nabla) y \circ F_t^{-1}, \lambda \circ F_t^{-1}\big)_{\Omega_t} - (\operatorname{div}(\lambda \circ F_t^{-1}), p \circ F_t^{-1})_{\Omega_t} \\ &- (f, \lambda \circ F_t^{-1})_{\Omega_t} + (\operatorname{div}(y \circ F_t^{-1}), q \circ F_t^{-1})_{\Omega_t}.\end{aligned}$$

To verify conditions (H1)–(H4) we introduce the continuous trilinear form $c\colon H_0^1(\Omega)^d \times H_0^1(\Omega)^d \times H_0^1(\Omega)^d$ by $c(y, v, w) = ((y \cdot \nabla)v, w)$ and assume that

$$\nu^2 > \mathcal{N} |f|_{H^{-1}} \quad \text{and} \quad \nu > \mathcal{M}, \tag{11.3.21}$$

where

$$\mathcal{N} = \sup_{y,v,w\in H_0^1} \frac{c(y,v,w)}{|y|_{H_0^1}|v|_{H_0^1}|w|_{H_0^1}} \quad \text{and} \quad \mathcal{M} = \sup_{v\in H_0^1} \frac{c(v,v,y)}{|v|^2_{H_0^1}},$$

with y the solution to (11.3.16). Condition (H1) is satisfied by construction. If ν is sufficiently large so that the first inequality in (11.3.21) is satisfied, existence of a unique solution $(y, p) \in H_0^1(\Omega)^d \times L^2(\Omega)/\mathbb{R}$ to (11.3.16) is guaranteed; see, e.g., [Te]. The second condition in (11.3.21) ensures the bijectivity of the linearized operator $e'(y, p)$, and thus (H6) holds. In particular this implies that (H2) holds and that the adjoint equation admits a unique solution. To verify (H3) we consider for arbitrary $(v, q) \in X$ and $(\psi, \chi) \in X$

$$\begin{aligned}
&\langle e((v,q),\Omega) - e((y,p),\Omega) - e'((y,p),\Omega)((v,q)-(y,p)), (\psi,\chi)\rangle_{X^*,X} \\
&\quad = ((v-y)\cdot\nabla(v-y), \psi) \le K|\psi|_{H_0^1(\Omega)}|v-y|^2_{H_0^1(\Omega)},
\end{aligned}$$

where K is an embedding constant, independent of $(v, q) \in X$ and $(\psi, \chi) \in X$. Verifying (H4) requires us to consider the quotient of the following expression with t and taking the limit as $t \to 0$:

$$\begin{aligned}
&\nu[(I_t A_t\nabla(y^t - y), A_t\nabla\psi) - (\nabla(y^t - y), \nabla\psi)] \\
&\quad + [((y^t\cdot A_t\nabla)y^t, I_t\psi) - ((y^t\cdot\nabla)y^t, \psi) - ((y\cdot A_t\nabla)y, I_t\psi) + ((y\cdot\nabla)y,\psi)] \\
&\quad - [(I_t(A_t)_k\nabla\psi_k, p^t - p) + (\operatorname{div}\psi, p^t - p)] \\
&\quad + [(I_t(A_t)_k\nabla\,(y_k^t - y_k), \chi) - (\operatorname{div}(y^t - y), \chi)]
\end{aligned}$$

for $(\psi, \chi) \in X$. The first two terms in square brackets can be treated by analogous estimates as in Sections 11.3.1 and 11.3.2. Noting that the third and fourth square brackets can be estimated quite similarly to each other, we give the estimate for the last one:

$$\begin{aligned}
&((I_t - 1)(A_t)_k\nabla(y_k^t - y_k), \chi) + (((A_t)_k - e_k)\nabla(y_k^t - y_k), \chi) \\
&\quad (e_k\nabla(y_k^t - y_k) - \operatorname{div}(y^t - y), \chi),
\end{aligned}$$

which, upon division by t, tends to 0 for $t \to 0$.

In the following calculation we utilize that $(y, p), (\lambda, q) \in H^2(\Omega)^d \times H^1(\Omega)$, which is satisfied if Γ is C^2. Applying Lemma 11.5 results in

$$\begin{aligned}
&\frac{d}{dt}\langle \tilde{e}((y,p),t), (\lambda,q)\rangle_{X^*\times X}|_{t=0} \\
&\quad = \nu(\nabla(-\nabla y^T h), \nabla\lambda)_\Omega + \nu(\nabla y, \nabla(-\nabla\lambda^T h))_\Omega + \nu\int_\Gamma (\nabla y, \nabla\lambda)\,(h,n)\,ds \\
&\quad + \Big(((-\nabla y^T h)\cdot\nabla)y, \lambda\Big)_\Omega + \big((y\cdot\nabla)(-\nabla y^T h), \lambda\big)_\Omega \\
&\quad + \big((y\cdot\nabla)y, -\nabla\lambda^T h\big)_\Omega + \int_\Gamma \big((y\cdot\nabla)y, \lambda\big)\,(h,n)\,ds \\
&\quad - (-\nabla p^T h, \operatorname{div}\lambda)_\Omega - (p, \operatorname{div}(-\nabla\lambda^T h))_\Omega - \int_\Gamma p\operatorname{div}\lambda\,(h,n)\,ds \\
&\quad - (f, -\nabla\lambda^T h)_\Omega - \int_\Gamma f\lambda\,(h,n)\,ds \\
&\quad + (\operatorname{div}(-\nabla y^T h), q)_\Omega + (\operatorname{div} y, -\nabla q^T h)_\Omega + \int_\Gamma q\operatorname{div} y\,(h,n)\,ds.
\end{aligned}$$

Since $\operatorname{div} y = \operatorname{div} \lambda = 0$ and $y, \lambda \in H_0^1(\Omega)^d$, this expression simplifies to

$$\begin{aligned}
&\frac{d}{dt}\langle \tilde{e}\big((y,p),t\big),(\lambda,q)\rangle_{X^*\times X}|_{t=0} \\
&\quad = \nu(\nabla y, \nabla\psi_\lambda)_\Omega + ((y\cdot\nabla)y,\psi_\lambda)_\Omega - (p, \operatorname{div}\psi_\lambda)_\Omega - (f,\psi_\lambda)_\Omega \\
&\quad + \nu(\nabla\psi_y,\nabla\lambda)_\Omega + ((\psi_y\cdot\nabla)y + (y\cdot\nabla)\psi_y,\lambda)_\Omega \\
&\quad + (\operatorname{div}\psi_y, q)_\Omega + \nu\int_\Gamma (\nabla y,\nabla\lambda)\,(h,n)\,ds,
\end{aligned}$$

where we have used the abbreviation

$$\psi_y = -(\nabla y)^T h, \qquad \psi_\lambda = -(\nabla\lambda)^T h.$$

Note that $\psi_y, \psi_\lambda \in H^1(\Omega)^d$ but not in $H_0^1(\Omega)^d$. Green's formula, together with (11.3.16), (11.3.20), entails

$$\begin{aligned}
&\frac{d}{dt}\langle \tilde{e}\big((y,p),t\big),(\lambda,q)\rangle_{X^*\times X}|_{t=0} \\
&= (-\nu\Delta y + (y\cdot\nabla y)y + \nabla p - f, \psi_\lambda)_\Omega + \int_\Gamma \nu\left(\frac{\partial y}{\partial n},\psi_\lambda\right)ds + \int_\Gamma p\,(\psi_\lambda,n)\,ds \\
&+ (\psi_y, -\nu\Delta\lambda + (\nabla y)\lambda - (y\cdot\nabla)\lambda - \nabla q)_\Omega + \int_\Gamma \nu\left(\frac{\partial\lambda}{\partial n},\psi_y\right)ds \\
&+ \int_\Gamma q\,(\psi_y,n)\,ds + \nu\int_\Gamma(\nabla y,\nabla\lambda)\,(h,n)\,ds \\
&= -\int_\Gamma \nu\left(\frac{\partial y}{\partial n},(\nabla\lambda)^T h\right)ds - \int_\Gamma p\,((\nabla\lambda)^T h,n)\,ds - \int_\Gamma \nu\left(\frac{\partial\lambda}{\partial n},(\nabla y)^T h\right)ds \\
&- \int_\Gamma q\,((\nabla y)^T h,n)\,ds + \nu\int_\Gamma(\nabla y,\nabla\lambda)\,(h,n)\,ds - (j_1'(y),(\nabla y)^T h)_\Omega \\
&= -\int_\Gamma\left(\nu\left(\frac{\partial y}{\partial n},\frac{\partial\lambda}{\partial n}\right) + p\left(\frac{\partial\lambda}{\partial n},n\right) + q\left(\frac{\partial y}{\partial n},n\right)\right)(h,n)\,ds - (j_1'(y),(\nabla u)^T h)_\Omega.
\end{aligned}$$

Arguing as in Section 11.3.3 one eventually obtains by Theorem 11.3

$$\begin{aligned}
\hat{J}'(\Omega)h &= \int_\Gamma\left(\nu\left(\frac{\partial y}{\partial n},\frac{\partial\lambda}{\partial n}\right) + p\left(\frac{\partial\lambda}{\partial n},n\right) + q\left(\frac{\partial y}{\partial n},n\right)\right)(h,n)\,ds \\
&+ \int_\Omega (j_1(y)\operatorname{div} h + j_1'(y)\nabla y^T h)\,dx \\
&= \int_\Gamma\left(\nu\left(\frac{\partial y}{\partial n},\frac{\partial\lambda}{\partial n}\right) + p\left(\frac{\partial\lambda}{\partial n},n\right) + q\left(\frac{\partial y}{\partial n},n\right) + j_1(y)\right)(h,n)\,ds.
\end{aligned}$$

Bibliography

[ACFK] J. Albert, C. Carstensen, S. A. Funken, and R. Close, *MATLAB implementation of the finite element method in elasticity*, Computing 69(2002), 239–263.

[AcVo] R. Acar and C. R. Vogel, *Analysis of bounded variation penalty methods for ill-posed problems*, Inverse Problems 10(1994), 1217–1229.

[Ad] R. Adams, *Sobolev Spaces*, Academic Press, Boston, 1975.

[AlMa] W. Alt and K. Malanowski, *The Lagrange-Newton method for nonlinear optimal control problems*, Comput. Optim. Appl. 2(1993), 77–100.

[Alt1] W. Alt, *Stabilität mengenwertiger Abbildungen mit Anwendungen auf nichtlineare Optimierungsprobleme*, Bayreuther Mathematische Schriften 3, 1979.

[Alt2] W. Alt, *Lipschitzian perturbations in infinite optimization problems*, in Mathematical Programming with Data Perturbations II, A.V. Fiacco, ed., Lecture Notes in Pure and Appl. Math. 85, Marcel Dekker, New York, 1983, 7–21.

[Alt3] W. Alt, *Stability of Solutions and the Lagrange-Newton Method for Nonlinear Optimization and Optimal Control Problems*, Habilitation thesis, Bayreuth, 1990.

[Alt4] W. Alt, *The Lagrange-Newton method for infinite-dimensional optimization problems*, Numer. Funct. Anal. Optimiz. 11(1990), 201–224.

[Alt5] W. Alt, *Sequential quadratic programming in Banach spaces*, in Advances in Optimization, W. Oettli and D. Pallaschke, eds., Lecture Notes in Econom. and Math. Systems 382, Springer-Verlag, Berlin, 1992, 281–301.

[Ba] V. Barbu, *Analysis and Control of Nonlinear Infinite Dimensional Systems*, Academic Press, Boston, 1993.

[BaHe] A. Battermann and M. Heinkenschloss, *Preconditioners for Karush-Kuhn-Tucker matrices arising in the optimal control of distributed systems*, in Control and Estimation of Distributed Parameter Systems, Vorau (1996), Internat. Ser. Numer. Math. 126, Birkhäuser, Basel, 1998, 15–32.

[BaK1] V. Barbu and K. Kunisch, *Identification of nonlinear elliptic equations*, Appl. Math. Optim. 33(1996), 139–167.

[BaK2] V. Barbu and K. Kunisch, *Identification of nonlinear parabolic equations*, Control Theory Adv. Tech. 10(1995), 1959–1980.

[BaKRi] V. Barbu, K. Kunisch, and W. Ring, *Control and estimation of the boundary heat transfer function in Stefan problems*, RAIRO Modél Math. Anal. Numér. 30(1996), 1–40.

[BaPe] V. Barbu and Th. Percupanu, *Convexity and Optimization in Banach Spaces*, Reidel, Dordrecht, 1986.

[BaSa] A. Battermann and E. Sachs, *An indefinite preconditioner for KKT systems arising in optimal control problems*, in Fast Solution of Discretized Optimization Problems, Berlin, 2000, Internat. Ser. Numer. Math. 138, Birkhäuser, Basel, 2001, 1–18.

[Be] D. P. Bertsekas, *Constrained Optimization and Lagrange Multiplier Methods*, Academic Press, Paris, 1982.

[BeK1] M. Bergounioux and K. Kunisch, *Augmented Lagrangian techniques for elliptic state constrained optimal control problems*, SIAM J. Control Optim. 35(1997), 1524–1543.

[BeK2] M. Bergounioux and K. Kunisch, *Primal-dual active set strategy for state constrained optimal control problems*, Comput. Optim. Appl. 22(2002), 193–224.

[BeK3] M. Bergounioux and K. Kunisch, *On the structure of the Lagrange multiplier for state-constrained optimal control problems*, Systems Control Lett. 48(2002), 16–176.

[BeMeVe] R. Becker, D. Meidner, and B. Vexler, *Efficient numerical solution of parabolic optimization problems by finite element methods*, Optim. Methods Softw. 22(2007), 813–833.

[BePl] A. Berman and R. J. Plemmons, *Nonnegative Matrices in the Mathematical Sciences*, Computer Science and Scientific Computing Series, Academic Press, New York, 1979.

[Ber] M. Berggren, *A Unified Discrete-Continuous Sensitivity Analysis Method for Shape Optimization*, Lecture at the Radon Institut, Linz, Austria, 2005.

[Bew] T. Bewley, *Flow control: New challenges for a new renaissance*, Progress in Aerospace Sciences 37(2001), 24–58.

[BGHW] L. T. Biegler, O. Ghattas, M. Heinkenschloss, and B. van Bloemen Waanders, eds., *Large-Scale PDE-Constrained Optimization*, Lecture Notes in Comput. Sci. Eng. 30, Springer-Verlag, Berlin, 2003.

[BGHKW] L. T. Biegler, O. Ghattas, M. Heinkenschloss, D. Keyes, and B. van Bloemen Waanders, eds., *Real-Time PDE-Constrained Optimization*, SIAM, Philadelphia, 2007.

[BHHK] M. Bergounioux, M. Haddou, M. Hintermüller, and K. Kunisch, *A comparison of a Moreau–Yosida-based active set strategy and interior point methods for constrained optimal control problems*, SIAM J. Optim. 11(2000), 495–521.

[BIK] M. Bergounioux, K. Ito, and K. Kunisch, *Primal-dual strategy for constrained optimal control problems*, SIAM J. Control Optim. 37(1999), 1176–1194.

[BKW] A. Borzì, K. Kunisch, and D. Y. Kwak, *Accuracy and convergence properties of the finite difference multigrid solution of an optimal control optimality system*, SIAM J. Control Optim. 41(2003), 1477–1497.

[BoK] A. Borzì and K. Kunisch, *The numerical solution of the steady state solid fuel ignition model and its optimal control*, SIAM J. Sci. Comput. 22(2000), 263–284.

[BoKVa] A. Borzì, K. Kunisch, and M. Vanmaele, *A multigrid approach to the optimal control of solid fuel ignition problems*, in Multigrid Methods VI, Lect. Notes Comput. Sci. Eng. 14, Springer-Verlag, Berlin, 59–65.

[Bre] H. Brezis, *Opérateurs Maximaux Monotones et Semi-groupes de Constraction das le Espaces de Hilbert*, North–Holland, Amsterdam, 1973.

[Bre2] H. Brezis, *Problèmes unilatéraux*, J. Math. Pures Appl. 51(1972), 1–168.

[CaTr] E. Casas and F. Tröltzsch, *Second-order necessary and sufficient optimality conditions for optimization problems and applications to control theory*, SIAM J. Optim. 13(2002), 406–431.

[ChK] G. Chavent and K. Kunisch, *Regularization of linear least squares problems by total bounded variation*, ESAIM Control Optim. Calc. Var. 2(1997), 359–376.

[ChLi] A. Chambolle and P.-L. Lions, *Image recovery via total bounded variation minimization and related problems*, Numer. Math. 76(1997), 167–188.

[ChZo] Z. Chen and J. Zou, *Finite element methods and their convergence for elliptic and parabolic interface problems*, Numer. Math. 79(1998), 175–202.

[Ci] P. G. Ciarlet, *Mathematical Elasticity, Vol.* 1, North–Holland, Amsterdam, 1987.

[CKP] E. Casas, K. Kunisch, and C. Pola, *Regularization by functions of bounded variation and applications to image enhancement*, Appl. Math. Optim. 40(1999), 229–258.

[Cla] F. H. Clarke, *Optimization and Nonsmooth Analysis*, John Wiley and Sons, New York, 1983.

[CNQ] X. Chen, Z. Nashed, and L. Qi, *Smoothing methods and semismooth methods for nondifferentiable operator equations*, SIAM J. Numer. Anal. 38(2000), 1200–1216.

[CoKu] F. Colonius and K. Kunisch, *Output least squares stability for parameter estimation in two point value problems*, J. Reine Angew. Math. 370(1986), 1–29.

[DaLi] R. Dautray and J. L. Lions, *Mathematical Analysis and Numerical Methods for Science and Technology, Vol.* 3, Springer-Verlag, Berlin, 1990.

[Dei] K. Deimling, *Nonlinear Functional Analysis*, Springer-Verlag, Berlin, 1985.

[DeZo] M. C. Delfour and J.-P. Zolésio, *Shapes and Geometries: Analysis, Differential Calculus, and Optimization*, SIAM, Philadelphia, 2001.

[Don] A. L. Dontchev, *Local analysis of a Newton-type method based on partial elimination*, in The Mathematics of Numerical Analysis, Lectures in Appl. Math. 32, AMS, Providence, RI, 1996, 295–306.

[DoSa] D. C. Dobson and F. Santosa, *Recovery of blocky images from noisy and blurred data*, SIAM J. Appl. Math. 56(1996), 1181–1198.

[dReKu] C. de los Reyes and K. Kunisch, *A comparison of algorithms for control constrained optimal control of the Burgers equation*, Calcolo 41(2004), 203–225.

[EcJa] C. Eck and J. Jarusek, *Existence results for the static contact problem with Coulomb friction*, Math. Models Methods Appl. Sci. 8(1998), 445–468.

[EkTe] I. Ekeland and R. Temam, *Convex Analysis and Variational Problems*, North–Holland, Amsterdam, 1976.

[EkTu] I. Ekeland and T. Turnbull, *Infinite Dimensional Optimization and Convexity*, The University of Chicago Press, Chicago, 1983.

[Fat] H. O. Fattorini, *Infinite Dimensional Optimization and Control Theory*, Cambridge University Press, Cambridge, 1999.

[FGH] A. V. Fursikov, M. D. Gunzburger, and L. S. Hou, *Boundary value problems and optimal boundary control for the Navier–Stokes systems: The two-dimensional case*, SIAM J. Control Optim. 36(1998), 852–894.

[FoGl] M. Fortin and R. Glowinski, *Augmented Lagrangian Methods: Applications to Numerical Solutions of Boundary Value Problems*, North–Holland, Amsterdam, 1983.

[Fr] A. Friedman, *Variational Principles and Free Boundary Value Problems*, John Wiley and Sons, New York, 1982.

[Geo] V. Georgescu, *On the unique continuation property for Schrödinger Hamiltonians*, Helv. Phys. Acta 52(1979), 655–670.

[GeYa] D. Geman and C. Yang, *Nonlinear image recovery with half-quadratic regularization*, IEEE Trans. Image Process. 4(1995), 932–945.

[GHS] M. D. Gunzburger, L. Hou, and T. P. Svobodny, *Finite element approximations of optimal control problems associated with the scalar Ginzburg-Landau equation*, Comput. Math. Appl. 21(1991), 123–131.

[GiRa] V. Girault and P.-A. Raviart, *Finite Element Methods for Navier–Stokes Equations*, Springer-Verlag, Berlin, 1986.

[Giu] E. Giusti, *Minimal Surfaces and Functions of Bounded Variation*, Birkhäuser, Boston, 1984.

[Glo] R. Glowinski, *Numerical Methods for Nonlinear Variational Problems*, Springer-Verlag, Berlin, 1984.

[GLT] R. Glowinski, J. L. Lions, and R. Tremoliers, *Numerical Analysis of Variational Inequalities*, North–Holland, Amsterdam, 1981.

[Gri] P. Grisvard, *Elliptic Problems in Nonsmooth Domains*, Monographs Stud. Math. 24, Pitman, Boston, MA, 1985.

[Grie] A. Griewank, *The local convergence of Broyden-like methods on Lipschitzian problems in Hilbert spaces*, SIAM J. Numer. Anal. 24(1987), 684–705.

[GrVo] R. Griesse and S. Volkwein, *A primal-dual active set strategy for optimal boundary control of a nonlinear reaction-diffusion system*, SIAM J. Control Optim. 44(2005), 467–494.

[Gu] M. D. Gunzburger, *Perspectives of Flow Control and Optimization*, SIAM, Philadelphia, 2003.

[Han] S.-P. Han, *Superlinearly convergent variable metric algorithms for general nonlinear programming problems*, Math. Programming 11(1976), 263–282.

[HaMä] J. Haslinger and R. A. E. Mäkinen, *Introduction to Shape Optimization*, SIAM, Philadelphia, 2003.

[HaNe] J. Haslinger and P. Neittaanmäki, *Finite Element Approximation for Optimal Shape, Material and Topology Design*, 2nd ed, Wiley, Chichester, 1996.

[HaPaRa] S.-H. Han, J.-S. Pang, and N. Rangaray, *Glabally convergent Newton methods for nonsmooth equations*, Math. Oper. Res. 17(1992), 586–607.

[Har] A. Haraux, *How to differentiate the projection on a closed convex set in Hilbert space. Some applications to variational inequalities*, J. Math. Soc. Japan 29(1977), 615–631.

[Has] J. Haslinger, *Approximation of the Signorini problem with friction, obeying the Coulomb law*, Math. Methods Appl. Sci. 5(1983), 422–437.

[Hei] M. Heinkenschloss, *Projected sequential quadratic programming methods*, SIAM J. Optim. 6(1996), 373–417.

[Hes1] M. R. Hestenes, *Optimization Theory. The Finite Dimensional Case*, John Wiley and Sons, New York, 1975.

[Hes2] M. R. Hestenes, *Multiplier and gradient methods*, J. Optim. Theory Appl. 4(1968), 303–320.

[HHNL] I. Hlavacek, J. Haslinger, J. Necas, and J. Lovisek, *Solution of Variational Inequalities in Mechanics*, Appl. Math. Sci. 66, Springer-Verlag, New York, 1988.

[HiIK] M. Hintermüller, K. Ito, and K. Kunisch, *The primal-dual active set strategy as a semismooth Newton method*, SIAM J. Optim. 13(2002), 865–888.

[HiK1] M. Hinze and K. Kunisch, *Three control methods for time-dependent fluid flow*, Flow Turbul. Combust. 65(2000), 273–298.

[HiK2] M. Hinze and K. Kunisch, *Second order methods for optimal control of time-dependent fluid flow*, SIAM J. Control Optim. 40(2001), 925–946.

[HinK1] M. Hintermüller and K. Kunisch, *Total bounded variation regularization as a bilaterally constrained optimization problem*, SIAM J. Appl. Math. 64(2004), 1311–1333.

[HinK2] M. Hintermüller and K. Kunisch, *Feasible and noninterior path-following in constrained minimization with low multiplier regularity*, SIAM J. Control Optim. 45(2006), 1198–1221.

[HPR] S.-P. Han, J.-S. Pang, and N. Rangaraj, *Globally convergent Newton methods for nonsmooth equations*, Math. Oper. Res. 17(1992), 586–607.

[HuStWo] S. Hüeber, G. Stadler, and B. I. Wohlmuth, *A primal-dual active set algorithm for three-dimensional contact problems with Coulomb friction*, SIAM J. Sci. Comput. 30(2008), 572–596.

[IK1] K. Ito and K. Kunisch, *The augmented Lagrangian method for equality and inequality constraints in Hilbert space*, Math. Programming 46(1990), 341–360.

[IK2] K. Ito and K. Kunisch, *The augmented Lagrangian method for parameterization in elliptic systems*, SIAM J. Control Optim. 28(1990), 113–136.

[IK3] K. Ito and K. Kunisch, *The augmented Lagrangian method for estimating the diffusion coefficient in an elliptic equation*, in Proceedings of the 26th IEEE Conference on Decision and Control, Los Angeles, 1987, 1400–1404.

[IK4] K. Ito and K. Kunisch, *An augmented Lagrangian technique for variational inequalities*, Appl. Math. Optim. 21(1990), 223–241.

[IK5] K. Ito and K. Kunisch, *Sensitivity analysis of solutions to optimization problems in Hilbert spaces with applications to optimal control and estimation*, J. Differential Equations 99(1992), 1–40.

[IK6] K. Ito and K. Kunisch, *On the choice of the regularization parameter in nonlinear inverse problems*, SIAM J. Optim. 2(1992), 376–404.

[IK7] K. Ito and K. Kunisch, *Sensitivity measures for the estimation of parameters in 1-D elliptic boundary value problems*, J. Math. Systems Estim., Control 6(1996), 195–218.

[IK8] K. Ito and K. Kunisch, *Maximizing robustness in nonlinear illposed inverse problems*, SIAM J. Control Optim. 33(1995), 643–666.

[IK9] K. Ito and K. Kunisch, *Augmented Lagrangian-SQP-methods in Hilbert spaces and application to control in the coefficients problems*, SIAM J. Optim. 6(1996), 96–125.

[IK10] K. Ito and K. Kunisch, *Augmented Lagrangian-SQP methods for nonlinear optimal control problems of tracking type*, SIAM J. Control Optim. 34(1996), 874–891.

[IK11] K. Ito and K. Kunisch, *Augmented Lagrangian methods for nonsmooth convex optimization in Hilbert spaces*, Nonlinear Anal. 41(2000), 573–589.

[IK12] K. Ito and K. Kunisch, *An active set strategy based on the augmented Lagrangian formulation for image restoration*, MZAN Math. Model Numer. Anal. 33(1999), 1–21.

[IK13] K. Ito and K. Kunisch, *Augmented Lagrangian formulation of nonsmooth, convex optimization in Hilbert spaces*, in Control of Partial Differential Equations, E. Casas, ed., Lecture Notes in Pure and Appl. Math. 174, Marcel Dekker, New York, 1995, 107–117.

[IK14] K. Ito and K. Kunisch, *Estimation of the convection coefficient in elliptic equations*, Inverse Problems 13(1997), 995–1013.

[IK15] K. Ito and K. Kunisch, *Newton's method for a class of weakly singular optimal control problems*, SIAM J. Optim. 10(1999), 896–916.

[IK16] K. Ito and K. Kunisch, *Optimal control of elliptic variational inequalities*, Appl. Math. Optim. 41(2000), 343–364.

[IK17] K. Ito and K. Kunisch, *Optimal control of the solid fuel ignition model with H^1-cost*, SIAM J. Control Optim. 40(2002), 1455–1472.

[IK18] K. Ito and K. Kunisch, *BV-type regularization methods for convoluted objects with edge, flat and grey scales*, Inverse Problems 16(2000), 909–928.

[IK19] K. Ito and K. Kunisch, *Optimal control*, in Encyclopedia of Electrical and Electronic Engineering, J. G. Webster, ed., 15, John Wiley and Sons, New York, 1999, 364–379.

[IK20] K. Ito and K. Kunisch, *Semi-smooth Newton methods for variational inequalities of the first kind*, MZAN Math. Model. Numer. Anal. 37(2003), 41–62.

[IK21] K. Ito and K. Kunisch, *The primal-dual active set method for nonlinear optimal control problems with bilateral constraints*, SIAM J. Control Optim. 43(2004), 357–376.

[IK22] K. Ito and K. Kunisch, *Semi-smooth Newton methods for state-constrained optimal control problems*, Systems Control Lett. 50(2003), 221–228.

[IK23] K. Ito and K. Kunisch, *Parabolic variational inequalities: The Lagrange multiplier approach*, J. Math. Pures Appl. (9) 85(2006), 415–449.

[IKK] K. Ito, M. Kroller, and K. Kunisch, *A numerical study of an augmented Lagrangian method for the estimation of parameters in elliptic systems*, SIAM J. Sci. Statist. Comput. 12(1991), 884–910.

[IKP] K. Ito, K. Kunisch, and G. Peichl, *Variational approach to shape derivatives for a class of Bernoulli problems*, J. Math. Anal. Appl. 314(2006), 126–149.

[IKPe] K. Ito, K. Kunisch, and G. Peichl, *On the regularization and the numerical treatment of the inf-sup condition for saddle point problems*, Comput. Appl. Math. 21(2002), 245–274.

[IKPe2] K. Ito, K. Kunisch, and G. Peichl, *A variational approach to shape derivatives*, to appear in ESAIM Control Optim. Calc. Var.

[IKSG] K. Ito, K. Kunisch, V. Schulz, and I. Gherman, *Approximate Nullspace Iterations for KKT Systems in Model Based Optimization*, preprint, University of Trier, 2008.

[ItKa] K. Ito and F. Kappel, *Evolution Equations and Approximations*, World Scientific, River Edge, NJ, 2002.

[Ja] J. Jahn, *Introduction to the Theory of Nonlinear Optimization*, Springer-Verlag, Berlin, 1994.

[JaSa] H. Jäger and E. W. Sachs, *Global convergence of inexact reduced SQP methods*, Optim. Methods Softw. 7(1997), 83–110.

[Ka] K. Kato, *Perturbation Theory for Linear Operators*, Springer-Verlag, Berlin, 1980.

[KaK] A. Kauffmann and K. Kunisch, *Optimal control of the solid fuel ignition model*, ESAIM Proc. 8(2000), 65–76.

[Kan] C. Kanzow, *Inexact semismooth Newton methods for large-scale complementarity problems*, Optim. Methods Softw. 19(2004), 309–325.

[KeSa] C. T. Kelley and E. W. Sachs, *Solution of optimal control problems by a pointwise projected Newton method*, SIAM J. Control Optim. 33(1995), 1731–1757.

[KO] N. Kikuchi and J. T. Oden, *Contact Problems in Elasticity: A Study of Variational Inequalities and Finite Element Methods*, SIAM Stud. Appl. Math. 8, SIAM, Philadelphia, 1988.

[Kou] S. G. Kou, *A jump-diffusion model for option pricing*, Management Sci. 48(2002), 1086–1101.

[KPa] K. Kunisch and X. Pan, *Estimation of interfaces from boundary measurements*, SIAM J. Control Optim. 32(1994), 1643–1674.

[KPe] K. Kunisch and G. Peichl, *Estimation of a temporally and spatially varying diffusion coefficient in a parabolic system by an augmented Lagrangian technique*, Numer. Math. 59(1991), 473–509.

[KRe] K. Kunisch and F. Rendl, *An infeasible active set method for quadratic problems with simple bounds*, SIAM J. Optim. 14(2003), 35–52.

[KRo] K. Kunisch and A. Rösch, *Primal-dual active set strategy for a general class of optimal control problems*, to appear in SIAM J. Optim.

[KSa] K. Kunisch and E. W. Sachs, *Reduced SQP methods for parameter identification problems*, SIAM J. Numer. Anal. 29(1992), 1793–1820.

[Kup] F.-S. Kupfer, *An infinite-dimensional convergence theory for reduced SQP methods in Hilbert space*, SIAM J. Optim. 6(1996), 126–163.

[KuSa] F.-S. Kupfer and E. W. Sachs, *Numerical solution of a nonlinear parabolic control problem by a reduced SQP method*, Comput. Optim. Appl. 1(1992), 113–135.

[KuSt] K. Kunisch and G. Stadler, *Generalized Newton methods for the 2D-Signorini contact problem with friction in function space*, MZAN Math. Model. Numer. Anal. 39(2005), 827–854.

[KuTa] K. Kunisch and X-. Tai, *Sequential and parallel splitting methods for bilinear control problems in Hilbert spaces*, SIAM J. Numer. Anal. 34(1997), 91–118.

[KVo1] K. Kunisch and S. Volkwein, *Augmented Lagrangian-SQP techniques and their approximation*, in Optimization Methods in Partial Differential Equations, Contemp. Math. 209, AMS, Providence, RI, 1997, 147–160.

[KVo2] K. Kunisch and S. Volkwein, *Galerkin proper orthogonal decomposition methods for parabolic systems*, Numer. Math. 90(2001), 117–148.

[KVo3] K. Kunisch and S. Volkwein, *Galerkin proper orthogonal decomposition methods for a general equation in fluid dynamics*, SIAM J. Numer. Anal. 40(2002), 492–515.

[LeMa] F. Lempio and H. Maurer, *Differential stability in infinite-dimensional nonlinear programming*, Appl. Math. Optim. 6(1980), 139–152.

[LiMa] J.-L. Lions and E. Magenes, *Non-Homogeneous Boundary Value Problems and Applications*, Springer-Verlag, Berlin, 1972.

[Lio1] J.-L. Lions, *Control of Distributed Singular Systems*, Gauthier-Villars, Paris, 1985.

[Lio2] J.-L. Lions, *Optimal Control of Systems Governed by Partial Differential Equations*, Springer-Verlag, Berlin, 1971.

[Lio3] J.-L. Lions, *Quelques Methodes de Resolution de Problemes aux Limites Non Lineares*, Dunod, Paris, 1969.

[MaZo] H. Maurer and J. Zowe, *First and second-order necessary and sufficient optimality conditions for infinite-dimensional programming problems*, Math. Programming 16(1979), 98–110.

[Mif] R. Mifflin, *Semismooth and semiconvex functions in constrained optimization*, SIAM J. Control Optim. 15(1977), 959–972.

[MuSi] F. Murat and J. Simon, *Sur le controle par un domain geometrique*, Rapport 76015, Universite Pierre et Marie Curie, Paris, 1976.

[Pan1] J. S. Pang, *Newton's method for B-differentiable equations*, Math. Oper. Res. 15(1990), 311–341.

[Pan2] J. S. Pang, *A B-differentiable equation-based, globally and locally quadratically convergent algorithm for nonlinear programs, complementarity and variational inequality problems*, Math. Programming 51(1991), 101–131.

[Paz] A. Pazy, *Semi-groups of nonlinear contractions and their asymptotic behavior*, in Nonlinear Analysis and Mechanics: Heriot Watt Symposium, Vol III, R. J. Knops, ed., Res. Notes in Math. 30, Pitman, Boston, MA, 1979, 36–134.

[PoTi] E. Polak and A. L. Tits, *A globally convergent implementable multiplier method with automatic limitation*, Appl. Math. Optim. 6(1980), 335–360.

[PoTr] V. T. Polak and N. Y. Tret'yakov, *The method of penalty estimates for conditional extremum problems*, Žurnal Vyčislitel'noĭ Matematiki i Matematičeskoĭ Fiziki 13(1973), 34–46.

[Pow] M. J. D. Powell, *A method for nonlinear constraints in minimization problems*, in Optimization, R. Fletcher, ed., Academic Press, New York, 1968, 283–298.

[Pow1] M. J. D. Powell, *The convergence of variable metric methods for nonlinearly constrained optimization problems*, Nonlinear Programming 3, O. L. Mangasarian, ed., Academic Press, New York, 1987, 27–63.

[Qi] L. Qi, *Convergence analysis of some algorithms for solving nonsmooth equations*, Math. Oper. Res. 18(1993), 227–244.

[QiSu] L. Qi and J. Sun, *A nonsmooth version of Newton's method*, Math. Programming 58(1993), 353–367.

[Rao] M. Raous, *Quasistatic Signorini problem with Coulomb friction and coupling to adhesion*, in New Developments in Contact Problems, P. Wriggers and Panagiotopoulos, eds., CISM Courses and Lectures 384, Springer-Verlag, New York, 1999, 101–178.

[Ro1] S. M. Robinson, *Regularity and stability for convex multivalued functions*, Math. Oper. Res. 1(1976), 130–143.

[Ro2] S. M. Robinson, *Stability theory for systems of inequalities, Part* II*: Differentiable nonlinear systems*, SIAM J. Numer. Anal. 13(1976), 497–513.

[Ro3] S. M. Robinson, *Strongly regular generalized equations*, Math. of Oper. Res. 5(1980), 43–62.

[Ro4] S. M. Robinson, *Local structure of feasible sets in nonlinear programming, Part* III*: Stability and sensitivity*, Math. Programming Stud. 30(1987), 45–66.

[Roc1] R. T. Rockafeller, *Local boundedness of nonlinear monotone operators*, Michigan Math. J. 16(1969), 397–407.

[Roc2] R. T. Rockafeller, *The multiplier method of Hestenes and Powell applied to convex programming*, J. Optim. Theory Appl. 12(1973), 34–46.

[ROF] L. I. Rudin, S. Osher, and E. Fatemi, *Nonlinear total variation based noise removal algorithms*, Phys. D 60 (1992), 259–268.

[RoTr] A. Rösch and F. Tröltzsch, *On regularity of solutions and Lagrange multipliers of optimal control problems for semilinear equations with mixed pointwise control-state constraints*, SIAM J. Control Optim. 46(2007), 1098–1115.

[Sa] E. Sachs, *Broyden's method in Hilbert space*, Math. Programming 35(1986), 71–82.

[Sey] R. Seydel, *Tools for Computational Finance*, Springer-Verlag, Berlin, 2002.

[Sha] A. Shapiro, *On concepts of directional differentiability*, J. Optim. Theory Appl. 13(1999), 477–487.

[SoZo] J. Sokolowski and J. P. Zolesio, *Introduction to Shape Optimization*, Springer-Verlag, Berlin, 1991.

[Sta1] G. Stadler, *Infinite-Dimensional Semi-Smooth Newton and Augmented Lagrangian Methods for Friction and Contact Problems in Elasticity*, Ph.D. thesis, University of Graz, 2004.

[Sta2] G. Stadler, *Semismooth Newton and augmented Lagrangian methods for a simplified problem friction*, SIAM J. Optim. 15(2004), 39–62.

[Sto] J. Stoer, *Principles of sequential quadratic programming methods for solving nonlinear programs*, in Computation Mathematical Programming, K. Schittkowski, ed., NATO Adv. Sci. Inst. Ser F Comput. Systems Sci. 15, Springer-Verlag, Berlin, 1985, 165–207.

[StTa] J. Stoer and R. A. Tapia, *The Local Convergence of Sequential Quadratic Programming Methods*, Technical Report 87–04, Rice University, 1987.

[Tan] H. Tanabe, *Equations of Evolution*, Pitman, London, 1979.

[Te] R. Temam, *Navier Stokes Equations: Theory and Numerical Analysis*, North–Holland, Amsterdam, 1979.

[Tem] R. Temam, *Infinite-Dimensional Dynamical Systems in Mechanics and Physics*, Springer-Verlag, Berlin, 1988.

[Tin] M. Tinkham, *Introduction to Superconductivity*, McGraw–Hill, New York, 1975.

[Tr] G. M. Troianiello, *Elliptic Differential Equations and Obstacle Problems*, Plenum Press, New York, 1987.

[Tro] F. Troeltzsch, *An SQP method for the optimal control of a nonlinear heat equation*, Control Cybernet. 23(1994), 267–288.

[Ulb] M. Ulbrich, *Semismooth Newton methods for operator equations in function spaces*, SIAM J. Optim. 13(2003), 805–841.

[Vog] C. R. Vogel, *Computational Methods for Inverse Problems*, Frontiers Appl. Math. 23, SIAM, Philadelphia, 2002.

[Vol] S. Volkwein, *Mesh-independence for an augmented Lagrangian-SQP method in Hilbert spaces*, SIAM J. Control Optim. 38(2000), 767–785.

[We] J. Werner, *Optimization—Theory and Applications*, Vieweg, Braunschweig, 1984.

[Zar] E. M. Zarantonello, *Projection on convex sets in Hilbert space and spectral theory*, in Contributions to Nonlinear Functional Analysis, E. H. Zarantonello, ed., Academic Press, New York, 1971, 237–424.

[Zo] J. P. Zolesio, *The material derivative (or speed) method for shape optimization*, in Optimization of Distributed Parameter Structures, Vol II, E. Haug and J. Cea, eds., Sijthoff & Noordhoff, Alphen aan den Rijn, 1981, 1089–1151.

[ZoKu] J. Zowe and S. Kurcyusz, *Regularity and stability for the mathematical programming problem in Banach spaces*, Appl. Math. Optim. 5(1979), 49–62.

Index